Advances in Geological Science

Studies in the twentieth century uncovered groundbreaking facts in geophysics and produced a radically new picture of the Earth's history. However, in some respects it also created more puzzles for the research community of the twenty-first century to tackle. This book series aims to present the state of the art of contemporary geological studies and offers the opportunity to discuss major open problems in geosciences and their phenomena. The main focus is on physical geological features such as geomorphology, petrology, sedimentology, geotectonics, volcanology, seismology, glaciology, and their environmental impacts. The monographs in the series, including multi-authored volumes, will examine prominent features of past events up to their current status, and possibly forecast some aspects of the foreseeable future. The guiding principle is that understanding the fundamentals and applied methodology of overlapping fields will be key to paving the way for the next generation.

Junzo Kasahara · Ko Sato · Hirofumi Muraoka ·
Koichi Tagomori · Narimitsu Ito
Editors

Challenges in Geothermal Development

Geological Investigation and Well Targeting in Japan

Springer

Editors
Junzo Kasahara
Shizuoka University
Shizuoka, Japan

Hirofumi Muraoka
Institute of Regional Innovation
Hirosaki University
Hirosaki, Aomori, Japan

Narimitsu Ito
NEWJEC Inc.
Osaka, Japan

Ko Sato
Socio Engineering Co., Ltd.
Morioka, Japan

Koichi Tagomori
West Japan Engineering Consultants, Inc.
Fukuoka, Japan

ISSN 2524-3829 ISSN 2524-3837 (electronic)
Advances in Geological Science
ISBN 978-981-95-0552-4 ISBN 978-981-95-0553-1 (eBook)
https://doi.org/10.1007/978-981-95-0553-1

This Springer imprint is published by the registered company Springer Nature Singapore Pte Ltd.
The registered company address is: 152 Beach Road, #21-01/04 Gateway East, Singapore 189721, Singapore

Foreword

The global energy landscape is undergoing a profound transformation. As nations grapple with the imperative to decarbonize their economies and secure sustainable energy futures, geothermal energy stands as a silent giant, holding immense promise. Unlike intermittent renewables such as solar and wind, geothermal offers a baseload power source, available 24/7, harnessing the Earth's natural heat. This book, *Challenges in Geothermal Development—Geological Investigation and Well Targeting in Japan*, arrives at a critical juncture, providing an invaluable contribution to the worldwide pursuit of this vital resource.

Japan, a nation uniquely positioned atop the Pacific Ring of Fire, possesses one of the world's richest geothermal endowments. Yet, despite this geological advantage, its geothermal potential remains only partly developed. This book embarks on a meticulous exploration of Japan's geothermal journey, dissecting both triumphs and setbacks from actual case studies. By systematically documenting the historical context and the specific details of various sites, it offers a vision not only for future domestic endeavors but also for other countries embarking on their own geothermal development paths. The lessons learned from Japan's diverse experiences—the geological complexities encountered, the successes achieved, and the challenges overcome—provide a pragmatic roadmap for others navigating the intricate world of geothermal exploration and development.

While acknowledging the foundational role of conventional exploration and survey methods, this text also includes the cutting edge of geothermal technology. It shines a spotlight on Distributed Temperature Sensing (DTS) transformative technology that helps how we understand and exploit geothermal wells. Through detailed practical case studies, the book illustrates the immense potential of DTS to provide insights into subsurface conditions, enabling more precise well targeting and optimized resource management. This focus carries the book beyond a general overview, providing a uniquely specialized technical resource for geothermal professionals seeking to leverage the latest advancements.

Ultimately, this book is a testament to the hands-on experience of the engineers who have grappled with the realities of geothermal development in Japan and the

World. It tackles the fundamental questions of how to approach geothermal reservoirs with exploratory techniques and, critically, how to identify viable development targets. Drawing directly from real-world Japanese case studies, as well as others in the World, this book provides a practical perspective that is less often seen in academic texts. For anyone involved in the complex, challenging, yet rewarding field of geothermal energy, this book offers not just knowledge, but also the wisdom born from direct experience, making it an essential guide for unlocking the Earth's hidden power.

Roland N. Horne
Director
Stanford Geothermal Program
Stanford University
Stanford, California, USA

The original version of the book has been revised. A correction to this book can be found at https://doi.org/10.1007/978-981-95-0553-1_15

Preface

Fossil energy resources such as oil, gas, and coal emit significant amounts of CO_2 into the atmosphere, contributing to the greenhouse effect and the rise in global temperatures. As the urgency to address global warming grows, renewable energy sources are becoming increasingly vital. Among these, geothermal energy stands out as a promising solution. However, developing geothermal power plants presents notable challenges, including high drilling costs and a low success rate in accessing viable geothermal resources. Additionally, many geothermal sites are in natural parks, where development poses environmental concerns.

To overcome these hurdles, it is essential to innovate exploration methods, to seek deep geothermal resources, to improve the success rate of geothermal drilling and reduce associated costs. This book emphasizes gaining a deeper understanding of geothermal fields and improving the precision of geothermal exploration through innovative technologies and methodologies.

Geothermal fields around the world exhibit significant variations in their geological structures, rock types, heat conduction mechanisms, temperature gradients, permeability structures, brittle-ductile transitions, reservoir sizes, acidity levels, and fluid-to-gas ratios. Factors such as population density, accessibility, and infrastructure also vary widely. This diversity underscores the importance of sharing insights and experiences from specific regions. In this context, we aim to contribute to the global geothermal community by presenting Japan's unique geothermal characteristics and development case studies.

Furthermore, this book explores state-of-the-art technologies in geothermal exploration, including Distributed Temperature Sensing (DTS), Distributed Acoustic Sensing (DAS), Full Waveform Inversion (FWI) methods, seismicity analysis, and gravity tensor measurements. These advancements enhance our ability to locate and characterize geothermal reservoirs with greater precision.

The book is organized into two parts:

- **Part I**: Development in High-Temperature Geothermal Systems in the World
- **Part II**: Geological Investigation and Well Targeting for Geothermal Development, Japan

Part I consists of eight chapters:

1. "Overview of High-Temperature Geothermal Systems in the World"—Summarizes the global characteristics of geothermal fields.
2. "Tectonic Backgrounds and Advances of Geothermal Developments in Japan"—Explores Japan's tectonic framework.
3. "High-Temperature Hydrothermal Systems Induced from Shallow-Depth Magma Intrusions in Japan"—Discusses geothermal systems influenced by magma.
4. "Fundamentals of the Brittle-Ductile Transition in the Crust"—Examines its role in heat transfer and thermal gradients.
5. "General Exploration Methods Applied for Geothermal Development"—Provides an overview of exploration technologies.
6. "Case Studies I: High- and Ultra-High-Temperature Geothermal Fields in Japan and the World"—Shares insights into exceptional geothermal fields.
7. "Geothermal Exploration Using the Optical Fiber Sensing: Distributed Temperature Sensing (DTS) Technology"—Highlights the significance of DTS technology using optical fibers.
8. "Latest Geothermal Exploration and Development"—Explores DAS applications and other innovative techniques such as FWI and gravity tensor analysis.

Part II includes six chapters, grounded in fieldwork conducted in Japan:

9. "Geothermal Systems and Geothermal Reservoirs"—Provides a description of geothermal systems and reservoirs, explaining how water is heated, rises, and forms exploitable heat or fluid sources.
10. "Understanding of High-Temperature Zones in Geothermal Reservoirs"—Outlines an overview of geochemical surveys, borehole temperature classifications, and fluid inclusion surveys in geothermal exploration.
11. "Understanding of High Permeable Zones in Geothermal Reservoirs"—Explains that fracture permeability, governed by Cubic Law, is key for steam production, with minimum kh of 10^{-12} m^3.
12. "Production Well Targeting Based on Geological Analyses"—Outlines that targeting production wells involves geological surveys, including thermal manifestations, hydrothermal rock alteration, and geomorphology.
13. "Reinjection Wells"—Discusses reinjected hot water flow mechanisms in reservoirs, including cold sweep processes, fracture spacing, breakthrough, spread-out, and temperature influence. Six general plans for setting reinjection wells are also illustrated.
14. "Case Studies II: Well-Targeting Field Examples in Japan"—Describes case studies for well-targeting in the early stages of investigation, using literature-based methods in five geothermal areas.

By sharing these studies and technologies, we aim to advance the understanding and development of geothermal energy worldwide.

We sincerely thank the engineers and scientists involved in geothermal development in Japan for sharing valuable insights from their challenging work. We would also like to express our great thanks to Mr. Yosuke Nishida Editorial Chief of Earth Sciences and Geography for encouraging our publication.

Tokyo, Japan
May 2025

Junzo Kasahara

Acknowledgements

We, the authors, would like to express our heartfelt gratitude to all the individuals and organizations whose support and contributions made the completion of this book possible.

In particular, we extend our deep appreciation to the New Energy and Industrial Technology Development Organization (NEDO), Hokkaido Electric Power Co., Inc., Mitsubishi Materials Corporation, Japan Metals and Chemicals Co., Ltd., Tohoku Sustainable and Renewable Energy Co., Inc., Kyuden Mirai Energy Co., Inc., Idemitsu Kosan Co., Ltd., WELMA Co., Ltd., West Japan Engineering Consultants, Inc., NEWJEC Inc., Nippon Koei Co., Ltd., and The Engineering Advancement Association of Japan (ENAA). Their invaluable assistance, provision of critical information, and expertise were instrumental in shaping this work.

Additionally, in the process of creating this book, ChatGPT was utilized as a supplementary tool. It contributed particularly to improving the conciseness of expressions and enhancing the overall structure, thereby supporting the refinement of our ideas.

Lastly, we sincerely thank all our readers. We hope this book serves as a valuable resource and contributes meaningfully to advancements and insights in the field.

Junzo Kasahara
Ko Sato
Hirofumi Muraoka
Koichi Tagomori
Narimitsu Ito

Contents

Part I Development in High-Temperature Geothermal Systems in the World

1 Overview of High-Temperature Geothermal Systems in the World 3
Hirofumi Muraoka, Koichi Tagomori, and Junzo Kasahara

2 Tectonic Backgrounds and Advances of Geothermal Developments in Japan 75
Hirofumi Muraoka and Koichi Tagomori

3 High-Temperature Hydrothermal Systems Induced from Shallow-Depth Magma Intrusions in Japan 127
Hirofumi Muraoka

4 Fundamentals of the Brittle–Ductile Transition in the Crust 161
Toshiaki Masuda and Yasutomo Omori

5 General Exploration Methods Applied for Geothermal Development 181
Koichi Tagomori and Hirofumi Muraoka

6 Case Studies I: High- and Ultra-High-Temperature Geothermal Fields in Japan and the World 201
Koichi Tagomori, Naotsugu Ikeda, Ko Sato, Hirofumi Muraoka, and Junzo Kasahara

7 Geothermal Exploration Using the Optical Fiber Sensing: Distributed Temperature Sensing (DTS) Technology 263
Naotsugu Ikeda

8 Latest Geothermal Exploration and Development 295
Junzo Kasahara and Yoko Hasada

Part II Geological Investigation and Well Targeting for Geothermal Development, Japan

9 Geothermal Systems and Geothermal Reservoirs 337
Ko Sato

10 Understanding of High-Temperature Zones in Geothermal Reservoirs 345
Narimitsu Ito and Ko Sato

11 Understanding of High Permeable Zones in Geothermal Reservoirs 359
Ko Sato and Narimitsu Ito

12 Production Well Targeting Based on Geological Analyses 393
Ko Sato

13 Reinjection Wells 415
Ko Sato

14 Case Studies II: Well Targeting Field Examples in Japan 429
Ko Sato, Junzo Kasahara, and Narimitsu Ito

Correction to: Challenges in Geothermal Development C1
Junzo Kasahara, Ko Sato, Hirofumi Muraoka, Koichi Tagomori, and Narimitsu Ito

Epilogue 497

Index 499

Contributors

Yoko Hasada The Engineering Advancement Association of Japan, Tokyo, Japan

Naotsugu Ikeda Welma Co., Ltd., Fukuoka, Japan

Narimitsu Ito NEWJEC Inc., Osaka, Japan

Junzo Kasahara Shizuoka University, Shizuoka, Japan

Toshiaki Masuda Shizuoka University, Shizuoka, Japan

Hirofumi Muraoka Hirosaki University, Hirosaki, Japan

Yasutomo Omori Shizuoka University, Shizuoka, Japan

Ko Sato Socio Engineering Co., Ltd., Tokyo, Japan

Koichi Tagomori West Japan Engineering Consultants, Inc., Fukuoka, Japan

Abbreviations

ΔP	Pressure difference of fluid flow
3D	Three-Dimensional
3KEYGS	Three Key Elements for Geothermal Systems
AD	Annual dose by measurement of U, Th, and K_{40}
AI	Anion Index (major anion composition)
AIST	National Institute of Advanced Industrial Science and Technology
AMT	Audio-Magnetotelluric survey
API	American Petroleum Institute
Ar-Ar	Argon-Argon (Radiometric Dating)
BHA	Bottom Hole Assembly
BHT	Bottom Hole Temperature
BHTV	Borehole Televiewer
BOP	Blowout Preventer
ca.	Circa = about
CGS	Unit of centimeters, gram, and second
CNR	National Research Council (Italy)
cp	Centipoise (viscosity unit, cP)
Cr	Rock-specific heat capacity
CSAMT	Controlled Source Audio-frequency Magneto-Telluric method
CT	Computed Tomography
Cw	Fluid-specific heat capacity
d	Width of the fracture aperture
DAS	Distributed Acoustic Sensing
Dc	Annual dose of cosmic ray
DC	Direct Current
DEM	Digital Elevation Model
DFOS	Distributed Fiber Optic Sensing
DGE	Dohnan Geothermal Energy Co. Ltd.
DSCA	Differential Strain Curve Analysis
DTF	Drilling-induced Tensile Fracture

DTS	Distributed Temperature Sensing
Dα	Annual dose of alpha ray
Dβ	Annual dose of beta ray
Dγ	Annual dose of gamma ray
EGS	Enhanced or Engineered Geothermal System
EL	Elevation
EPC	Electric Power Company
EQ	Earthquake
FIT	Feed-in Tariff
FMI	Formation Micro-Imager (a SLB's trademark)
FP	Feed point: thermal outflow point from geological layers
FREA	Fukushima Renewable Energy Institute, AIST
FT	Average formation temperature
FWI	Full Waveform Inversion
GI-Fiber	Graded-Index Fiber
GIS	Geographical Information System
GPP	Geothermal Power Plant
Gs	Steam flow rate
GSJ	Geological Survey of Japan
Gw	Hot water flow rate
GWeh	Gigawatt hours of electricity generation
Gy	J/kg absorbed radiation dose
h'	Specific enthalpy of hot water
h''	Specific enthalpy of steam
$\hbar$	Reduced Planck constant which is the Planck constant divided by 2π, read as "h-bar"
HEP	Hokkaido Electric Power Co., Inc.
HFEs	Ordinary High-Frequency Earthquakes
HHL	Hita-Hakuso line: western fault line in Hohi VTD
HQ	Outer diameter of about 98 mm on boring (drilling)
IDDP	Iceland Deep Drilling Project
IVI	Industrial Vehicles International
JMA	Japan Meteorological Agency
JMC	Japan Metals and Chemicals Co. Ltd.
JOGMEC	Japan Organization for Metals and Energy Security
k	Permeability
ka	Kilo annum: thousand years before the present
K-Ar	Kalium (potassium)-Argon age determinaton
kh	Permeability-thickness
kh/μ	Transmissibility
KHL	Kurume-Hiji line: northern fault line in Hohi VTD
K-horizon	A seismic marker horizon in the Larderello geothermal field
KI	Potassium Iodide
L/C	Lost circulation during drilling
L/min	Liters per minute

LC^2	Total volume of the layer
LFEs	Deep Low-Frequency Earthquakes
LiDAR	Laser Imaging Detection and Ranging
LOT	Leak-Off Test
Ma	Million years before the present
METI	Ministry of Economy, Trade and Industry
mgal	10^{-3} m/s^2 using gravity analysis
mGy/a	10^{-3} J/kg in radiation dose per year
MHTF	Minimum homogenization temperature in fluid inclusion
MINC	Multiple interacting continue method in reservoir numerical simulation
Mise-À-La-Masse	Fluid flow potential method in electrical survey
MITI	Ministry of International Trade and Industry
Moho	Mohorovičić Discontinuity (boundary between Earth's crust and mantle)
MPaG	Wellhead pressure in gauge
MT	Magneto-Telluric method
MTL	Median Tectonic line
MWD	Measurement While Drilling
MWe	Megawatt electrical
Na-K	Sodium-Potassium (commonly used in geothermometers)
ndLC	Total volume of porosity
NEDO	New Energy and Industrial Technology Development Organization
ng	10^{-6} mg, 10^{-9} g
NQ	Outer diameter of about 75 mm on boring (drilling)
OKL	Oita-Kumamoto Tectonic line: southern fault line in Hohi VTD
OTDR	Optical Time Domain Reflectometry
P	Pressure
P0	Thermal fluid production pressure in the wellhead
PD	Paleodose: crystal mineral underwent radiation exposure during geologic time
PDC	Polycrystalline Diamond Compact (drill bit type)
PTS	Pressure-Temperature-Spinner Logs
PVT	Pressure-Volume-Temperature
P-wave	Primary Wave, a type of compressional seismic wave
q	Flow rate per fracture
Q	Total flow rate
R&D	Research and Development
R.A.	Radioactive (Tracer Logs)
S.M-Fiber	Single-Mode Fiber
SAR	Synthetic Aperture Radar
SEI	Sumitomo Electric Industry

SI	International System of Units (from French Système international d'unités)
SINTEF	Foundation for Scientific and Industrial Research (Norway)
SS	Sand Stone
ST	Standing time
TDS	Top Drive System or Total Dissolved Solids
TEM	Time Domain Electromagnetic
Tf	Formation temperature of wells
TGE	Tohoku Geothermal Energy Co., Ltd.
TL	Thermolumienscence age determination
$T_{Na\text{-}K}$	Temperature measurement by sodium-potassium in geochemical thermometers
Tr	Minimum homogenization temperature in fluid inclusion (=MHTF)
U	Darcy's velocity
UD	Up-down
U-Pb	Uranium-lead age determination
USGS	United States Geological Survey
UTM	Universal Transverse Mercator (projection system)
V	Viscosity of fluid flow
Vp	Primary wave velocity: longitudinal, compressive, dilatational wave
Vs	Secondary wave velocity: transverse, shear, torsion, skew wave
VTD	Volcano-Tectonic Depression
Vu	Vulnerable portion strata at valley
W	water level
WL	Water Level
X	Dryness
XRD	X-ray Diffraction
γ_m	Inclination in DTF
δ	Error in TL or aperture of fracture (= d)
θ	Cessation time of mud circulation to the temperature measurement (= standing time)
θ_{md}	Circumferential position in DTF
λ	Fracture spacing
μ	Velocity of fluid flow
μgal	Microgal (unit of gravity measurement) = 10^{-6} cm/s^2
μGy	10^{-6} J/kg in radiation dose per year
ρ	Gravity density
ρ_3	Minimum compressive principal stress in a regional stress field or tensional principal stress
ρ_r	Rock density
ρ_w	Fluid density
σ	Standard deviation

σ_1	Maximum compressive principal stress in a regional stress field
σ_1'	Maximum compressive principal stress in a local stress field
σ_2	Intermediate compressive principal stress in a regional stress field
σ_2'	Intermediate compressive principal stress in a local stress field
σ_3	Minimum compressive principal stress in a regional stress field
σ_3'	Minimum compressive principal stress in a local stress field
σ_H	Maximum compressive principal stress in a horizontal plane
σ_h	Minimum compressive principal stress in a horizontal plane
σ_{mD}	Maximum compressive principal stress in a circumferential direction
σ_v	Vertical stress
ϕ	Porosity

Part I
Development in High-Temperature Geothermal Systems in the World

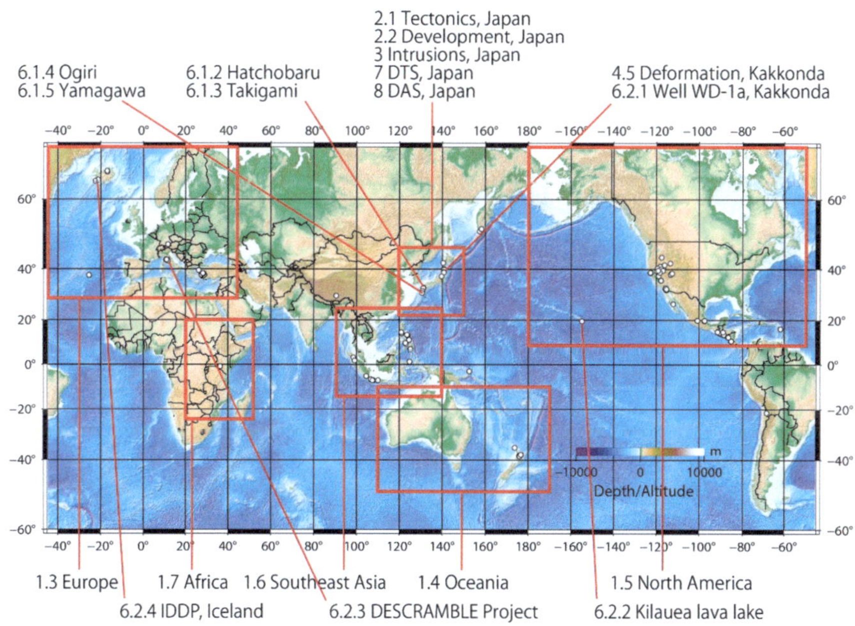

The site index map of Part I, the open circle shows more than 7 MWe geothermal power plants. Numbers indicate those of a chapter, section, or subsection.

Chapter 1
Overview of High-Temperature Geothermal Systems in the World

Hirofumi Muraoka, Koichi Tagomori, and Junzo Kasahara

Abstract This book focuses on high-temperature hydrothermal convection systems, including the challenging efforts for next-generation high-temperature geothermal explorations. In this chapter, the thermal structure of the Earth's interior and thermal energy emitted from the Earth's interior are first overviewed. Then, the world's high-temperature geothermal systems are briefly described. General aspects of the high-temperature geothermal systems are also summarized.

Keywords Thermal structure of the Earth's interior · Thermal energy emitted from the Earth's interior · The world's high-temperature geothermal systems · General aspects of the high-temperature geothermal systems

1.1 Thermal Energy of the Earth's Interior

1.1.1 Geothermal Resources

Geothermal resources can be defined as the economically utilized part of thermal energy supplied from the Earth's interior. The most common geothermal resource is the direct use of hot spring water for bathing. The economic utilization of hydrothermal resources varies depending on the direct use, such as bathing, hot pools, balneology, space heating, hot water supply, greenhouse, fish farming, and snow melting. In addition, one of the most valuable modern utilization categories is

H. Muraoka (✉)
Hirosaki University, Hirosaki, Japan
e-mail: hiro@hirosaki-u.ac.jp

K. Tagomori
West Japan Engineering Consultants, Inc., Fukuoka, Japan
e-mail: k-tagomori@wjec.co.jp

J. Kasahara
Shizuoka University, Shizuoka, Japan
e-mail: kasahara.junzo@shizuoka.ac.jp

J. Kasahara et al. (eds.), *Challenges in Geothermal Development*, Advances in Geological Science, https://doi.org/10.1007/978-981-95-0553-1_1

electric power generation, which includes dry steam, steam-flash, and binary-cycle power generation. Hydrothermal resources can be classified by temperature into high-temperature (>150 °C), intermediate-temperature (90–150 °C), and low-temperature (<90 °C) hydrothermal convection systems (White and Williams 1975; Muffler and Guffanti 1975).

1.1.2 Thermal Structure of the Earth's Interior

The Earth's interior is divided into three layers, crust, mantle, and core, from the surface to the center (Fig. 1.1). This division is based on the difference in the lithologic compositions of the layers and their physical states; for instance, the outer core is considered only an almost entirely molten layer. The observation of the seismic travel time curves recognizes these layers. The crust ranges in thickness from the surface to approximately 7 km in oceans and 35 km in continents. However, it occasionally reaches 70 km, like Tibet, by duplicating the crust through thrusting or collision tectonics. The oceanic crust comprises mafic rocks such as gabbro (plutonic rock) and basalt (volcanic rock) with overlying thin pelagic sediments. The continental crust is subdivided into the lower crust and upper crust. The continental lower crust is composed of mafic rocks such as gabbro (plutonic rock) and basalt (volcanic rock), the same as the oceanic crust. The continental upper crust is distinctively composed of silicic rocks such as granodiorite, granite (plutonic rock), andesite, dacite, and rhyolite (volcanic rock) with thick sedimentary rocks mainly accreted as accretionary prisms from the oceanic crust through the plate tectonic processes of subduction, collision, and obduction. The low density of the silicic rocks composes the high-altitude mountains on continents by isostasy (gravity equilibria). Therefore, whether a given place becomes an ocean or a continent is essentially determined by the rock composition of the crust, and the seawater filling is only a result (Uyeda 1971).

The crust–mantle boundary is called the Mohorovičič discontinuity or briefly called the Moho discontinuity, where P- and S-wave velocities abruptly increase toward the mantle. The mantle is subdivided into the upper mantle at a depth from 35 to 660 km and the lower mantle at a depth from 660 to 2,900 km (Fig. 1.1). The mantle is composed of silica-poor and high-density ultramafic rocks such as dunite, lherzolite, harzburgite, and wehrlite, collectively called peridotite, or their high-pressure phase change equivalents. Obduction fragments of the oceanic upper mantle and crust are sometimes exposed to the surface as ophiolites, as seen in Cyprus, Oman, New Caledonia, and many other subduction zones. The core is subdivided into the outer core at a depth from 2,900 to 5,100 km and the inner core at a depth from 5,100 to 6,378 km (a mean equatorial radius; Fig. 1.1). The density of the Earth's core has long been known to be much greater than the Earth's mantle having been ascribed to iron–nickel alloy. The outer core is impervious to the S-wave, and the inner core is pervious to the S-wave, so that the outer core is considered a molten

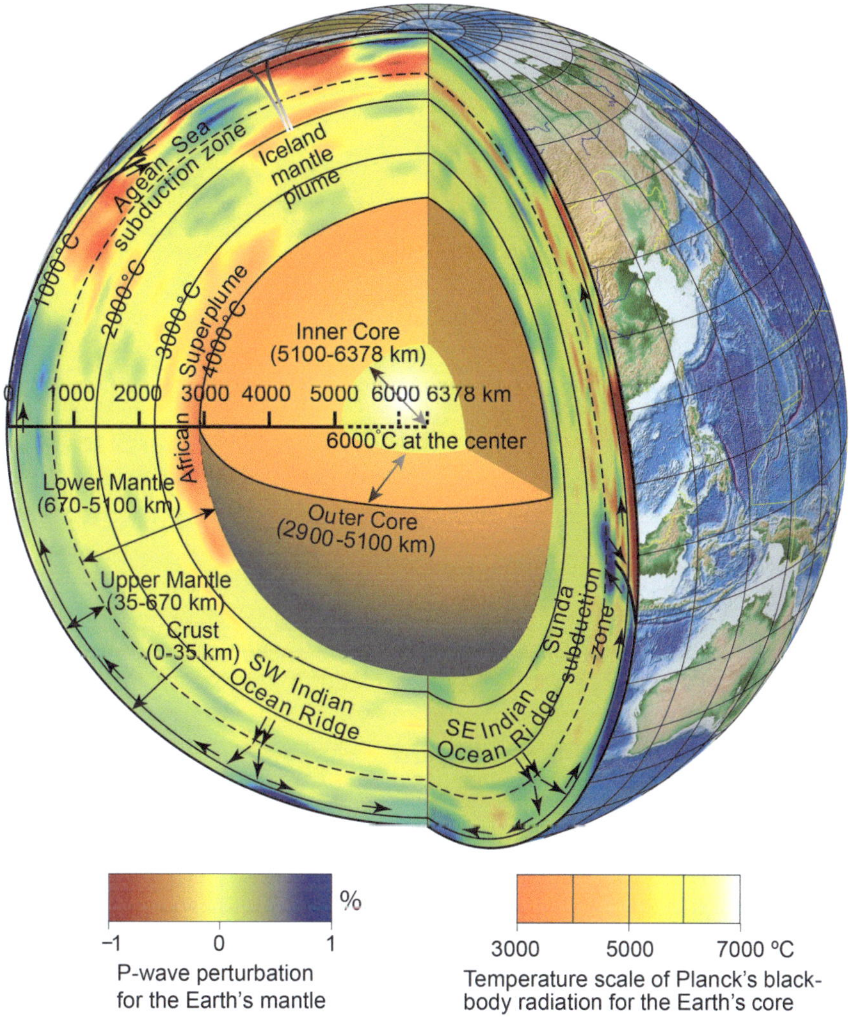

Fig. 1.1 Thermal structure of the Earth's interior. Concentric isotherms are modified from the temperature profile by Kanamori (1978). Seismic tomography for the mantle is modified from Ichikawa (2017). Topographic and bathymetric maps are drawn by the ETOPO1 1 arc-minute global relief model (Amante and Eakins 2009) and GMT 6.4.0 (Wessel et al. 2019)

state of iron–nickel alloy, and the inner core is considered a solid state of iron–nickel alloy.

We are blessing geothermal energy resources from the Earth. To visualize geothermal energy released from the Earth, the thermal structure of the Earth's interior is illustrated in Fig. 1.1. Concentric isotherms with an every 1,000 °C increment in Fig. 1.1 reveal the most commonly accepted classical thermal structure of the Earth's interior as shown in Fig. 1.2 (Kanamori 1978). The isotherms indicate only

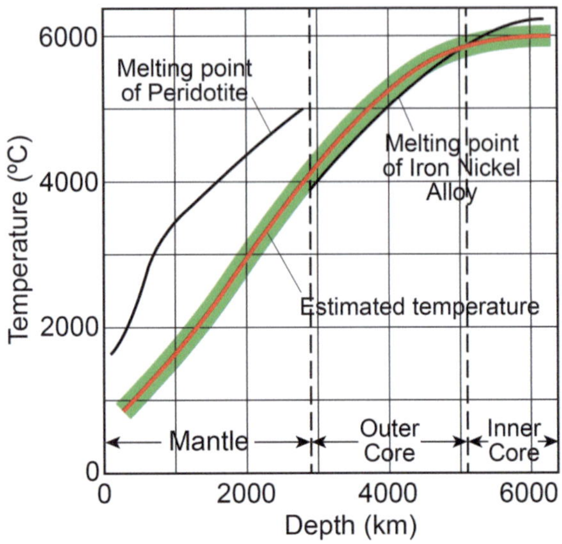

Fig. 1.2 Estimated depth–temperature curve in the Earth's interior (redrawn from Kanamori 1978). The melting point of peridotite is a dry state

average temperature distribution with depth, neglecting lateral temperature variations. However, as quickly speculated from the P-wave perturbation of current mantle seismic tomography in Fig. 1.1 (Obayashi et al. 2013; Fukao and Obayashi 2013; Ichikawa 2017), the actual temperature distribution must be much more varied laterally from place to place, particularly in mid-ocean ridges and subduction zones. The deepest drill hole ever before was the well SG-3 drilled to a depth of 12,262 m in the Kola Peninsula, Russia, in 1989 (Trčková et al. 2002), and this depth is only 0.19% of the mean equatorial radius of 6,378 km of the Earth. Therefore, the temperature of the Earth's interior cannot be directly measured in almost the entire part of the depth range, except for the very shallow surface of the Earth's crust.

Because the drilling can access only a very shallow depth of the Earth's crust, the temperature of the Earth's interior is indirectly estimated from a variety of geophysical methods, such as the materials of the Earth's interior and their physical states deduced from the velocity observation of seismic waves, the materials of the Earth's interior and their melting temperatures deduced from high-temperature and high-pressure experiments in the laboratories, the temperature increase deduced from the adiabatic compression of the materials of the Earth's interior, and the electrical conduction models of the materials of the Earth's interior. These indirect estimations still contain uncertainty with very large errors. The mantle–core boundary is estimated at 4,000 °C from the melting temperature of the iron–nickel alloy, which is one of the most reliable estimations (Fig. 1.2). The temperature of the Earth's center is generally estimated at around 6,000 °C (Figs. 1.1 and 1.2). However, its error may probably reach plus or minus 1,000 °C. The water contents in minerals decrease the melting temperature, and how the H_2O is included in minerals has been investigated.

The high-temperature thermal structure of the Earth's interior is mainly caused by the release of the accretion (accumulation) energy of planetesimals and the release of gravity energy of differentiation of core–mantle layering during the founding period of the Earth (Takahashi 1996). A significant portion of the Earth was under molten states called "magma oceans" by the released energy during the founding period of the Earth. The lighter silicate melts ascended, and heavier iron–nickel alloy melts (only solid under the low-pressure state) descended by the gravity sorting in the magma oceans stage. This process is considered to cause the differentiation of the core–mantle layers (Takahashi 1996). Therefore, the Earth has been cooling since the founding period. Nevertheless, the Earth's interior is still significantly hot after 4.5 billion years since the founding period. This is partly ascribed to the fact that the decay of radioactive elements such as uranium, thorium, and potassium plays an important role as heat sources for the thermal budget of the Earth's interior.

1.1.3 Thermal Energy Budget of the Earth's Interior

Heat is transferred by three modes of processes: thermal radiation, thermal conduction, and thermal convection. Thermal radiation is not significant in the solid-dominant Earth's interior. Thermal conduction is a diffusive process in which molecules transmit their kinetic energy to other molecules by colliding (Turcotte and Schubert 2014). This is an invisible process without any motion of a medium. Thermal convection is caused by the motion of a medium, such as hydrothermal convection, magma convection, mantle convection, gas plume, and mantle plume.

Thermal energy is always transported from the Earth's interior to the surface by thermal conduction. The heat flux per unit area and unit time is called terrestrial heat flow as follows:

$$Q = K\left(\frac{dT}{dz}\right) \tag{1.1}$$

where K is the thermal conductivity, and dT/dz is the geothermal gradient. To obtain the terrestrial heat flow, measurements of the thermal conductivity of borehole core samples in the laboratory and the geothermal gradient in the boreholes are required. The shallower depths in the continental crust are exclusively subject to the solar radiant heat. Therefore, the steady-state geothermal gradient is better measured deeper than 300 m to avoid the solar radiance effect. Hence, ground-sourced heat pump systems can be used seasonally as heat sinks or heat sources in the depth of solar radiance effect zones. This category of utilization is usually not included in geothermal utilization in a strict sense because of the utilization of the seasonal change of the solar radiant heat.

The mean terrestrial heat flow in continents is 65 ± 1.6 mW/m^2, and that in oceans is 101 ± 2.2 mW/m^2 (Turcotte and Schubert 2014). The value in continents includes not only the heat flow from the Earth's interior but also the decay heat from radioactive elements predominant in the continental crust itself (Table 1.1; Turcotte and Schubert 2014). Some calculation shows that the fraction of radioactive heat is

37 mW/m^2 from the continental crust, and the remaining fraction of 28 mW/m^2 is supplied from the mantle among the total 65 mW/m^2 (Turcotte and Schubert 2014). The terrestrial heat flow from the continents is, in fact, known to correlate with the concentration of radioactive elements in rocks taken from the given area. The terrestrial heat flow from the oceans is fundamentally proportional to the formation age of the oceanic crust because the oceanic crust is newly produced by cumulate gabbro, sheeted dikes, and pillow lava from mid-ocean ridge basalt (MORB) magma at the mid-ocean ridges (Fig. 1.3) (Ryan 1994).

The terrestrial heat flow is the largest energy category that escapes from the Earth's surface. The continents and their margins have an area of 2.0×10^8 km^2, and the oceans and their marginal basins have an area of 3.1×10^8 km^2. Weighted by their area ratios, we obtain the mean surface terrestrial heat flow 87 ± 2.2 mW/m^2 (Turcotte and Schubert 2014). Integrating this value by the total Earth's surface 5.1×10^8 km^2, we obtain 4.4×10^{13} W as a total terrestrial surface heat flow from the Earth (Table 1.2).

The second largest energy category that escapes from the Earth's surface is the discharge heat of volcanic eruption ejecta. Including the volcanic eruption ejecta

Table 1.1 Average concentrations of heat-producing radiogenic elements in typical rocks and chondrites (Turcotte and Schubert 2014)

Rock type	U (ppm)	Th (ppm)	K (wt %)
Reference undepleted (fertile) mantle	0.031	0.124	0.031
"Depleted" peridotite	0.001	0.004	0.003
Tholeiitic basalt	0.07	0.19	0.088
Granite	4.7	20	4.2
Shale	3.7	12	2.7
Average continental crust	1.42	5.6	1.43
Chondritic meteorites	0.008	0.029	0.056

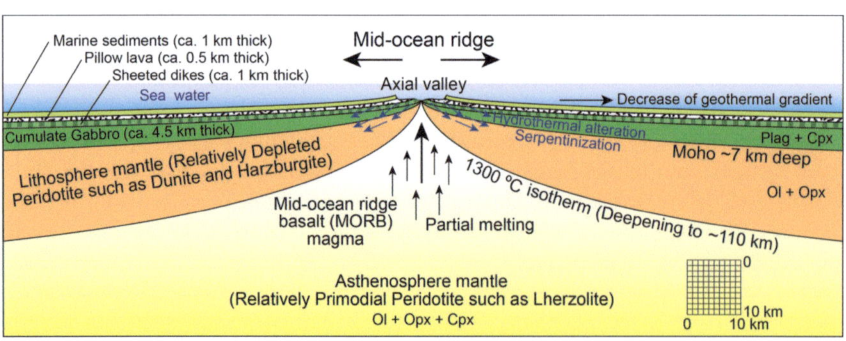

Fig. 1.3 Cross-section model of mid-ocean ridges

Table 1.2 Energy categories that escape from the Earth's surface or are produced in the Earth's interior

	In J/yr	In W (J/s)	In cal/yr
Total terrestrial surface heat flow from the Earth	1.4×10^{21} J/yr	4.4×10^{13} W	3.3×10^{20} cal/yr
Discharge heat of volcanic eruption ejecta from the Earth	3.0×10^{19} J/yr	9.5×10^{11} W	7.2×10^{18} cal/yr
Discharge heat of steam and hot springs from the Earth	5.1×10^{18} J/yr	1.6×10^{11} W	1.2×10^{17} cal/yr
Earthquake energy released from the Earth	7.0×10^{17} J/yr	2.2×10^{10} W	1.7×10^{16} cal/yr
Energy categories produced in the Earth's interior	In J/yr	In W (J/s)	In cal/yr
Present-day radioactive heat generation in the Earth	1.1×10^{21} J/yr	3.5×10^{13} W	2.5×10^{20} cal/yr

from mid-ocean ridges, the total discharge heat was estimated to be 9.5×10^{11} W (Table 1.2; Mizutani and Watanabe 1978).

The third largest energy category that escapes from the Earth's surface is the discharge heat of steam and hot springs. This value was estimated to be 6×10^{10} W from continents (Mizutani and Watanabe 1978). Weighted by the area ratio of the whole globe 5.1×10^8 km^2 to continents 2.0×10^8 km^2, we obtain 1.6×10^{11} W for the discharge heat of steam and hot springs from the entire Earth (Table 1.2).

The fourth largest energy category that escapes from the Earth's shallow crust is the seismic wave energy released from the earthquake hypocenters. One magnitude larger earthquake has about 31.6 times larger energy. Therefore, this energy category can be virtually determined by the great earthquakes with magnitudes larger than 7. This value was estimated to be 2.2×10^{10} W from the historical records of the great earthquakes (Mizutani and Watanabe 1978).

The largest energy category that escapes from the Earth's surface is transferred by thermal conduction and escapes ubiquitously from the entire surface of the Earth. Hence, the second, third, and fourth largest energy categories are transferred by thermal convection and exclusively escape from the active tectonic zones such as the subduction zones, rift zones, and hot spots. However, the summation of the second, third, and fourth largest energy categories is only 2.6% of the largest energy category, and the terrestrial heat flow plays the most important role in escaping energy from the Earth's surface.

On the contrary, the major heat sources of the Earth's interior are the accretion (accumulation) energy of planetesimals and the gravitational energy of differentiation of core–mantle layering during the early period of the Earth. These huge energy categories were estimated to be 2.5×10^{32} J and 2.5×10^{31} J, respectively (Matsui 1996). Most of these energy categories might have been released to outer space in the magma oceans stage. Still, it is important that they contributed to the formation of the present-day high-temperature thermal structure of the Earth's interior. Radioactive heat generation is the most important present-day energy category produced in the

Earth's interior. There are several estimations for this value. Among them, we here refer to the concise conclusions: "about 80% of the present-day surface heat flow can be attributed to the decay of radioactive isotopes presently in the Earth, and about 20% comes from the cooling of the Earth" (Turcotte and Schubert 2014). Then, we estimate the radioactive heat generation to be 3.5×10^{13} W as 80% of the terrestrial heat flow (Table 1.2). This category prolongs the Earth's heat life against its simple, rapid cooling.

1.2 High-Temperature Geothermal Systems in the World

High-temperature geothermal systems are usually generated near high-level magma systems beneath relatively young volcanoes. Active magmatism occurs along volcanic zones near plate boundaries, such as the mid-ocean ridges (Fig. 1.3), subduction zones, continental rift zones, and collision zones. One of the most reliable criteria for defining the plate boundaries is the earthquake epicenter concentrations. The 42,977 events with a magnitude higher than 5.0 earthquake epicenters from January 2000 to February 2024 exhibit the plate boundary in the world (Fig. 1.4; US Geological Survey 2024). Based on the published literature, earthquake hypocenters, and plate motions, plate boundaries of 52 plates on the Earth were compiled in a digital form by Bird (2003). According to the data, major plates are illustrated on the Earth (Figs. 1.5, 1.6 and 1.7). Holocene volcanoes and representative high-temperature geothermal systems are closely associated with these plate boundaries.

The mid-ocean ridges are undoubtedly the most significant volcanic expressions on the Earth because they are spreading centers for plate tectonics where the oceanic lithospheres are produced through the formation of the layers of peridotite, cumulate gabbro, sheeted dikes, and pillow lava (Fig. 1.3). They are some line source volcanoes. Nevertheless, despite the shallow-depth crests along the mid-ocean ridges close to a relatively shallow depth of 2 km, conventional geothermal power developments in submarine environments are still demanding from technological and economic viewpoints. Geothermal power developments are only performed in Iceland on the Mid-Atlantic Ridge and in the Azores Islands, Portugal, on the Azores-Gibraltar Ridge by this time.

Numbers of the Holocene volcanoes are widely observed in the subduction zones (Fig. 1.8; van Keken and Wilson 2023). Geothermal powers are commonly developed in subduction zones such as Italy, Indonesia, the Philippines, New Zealand, and Japan. The first geothermal power generation in the world was initiated in Larderello, Italy, in 1913. Submarine ocean ridges are challenging to develop, but geothermal powers in the Great Rift Valley in eastern Africa were successfully developed. Kenya is the leading country in Africa in terms of geothermal power development.

Many people have tried to compile the locations of geothermal power plants worldwide, including the longitudes and latitudes. We here refer to one of those efforts by the European Commission: opening the data on the Website (Uihlein 2018). We slightly simplified the list of geothermal power plants, where the sizes

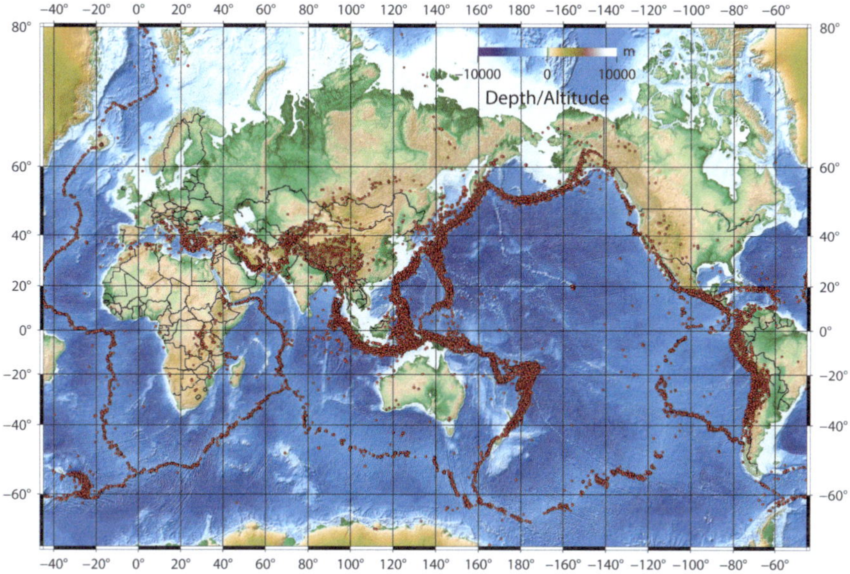

Fig. 1.4 42,977 earthquake epicenters (red dot) in the world with a magnitude of 5.0–9.5 and a depth of −4–750 km were observed from January 2000 to February 2024. Hypocenter information is quoted from the Search Earthquake Catalog of the Earthquake Hazard Program (US Geological Survey 2024). The Mercator projection map was drawn by the ETOPO1 1 arc-minute global relief model (Amante and Eakins 2009) and GMT 6.4.0 (Wessel et al. 2019)

Fig. 1.5 Ocean lithospheres produced from the mid-ocean ridges, such as the Mid-Atlantic Ridge and the East Pacific Rise. Red triangle: Holocene volcanoes (Global Volcanism Program 2024), thick line: plate boundaries (Bird 2003), and thin line: borders of countries (Wessel et al. 2019)

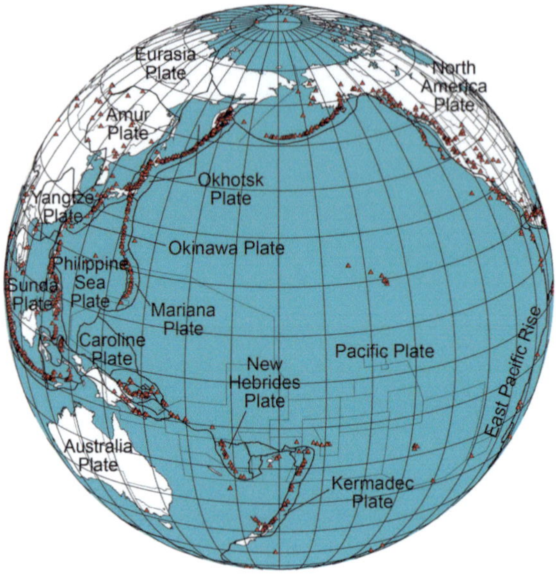

Fig. 1.6 Volcanic zones in the surroundings of the Pacific. Red triangle: Holocene volcanoes (Global Volcanism Program 2024), thick line: plate boundaries (Bird 2003), and thin line: borders of counties (Wessel et al. 2019)

of power plants were restricted to relatively large-scale plants of more than 7 MWe. Multiple units were unified into a single category if the longitudes and latitudes were exactly the same. Then, we obtain 240 major geothermal power plants worldwide as shown in Table 1.3. The dataset will be used to plot the following maps.

The top ten geothermal power production countries change year by year. Table 1.4 shows the top ten geothermal power production countries as of 2020 (Huttrer 2021). Active changes among these ten countries can be seen in Fig. 1.9. The USA is the stable top production country in the world throughout the period from 1995 to 2020. The Philippines used to rank second until 2015, but was overtaken by Indonesia in 2020. New Zealand and Iceland were marked by rapid increases until 2015. However, Turkey and Kenya further rapidly increased in 2020 and reached rankings fourth and fifth, respectively. Japan used to rank fifth in 1995 but has gradually fallen to tenth in 2020. This was mainly due to the less political support during "the lost 15 years" from 1997 to 2010. After the Great East Japan Earthquake in 2011, political support for geothermal energy developments was revived.

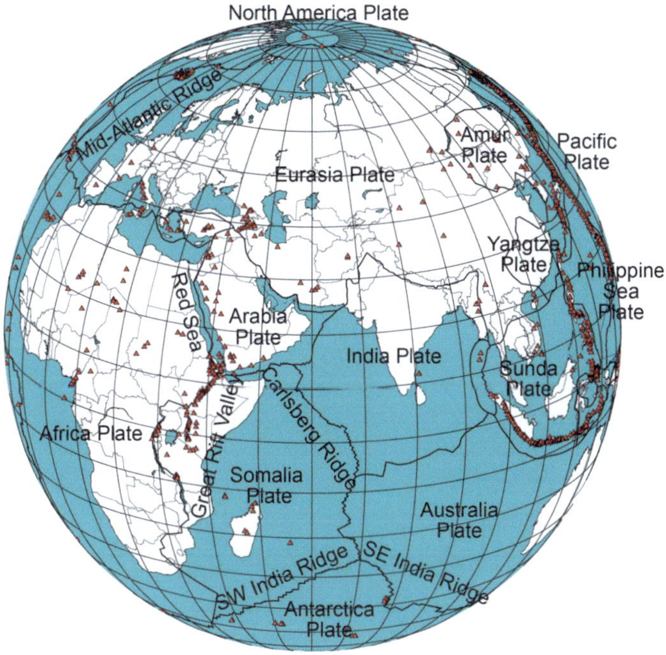

Fig. 1.7 The Great Rift Valley, a rift system for volcanic zones in eastern Africa. Red triangle: Holocene volcanoes (Global Volcanism Program 2024), thick line: plate boundaries (Bird 2003), and thin line: borders of counties (Wessel et al. 2019)

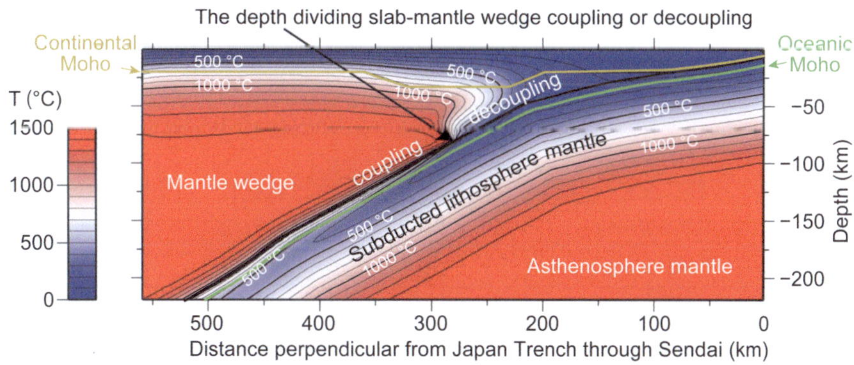

Fig. 1.8 Thermal structure model of subduction zones along the cross-section perpendicular to the Japan Trench through Sendai City (modified from van Keken and Wilson 2023). The slab changes from decoupling at shallower depths to full coupling with the overlying mantle wedge at the decoupling depth. The terms coupling and decoupling are used here in their long-term geodynamic context. This cross-section shows the predicted thermal structure for an end-member cold subduction zone, which is caused by the rapid (8.3 cm/yr) subduction of old (130 Myr) oceanic lithosphere (van Keken and Wilson 2023)

Table 1.3 Representative geothermal power plants in the world. Simplified from Uihlein (2018), where power plants with more than seven MWe were selected, and units were unified if the latitudes and longitudes were the same

Country	Name	Capacity	Longitude	Latitude
Indonesia	Gunung Salak I–III	180.0	−6.74030	106.67300
	Gunung Salak IV–VI	195.0	−6.74170	106.64700
	Wayang Windu I–II	227.0	−7.19780	107.62600
	Kamojang I–V	238.0	−7.13860	107.78500
	Darajat I	55.0	−7.21625	107.73800
	Darajat II–III	216.0	−7.21803	107.72700
	Dieng	60.0	−7.22280	110.01000
	Ulubelu I	55.0	−5.30879	104.57300
	Ulubelu II	55.0	−5.30547	104.57700
	Lahendong I–II	40.0	1.25462	124.82200
	Lahendong III–IV	47.5	1.27124	124.83600
	Sibayak	12.0	3.19250	98.50190
	Sarulla I–II	220.0	1.82089	99.08900
	Patuha	60.0	−7.18391	107.42200
Philippines	Bacman I	30.0	13.05490	123.96800
	Bacman II Cawayan	31.0	13.03490	123.93700
	Mindanao I	52.0	7.01071	125.09200
	Mindanao II	54.0	6.99900	125.19500
	Palinpinon I	112.5	9.31381	123.20100
	Palinpinon II	80.0	9.73670	122.83200
	Tongonan I	112.5	11.17290	124.63200
	Tongonan I Binary	19.5	11.14780	124.64100
	Upper Mahiao	136.5	11.17260	124.63000
	Malitbog	247.1	11.15140	124.64900
	Mahandong	199.5	11.15310	124.65000
	Unit 1 and 2	120.0	13.46550	123.64900
	Unit 5 and 6	114.0	13.45940	123.64000
	Macban A–D	402.8	14.08810	121.22800
	Macban E	40.0	14.09030	121.21300
	Maibarara	20.0	14.60920	121.02200

(continued)

Table 1.3 (continued)

Country	Name	Capacity	Longitude	Latitude
China	Yangbajain	26.2	30.08780	90.54060
Russia	Pauzhetskaya	8.0	51.91330	157.35200
	Verkhne-Mutnovskaya	12.0	52.49170	158.24600
	Mutnovskaya	50.0	52.53640	158.20300
Papua New Guinea	Lihir	50.0	−3.12440	152.64600
Japan	Mori	25.0	42.13310	140.45500
	Onuma	9.5	39.98140	140.78200
	Sumikawa	50.0	39.98130	140.78200
	Matsukawa	23.5	39.92710	140.93300
	Kakkonda	80.0	39.82910	140.86500
	Uenotai	28.8	39.00120	140.60400
	Onikobe	15.0	38.80980	140.70900
	Yanaizu-Nishiyama	30.0	37.44000	139.69400
	Takigami	25.0	33.21150	131.27300
	Otake	12.5	33.12170	131.19000
	Hatchobaru	112.0	33.10530	131.18700
	Ogiri	30.0	31.93180	130.79800
	Yamagawa	30.0	31.19080	130.61400
Italy	Carboli 1	15.4	43.13690	10.82910
	Carboli 2	15.4	43.13610	10.82900
	Cornia 2	12.0	43.16600	10.80600
	Nuova Lagoni Rossi	12.7	43.16670	10.78460
	Le Prata	18.0	43.19620	10.83220
	Monteverdi 1	17.8	43.18610	10.76550
	Monteverdi 2	15.6	43.19730	10.76520
	Nuova Lago	10.9	43.15570	10.81450
	Nuova Monterotondo	8.0	43.15170	10.84970
	Nuova San Martino	36.2	43.14520	10.83210
	Nuova Sasso	14.0	43.17550	10.85300
	Sasso 2	16.7	43.17550	10.85400

(continued)

Table 1.3 (continued)

Country	Name	Capacity	Longitude	Latitude
Italy	Nuova Serrazzano	47.5	43.20420	10.80020
	Selva 1	18.3	43.18090	10.88320
	Farinello	52.4	43.23850	10.90450
	Nuova Castelnuovo	14.9	43.21690	10.90510
	Nuova Gabbro	19.1	43.26110	10.88540
	Nuova Lardarello 3	16.6	43.20420	10.80020
	Nuova Molinetto	14.5	43.21150	10.85860
	Sesta 1	13.9	43.22960	10.94750
	Valle Secolo 1–2	110.6	43.23620	10.86750
	Bagnore 3	19.9	42.84220	11.55900
	Bagnore 4	40.0	42.84070	11.55750
	Piancastagnaio 3	20.0	42.83320	11.70030
	Piancastagnaio 4	20.0	42.85800	11.70560
	Piancastagnaio 5	20.0	42.85690	11.70290
	Chiusdino 1	19.4	43.16060	11.06380
	Nuova Radicondoli	39.5	43.19340	11.04500
	Nuova Radicondoli 2	19.4	43.19310	11.04510
	Pianacce	14.1	43.21040	11.03110
	Rancia 1	19.1	43.21400	11.04680
	Rancia 2	19.1	43.21030	11.03540
	Travale 3	16.5	43.17110	11.03890
	Travale 4	38.9	43.16860	11.03800
Turkey	Efeler 1	47.4	37.87870	27.60800
	Efeler 2	24.0	37.86270	27.62530
	Efeler 3	24.0	37.86130	27.64710
	Efeler 4	24.0	37.86770	27.64000
	Deniz	24.0	37.86930	27.60020
	Irem	22.0	37.88610	27.58130
	Sinem	24.0	37.86560	27.59480
	Kerem 3	25.0	37.86560	27.58530

(continued)

Table 1.3 (continued)

Country	Name	Capacity	Longitude	Latitude
Turkey	Ken	24.0	37.85080	27.90430
	Gürmat 1	47.4	37.88480	27.63130
	Kubilay 1	24.0	37.82800	27.54210
	Pamukören 1–5	112.5	37.92280	28.53180
	Gumuskoy 1	13.2	37.85920	27.47700
	Dora 1	7.5	37.87650	28.09530
	Dora 2	9.5	37.85780	28.09210
	Dora 3	34.0	37.85690	28.07260
	Dora 4	17.0	37.87380	28.10280
	Sultanhisar 1	14.0	37.89590	28.17350
	Umurlu 1	12.0	37.85260	27.92760
	Umurlu 2	12.0	37.85350	27.92750
	Ken Kipas 3	25.0	37.83680	27.90380
	Mehmethan	25.0	37.89770	27.60650
	Greeneco 1	12.0	37.93430	28.79710
	Greeneco 2	12.0	37.93400	28.79790
	Kizildere 1	15.0	37.96000	28.87120
	Kizildere 2	80.0	37.94780	28.86700
	Türkeler 1–2	48.0	38.34250	28.55690
	Kemaliye 1–2	24.0	38.44830	28.42830
	Alasehir JES	45.0	38.40140	28.50420
	Babadere JES	8.0	39.60640	26.18530
	Dardanel Tuzla	7.5	39.55390	26.16120
	Efeler 6	27.0	37.86780	27.62360
	Kuyucak	22.0	37.93970	28.52660
	Kizildere 3	150.8	37.96980	28.80550
	Maspo	10.0	38.43730	28.39660
	Özmen 1 (SIS)	24.0	38.35950	28.40830
	Caferbey Salihil	16.0	38.47350	28.09590
	Sanko	15.0	38.49210	28.09200

(continued)

Table 1.3 (continued)

Country	Name	Capacity	Longitude	Latitude
Portugal	Ribeira Grande 1–2	13.0	37.78230	−25.48920
	Pico Vermelho	13.0	37.80170	−25.50460
France	Bouillante II	11.0	16.13710	−61.76740
Germany	Dürrnhaar	7.0	47.99610	11.72590
	Kirchstockach	7.0	48.02720	11.68780
Iceland	Hellisheidi 1–2	303.0	64.03720	−21.40080
	Svartsengi	74.4	63.87860	−22.43280
	Krafla	60.0	65.70290	−16.77440
	Nesjavellir	120.0	64.10820	−21.25660
	Reykjanesvirkjun	100.0	63.82640	−22.68190
	Theistareykir	90.0	65.89000	−16.96180
Kenya	Olkaria I	195.0	−0.89302	36.30850
	Olkaria II	105.0	−0.86649	36.29920
	Olkaria III	139.0	−0.89061	36.28910
	Olkaria IV	140.0	−0.91868	36.33390
	Olkaria I Wellhead	38.3	−0.89128	36.29540
	Olkaria IV Wellhead	42.8	−0.91820	36.33430
New Zealand	Ohaaki	40.0	−38.52770	176.29400
	Rotokawa	34.0	−38.61210	176.19300
	Nga Awa Purua	140.0	−38.61390	176.18400
	Mokai	113.0	−38.50850	175.92400
	Ngatamariki	82.0	−38.54700	176.19600
	Kawerau (MRP)	100.0	−38.06330	176.72600
	Ngawha	25.0	−35.41750	173.85200
	Wairakei	136.0	−38.62580	176.10400
	Poihipi Road	55.0	−38.63030	176.04100
	KA24	9.0	−38.06080	176.72200
	Te Mihi	166.0	−38.61910	176.04800
	Te Huka	23.0	−38.66840	176.11500
	TOPP1	25.0	−38.06530	176.72100

(continued)

Table 1.3 (continued)

Country	Name	Capacity	Longitude	Latitude
Costa Rica	Las Pailas I	41.5	10.75760	−85.35740
	Miravalles I–II	110.0	10.70060	−85.19390
	Miravalles III	29.5	10.71790	−85.18140
	Miravalles V	19.0	10.68380	−85.19330
Mexico	Cerro Prieto I–IV	570.0	32.39350	−115.25200
	Los Azufres I–II	191.0	19.78970	−100.66200
	Los Humeros II	53.4	19.64600	−97.43740
	Las Tres Virgenes	10.0	26.89500	−111.99000
	Los Humeros III—Phase A	25.0	19.81250	−97.38780
El Salvador	Ahuachapan	84.0	13.92030	−89.81750
	Berlin El Salvador I–II	108.0	13.52500	−88.50890
Guatemala	Amatitlan	25.2	14.40890	−90.60680
	Orzunil	28.0	14.77380	−91.50630
Honduras	Platanares	35.0	14.76090	−88.92720
Nicaragua	San Jacinto—Tizate 1–3	82.0	12.60860	−86.77400
	Momotombo I–II	25.0	12.39390	−86.54140
United States	McCabe (Calpine 5&6)	78.0	38.80540	−122.80900
	Ridge Line (Calpine 7&8)	69.0	38.81510	−122.80100
	Quicksilver (Calpine 16)	53.0	38.76620	−122.70700
	Lake View (Calpine 17)	52.0	38.82360	−122.78200
	Sulphur Springs (Calpine 14)	51.0	38.78570	−122.78300
	Socrates (Calpine 18)	50.0	38.76850	−122.74800
	Grant (Calpine 20)	43.0	38.77870	−122.75500
	Sonoma (Calpine 3)	42.0	38.79040	−122.75600
	West Ford Flat (Calpine 4)	24.0	38.78790	−122.72200
	Eagle Rock (Calpine 11)	66.0	38.82690	−122.80000
	Cobb Creek (Calpine 12)	52.0	38.80460	−122.78400
	Big Geysers (Calpine 13)	48.0	38.77260	−122.72900
	Calistoga (Calpine 19)	66.0	38.79000	−122.74500
	Bear Canyon (Calpine 2)	14.0	38.76290	−122.69300

(continued)

Table 1.3 (continued)

Country	Name	Capacity	Longitude	Latitude
United States	Aidlin (Calpine 1)	17.0	38.83360	−122.88100
	NCPA1	110.0	38.75160	−122.71900
	NCPA2	110.0	38.74860	−122.71100
	Bottle Rock	11.0	38.83480	−122.76800
	Ormesa I	14.0	32.81520	−115.25700
	Ormesa II	18.0	32.78760	−115.24800
	GEM Resources II	9.0	32.77560	−115.26300
	GEM Resources III	12.0	32.77600	−115.26300
	CE Turbo	10.0	33.16400	−115.61700
	Del Ranch (Hoch)	38.0	33.16360	−115.61800
	Elmore	38.0	33.17810	−115.60300
	Leather	38.0	33.17780	−115.56500
	Salton Sea I	10.0	33.15810	−115.64700
	Salton Sea II	20.0	33.15700	−115.64700
	Salton Sea III	50.0	33.15750	−115.63900
	Salton Sea IV	40.0	33.15680	−115.63900
	Salton Sea V	49.0	33.15380	−115.63900
	Vulcan	35.0	33.16310	−115.62000
	BLM	90.0	36.00230	−117.78900
	Navy I	100.0	36.03760	−117.79800
	Navy II	102.0	36.01960	−117.79200
	North Brawley	50.0	33.01440	−115.54100
	Gould I	10.0	32.71590	−115.51700
	Heber I	37.0	32.71590	−115.51800
	Heber South (Gould II)	14.5	32.71370	−115.51800
	Mammoth Pacific I	7.5	37.64510	−118.91400
	Mammoth Pacific II	11.0	37.64620	−118.91000
	Mammoth III (PLES 1)	15.0	37.64590	−118.90900
	SIGC	42.0	33.03990	−115.36700
	Beowave	19.6	40.55470	−116.61700

(continued)

Table 1.3 (continued)

Country	Name	Capacity	Longitude	Latitude
United States	Brady Hot Springs	12.0	39.79600	−119.01000
	Desert Peak	11.0	39.75390	−118.95300
	Galena II	15.0	39.38880	−119.74900
	Galena III	20.0	39.39030	−119.75500
	Galena I (Richard Burdett)	30.0	39.39050	−119.75500
	Steamboat Hills I	14.4	39.37320	−119.76500
	Steamboat II	29.0	39.39490	−119.74700
	Steamboat III	24.0	39.39430	−119.74800
	Soda Lake II	18.0	39.55640	−118.85000
	Stillwater	47.3	39.54640	−118.55600
	Dixie Valley I–II	73.4	39.96740	−117.85600
	Salt Wells	23.6	39.29500	−118.57300
	Faulkner I	39.5	40.99500	−118.14300
	San Emidio Repower	12.8	40.38070	−119.40000
	Jersey Valley	22.5	40.18060	−117.47600
	Tuscarora	32.0	41.46710	−116.15100
	McGinness Hills	52.0	39.59070	−116.91300
	Hudson Ranch I	49.5	33.20420	−115.57400
	Neal	23.0	44.02250	−117.46700
	Blundell I	26.0	38.48890	−112.85200
	Blundell II	11.0	38.48980	−112.85300
	Thermo Hot Springs	10.0	38.16120	−113.19500
	Don Campbell	16.0	38.83700	−118.32500
	Puna	30.0	19.47810	−154.88800
	Cove Fort	25.0	38.56130	−112.58000
	GEM Bottoming Unit	9.0	32.77530	−115.26400
	Puna Expansion	8.0	19.47680	−154.88900
	Patua Phase I	30.0	39.58240	−119.07600
	Raft River	15.8	42.09920	−113.38200
Chile	Cerro Pabellon	48.0	−21.44950	−68.33780

Table 1.4 Installed geothermal power capacity of the top ten geothermal power countries as of 2020 (Huttrer 2021)

Rank	Country	Installed Capacity in 2020 (MWe)	Rank	Country	Installed Capacity in 2020 (MWe)
1	USA	3,700	6	Mexico	1,105
2	Indonesia	2,289	7	New Zealand	1,064
3	Philippines	1,918	8	Italy	916
4	Turkey	1,549	9	Iceland	755
5	Kenya	1,193	10	Japan	550

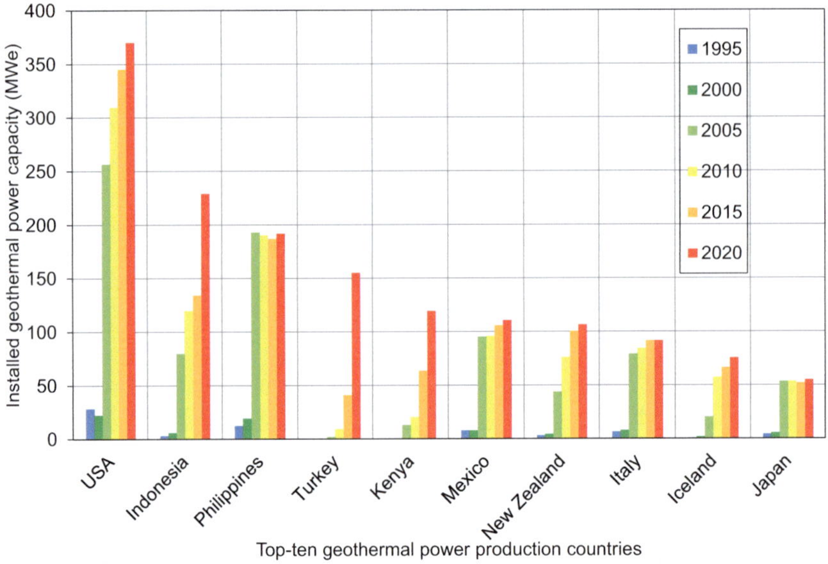

Fig. 1.9 Historical trends of the installed geothermal power capacities among the top ten geothermal power production countries from 1995 to 2020 (Huttrer 2001, 2021; Bertani; 2005; Bertani 2016)

1.3 High-Temperature Geothermal Systems in Europe

1.3.1 Geothermal Settings of Europe

High-temperature geothermal systems are found in Italy, Iceland, Greece, Turkey, and Portugal in Europe (Fig. 1.10). In particular, Italy first performed geothermal power generation in the world, experimentally in 1904 and commercially in 1913, opening the world geothermal power development history. Iceland is not only situated on the Mid-Atlantic Ridge but also on the Iceland Hot Spot. Geothermal resources

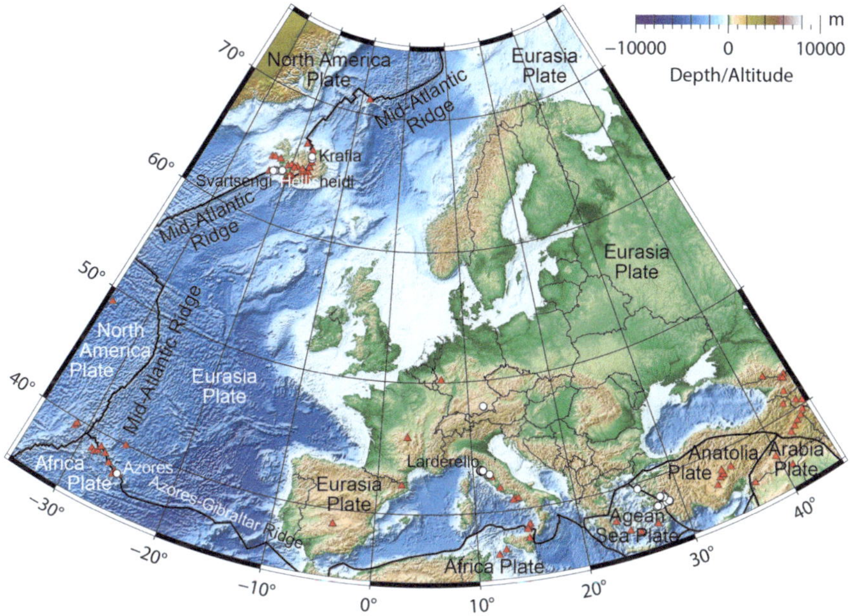

Fig. 1.10 Topographic relief map of Europe. White circle: geothermal power plants, red triangle: Holocene volcanoes, thick line: plate boundaries, and thin line: borders of counties. Topographic and bathymetric maps are drawn by the ETOPO1 1 arc-minute global relief model (Amante and Eakin 2009) and GMT 6.4.0 (Wessel et al. 2019)

are exceeding the energy demands of the population of Iceland, 380,000, and Iceland has built a geothermal utopia using abundant geothermal resources. The refining of aluminum requires abundant electricity, and Iceland performed this with excessive geothermal power, which brings excellent income from aluminum trading.

Europe is composed of two types of volcanic settings: the subduction zones in the Mediterranean Sea, like Italy, Greece, and Turkey, the ocean ridges in the Mid-Atlantic Ridge, like Iceland, and the Azores-Gibraltar Ridge, like the Azores Islands, Portugal.

1.3.2 Italy

It is known that the Africa Plate moves down beneath the Eurasia Plate in the Mediterranean Sea area (Fig. 1.10). The tectonic process is, however, complicated in the surroundings of Italy, which includes not only the oceanic plate subduction but also the continental collision and back-arc basin formation (Fig. 1.11; Minissale et al. 2000; Conti et al. 2015). A typical Wadati-Benioff zone is only recognized from the Calabrian Arc to the Tyrrhenian Sea in the southeastern sector of Italy by the seismological study (Palano 2015). As a result, most of the active volcanoes such as Etna, Stromboli, and Vesuvius are concentrated in the southeastern sector of Italy (Fig. 1.11), and the northernmost Quaternary volcano is Monte Amiata in Italy, which is about 75 km southeast of the Larderello geothermal field (Fig. 1.12).

A volcano in Monte Amiata is classified as a Pleistocene volcano, not a Holocene volcano. Distributions of the Pliocene–Pleistocene volcanic rocks are also described

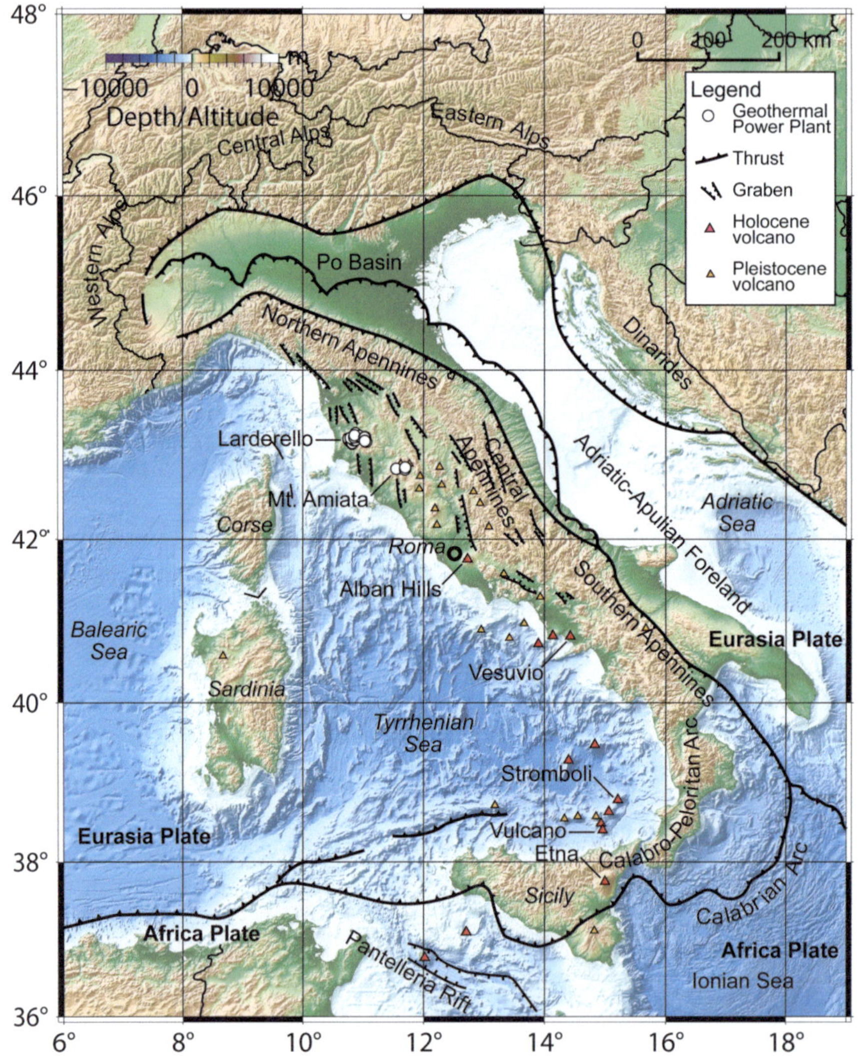

Fig. 1.11 Tectonic map of Italy with thrusts, grabens, Quaternary volcanoes, and geothermal power plants (Minissale et al. 2000; Palano 2015; Conti et al. 2015). The UTM projection map was drawn by the ETOPO2022 15 arc-second global relief model (NOAA National Centers for Environmental Information 2022) and GMT 6.4.0 (Wessel et al. 2019)

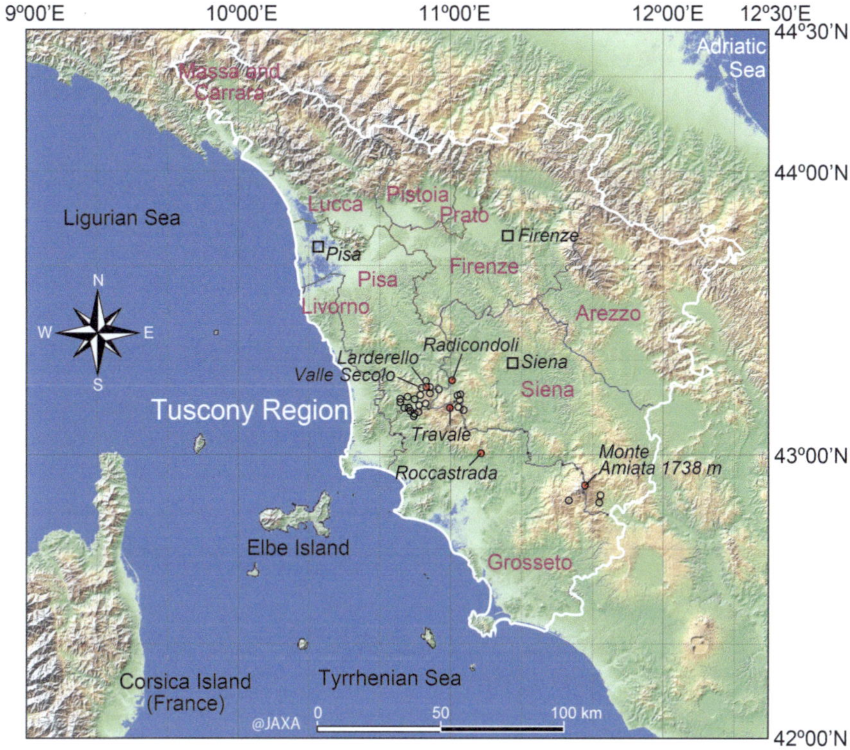

Fig. 1.12 Map of localities for geothermal power development in the southern Tuscany Region, Italy. Small open circle: geothermal power plants in Table 1.3, and circle filled with red: key locality names. Topography was drawn by the authors using the ALOS World 3D data with 30 m resolution provided by the Japan Aerospace Exploration Agency (JAXA) and the shading software KASHMIL 3D

in Roccastrada, 35 km southeast of Larderello (Fig. 1.12; Arias et al. 2010), but the volcanic edifice is not necessarily clear and seems deeply dissected (Fig. 1.12). Thus, active volcanoes are not recognized in the Larderello-Travale geothermal field. The young magmatism is supposed in the Larderello-Travale geothermal field where Quaternary granitic intrusions are described at a depth greater than the K-seismic reflection horizon at a depth of 3–5 km (Fig. 1.13; Romagnoli et al. 2010; Arias et al. 2010). The meaning of the intensive seismic reflection at the K-horizon is still controversial. One possibility is the brittle-plastic transition, and the other is the top of the Quaternary granitic intrusions. Both depths are probably close, as seen in Kakkonda, Japan (See Chap. 6, Sect. 2.1), and both possibilities are broadly compatible. A shallow zone of the K-horizon in Fig. 1.13 shows a 25 km (NW–SE) by 38 km (NE-SW) area, which encompasses the 950 km^2 area. If this zone exhibits a Quaternary granitic intrusion complex, the area is three times wider than the Felsite intrusion in The Geysers, USA (Thompson 1991). Thus, a geothermal heat source in the Larderello-Travale geothermal field is probably ascribed to the Quaternary granitic intrusion complex on a large-batholith scale.

It is well known that the Larderello-Travale geothermal field is characterized by a vapor-dominated geothermal system (White et al. 1971). From a geological point of view, two types of geothermal reservoirs are utilized in the Larderello-Travale geothermal field. Until the 1970s, the main reservoirs for geothermal power developments were carbonate rocks at a shallow depth, mainly less than 1,500 m which compose a principal constituent of the late-Triassic to Cretaceous Tuscan Sequence or Tuscan Unit (Fig. 1.14). The Tuscan Sequence is exposed on the surface to the southeast of the Larderello-Travale geothermal field (Fig. 1.14). Limestone is easily dissolved by the carbon dioxide-bearing meteoric water and easily forms solution caves. Therefore, limestone plays the role of a non-fracture type porous reservoir. The shallow carbonate reservoir originally has a temperature of about 220–250 °C and a pressure of 2–4 MPa. During the 1980s, a deep exploration program

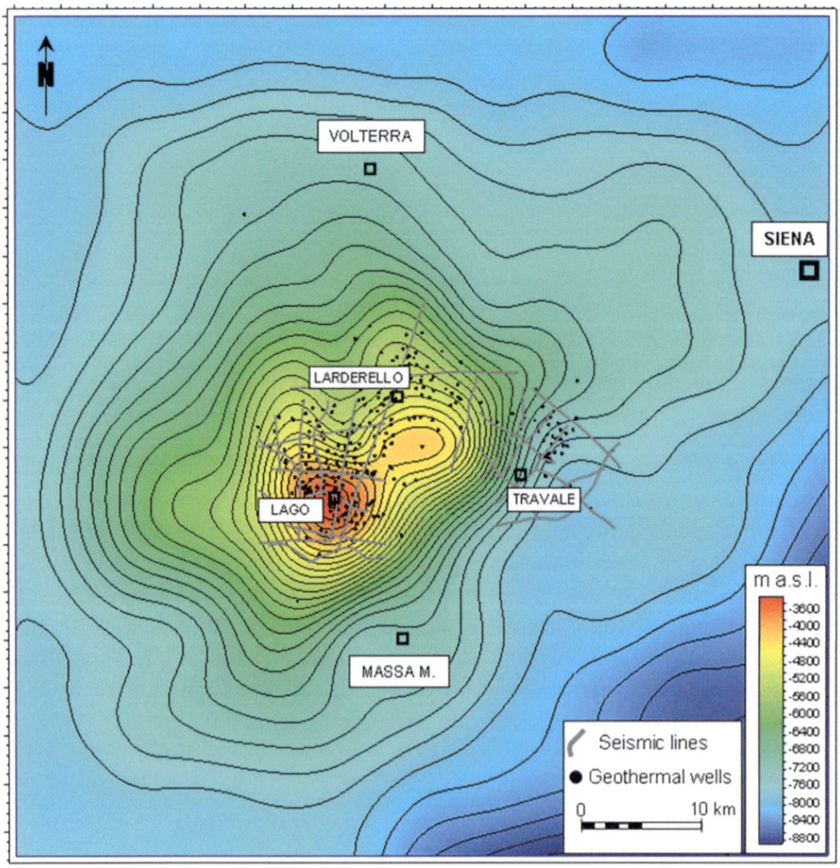

Fig. 1.13 Map of the depth of "K-horizon" (See Fig. 1.15) in m above sea level (Arias et al. 2010)

was conducted into the early Paleozoic metamorphic basements at depths between 2,500 m and 3,500 m. This deep reservoir is characterized by a temperature greater than 300–350 °C and pressure around 60–70 MPa (Arias et al. 2010). The geological age of the shallow reservoirs is from the Triassic to Cretaceous, and that of the deep reservoirs is from the Precambrian to early Paleozoic (Fig. 1.15; Romagnoli et al. 2010). The age of the deep reservoirs is one of the oldest geothermal reservoirs in the world.

Boric acid was discovered in natural geothermal steam in Larderello in 1777, and industrial extraction of the boric acid with the utilization of the geothermal steam boiler has been operated since 1818. Then, geothermal power generation was experimentally invented by Prince Piero Ginori Conti in 1904 and has been commercially operated since 1913 (Nakamura 1964). Italy initiated geothermal power developments worldwide throughout this history.

It should be noted that almost the entire geothermal power developments of Italy were achieved in a very compact area, southern Tuscany which is 30 km (NE–SW) by 100 km (NW–SE) including the Larderello-Travale geothermal field and the

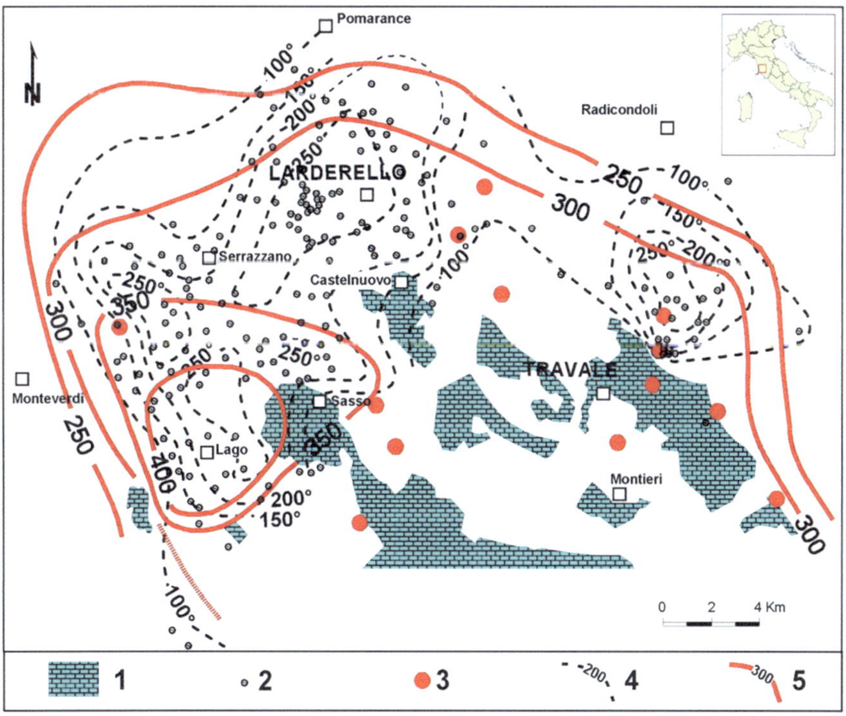

Fig. 1.14 Temperature distribution in the Larderello-Travale geothermal field. (1) Outcrop of permeable formations (mainly carbonate formation of Tuscan Units, Late Triassic-Late Jurassic); (2) Shallow wells; (3) Deep and recent wells; (4) Temperature at the top of the shallow reservoir; (5) Temperature at 3000 m b.s.l. (Arias et al. 2010)

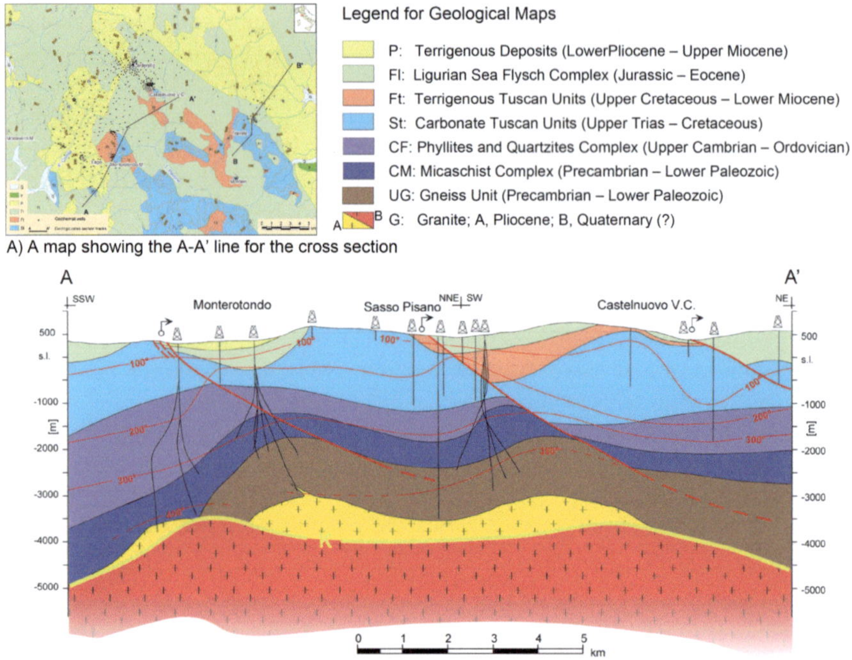

B) Geological cross section along the A-A' line described in A)

Fig. 1.15 Geological cross-section B) along the SW-NE (A-A′) line in A) in the Larderello geothermal field (Romangnoli et al. 2010). The thick yellow line indicates seismically identified "K-horizon"

Monte Amiata geothermal field (Fig. 1.12). As of 2020, this area is producing 916 MWe geothermal power in the installed capacity (Huttrer 2021). Remarkable is the fact that the area density of geothermal power generation capacity is very high in Italy. This efficiency was enabled by the numerous shallow geothermal wells in the vapor-dominated geothermal systems (Fig. 1.14) that were probably provided by a large-scale geothermal heat source, the Quaternary granitic batholith (Figs. 1.13 and 1.15).

1.3.3 Iceland

Iceland is situated on the Mid-Atlantic Ridge and forms one of the large-scale hot spots, called the Iceland Hot Spot (Fig. 1.16). The boundary between the America Plate and Eurasia Plate spreads 3 cm a year from west to east. The simplified plate boundary drawn in Fig. 1.16 was quoted from the global data by Bird (2003), indicating that the southern boundary jumps 100 km from west to east by a transform fault going to the north. However, part of the south boundary is straightforwardly continued from the Reykjanes peninsula to the further northeast, such as the Hellisheidi geothermal field, Nesjavellir geothermal field, and Þingvellir (Thingvellir) National Park (famous lookout for the number of fissures called gjá) without any significant transform faults (Fig. 1.17). Therefore, the spreading center is probably duplicated from west to east with a 100 km gap or has a total of 100 km width in southern Iceland.

The answer to this view could probably be given as the former solution as seen in the geological age map of basaltic lava (Fig. 1.17). Two zones of basalt younger than 0.8 million years are separately described from west to east by the intervention of the zone of somewhat older basalt of 3.3 to 0.8 million years in southern Iceland. Therefore, two lines of spreading centers are probably distributed separately in southern Iceland.

As of 2020, the installed capacity of geothermal power plants is 755 MWe in Iceland (Ragnarsson et al. 2021; Table 1.4). In addition to this, about 90% of all houses in Iceland are heated by geothermal water (Ragnarsson et al. 2021). The population of Iceland is only 380,000. If you think only about the living demands, the installed capacity of 150 MWe is probably more than enough. However, the excess electricity

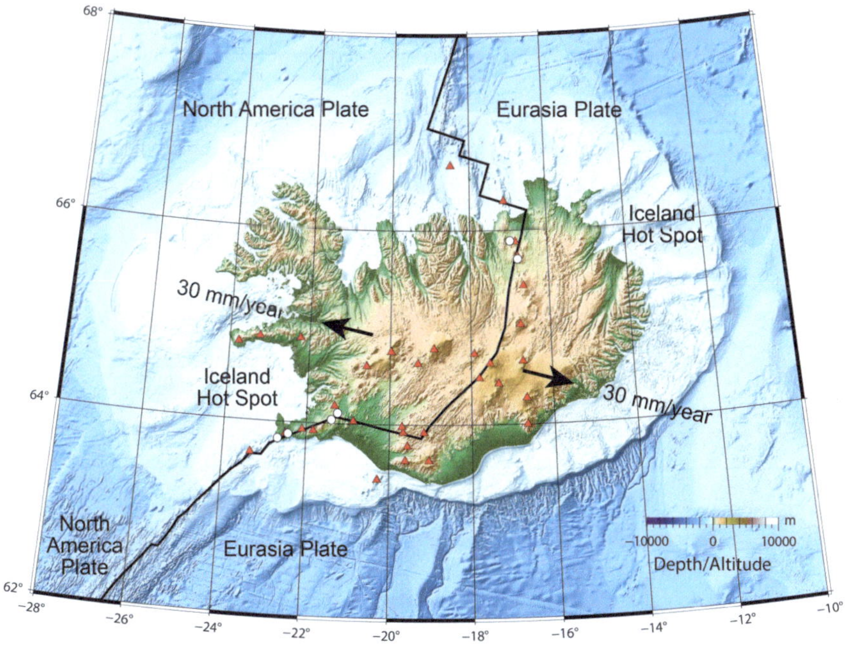

Fig. 1.16 Tectonic map of Iceland with a plate boundary (thick line, Bird 2003), Holocene volcanoes (red triangle), and geothermal power plants (open circle). The Lambert projection map was drawn by the ETOPO2022 15 arc-second global relief model (NOAA National Centers for Environmental Information 2022) and GMT 6.4.0 (Wessel et al. 2019)

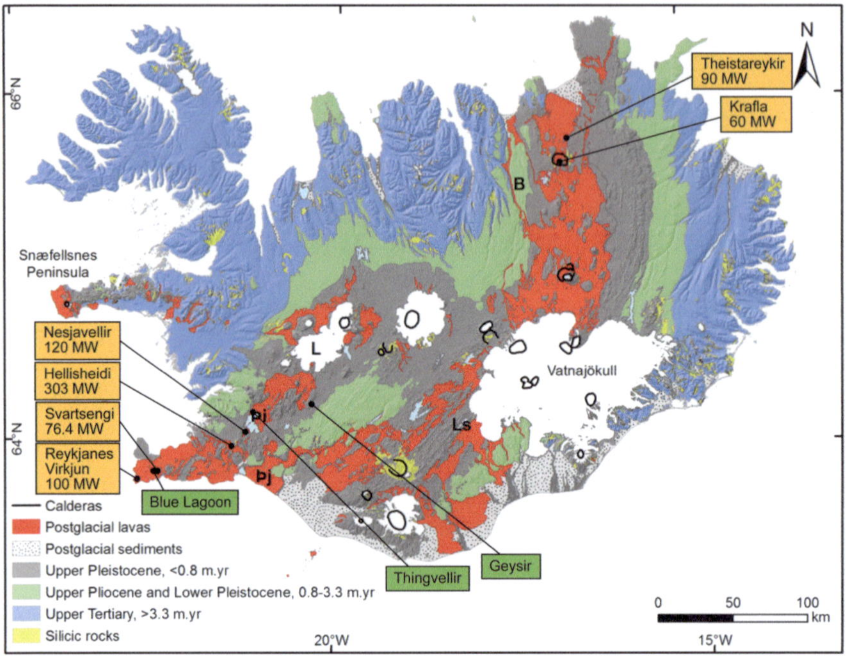

Fig. 1.17 Geologic map of Iceland (Sigmundsson et al. 2020), and the capacities of geothermal power plants are according to the data as of 2019 (Ragnarsson et al. 2021)

is efficiently used for the electricity-consuming aluminum refinement in Iceland, and then Iceland is a rich country in aluminum exports. Iceland is a geothermal utopia.

1.4 High-Temperature Geothermal Systems in Oceania

1.4.1 Geothermal Settings of Oceania

Many Holocene volcanoes are known along the subduction zones of the Pacific Plate to the Australia Plate in the Oceania region. Most of them form isolated small volcanic islands (Fig. 1.18), providing valuable geothermal resources for near-future green power developments. A present center for geothermal power developments is the Taupo Volcanic Zone, North Island in New Zealand, where clustered calderas are formed under the extension tectonics (Wilson et al. 1984).

If the subduction angle is low, as in the Chilean-type subduction, contraction tectonics typically occurs. If the subduction angle is high, as in the Mariana-type subduction, extension tectonics tends to form back-arc spreading zones (Uyeda and

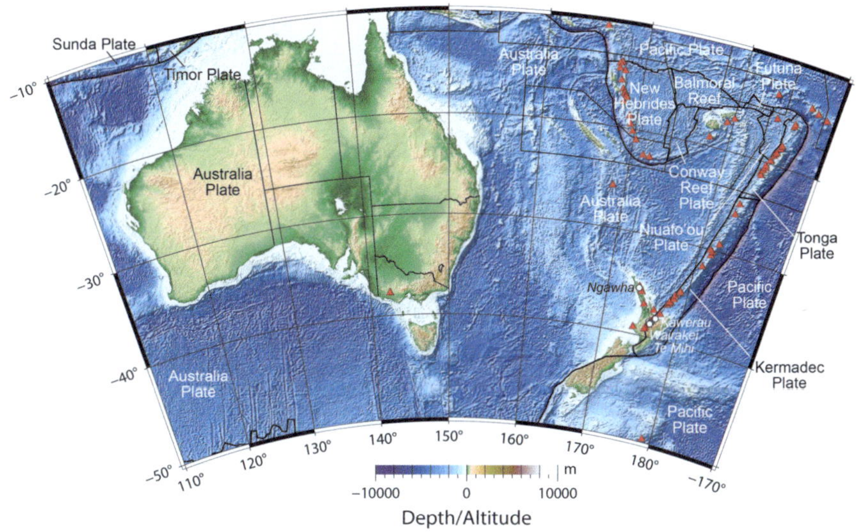

Fig. 1.18 Topographic relief map of Oceania. White circle: geothermal power plants, red triangle: Holocene volcanoes, thick line: plate boundaries, and thin line: borders of counties. Topographic and bathymetric maps are drawn by the ETOPO1 1 arc-minute global relief model (Amante and Eakins 2009) and GMT 6.4.0 (Wessel et al. 2019)

Kanamori 1978). The Taupo Volcanic Zone is a typical back-arc spreading zone by the high-angle Kermadec–Hikurangi subduction.

1.4.2 New Zealand

The earlier geothermal power developments in the world were biased toward the vapor-dominated geothermal systems such as Larderello in Italy and The Geysers in the USA. In 1958, New Zealand opened the two-phase separation technology using water-dominated geothermal systems in the Wairakei geothermal power plant. This breakthrough expanded the applicability of geothermal power developments to the widespread water-dominated geothermal systems in the world.

The South Island in New Zealand mainly comprises metamorphic rocks, and the Holocene volcanoes and high-temperature geothermal resources are only found on the North Island. Notably, the Taupo Volcanic Zone is a volcanic rift system formed in the back-arc spreading by the high-angle subduction of the Pacific Plate from the Hikurangi Trough (Fig. 1.19). Under the extension tectonics, several calderas are recognized in the Taupo Volcanic Zone (Wilson et al. 1984). Most of the geothermal power developments in New Zealand were conducted in the Taupo Volcanic Zone. Numerous monogenic volcanoes composed of basalts, such as Rangitoto Island, One Tree Hill, Mt. Eden, and Mt. Wellington, are recognized in Auckland City and its vicinity (Heming and Barnet 1986). Rangitoto Island is the largest and youngest volcano, estimated to be 770 ± 50 year B.P. in age (Heming and Barnet 1986). However, they do not have any geothermal and hydrothermal manifestations. One exception is the Ngawha geothermal power plants with 25 MWe in North Land (Tabel 1.2.1; Fig. 1.19). This field may have a long-lived magma system beneath the volcano. As of 2020, the installed capacity of geothermal power plants is 1,064 MWe in New Zealand (Table 1.4; Huttrer 2021).

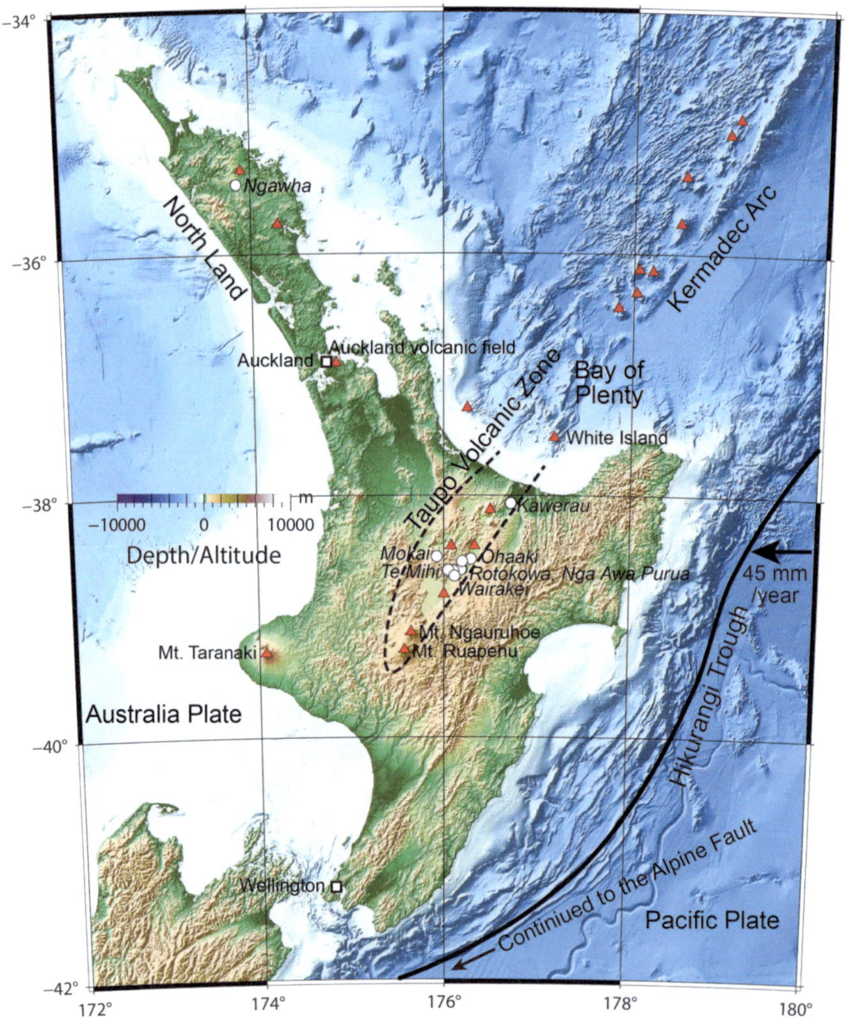

Fig. 1.19 Tectonic map of North Island, New Zealand, with Holocene volcanoes (red triangle) and geothermal power plants (open circle). The Lambert projection map was drawn using the ETOPO2022 15 arc-second global relief model (NOAA National Centers for Environmental Information 2022) and GMT 6.4.0 (Wessel et al. 2019)

1.5 High-Temperature Geothermal Systems in North America

1.5.1 Geothermal Settings of North America

The western United States and western Mexico have unique geological settings where, for the most part, it is a northern extension of the East Pacific Rise and accompanied by transform faults in the onshore areas. It should be noted that a zone with about 1,500 km distance between the Gorda Ridge in the west off California, the USA, and segmented small ridges near the border of the USA-Mexico can be represented as the San Andreas transform fault, probably one of the longest transform faults on the Earth (Fig. 1.20). This zone involves large-scale geothermal fields such as Cerro Prieto, Salton Sea, Coso, and probably The Geysers, the largest geothermal development field in the world. Although this region is situated on the active continental margin, it is also represented as the ocean ridge from a plate tectonic point of view. This special situation occurred when an ocean ridge system, such as the Juan de Fuca Ridge, Gorda Ridge, and East Pacific Rise, moved eastward and subducted to the western subduction zone of the North America Plate through the Cenozoic era. As a result, the San Andreas fault grew by the subduction of the ocean ridge as a transform fault (Atwater 1970). Horst-graben physiography and volcanisms in the Basin and Range Province east of the San Andreas fault were also formed by extension tectonics by the subduction of the East Pacific Rise beneath the western margins of the North America Plate since the Middle Miocene age. Therefore, many geothermal power plants in the western USA and northwestern Mexico are broadly explained by the extension tectonics due to the subduction of the East Pacific Rise.

Two exceptions are known: one is the Yellowstone Hot Spot at the eastern edge along the flood basalt lava distribution on the Snake River Plain, and the other is the Hawaii Hot Spot at the east edge of the volcanic Hawaiian Islands where volcano comes to be young eastwards (Fig. 1.20). These hot spots are laterally a stable anchor fixed to the lower mantle, but their overlying plates are constantly moving westwards. Then, the Snake River Plain and Hawaiian Islands were formed as renewable volcanic trajectories of the mantle plumes from the west to the east.

1.5.2 The USA

The USA's installed geothermal power capacity is 3,700 MWe as of 2020 (Table 1.4; Huttrer 2021), the highest in the world. The Geysers is not only the largest geothermal field in the USA but also in the world, encompassing an area of Sonoma County, Lake County, and Mendocino County in California (Fig. 1.21; Dobson et al. 2020). The Geysers geothermal field is a vapor-dominated geothermal system characterized by producing dry steam.

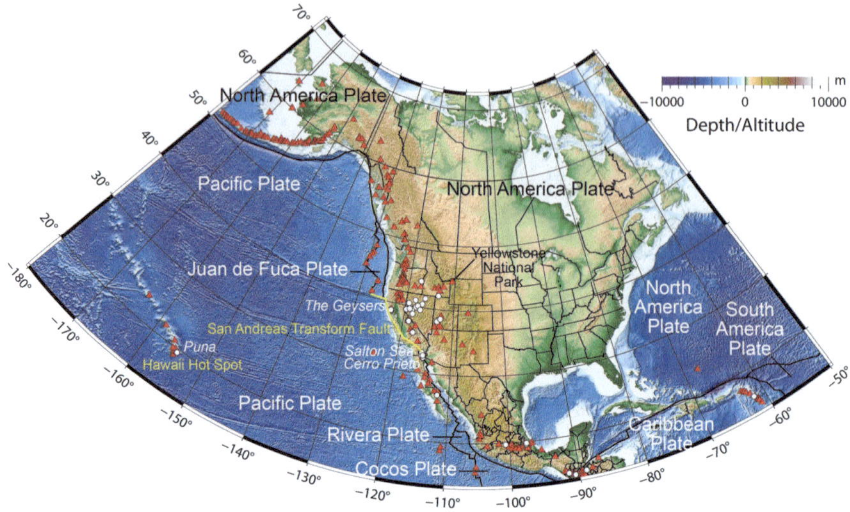

Fig. 1.20 Topographic and tectonic map of North America. White circle: geothermal power plants, red triangle: Holocene volcanoes, thick line: plate boundaries, and thin line: borders of counties. Topographic and bathymetric maps are drawn by the ETOPO1 1 arc-minute global relief model (Amante and Eakins 2009) and GMT 6.4.0 (Wessel et al. 2019)

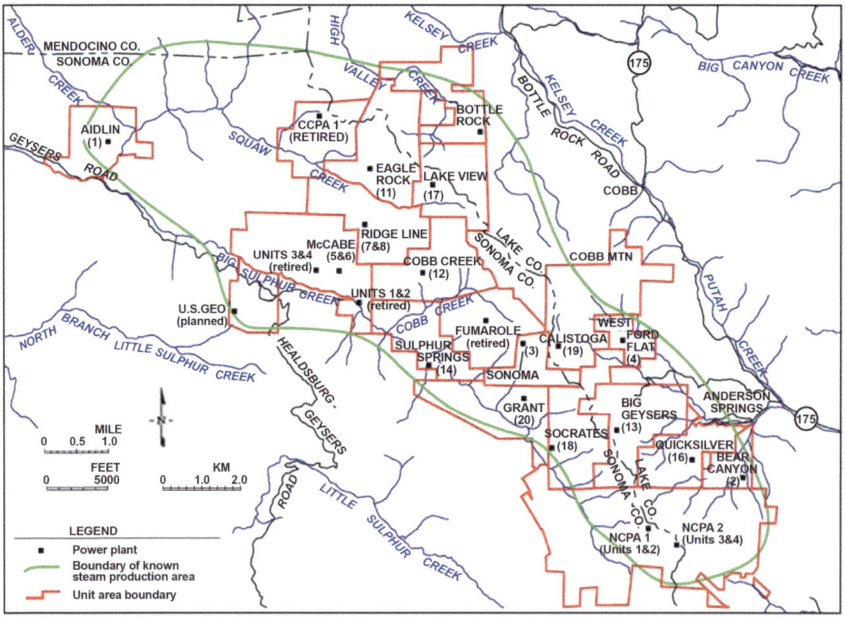

Fig. 1.21 Map of The Geysers geothermal field (Dobson et al. 2020). Note the steam production zone indicated by the green line

Electric power generation started at The Geysers in 1960 with 12 MWe of the installed gross capacity in the PG&E's Unit 1 plant. Since then, steam production has rapidly increased by 23 million tons per year in 1974, 46 million tons in 1981, and 249 million tons in 1988 (Endy and Butler 2010). If steam production represents the installed gross capacity, the number is about 400 MWe in 1974, 943 MWe in 1984, and 2,100 MWe in 1988 (Barker et al. 1991). The number one ranking of the USA on the geothermal power developments in the world is mainly owing to The Geysers geothermal field. The dry steam production rose from 70 kg/s in 1965 to a peak of 3546 kg/s in 1987. However, the decline started at The Geysers from 1988 (Barker et al. 1991; Endy and Butler 2010; Goyal and Conant 2010).

The decline was derived from the essential characteristics of the vapor-dominated geothermal systems, which made it difficult to collect amounts of water for reinjection, except for the small amounts of steam condensate. Injection of steam condensate began in April 1969 into the well sulfur Bank 1, with the startup of PG&E's Unit 4 (Endy and Buttler 2010). Steam condensate injection alone replaces about 22% of the mass of steam withdrawn from the reservoir. The net loss of mass evaporating into the air from the cooling towers in geothermal power plants at The Geysers is typically between 70 and 80% of the produced mass (Endy and Buttler 2010).

Gravity surveys, seismic studies, and drillhole data suggest that the heat source for the steam reservoir is a large-scale intrusion called the Felsite intrusion. The top is 1.5 km beneath the surface (0.61 km below sea level), and the areal extent is greater than 13 km in diameter. If the area of the steam field is roughly the same as that of the Felsite intrusion (Fig. 1.22), the area of the Felsite intrusion is estimated to be about 300 km^2, and this intrusion can be called a batholith, which is defined to be more than 100 km^2. The vapor-dominated reservoirs are thus heated by a silicic "Felsite" intrusion (Fig. 1.23). Steam is produced from the late Jurassic to late Cretaceous Franciscan meta-greywacke sandstone reservoir (Fig. 1.23). The reservoir is capped by the low-permeability ophiolite *mélange*, particularly capped by the serpentinite *mélange* (Fig. 1.23). Therefore, including Larderello in Italy and The Geysers in the USA, the pre-Tertiary reservoirs are common worldwide.

To overcome the decline of the dry steam production, relatively large-scale injection projects started (Goyal and Conant 2010). One project is called Southeast Geysers Effluent Pipeline (SEGEP), started in September 1997. It brought 341 L/s of lake water and secondary treated water from the treatment plants of Lake County Sanitation District (LACOSAN) to Calpine and NCPA through an 80 km pipeline. Another injection project is called Santa Rosa Geysers Recharge Project (SRGRP), started in October 2003, bringing 482 L/s of tertiary treated water from the city of Santa Rosa's treatment plant through a 72 km pipeline. By December 31, 2008, a gain of 397 kg/s in steam production was realized (Goyal and Conant 2010).

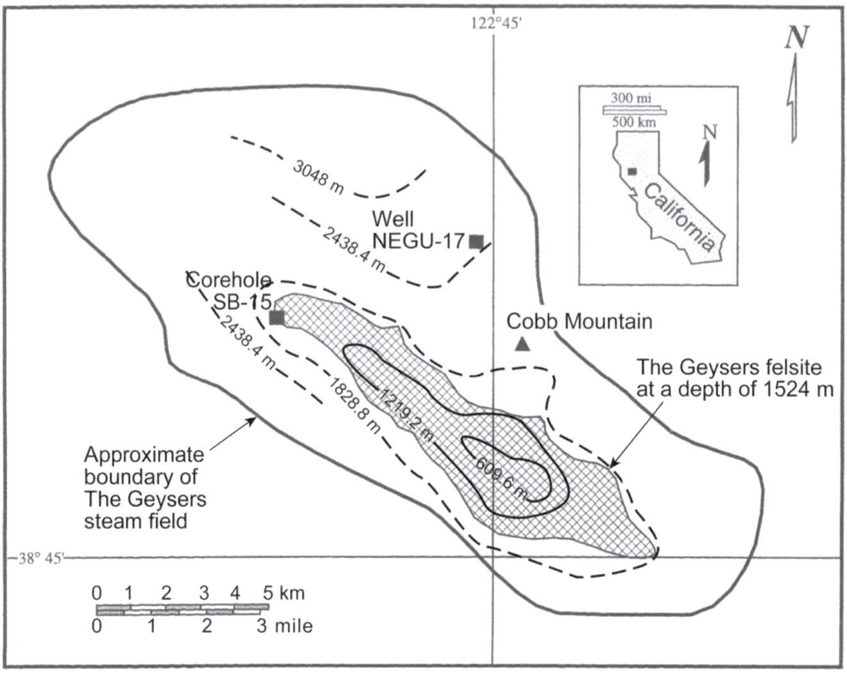

Fig. 1.22 The Geysers geothermal steam field and the depth contour map of the top of the Felsite intrusion below sea level in meters (non-rounded numbers derived from the original feet unit expression). Combined with Thompson (1991) and Persoff and Hulen (2001)

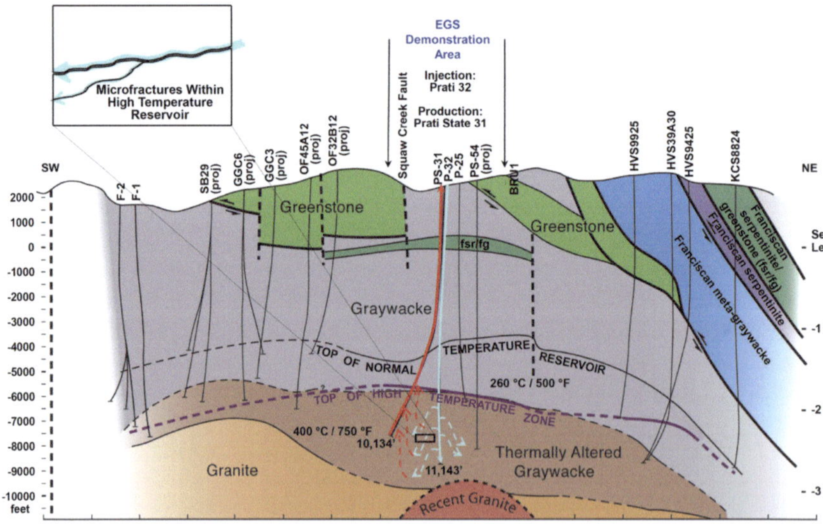

Fig. 1.23 SW-NE geological cross-section of the northwestern part of The Geysers geothermal fields (Rutqvist et al. 2016; Garcia et al. 2012)

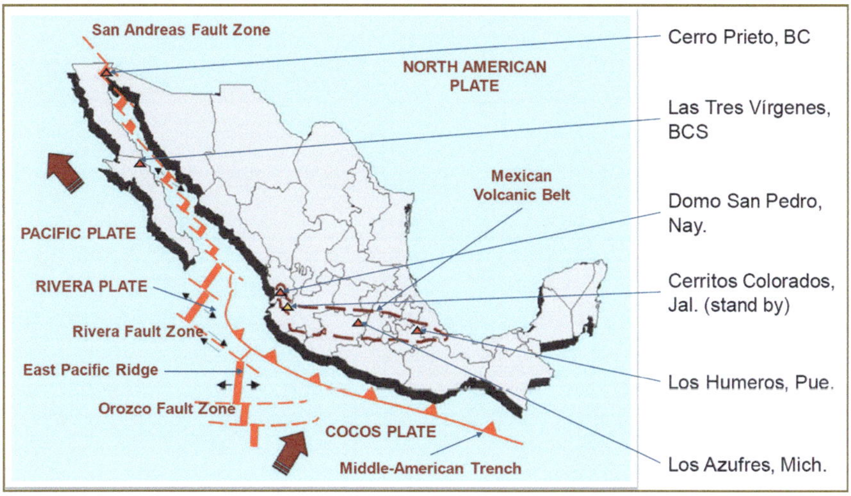

Fig. 1.24 Geothermal fields in Mexico and tectonic regimes (Gutiérrez-Negrín et al. 2021)

1.5.3 Mexico

Mexico's installed geothermal power capacity is 1,105 MWe as of 2020 and is number six in the world ranking (Table 1.4; Huttrer 2021). Tectonic settings of high-temperature geothermal systems are divided into two regimes (Fig. 1.24). One is a landward extension of the ocean ridge system consisting of the reputational combinations of the short axial valleys and the short transform faults of the East Pacific Rise. The largest Cerro Prieto geothermal field in Mexico, with an installed capacity of 570 MWe, is developed in this regime. The other is the Mexican volcanic belt by the subduction zone of the Cocos Plate (Fig. 1.24). The geothermal power plants in Los Azufres of 270.5 MWe and Los Humeros of 119.8 MWe were developed in this regime. These three geothermal power plants are producing most of the geothermal power in Mexico.

1.6 High-Temperature Geothermal Systems in Southeast Asia

1.6.1 Geothermal Settings of Southeast Asia

Southeast Asia geologically consists of the Sunda Plate (Fig. 1.25). Indo-China Peninsula, Malay Peninsula, Sumatra Island, Java Island, Kalimantan (Borneo) Island, and their broad continental shelf (shallow sea) areas are called "Sundaland" which is geologically a continent even though including broad shallow sea areas. The

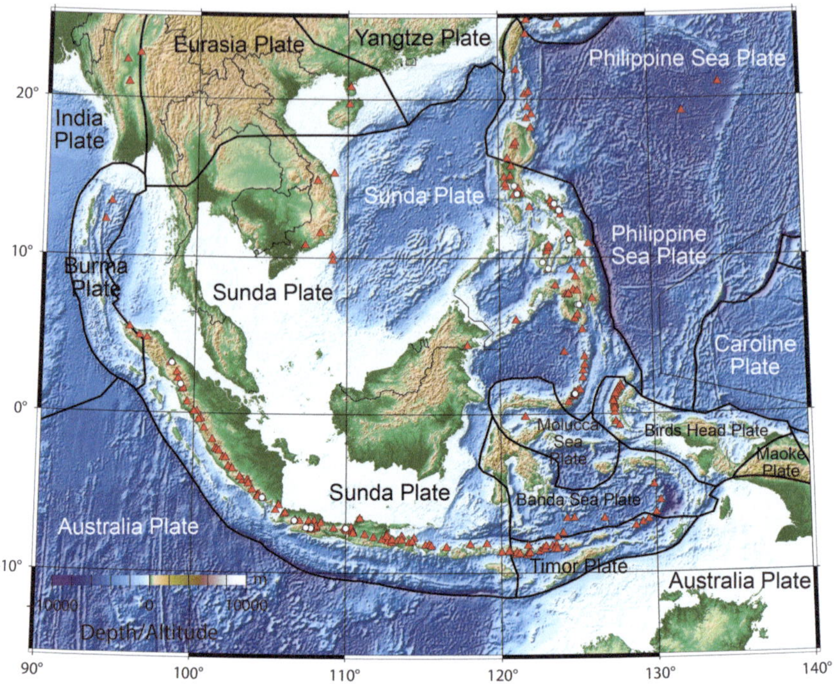

Fig. 1.25 Topographic and tectonic map of Southeast Asia. White circle: geothermal power plants, red triangle: Holocene volcanoes, thick line: plate boundaries, and thin line: borders of counties. Topographic and bathymetric maps are drawn by the ETOPO1 1 arc-minute global relief model (Amante and Eakins 2009) and GMT 6.4.0 (Wessel et al. 2019)

western central part of the Sunda Plate is composed of the "Sundaland" continent. The Philippine Sea Plate and the Carolina Plate are underplating the Sunda Plate from the east, and the Australia Plate is underplating the Sunda Plate from the south. The former subduction regime forms active volcanic regions in the Philippines, and the latter forms active volcanic areas in Indonesia.

In addition to the two outstanding geothermal countries, there are many other geothermal potential countries in Southeast Asia, such as Papua New Guinea, Malaysia, Vietnam, Thailand, Myanmar, and China.

1.6.2 *Indonesia*

Indonesia's installed geothermal power capacity is 2,289 MWe as of 2020 and ranks second in the world (Table 1.4; Huttrer 2021). The southern volcanic belt through Sumatra, Java, and the Lesser Sunda Islands (Nusa Tenggara) in Indonesia extends over 45 degrees along the equator from west to east, roughly 5,000 km on the great-circle distance. However, considering the curvature in Sumatra and tight bending in the Banda Sea, the volcanic belt reaches 6,000 km or more (Fig. 1.25). Active volcanic belts are also known in northern Sulawesi and Halmahera. Thus, the geothermal resource potentials in Indonesia are huge, probably comparable to the USA, and have recently been estimated to be 29 GWe (Darma et al. 2021).

In this subsection, an example of Indonesia's substantial geothermal resource potential is introduced through the relationship between the clustered volcanoes and the Sumatran fault in Sumatra Island in Fig. 1.26, by Muraoka et al. (2010b) with slight simplifications.

(1) *Introduction*

The Sumatran fault is the major trench-parallel strike–slip fault that bounds the fore-arc sliver plate and back-arc plate of Sundaland, as shown in Fig. 1.26. The Sumatran fault is a right-lateral strike–slip fault that is traced about 1,660 km over the entire Sumatra Island. A variety of observations such as a geologic unit offset, stream offset, and geodetic measurement, including the global positioning system (GPS), indicates that the slip rate of the Sumatran fault is large in northwestern Sumatra by 24 mm/yr and small in southeastern Sumatra by 6 mm/yr (McCaffrey 1991; Genrich et al. 2000; McCaffrey et al. 2000; Prawirodirdjo et al. 2000). This slip-rate regime inevitably leads to the along-arc stretching of the fore-arc sliver plate and the along-arc shortening of the back-arc plate. It should be noted that the major volcanic zone of Sumatra often crosses over the Sumatran fault, but most of the zone lies slightly in the back-arc side parallel to this fault. The Quaternary volcanoes tend to form *echelon*-shaped elliptical clusters commonly found in the oblique subduction zones (Muraoka et al. 2002; Muraoka and Nasution 2004; Fig. 1.26).

(2) *Distribution of Pull-Apart Basins Along the Sumatran Fault*

Thirteen pull-apart basins and one pressure ridge are recognized along the Sumatran fault by observing topographic maps, observing satellite images, and our field surveys (Fig. 1.27). Almost all the pull-apart basins along the Sumatran fault occur near the clustered volcanoes but rarely inside of the areas (Fig. 1.26). This observation indicates two contradictory roles of volcanic clusters in the formation of pull-apart basins: one is that the volcanic clusters macroscopically play the role of generators to pull-apart basins along the Sumatran fault, and the other is that the areas of the volcanic clusters themselves are inhibitors to pull-apart basins. The former role is typically observed in northwestern Sumatra. The Sumatran volcanic zone from 3.7°N to the north is away from the Sumatran fault to the back-arc side where pull-apart basins are not found (Fig. 1.26). The latter role is observed in the large volcanic cluster areas. A typical example is found in the fore-arc side of the largest Toba caldera volcanic cluster, where pull-apart basins are missing for 150 km along the Sumatran fault from 2.1°N to 3.2°N.

(3) *Geometry of the Fault Tips Against Clustered Volcanoes*

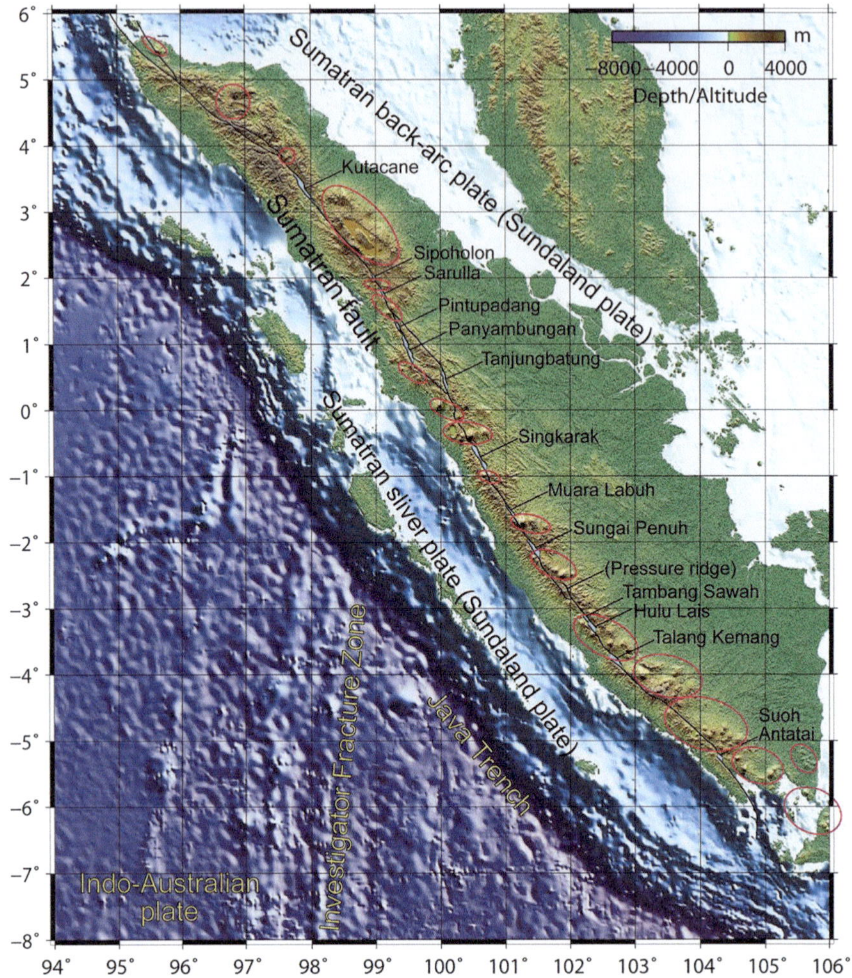

Fig. 1.26 Regional map of topography, fault segments (thick line), pull-apart basins (pale blue), a pressure ridge, and clustered volcanoes (elliptic red line) along the Sumatran fault (Muraoka et al. 2010b). Topography on land was made by the SRTM 3-s data from the U.S. Geological Survey, and the bathymetry was created by the data from the University of Alaska, respectively

The geometry of the fault tips against clustered volcanoes is described in two areas along the Sumatran fault. Figure 1.27a shows the shape of the fault tips against the Kerinci and Masurai volcanic clusters. The Muara Labuh pull-apart basin is situated north of the Kerinci volcanic cluster. This basin becomes wider southward to the nearest Kerinci volcanic cluster. Similarly, the Sungai Penuh pull-apart basin is situated north of the Masurai volcanic cluster. This basin also becomes wider southward to the nearest Masurai volcanic cluster. In general, forming a pull-apart basin needs stepping over two fault segments. The stepping over of the fault segments

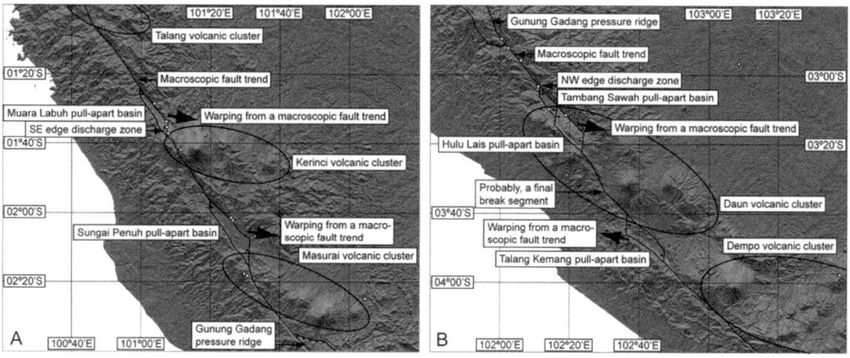

Fig. 1.27 Geometrical relation of the tips of the fault segment and clustered volcanoes (Muraoka et al. 2010b). The open circle shows natural hot springs.**a** Fault tips against the Kerinci and Masurai volcanic clusters. **b** Fault tips against the Daun volcanic cluster

in these pull-apart basins is performed by the eastward warping of the northwestern fault segment to the macroscopic fault trend as close to the clustered volcanoes. It is also noted that clockwise drags of the fault segment are observed in the inner areas of both the Kerinci and Masurai volcanic clusters, although the drag is weak.

Figure 1.27b shows the geometry of the fault tips against the Daun volcanic cluster. The Hulu Lais and Tambang Sawah pull-apart basins are located north of the Daun volcanic cluster. The Talang Kemang pull-apart basin is located south of the Daun volcanic cluster and becomes wider toward the nearest Daun volcanic cluster. It is formed by the eastward warping of the northwestern fault segment and the westward warping of the southeastern fault segment to the macroscopic fault trend as close to the clustered volcanoes. A weak clockwise drag of the fault segment is again observed in the inner area of the Daun volcanic cluster.

(4) *Geothermal Fluid Discharges in Pull-Apart Basins*

A close spatial relationship between pull-apart basins and geothermal fluid discharges is observed along the Sumatran fault. Figure 1.28 shows the distribution of 156 natural hot springs in Sumatra that are compiled from 23 sheets of the 1:250,000 geological map series in Indonesia with some additional observations by our field surveys (Muraoka et al. 2010b). It seems clear that about half of the hot springs are associated with volcanic edifices in volcanic clusters, and another half are associated with pull-apart basins. There still remains a high possibility of finding hot springs in the pull-apart basins.

A pull-apart basin is generally bounded by major strike–slip faults along its long axis and by normal faults along its short axis. When we observe the faults responsible for the discharges of geothermal fluids in pull-apart basins, the normal faults often play the most important role in major discharges of geothermal fluids. Figure 1.29 shows the NW boundary normal fault of the Sipoholon pull-apart basin at Ria-Ria-Sipoholon, which is situated at the top of a travertine terrace with a 35 m height, 500 m width in the ESE-WNW direction, and 1,000 m length in the NNW-SSE direction.

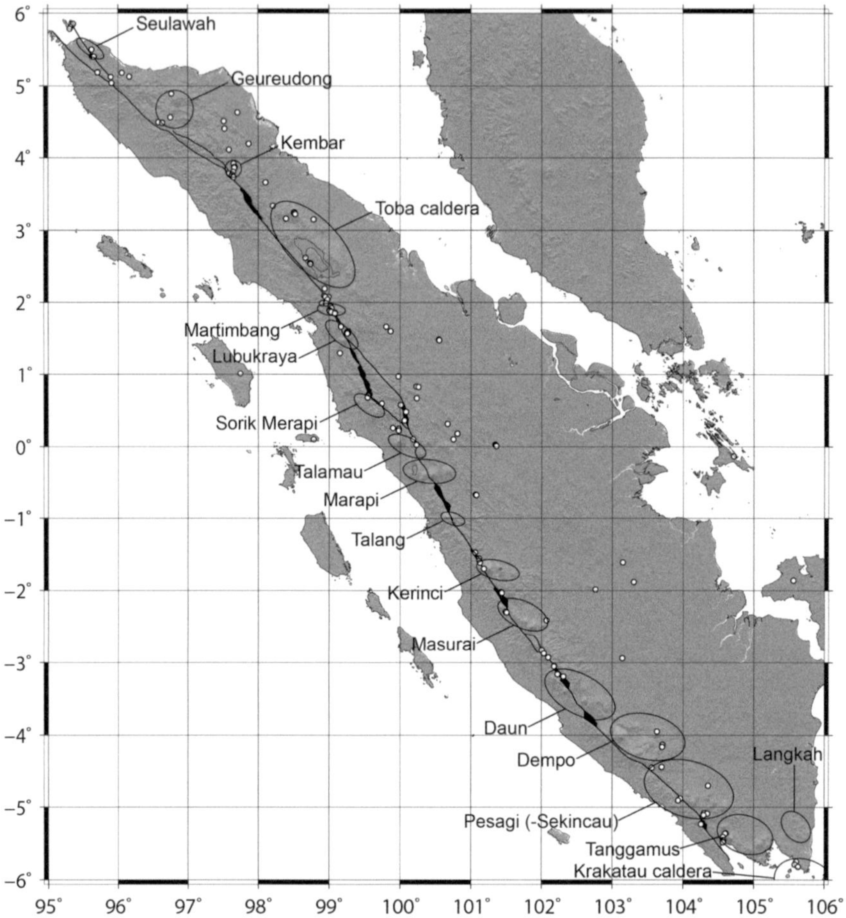

Fig. 1.28 Distribution of 156 hot springs on Sumatra Island (Compiled from 23 sheets of 1:250,000 geological map series in Indonesia by Muraoka et al. 2010b)

A 1-m-thick vertical travertine vein has an N5°E strike and 75°W dip. They cut the horizontal travertine terrace deposits where the thermal water discharges at a temperature of 64.9 °C and a rate of 6,000 L/minute (a total estimate from many discharge points near the fault). This could show a normal fault on the northwestern edge of the Sipoholon pull-apart basin.

(5) *Discussion: Localization of Pull-Apart Basins Along the Sumatran Fault*

Numerous studies have been done on the genesis of pull-apart basins along major strike–slip faults worldwide since the 1960s (Burchfiel et al. 1966). A geometrical concept of pull-apart basins is shown in Fig. 1.30. Faults usually reduce the displacement or slip rate toward their tips (Muraoka and Kamata 1983). However, a segment of the strike–slip fault in pull-apart basins is a special case. They do not reduce the

Fig. 1.29 Photograph of the hot spring discharges from the northwestern boundary normal fault in the Sipoholon pull-apart basin. Travertine generally forms horizontal terrace deposits, but the 1-m-thick high-angle travertine veins with a N5°E strike and 75°W dip penetrate them and are discharging thermal water at a temperature of 64.9 °C with a discharge rate of 6,000 L/minute (Muraoka et al. 2010b; Photograph by H. Muraoka on September 1, 2006)

displacement or slip rate at their tips because the excess displacement or slip rate accommodated at the tips is an essential driving force for the spreading of pull-apart basins (Fig. 1.30). This process requires the formation of a deep hole in the crust. However, it is not realistic to consider such a deep hole because of the large energy required for the spreading against the increasing stress of the crust with depth. It is more realistic that pull-apart basins are formed as shallow holes in the crust underlain by a horizontally decoupled plane at a relatively shallow depth. It is still unclear how to localize a pull-apart basin along a major strike–slip fault. Two key factors are critically important for generating pull-apart basins along a major strike–slip fault: one is the segmentation of the strike–slip fault, and the other is the stepping over between two segments across the fault direction.

Along the Sumatran fault, pull-apart basins usually occur near the volcanic clusters. The depth of the brittle-plastic transition in these areas is expected to be at a shallow depth because of the high-temperature thermal fields formed by the shallow-depth magma chambers (Fig. 1.31a). The depth of the brittle-plastic transition is known to be as shallow as 3.1 km at a temperature of 380 °C in the Kakkonda geothermal field of the Hachimantai volcanic cluster, Northeast Japan, where the well WD-1a exceeds 500 °C at a depth of 3,729 m (Muraoka et al. 1998). Even if this is the shallower case, the depth of the brittle-plastic transition in the volcanic clusters along the Sumatran fault may be several kilometers. On the other hand, the depth of the brittle-plastic transition at a nonvolcanic area along the Sumatran fault may be much deeper (Fig. 1.31a). The plastic layer does not yield any faults or fractures. The dehydration front will be physically formed in the brittle layer just above the brittle-plastic transition. The contact metamorphic dehydration reactions, such as the

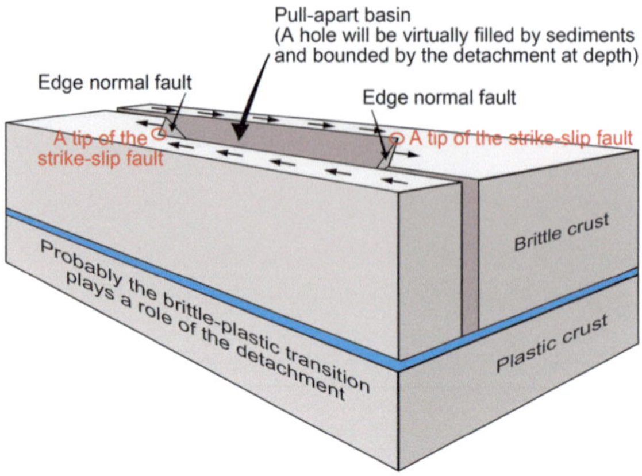

Fig. 1.30 Concept of a pull-apart basin along the strike–slip fault (modified from Muraoka et al. 2010b). Segmentation of the strike–slip fault and step-over of counter segments are key factors for generating the pull-apart basins

cordierite-forming reaction, will also chemically contribute to the formation of the dehydration front (Muraoka et al. 1998; Muraoka and Ohtani 2000). The dehydration front will yield a high pore pressure, resulting in numerous horizontal fractures (Muraoka et al. 1998; Muraoka and Ohtani 2000) and will provide the horizontally decoupled plane at the bottom of the pull-apart basins at a relatively shallow depth.

When we consider the development process of a major strike–slip fault, it should be initiated by the incipience of the oblique subduction (Fig. 1.31b). A paleomagnetic study demonstrated that a clockwise rotation of Sumatra Island relative to Java Island is ongoing at least for the last 2.1 Ma at the rate of 5° or 10°/m.y. (Nishimura et al. 1986). The nucleation of the fault segments needs a stress concentrator that should be expected in the thick, brittle layer area. The nucleated fault segments in the thick, brittle layer area will gradually expand their fault tips toward the volcanic cluster (Fig. 1.31b). However, the inside of the volcanic cluster is composed of a thin, brittle layer area that is free from the underlying plastic layer by the intervention of the dehydration front detachment (Fig. 1.31b). This thin, brittle layer area is easily deformed and inhibits the expansion of the fault segment at the earlier stage. This is the first step in forming pull-apart basins along the Sumatran fault, particularly for fault segmentation.

Before the fault segment penetrates the volcanic cluster, clockwise rotation deformation will occur in response to the dextral moment (Fig. 1.31b). A clockwise drag of the fault segment is observed in these volcanic clusters even after the fault segment appeared (Fig. 1.27). This dextral deformation causes lateral warping of the fault tips of the northwestern and southeastern segments to the macroscopic fault trend, providing across-fault space between the tips (Fig. 1.31b). Most pull-apart basins tend to be wider toward the nearest volcanic clusters (Fig. 1.27), demonstrating that the fault tips' warping occurred. This gives the second step for forming pull-apart basins, particularly for the across-fault stepping over between the tips of the fault segments.

When we consider the essential role of the major strike–slip fault decoupling on both the fore-arc sliver and back-arc plates, the slip rate should be partitioned over the

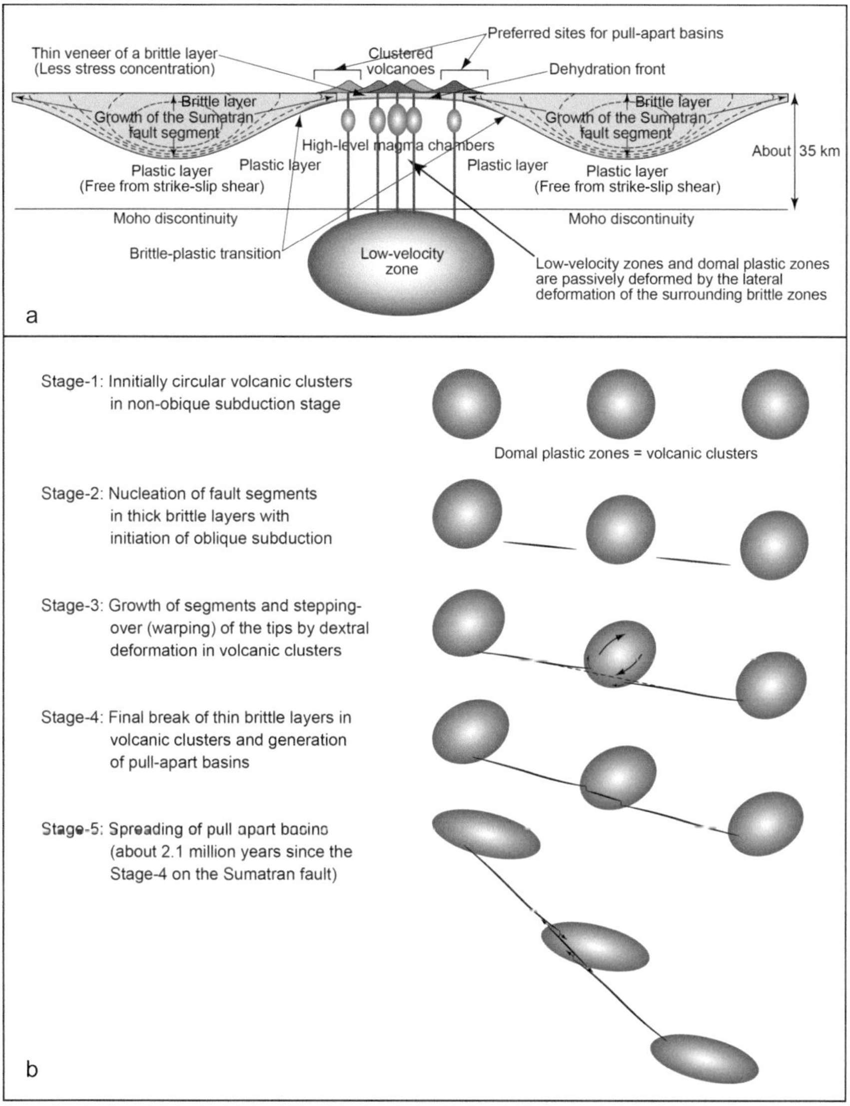

Fig. 1.31 Model for localization of pull-apart basins along the Sumatran fault (modified from Muraoka et al. 2010b). **a** Schematic cross-section model along the Sumatran fault showing the depth distribution of the brittle-plastic transition. **b** Schematic plan view model of the segmentation and step-over processes of the Sumatran fault with increasing obliquity of subduction

entire fault system. This is necessary for the macroscopic motions of both sides of the relatively rigid plates. Therefore, penetration of a fault segment is necessary even in the place of the volcanic clusters at the final stage (Fig. 1.31b). A new segment finally penetrates the volcanic cluster along the macroscopic fault trend. This proceeds to the final step for the completion of pull-apart basins. Spreading of pull-apart basins follows this process. Idealistically, two pull-apart basins are formed on opposite sides of the volcanic cluster (Fig. 1.31b), but only one basin occasionally appears on one side (Fig. 1.31a). This model, thus, explains a variety of the observed features on the pull-apart basins along the Sumatran fault, including the segmentation and stepping over, and could represent the localization process of the pull-apart basins.

(6) *Discussion: Estimate of Slip-Rate Partitioning by the Pull-Apart Basins*

If the excess displacement or slip rate accommodated at the fault tips is an essential driving force for spreading the pull-apart basins (Fig. 1.30), the length of pull-apart basins would provide the total displacement by the slips throughout the oblique subduction period (Aydin and Nur 1982). When the length of thirteen pull-apart basins is plotted against the latitude along the Sumatran fault (Fig. 1.32a), the length tends to be larger to the northwest and smaller to the southeast. This tendency is broadly consistent with the estimates of slip rates from other methods in the previous papers (McCaffrey 1991; Genrich et al. 2000; McCaffrey et al. 2000; Prawirodirdjo et al. 2000), but some pull-apart basins are shorter than the general trend. Because the Hulu Lais and Tambang Sawah pull-apart basins are exactly continued (Fig. 1.27b), if their lengths are summed up, their length comes close to the general trend (Fig. 1.32a). The Sarulla and Sipoholon pull-apart basins are very narrow pull-apart basins (Fig. 1.26), and there still remains a possibility that they may continue. If they are assumed to be continued, their total length comes close to the general trend (Fig. 1.32a).

The estimate becomes complex when looking at the part of the double fault zone from 0.3°N to 1.5°N along the Sumatran fault, where the total length of two pull-apart basins in the fore-arc branch is longer than the general trend, and the back-arc branch is shorter than this length (Fig. 1.32b). It should be noted that the long axes of these pull-apart basins are obviously oblique to the general trend of the Sumatran fault. This means that their spreading does not directly contribute to the displacement along the Sumatran fault. If these pull-apart basins spread further, the width between the double faults inevitably becomes larger, particularly to the fore-arc side (Fig. 1.32b). The double fault zone probably plays the role of a compensator to the along-arc stretching of the fore-arc sliver plate by this swelling process. The cause of the stretch absorber's existence in this place is not necessarily clear. However, a possible interpretation is that it may be related to the northern extension of the Investigator Fracture Zone (Fig. 1.26). Whatever the cause of the double fault zone, the length of pull-apart basins in this zone would be better excluded for the estimate of the integrated displacement.

As a result, we obtain a relatively reliable dataset on the length of pull-apart basins along the Sumatran fault (Fig. 1.32c). The period of the oblique subduction of Sumatra relative to Java is given to be the last 2.1 million years (Nishimura et al.

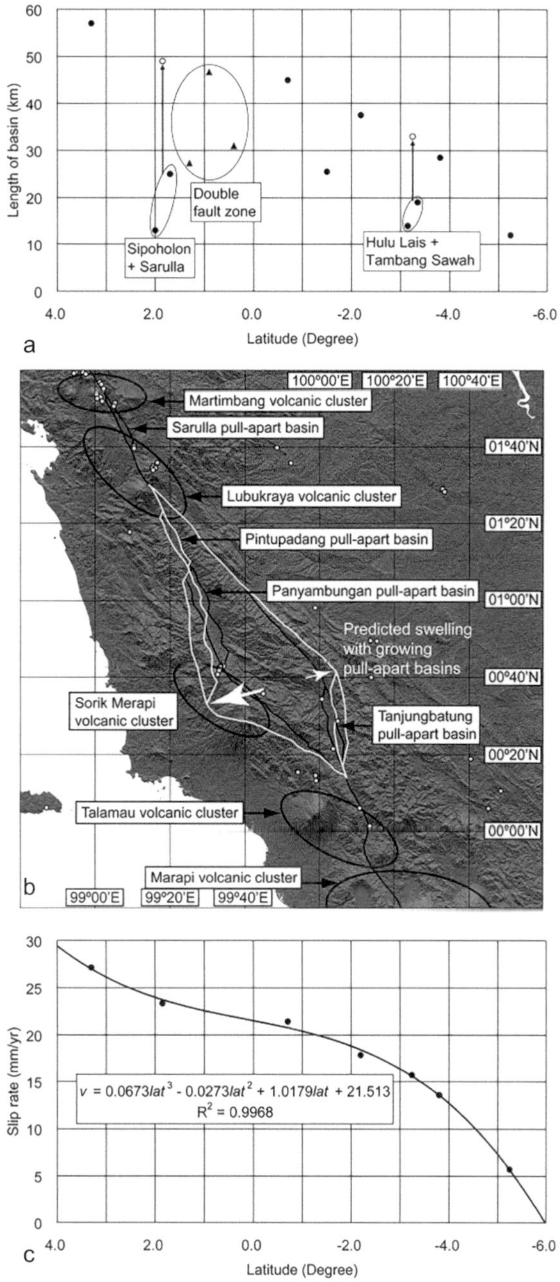

Fig. 1.32 Estimate of slip-rate partitioning along the Sumatran fault (Muraoka et al. 2010b). **a** Length of pull-apart basins along the Sumatran fault. **b** The double fault zone of the Sumatran fault from 0.3°N to 1.5°N as a stretch absorber. **c** Estimate of slip-rate partitioning along the Sumatran fault

1986). When we subdivide the integrated displacement by this period, we obtain a slip-rate partitioning along the Sumatran fault, as shown in Fig. 1.32c. This would provide a simple method to estimate the slip rate over almost the entire Sumatran fault and is well consistent with the previous estimates from other methods (McCaffrey 1991; Genrich et al. 2000; McCaffrey et al. 2000; Prawirodirdjo et al. 2000).

(7) *Tectonic Significance to Geothermal Systems in Sumatra*

Huge geothermal potentials in Sumatra could thus be attained by the spatial coincidence of the volcanic clusters and the Sumatran fault, particularly enhanced through the formation of pull-apart basins. This situation provides duplicated opportunities for extra geothermal reservoir potentials in Sumatra (Muraoka et al. 2010b).

(8) *Other Geothermal Fields in Indonesia*

Indonesia is an archipelagic nation composed of more than 17,000 islands. Rural electrification is earnestly required, particularly in remote islands such as the Lesser Sunda Islands (Nusa Tengara). Then, a five-year Indonesia-Japan bilateral research cooperation project named "Research Cooperation Project on the Exploration of Small-scale Geothermal Resources in the Eastern Part of Indonesia" (concisely called the Mataloko Project) was conducted in the central Flores Island from April 1997 to March 2002 (Muraoka and Uchida 2002; Fig. 1.33). The results were published as the special issue of 31 articles on the Bulletin of the Geological Survey of Japan (Muraoka and Uchida 2002). A flow test of the well MT-2 drilled in this project by the New Energy and Industrial Technology Development Organization of Japan was made in front of members of the Indonesian Parliament and several hundred residents in Mataloko, Flores Island, early in the morning of January 20, 2001. Since the drilling equipment was restricted in a remote island area, this well was only drilled to a depth of 162.35 m. However, about 15 tons per hour of dry steam was stably produced at a valve-full-open state (Sueyoshi et al. 2002), and the flow test was successful (Fig. 1.34).

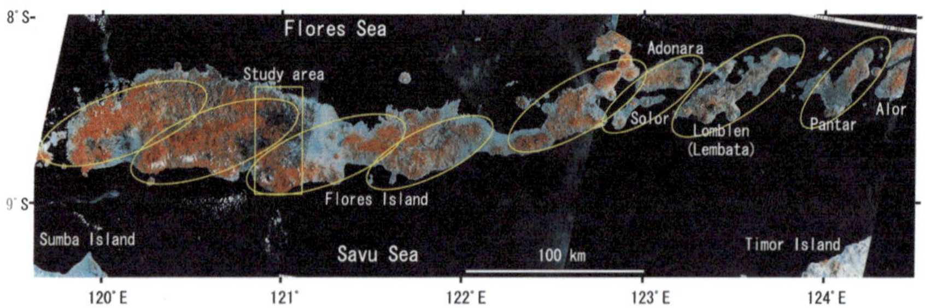

Fig. 1.33 Mosaic of three scenes of Landsat MSS imagery on the inner Lesser Sunda Islands. Note the *echelon*-shaped topographic structure in this sector of the Lesser Sunda Islands. Each element of the *echelon* structure consists of a culmination of the anticline and overlying clustered young volcanoes (Muraoka et al. 2002)

Fig. 1.34 Flow test of the well MT-2 in Mataloko, Flores Island, drilled by the New Energy and Industrial Technology Development Organization of Japan (Sueyoshi et al. 2002; Photograph by H. Muraoka on January 20, 2001)

Geothermal developments in Indonesia are gradually shifting from Java Island to the surrounding islands, particularly to Sumatra Island. The government of Indonesia is aiming at 10,000 MWe by 2030 (Darma et al. 2021).

It was established by the Minister of Energy in 2017 through Ministry Decree number 2268 K/30/MEM/2017 to develop and use all of the geothermal potential to power Flores Island (Darma et al. 2021). The total geothermal potential of this island is about 735 MWe in 18 locations. Then, Flores Island is now called the Geothermal Island. This new policy provides exquisite pleasure to all the participants of the Mataloko Project.

1.6.3 The Philippines

The installed geothermal power capacity in the Philippines is 1,918 MWe as of 2020 (Fig. 1.35). The Philippines ranks as the third largest geothermal power producer in the world (Table 1.4; Huttrer 2021).

However, the Philippines ranked as the second largest geothermal power producer from 1995 to 2000 (Fig. 1.9) and always led the world's geothermal developments. Geothermal energy production continuously contributed 11% of the total electricity requirements of 23,815 MWe in 2018 in the Philippines. The latest resource estimation for geothermal energy potential in the Philippines is 4,064 MWe, of which about 50% has already been developed (Fronda et al. 2021). This ratio is prominent. Thus, the Philippines is a well-matured geothermal country.

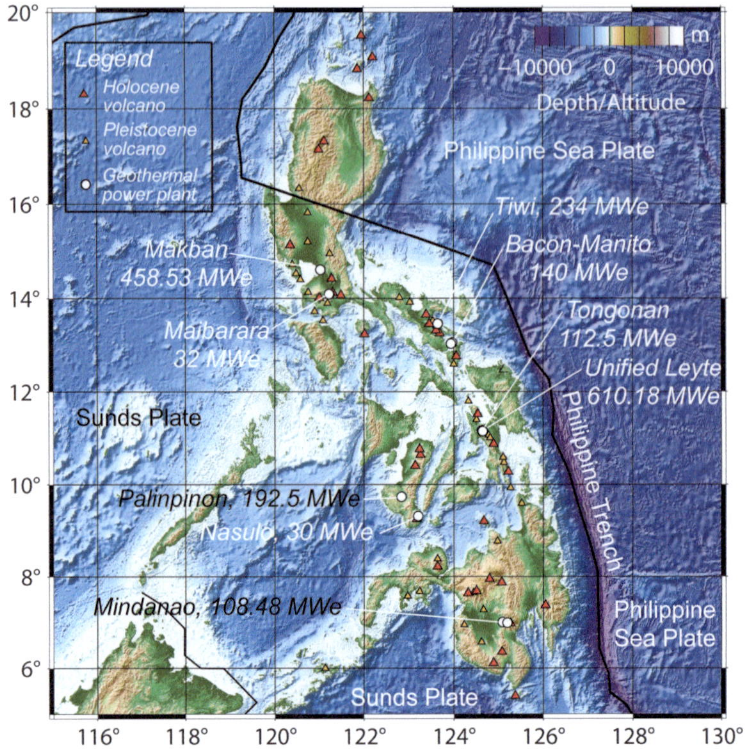

Fig. 1.35 Location map of existing geothermal power plants in the Philippines (Installed capacity was quoted from Fronda et al. 2021). The Lambert projection map was drawn using the ETOPO2022 15 arc-second global relief model (NOAA National Centers for Environmental Information 2022) and GMT 6.4.0 (Wessel et al. 2019)

1.7 High-Temperature Geothermal Systems in Africa

1.7.1 Geothermal Settings of Africa

High-temperature geothermal fields of Africa are concentrated along the Great Rift Valley in eastern Africa (Fig. 1.36), which is a spreading plate boundary between the Africa (Nubia) Plate and Somalia Plate. The spreading rate between the Africa (Nubia) Plate and Somalia Plate is larger in the north and smaller in the south because the Euler pole of the plate rotation between them is situated in the eastern Indian Ocean off South Africa (Bird 2003), which is nearly the southern end of the Somalia Plate. The spreading rate is estimated to be 6.9 mm/yr in the Afar Triangle (Fig. 1.36) and 1.9 mm/yr in the southeast of South Africa by the GPS measurements (Fernandes et al. 2004). The northern end of the Great Rift Valley forms the Afar Triple Junction, which is made of three plates: Africa (Nubia), Somalia, and Arabia Plates. The Gulf of Aden between the Arabia and Somalia Plates is an ocean-spreading ridge, and the Red Sea between the Africa and Arabia Plates is also an ocean-spreading ridge evolved from the continental rift valley. The spreading Afar Triple Junction forms a broad low-altitude land called the Afar Triangle or Afar Depression, extending over Djibouti, Eritrea, and Ethiopia.

The Great Rift Valley consists of the Eastern and Western Branches, where the Eastern Branch is more active in the current extension tectonics, volcanism, and geothermal phenomena. The Eastern Branch extends from Eritrea and Djibouti in the north, through Ethiopia and Kenya, and to Tanzania in the south. The Western Branch extends from Uganda in the north, through Rwanda, Congo, Burundi, Tanzania, Malawi, and Zambia, and to Mozambique in the south (Fig. 1.36). Among them, Ethiopia and Kenya are prominent high-potential geothermal resource countries. Mangi (2018) estimates that the geothermal resource potentials of Ethiopia and Kenya are 4,200–10,800 MWe and > 10,000 MWe, respectively. Although missing in

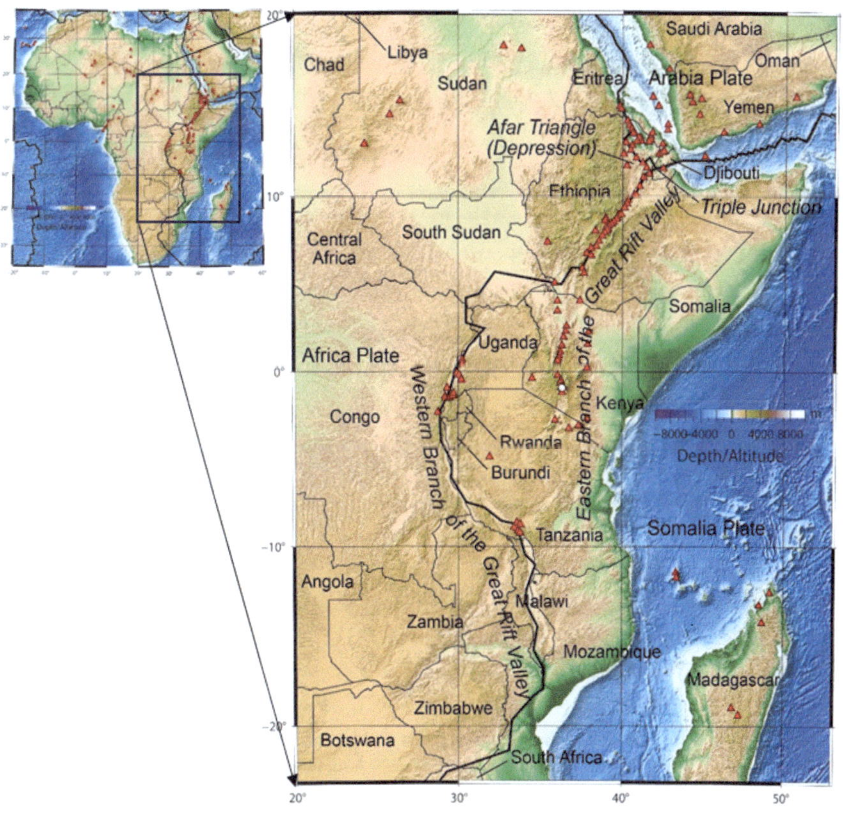

Fig. 1.36 Tectonic map of the Great Rift Valley in eastern Africa with Holocene volcanoes (red triangle) and geothermal power plants (open circle). The Lambert projection map was drawn by the ETOPO2022 15 arc-second global relief model (NOAA National Centers for Environmental Information 2022) and GMT 6.4.0 (Wessel et al. 2019)

Table 1.3, Ethiopia installed a 7.3 MWe geothermal power plant in the Aluto-Langano geothermal field as of 2010 (Bertani 2005).

1.7.2 Kenya

Kenya is a leading country in geothermal development in Africa. As of 2020, its installed geothermal power capacity was 1,193 MWe, ranking fifth in the world (Table 1.4; Huttrer 2021). The increase in the installed geothermal power capacity during the last 20 years was very speedy (Fig. 1.9). The installed geothermal power capacity was 865 MWe as of 2019 which is about 29% of the total installed electricity capacity in Kenya and about 47% of the electricity consumption in Kenya during 2019 (Omenda et al. 2021).

All six geothermal power plants in Table 1.3 are situated in the Olkaria geothermal field (Fig. 1.37), and geothermal power developments were concentrated in this field until recently. In addition, Menengai, Paka, Silali, Korosi, and many other geothermal fields are under exploration and exploitation (Fig. 1.37; Omenda et al. 2021).

1.8 Generalized Viewpoints on High-Temperature Geothermal Systems in the World

1.8.1 Vapor-Dominated Versus Water-Dominated Geothermal Systems

White et al. (1971) proposed a model of vapor-dominated geothermal systems found in relatively limited numbers of geothermal fields compared to the widely known water-dominated geothermal systems. The most conspicuous feature of the vapor-dominated geothermal systems is that the geothermal wells produce dry steam. The dry steam is derived from the near-vapor-static vapor-dominated reservoirs heated by the high-level magmatic intrusions, where the heat conduction rate exceeds the fluid recharge rate. White et al. (1971) considered that the vapor-dominated reservoir fluid was constrained by the prevailing heat supply into the maximum enthalpy point of the liquid-steam two-phase region at 2,804 kJ/kg, 31.19 bar (3.119 MPa), and 236 °C on the enthalpy–pressure diagram (Fig. 1.38). Coincidentally, the P–T conditions in the following well-known vapor-dominated geothermal fields in the world were very close to these numbers. Therefore, the maximum enthalpy point control theory has long been accepted.

Interesting is the fact that the earlier generations of geothermal power plants in the world were preferably operated in the vapor-dominated geothermal fields, such as the oldest Larderello plant in Italy, operated since 1913, The Geysers in the USA since 1960, Matsukawa in Japan since 1966, and Kamojang in Indonesia

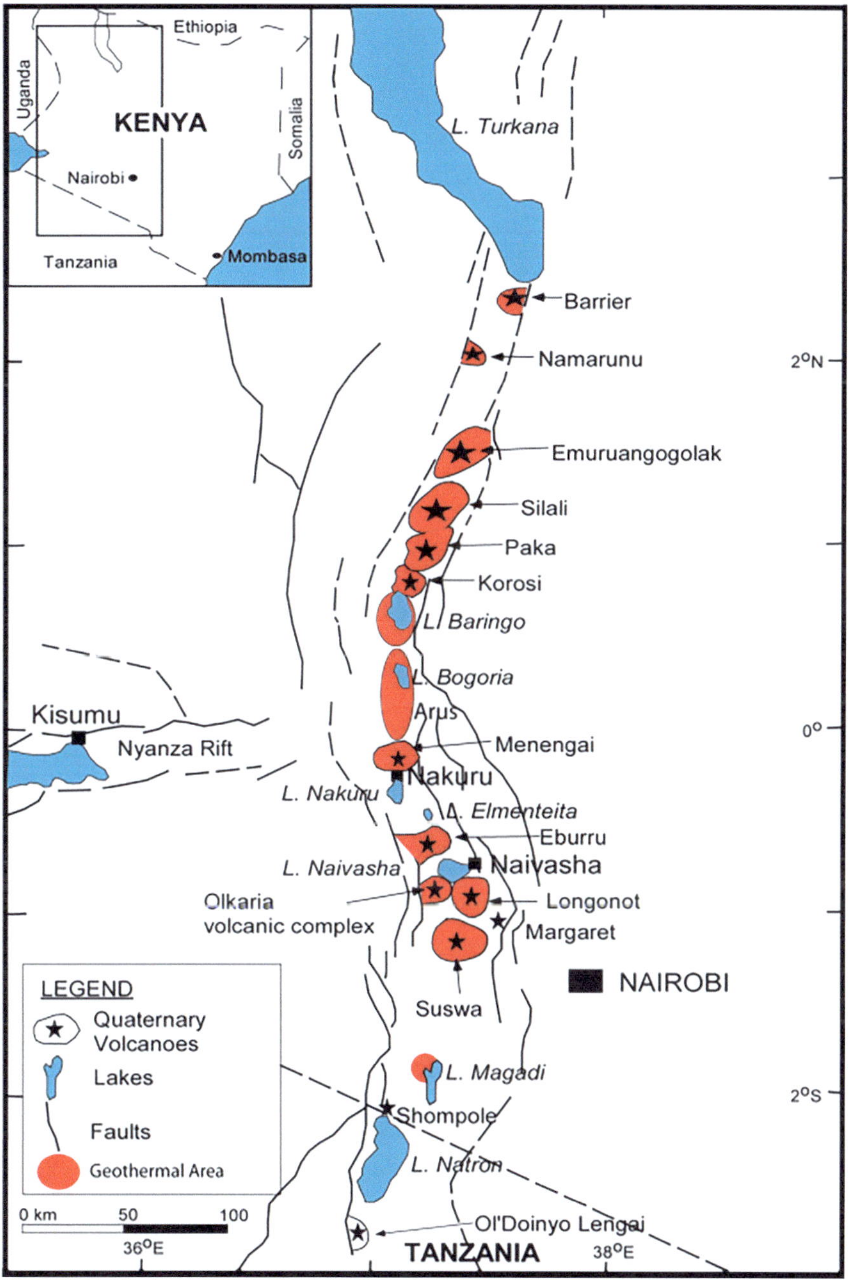

Fig. 1.37 Representative geothermal fields in Kenya (Omenda et al. 2021)

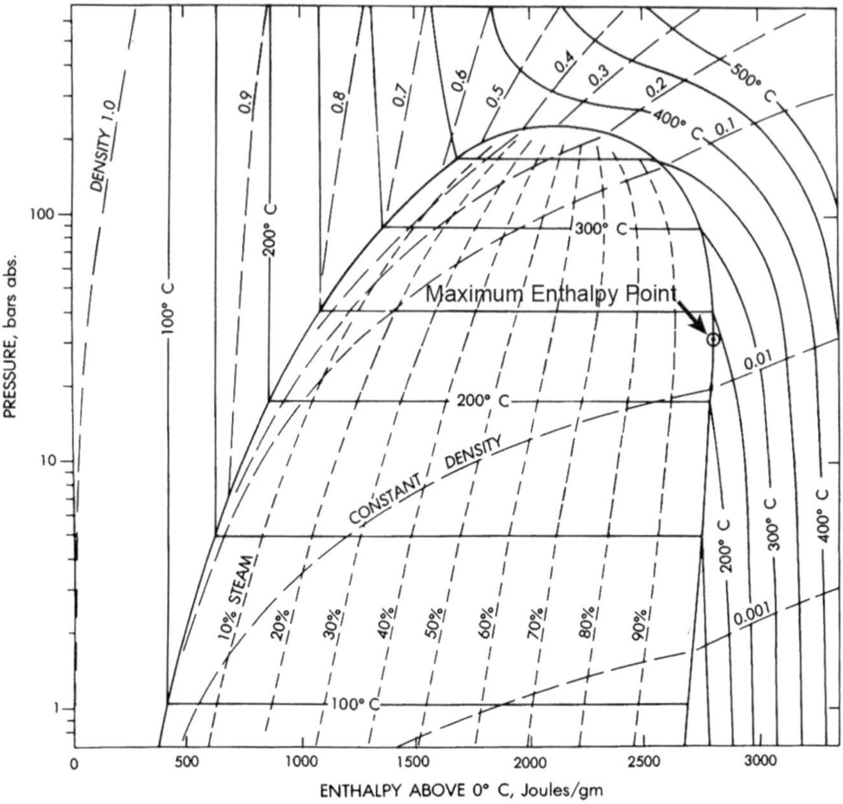

Fig. 1.38 Enthalpy–temperature–pressure–density relations for water to 500 °C and 600 bars (Truesdell 1984)

since 1978. The reason is quite simple that geothermal power generation is easily performed in the vapor-dominated geothermal systems because the single-phase dry steam apparently enables the generation of electricity without any two-phase separators, any complicated condenser systems, and any expensive reinjection wells for separated hot water. However, this advantage is only valid from a short-term viewpoint. Suppose we do not reinject amounts of used or separated liquid water into the surroundings of subsurface geothermal reservoirs. In that case, pressure drawdown easily occurs not only in water-dominated reservoirs but also in vapor-dominated reservoirs. However, the liquid water for reinjection is too difficult to collect in the vapor-dominated geothermal systems because of the dry steam production. We actually faced such a dilemma of declines in Valle Secolo, Larderello, during the 1950s (Fig. 1.39) and The Geysers during the 1980s (Fig. 1.40). Thus, efficient reinjection and sustainable operation are rather difficult in vapor-dominated geothermal systems from a long-term viewpoint. Then, geothermal development was widely extended to the water-dominated geothermal systems. The first geothermal power plant with a two-phase separator and condenser system in the water-dominated geothermal systems succeeded in Wairakei in New Zealand in 1958.

Ingebritsen and Sorey (1988) clarified that the pressure values measured in the vapor-dominated zones were not necessarily close to the maximum enthalpy point and often higher than that. The highest example reaches 100 bars (10 MPa) in The Geysers at a depth of more than 2,400 m. Their simulation results also indicate that the pressures at the top of vapor-dominated zones are proportional to the depth at the top of vapor-dominated zones (Fig. 1.41). Thus, they pointed out that the maximum enthalpy point control theory is not always true. The discrepancy is ascribed to the fact that the natural vapor-dominated zones are not single-phase steam reservoirs in a strict sense and two-phase zones with minor counter downflows of water (White et al. 1971; Ingebritsen and Sorey 1988). In other words, the physical state of the vapor-dominated zones might be situated slightly inside of the maximum enthalpy

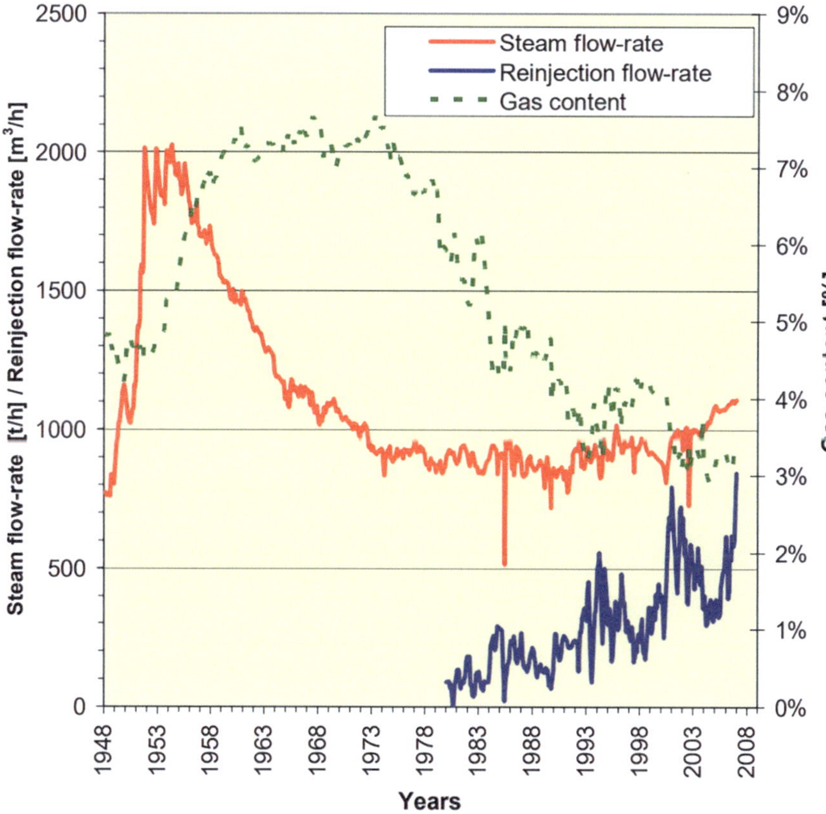

Fig. 1.39 Decline of steam production and effect of the reinjection in the Valle Secolo geothermal plant, Larderello, Italy. Arias et al. (2010) updated from Cappeti et al. (1995)

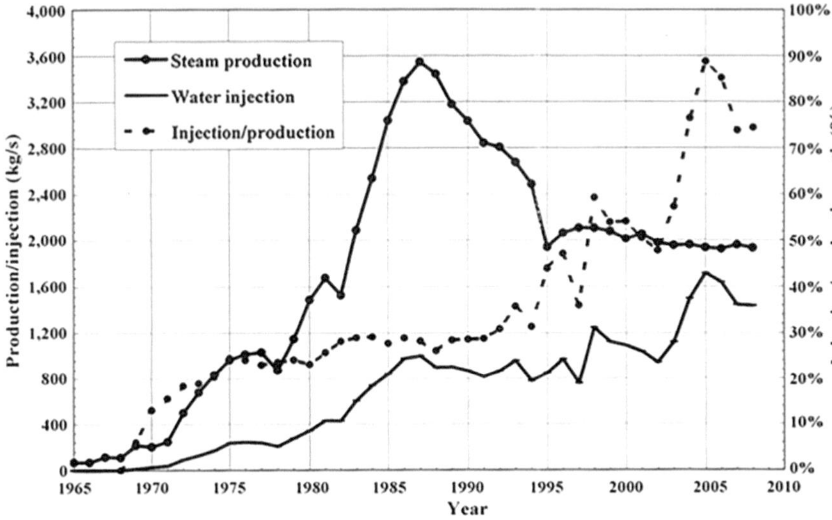

Fig. 1.40 Decline of steam production and effect of the reinjection in The Geysers geothermal field, USA (Goyal and Conant 2010)

point of the two-phase region on the enthalpy–pressure diagram, even in the vapor-dominated geothermal systems (Fig. 1.38), allowing the higher-pressure values with increasing depth up to the critical point at 22.06 MPa and 373.95 °C.

Fig. 1.41 Results of simulations on the relationship between near-steady state vapor-dominated zone pressures in the bar (vertical axis) and the depth from the upper-pressure boundary to the base of the caprock (lateral axis) from Ingebritsen and Sorey (1988)

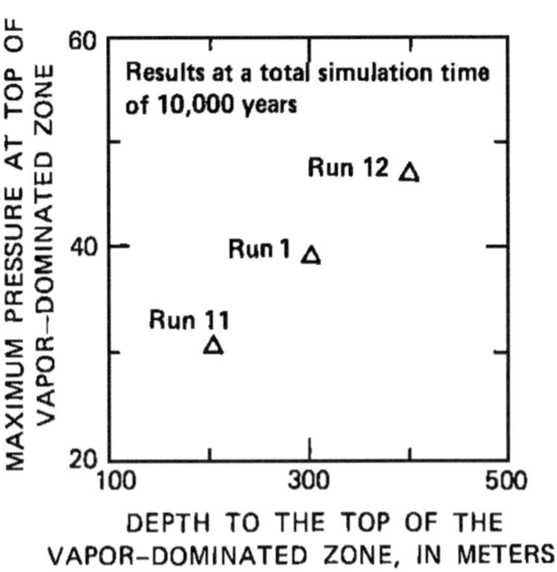

1.8.2 Hydrothermal Water Chemistry Induced from the Mafic Crust Versus Silicic Crust

To put it simply, we have two types of contrasting lithology on the surface of the Earth's crust: one is the mafic crust, and the other is the silicic crust. They are ultimately derived from mafic and silicic magmas, respectively, in terms of the essential replenishment materials. The former is represented by the mid-ocean ridge basalt (MORB) magma continuously producing the oceanic crust, and the latter is the granitic magma that drives the continental growth through the formation of large-scale batholiths such as the Mesozoic Sierra Nevada Batholith and the Hiroshima Batholith. Table 1.5 shows a variety of rock compositions, including the boron and chloride analyses from the mafic to silicic rocks taken from volcanic rocks and their affinities, because the rapid consolidation would better quench the original magma compositions. It can be easily speculated that these contrasting host rocks could closely control hydrothermal chemistry. Nevertheless, as the earlier geothermal developments are mainly performed in the silicic continental regions, we had not necessarily realized the contrasting hydrothermal chemistry, particularly from the mafic crust regions. Efforts for geothermal research and development in Iceland brought new insights into this subject.

Since Iceland is situated on the Mid-Atlantic Ridge, the crust is almost entirely composed of basalt with only minor amounts of Na_2O-rich rhyolite. Arnórsson and his collaborators clarified the conspicuous features of hot springs in Iceland by 2010 (Arnórsson 1992; Arnórsson and Andrésdóttir 1995; Kristmannsdóttir et al. 2010). One of the conspicuous features of natural hot water less than 100 °C in Iceland is the high pH values, typically from 8 to 10 (Fig. 1.42; Arnórsson and Andrésdóttir 1995). The pH of hot water for heating systems is also reported to be 8–11 (Kristmannsdóttir et al. 2010). Arnórsson and Andrésdóttir (1995) systematically analyzed the boron and chloride contents of hot springs and high-temperature geothermal water. Some hot springs show small mixing trends by the seawater of the Cl/B ratio 1330, and other hot springs show the effect of B enrichment from the magma degassing. However, most hot springs in Iceland reveal a Cl/B ratio of 30, which is ascribed to the dissolution from the host rock basalt. Therefore, the host rock basalt controls most of these hot springs. Then, Arnórsson (1992) considered chloride an important anion to dissolve solids into solutions as salts. However, the host rock basalt generally contains very limited chloride contents (Table 1.5), so that Icelandic geothermal water generally has low dissolved solids content. This is a reason why the Icelandic geothermal water is low H^+ or high OH^- (Arnórsson 1992). Although details of the high pH forming mechanism remain, these studies drew attention to the distinctive high pH values controlled by the basaltic host rocks in Iceland. Kristmannsdóttir et al. (2010) further added two interesting points: one is that the aluminum concentrations in the Icelandic hot waters have a positive correlation to the pH values in the relatively low aluminum concentration range of 0.0–0.6 mg/kg, and the other is that the higher-pH waters are categorized as the NaOH type.

Table 1.5 Reference rock compositions from mafic to silicic igneous rocks having the B and Cl analyses. Four data points are quoted from the US Geological Survey (2023), and four data points are from the Geological Survey of Japan (2023)

	DNC-1	BIR-1	JB-1a	W-2	JA-1	QLO-1	JR-1	JR-2
Rock name	N Carolina USA Diabase	Reykjavik Iceland Basalt	Sasebo Japan Basalt	Virginia USA Diabase	Hakone Japan Andesite	Oregon USA Quartz Latite	Wadatouge Japan Rhyolite	Wadatouge Japan Rhyolite
	wt %	wt %	wt %	wt %	wt %	wt %	wt %	wt %
SiO_2	47.15	47.96	52.41	52.68	63.97	65.55	75.45	75.69
TiO_2	0.484	0.96	1.28	1.062	0.85	0.624	0.11	0.07
Al_2O_3	18.34	15.53	14.45	15.45	15.22	16.18	12.83	12.72
Fe_2O_3	1.79	2.06	2.55	1.53	2.59	1.02	0.35	0.27
FeO	7.32	8.34	5.78	8.34	3.98	2.97	0.49	0.44
MnO	0.148	0.175	0.148	0.167	0.157	0.093	0.099	0.112
MgO	10.13	9.70	7.83	6.37	1.57	1.00	0.12	0.04
CaO	11.49	13.32	9.31	10.86	5.70	3.17	0.67	0.50
Na_2O	1.886	1.820	2.73	2.20	3.84	4.20	4.02	3.99
K_2O	0.234	0.030	1.40	0.626	0.77	3.60	4.41	4.45
P_2O_5	0.070	0.021	0.26	0.141	0.165	0.254	0.021	0.012
H_2O+			0.92		0.72		1.16	1.19
H_2O-			0.92		0.30		0.2	0.22
Total	99.042	99.916	99.988	99.426	99.832	98.661	99.93	99.704
	mg/kg	mg/kg	mg/kg	mg/kg	mg/kg	mg/kg	mg/kg	mg/kg
B	0.90	0.33	7.88	12.00	21.00	36.00	117.00	145.00
Cl	60.00	26.00	171.00	190.00	43.00	219.00	920.00	736.00

The Geothermal Implementing Agreement (GIA) of the International Energy Agency (IEA) provides a flexible framework for wide-ranging international cooperation. One of the operating Annexes was "Direct Use of Geothermal Energy", where Task A was conducted for "Resource Characterization" in the late 2000s. Participating countries were Iceland, Japan, South Korea, New Zealand, Switzerland, and the USA. The data collection depended on the publication state of each country. Under this opportunity, we have tried the global comparison of hot water chemistry (Muraoka et al. 2010a). The number of submitted hot water chemistry datasets as of November 2008 was 14 from South Korea, 352 from Iceland, 6,675 from the USA (Geo-Heat Center 2004), and 7,203 from Japan (Muraoka et al. 2007).

The temperature–pH diagram shows that the data from Iceland and the USA contains high-temperature geothermal fluids from geothermal wells (Fig. 1.43). However, we here focus on the data less than 100 °C for the cross-country comparison. The clusters from Japan are bimodal: one central cluster is the neutral pH

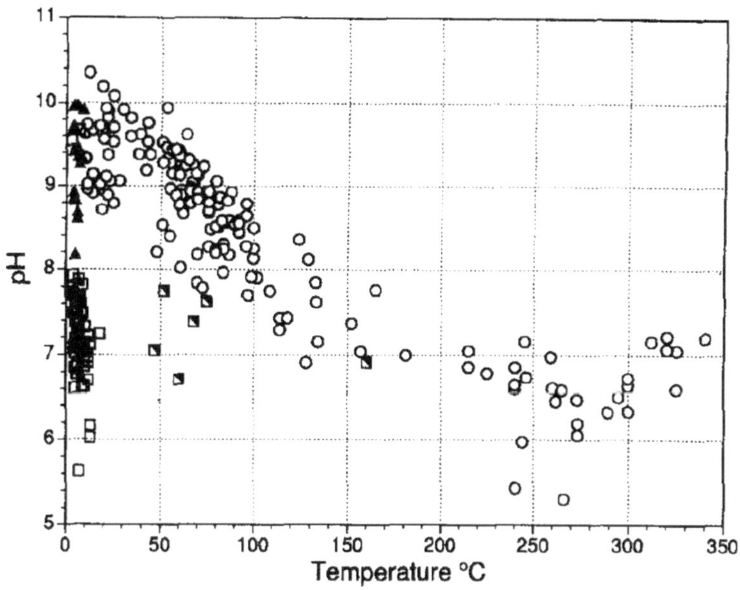

Fig. 1.42 Temperature–pH diagram of thermal waters in Iceland (Arnórsson and Andrésdóttir 1995). Data points are as follows: squares, surface waters; triangles, non-thermal groundwaters; circles, thermal waters, and half-filled squares, waters rich in CO_2. Quartz geothermometer was used for boiling hot springs in the high-temperature areas, and Na–K geothermometer was used for others

waters from 5.5 to 10.0, and the secondary cluster is the acidic pH waters from 1.0 to 3.5, derived from the near-surface condensation and oxidation of fumarolic gases commonly found in the volcanic centers of andesite volcanoes. Bimodal clustering is also recognized in the USA waters, although the acidic cluster is slightly fewer than in Japan. The cluster from South Korea shows neutral pH waters from 6.5 to 10.0, which are close to the main clusters of Japan and the USA. The acidic cluster is missing in South Korea due to fewer active volcanoes. As expected, a distinctive feature of Iceland can be seen in the cluster. The cluster from Iceland broadly ranges in pH from 6 to 11, but the highly dense portion is concentrated in pH from 8.5 to 10.5. Thus, it becomes clear through the global comparison that the silicic crust countries like Japan, the USA, and South Korea are characterized by neutral pH hot waters, and the mafic crust country like Iceland is characterized by alkaline pH hot waters (Muraoka et al. 2010a).

Figure 1.44 shows the Cl-B logarithmic plot data from Iceland, Japan, and the USA. The data from the USA contains an isolated small cluster of high-temperature geothermal fluids from geothermal wells close to the steam condensation line. Except for this, the main clusters from Japan and the USA almost overlap. The cluster from Iceland is significantly lower than the other two countries. The variation range of the central cluster of the chloride component in Japan and the USA expands from 2 to 20,000 mg/kg, which is about a four-magnitude range. The variation range of the

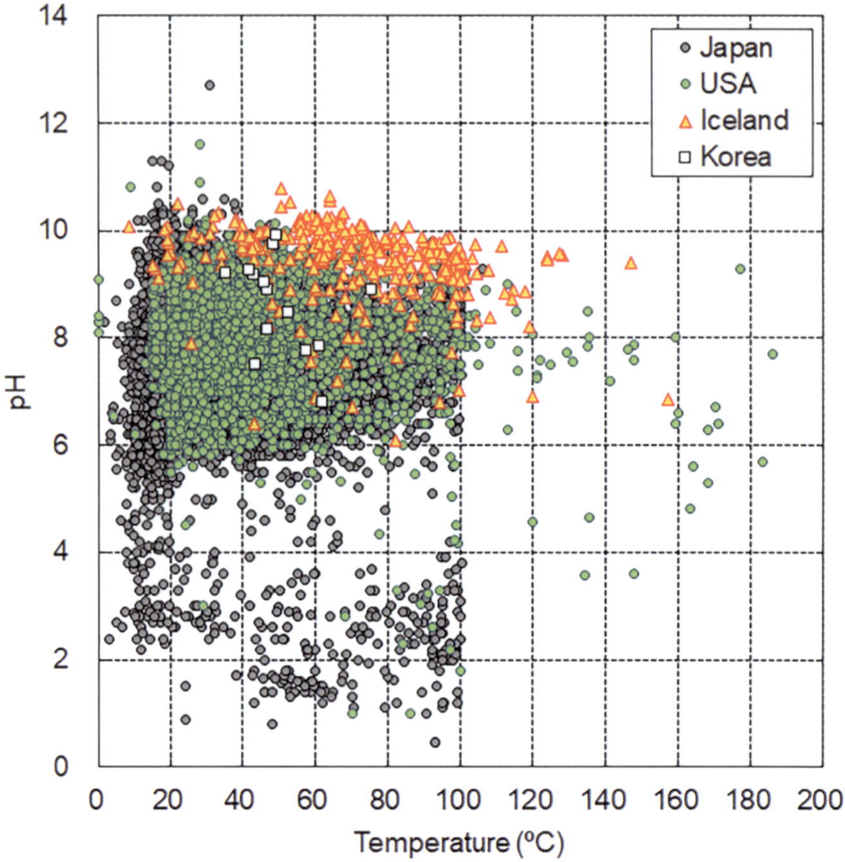

Fig. 1.43 Temperature–pH diagram of thermal water in South Korea, Iceland, the USA, and Japan (modified from Muraoka et al. 2010a)

chloride component in Iceland ranges from 3 to 300 mg/kg, which is about a two-magnitude range; therefore, the range is about half of the other two countries. The variation range of the main cluster of the boron component in Japan and the USA is expanding from 0.02 to 200 mg/kg, which is about a four-magnitude range. The variation range of the boron component in Iceland ranges from 0.02 to 2 mg/kg, which is about a two-magnitude range, and the range is about half of the other two countries, as is the case with the chloride component. Chloride and boron components are relatively incompatible with solid silicates and volatiles under hydrothermal temperatures. In other words, they are typically soluble components, preferably dissolved from the host rocks into liquid solutions. For this reason, they can be used as an indicator to host rock species (Ellis and Mahon 1964, 1967; Shigeno 1992).

When we plot the chloride and boron contents against SiO_2 contents of the host rocks using the reference rock chemistry in Table 1.5, it becomes clear that both

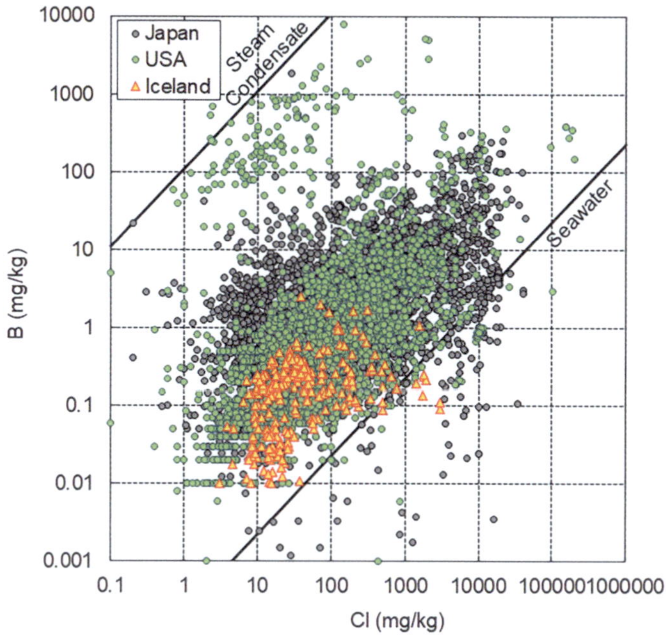

Fig. 1.44 Cl-B logarithmic plot diagram of thermal water in Iceland, the USA, and Japan (modified from Muraoka et al. 2010a)

components rapidly increase with SiO_2 contents (Figs. 1.45 and 1.46). If we assume that the SiO_2 contents of mafic and silicic crusts are 47.5 wt% and 67.5 wt%, the chloride contents of both crusts are 53.30 mg/kg and 287.16 mg/kg, respectively, from the regression line (Fig. 1.45). The chloride contents are about 5.4 times larger in silicic crust than in mafic crust. The boron contents of both crusts are 1.31 mg/kg and 39.57 mg/kg, respectively, from the regression line (Fig. 1.49). The boron contents are 30.2 times larger in silicic crust than in mafic crust.

The chemistry of hot springs is distinctive between the silicic crust and the mafic crust. One of the typical features is represented by the contrasting pH values: neutral to low in the silicic crust and high in the mafic crust. A simple explanation for the difference is ascribed to the abundance of soluble components like chloride, which is crucial for forming saline hot waters with high concentrations of dissolved solids. The chloride contents of the mafic magmas and the host rocks in the mafic crust are minimal; therefore, the chloride contents of hot waters are also minimal. This brings low-solubility waters in the mafic crust, which are high-pH waters such as the Na + -OH-type waters instead of the Na + -Cl-type waters.

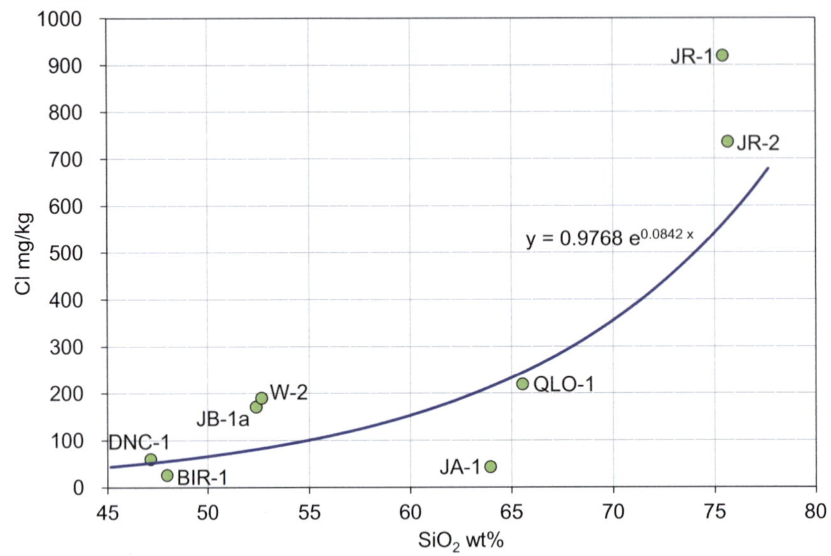

Fig. 1.45 SiO_2-Cl diagram of reference rock compositions from mafic to silicic igneous rocks (Table 1.5; modified from Muraoka et al. 2010a)

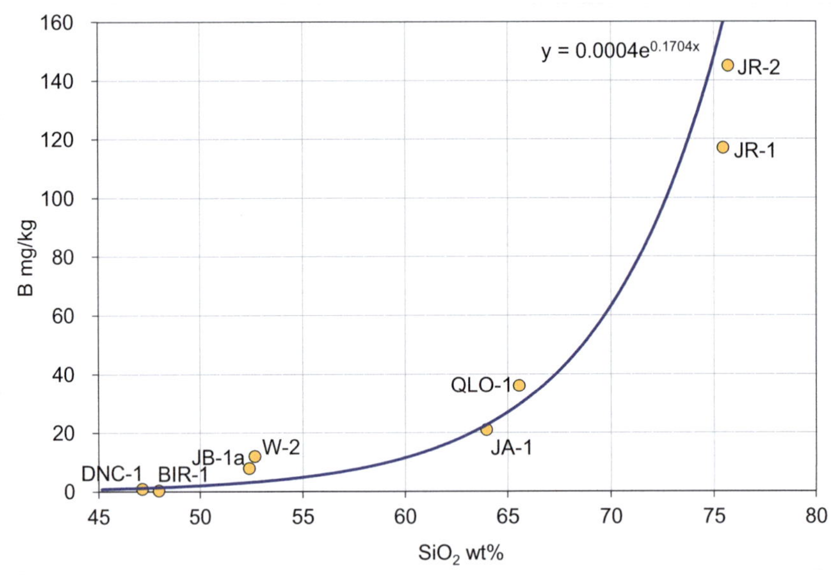

Fig. 1.46 SiO_2-B diagram of reference rock compositions from mafic to silicic igneous rocks (Table 1.5; modified from Muraoka et al. 2010a)

1.8.3 Steam Loss, Saline Brine Evolution, and Solubility of Solute in Hydrothermal Systems

Boiling or steamwater two-phase separation is common in high-temperature hydrothermal convection systems (Fournier 1987). Once a two-phase separation occurs, low-density steam rises upwards, accelerating steam loss in hydrothermal reservoirs. This is an enthalpy-decreasing process because steam has high enthalpy. At the same time, as chloride is almost entirely partitioned into a liquid phase, this is a salinity-increasing process in hydrothermal reservoirs. This is a common process for saline brine to evolve in the deeper part of the high-temperature hydrothermal convection systems. These processes are estimated by the enthalpy-chloride mixing diagram (Henley 1984; See Fig. 10.1.2 in Chap. 10).

It is well known that the high-temperature geothermal waters in the Salton Sea geothermal field in the USA have about ten times more saline water than seawater (Table 1.6; Henley 1984). The average chloride content of seawater is 18,980 mg/kg (Mason 1966), and that of geothermal water of the Salton Sea geothermal field shows 191,000 mg/kg (Table 1.6). Such hypersaline waters dissolve not only abundant cations, as shown in Fig. 1.47, but also a variety of base metals. Actually, zinc is extracted from geothermal wells in the Salton Sea geothermal field as a byproduct of the geothermal power generation. Currently, it is more focused on lithium extraction (McKibben et al. 2020).

The salinity drastically changes at a critical point of thermal water. A critical point of pure water is known to be 373.976 °C and 22.06 MPa (Driesner and Heinrich 2007). Pure water exceeding these PT conditions is defined as a supercritical fluid. Then, we wonder if we could easily access supercritical fluids through deeper wells, but this is not necessarily true. We seldom encounter the pure water in the deeper part of the high-temperature hydrothermal convection systems, because the saline brine easily evolves due to intense steam loss.

The relationship between the NaCl fraction and the PT conditions of the critical point of the H_2O-NaCl system has been well investigated. Figure 1.48 shows the NaCl fraction versus the critical point temperature of the H_2O-NaCl system (Driesner and Heinrich 2007). However, some attention is necessary in this diagram that the NaCl fraction is given by the NaCl mole fraction, but not the NaCl weight fraction. Then, the conversion factor between the two parameters is given here in Fig. 1.49.

According to this conversion, seawater with 0.032 in weight fraction (Fournier 1987) means 0.01 in mole fraction, the Salton Sea geothermal water with 0.32 in weight fraction (Table 1.6; Henley 1984) means 0.126 in mole fraction, and the Kakkonda WD-1a well geothermal water with 0.39 in weight fraction (Kasai et al. 1998) means 0.165 in mole fraction. The estimated pre-dilution Kakkonda WD-1a well geothermal water with 0.55 in weight fraction (Kasai et al. 1998) means 0.274 in mole fraction. The critical point temperatures for these four cases can be estimated from Fig. 1.48 to be 410 °C, 880 °C, more than 1000 °C, and more than 1000 °C, respectively. Among them, the critical point temperatures of three cases exceed the minimum melting point of the granitic magma system at about 670 °C (Luth et al.

Table 1.6 Concentration of major components in waters discharged from geothermal wells (Henley 1984)

Name of geothermal well	Country	Enthalpy	Separation Pressure	pH (at 20 °C)	Na	K	Ca	Mg	Cl	B	SO_4	HCO_3	SiO_2
		J/gm	bar		mg/kg	mg/kg	mg/kg	mg/kg	mg/kg	mg/kg	mg/kg	mg/kg	mg/kg
Wairakei W24	New Zealand	1085	1.00	8.30	1250	210	12.00	0.04	2210	28.8	28	23	670
Tauhara TH1	New Zealand	1120	1.00	8.00	1275	223	14.00	ns	2222	38.0	30	19	726
Broadland BR22	New Zealand	1160	1.00	8.40	1035	224	1.43	0.10	1705	51.0	2	233	848
Ngawha N4	New Zealand	1023	6.40	7.60	1025	90	2.90	0.11	1475	1080.0	27	298	464
Cerro Prieto CPM19A	Mexico	1203	7.55	7.27	7370	1660	438.00	0.35	13,800	14.4	18	52	808
Mahio-Tongonan 103	Philippines	1580	1.00	6.97	7155	2184	255.00	0.41	13,550	260.0	32	24	1010
Reykjanes 8	Iceland	1153	20.00	6.40	11,150	1720	1705.00	1.44	22,835	8.8	28	87	631
Salton Sea IID1	USA	1279	1.00	5.20	62,000	21,600	35,500.00	1690.00	191,000	481.2	6	220	1150

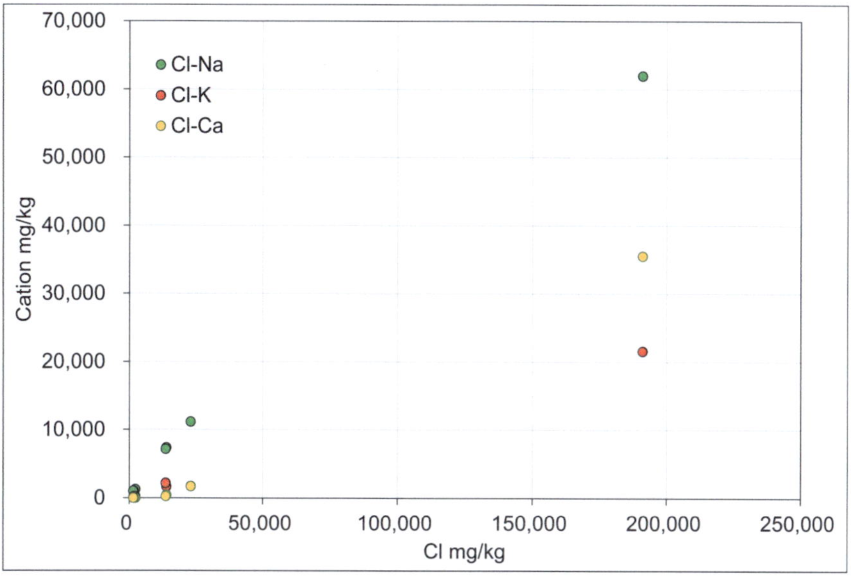

Fig. 1.47 Concentration of chloride versus major cations in waters discharged from geothermal wells (plotted from Table 1.6; Henley 1984)

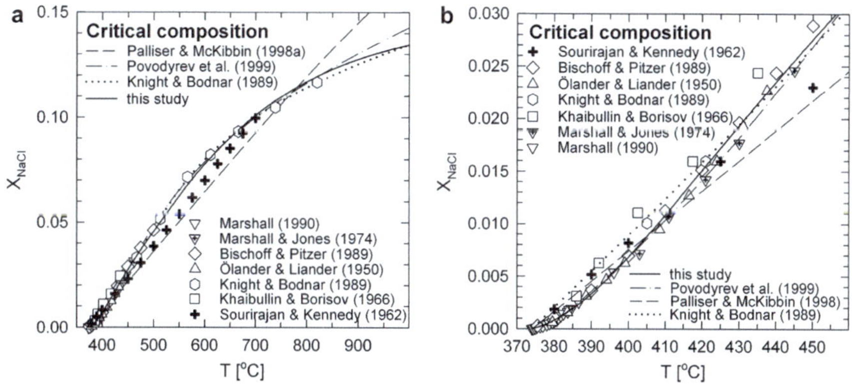

Fig. 1.48 Critical composition as a function of temperature. **a** Full range, **b** detail from a pure water critical point to 460 °C (Driesner and Heinrich 2007)

1964). Thus, we can only access supercritical fluids in very low-salinity water cases, less than seawater.

If we consider the hypersaline environments anticipated in the deeper part of the high-temperature hydrothermal convection systems, all the phase relations, such as the boiling point curve and critical point temperature, are dramatically shifted to the higher-temperature and higher-pressure regions by the salinity (Fig. 1.48; Bodnar

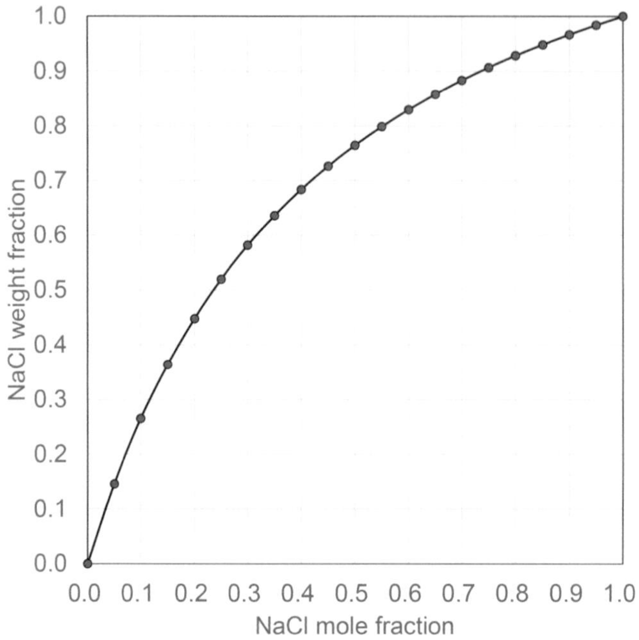

Fig. 1.49 Relation between the NaCl mole fraction and the NaCl weight fraction in the H_2O-NaCl system

and Costain 1991; Driesner and Heinrich 2007). Therefore, there is little probability of encountering the natural supercritical fluids at an easily accessible depth.

The mechanisms promoting chloride and silica concentration in hot water are different. Two-phase separation (boiling) and steam loss promote chloride concentration, while the rising enthalpy promotes the dissolution of silica from rocks into hot water. The former involves boiling liquid water, and the latter involves dissolving solid rocks. The silica and chloride contents of the Japanese hot springs show a more or less negative correlation, as shown in Fig. 1.50, which supports these different mechanisms.

It is experimentally recognized that the solubilities of silica minerals such as amorphous silica, opal, cristobalite, chalcedony, and quartz systematically increase with the temperature up to 300 °C at the given vapor pressure. For example, the solubility of quartz can often be used as a solute geothermometer to estimate the reservoir temperature (Fig. 1.51; Fournier 1985). However, the solubility of quartz decreases in the high-temperature region exceeding 300 °C at the given vapor pressure, which is called a region of retrograde solubility, as shown by the shaded area in Fig. 1.51 (Fournier 1985).

An empirical equation for the solubility of quartz in water from 25° to 900 °C at a pressure of up to 100 MPa based on the experimental data was given by Fournier and Potter (1982) as follows:

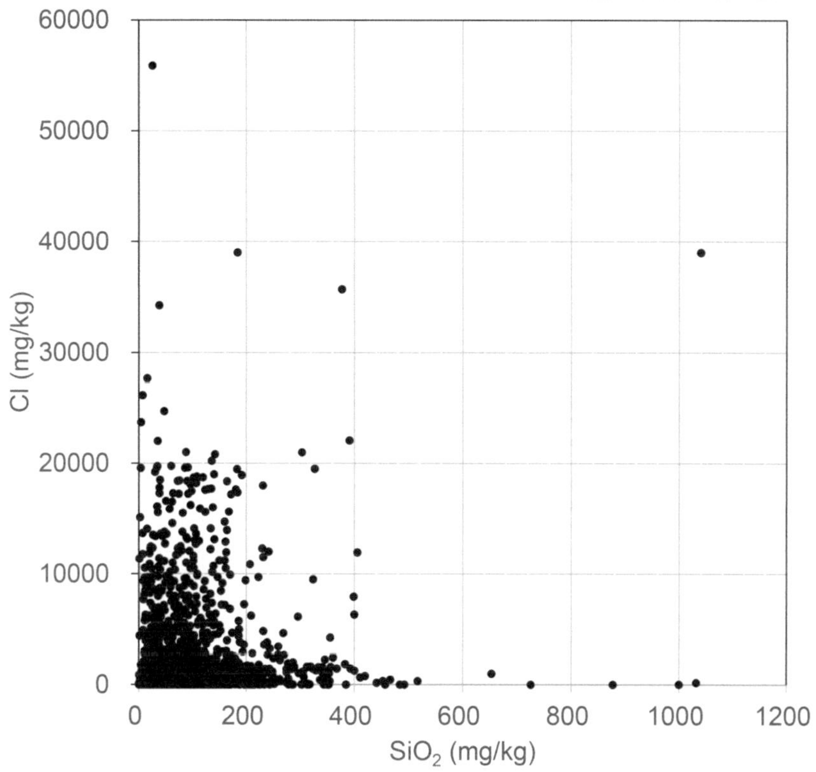

Fig. 1.50 Silica and chloride contents of 3,227 hot spring waters in Japan (Muraoka et al. 2007)

$$\log m = A + B(\log V) + C(\log V)^2 \tag{1.2}$$

where m is the molal silica concentration, V is the specific volume of pure water, and

$A = -4.66206 + 0.0034063T + 2179.7T^{-1} - 1.1292 \times 10^6 T^{-2} + 1.3543 \times 10^8 T^{-3}$

$B = -0.0014180T - 806.97T^{-1}$

$C = 3.9465 \times 10^{-4} T.$

T is the temperature in kelvins. Then, we can draw the solubility of quartz in the broad PT space as shown in Fig. 1.52.

As shown in Fig. 1.52, a region of retrograde solubility is typically controlled by the boiling point curve of pure water, where the solubility of quartz is higher in a liquid phase and lower in a vapor phase. Even if the pressure exceeds the critical point to the supercritical fluid region, this retrograde solubility relation by the pseudo-boiling point curve continues up to near 90 MPa.

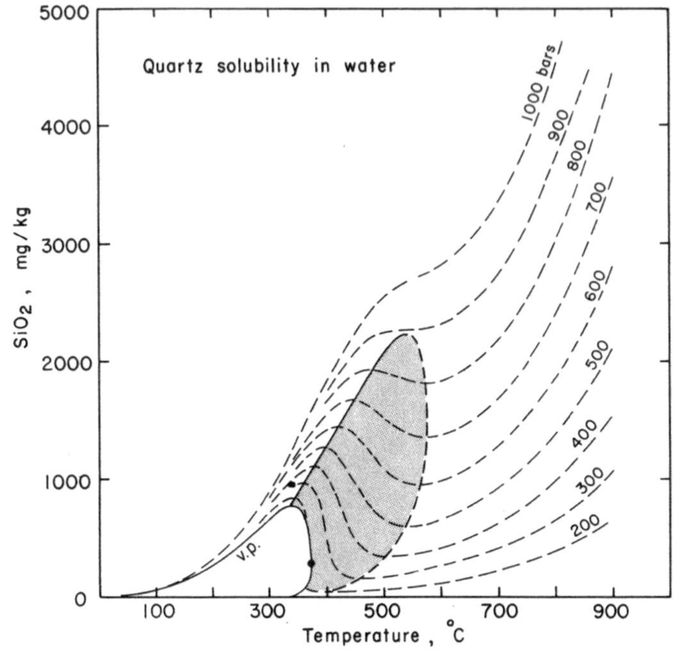

Fig. 1.51 Calculated solubilities of quartz in water up to 900 °C at the indicated pressures (Fournier 1985). The shaded area emphasizes a region of retrograde solubility

Below the boiling point curve and pseudo-boiling point curve, the solubility of quartz exclusively depends on temperature, so that this relation can be used for the quartz geothermometer. However, above 500 °C, the solubility of quartz depends more on pressure than temperature, so that it may play a possible role in the quartz geobarometer.

For example, hypersaline water taken from the depths of 2,500–3,220 m of the well S2-14 in the Salton Sea geothermal field contains a silica content of 588 mg/kg (McKibben et al. 2020), which is high but not extraordinarily high. The hypersaline water taken from the depth of 3,708 m of the well WD-1a by reverse circulation in Kakkonda, Japan (see Subsection 6.2.1 in detail), contains a silica content of 2,100 mg/kg (Table 6.2.1; Kasai et al. 1998; see Sect. 6.2.1 for details), one of the high-silica concentration records.

To simplify, the temperature of the high-silica brine of the well WD-1a is here exclusively assumed to be 550 °C (Fig. 6.2.5). According to the geobarometric concerns when we plot the data of 550 °C and 2,100 mg/kg to point A in Fig. 1.52, we obtain the pressure condition 86.2 MPa. However, the pressure of the high-silica brine of the well WD-1a was measured at about 36 MPa (Kasai et al. 1998). When we plot the data of 550 °C and 36 MPa into point B, the silica solubility contour indicates only 120 mg/kg, which is a far lower value. This discrepancy is easily solved if we consider the physical condition of the plastic zone at a depth of 3,708 m of the

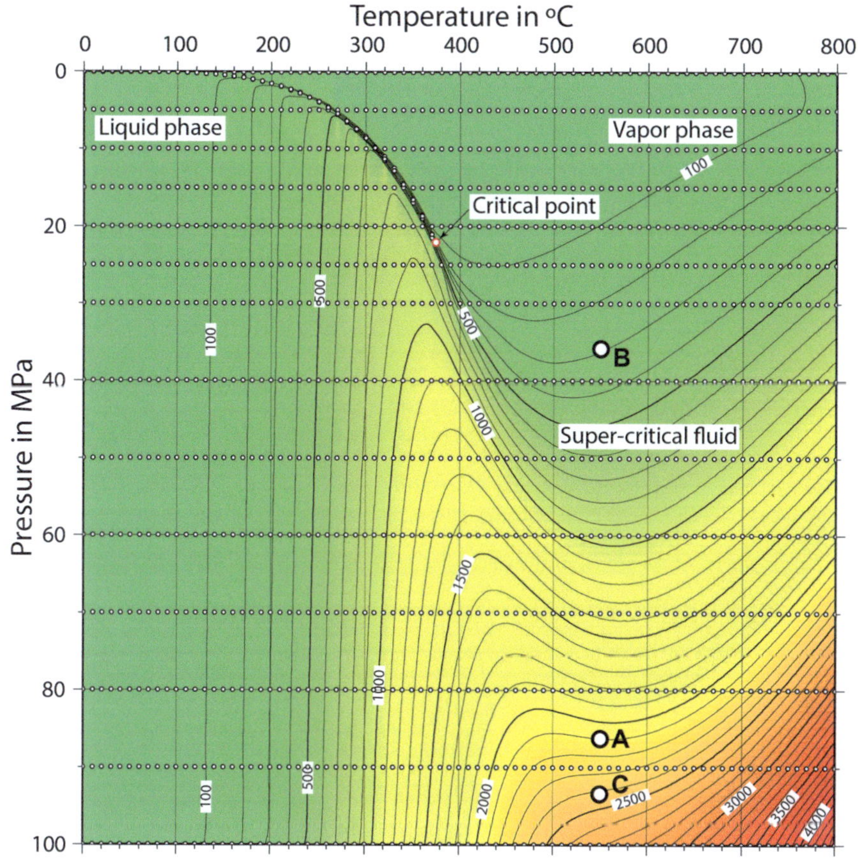

Fig. 1.52 Solubilities of quartz in the PT space calculated on minor circle points from the equations by Fournier and Potter (1982) and contoured by GMT 6.4.0 (Wessel et al. 2019). Numbers labeled on the contour lines show the SiO_2 concentration in mg/kg converted from molal concentration in Eq. (1.1)

well WD-1a. The pressure of 36 MPa was measured in nearly a hydrostatic physical condition after the pressure liberation by drilling. Before the drilling, the lithostatic condition probably controlled an impermeable plastic zone (see Subsection 6.2.1 for details). The lithostatic stress at a depth of 3,708 m is calculated to be 92.7 MPa by the 25 MPa/km conversion factor (see Subsection 3.2.5 for details). When we plot the data of 550 °C and 92.7 MPa into point C in Fig. 1.52, the silica solubility contour indicates 2,400 mg/kg. Point C is much closer to point A than to point B. The crustal strength analysis creep equation of the well WD-1a in Fig. 6.2.8 also shows 95.0 MPa at a depth of 3.7 km, roughly correlated to point A rather than to point B. It implies a possibility of quartz geobarometric application, even though it is still very rough.

References

Amante C, Eakins BW (2009) ETOPO1 1 arc-minute global relief model: procedures, data sources and analysis. NOAA technical memorandum NESDIS NGDC-24. National Geophysical Data Center, NOAA. https://doi.org/10.7289/V5C8276M. Access date: January 3, 2023

Arnórsson S (1992) Geochemistry and geothermal resources in Iceland. In: D' Amore D (ed) Application of geochemistry in geothermal reservoir development, UNITAR and UNDP, pp 145–196

Arnórsson S, Andrésdóttir A (1995) Processes controlling the distribution of boron and chlorine in natural waters in Iceland. Geochim Cosmochim Acta 59:4125–4146

Arias A, Dini I, Casini M, Fiordelisi A, Perticone I, Aiuto P (2010) Geoscientific feature update of the Larderello-Travale geothermal system (Italy) for a regional numerical modelling. *Proceedings World Geothermal Congress 2010*, Bali, Indonesia, 25–29 April 2010

Atwater T (1970) Implications of plate tectonics for the Cenozoic tectonic evolution of western North America. Geol Soc Am Bull 81:3513–3536

Aydin A, Nur A (1982) Evolution of pull-apart basins and their scale independence. Tectonics 1:91–105

Barker BJ, Gulati MS, Bryan MA, Riedel KL (1991) Geysers reservoir performance. Geothermal resources council transactions, monograph on the geysers geothermal field, special report 17:167–177

Bertani R (2005) World geothermal power generation in the period 2001–2005. Geothermics 34:651–690

Bertani R (2016) Geothermal power generation in the world 2010–2014 update report. Geothermics 60:31–43

Bird P (2003) An updated digital model of plate boundaries. Geochem Geophys Geosyst 4(3):1027, https://doi.org/10.1029/2001GC000252

Bodnar RJ, Costain JK (1991) Effect of varying fluid composition on mass and energy transport in the earth's crust. Geophys Res Lett 18:983–986

Burchfiel BC, Stewart JH (1966) 'Pull-apart' origin of the central segment of Death Valley, California. Geol Soc Am Bull 77:439–442

Conti P, Grassi W, Passaleva G, Cataldi R (2015) Geothermal direct uses in Italy: country update for WGC2015. In: Proceedings world geothermal congress 2015, Melbourne, Australia, 19–25 April 2015

Darma S, Imani YL, Naufal M, Shidqi A, Riyanto TD, Daud MY, (2021) Country update: the fast growth of geothermal energy development in Indonesia. Proceedings world geothermal congress 2020+1, Reykjavik, Iceland, 9p

Dobson P, Dwivedi D, Millstein D, Krishnaswamy N, Garcia J, Kiran M (2020) Analysis of curtailment at The Geysers geothermal Field, California. Geothermics 87:14. https://doi.org/10.1016/j.geothermics.2020.101871

Driesner T, Heinrich CA (2007) The system H_2O-NaCl. Part I: correlation formulae for phase relations in temperature-pressure-composition space from 0 to 1000 °C, 0 to 5000 bar, and 0 to 1 X_{NaCl}. Geochim Cosmochim Acta 71:4880–4901

Ellis AJ, Mahon WAJ (1964) Natural hydrothermal systems and experimental hot-water/rock interactions. Geochim Cosmochim Acta 28:1323–1357

Ellis AJ, Mahon WAJ (1967) Natural hydrothermal systems and experimental hot- water/rock interactions (part II). Geochim Cosmochim Acta 31:519–538

Endy SL, Butler SJ (2010) Numerical reservoir-modeling of forty years of injectate recovery at The Geysers geothermal field, California, USA. Geothermal Resour Council Trans 34:1221–1227

Fernandes RMS, Ambrosius BAC, Noomen R, Bastos L, Combrinck L, Miranda JM, Spakman W (2004) Angular velocities of Nubia and Somalia from continuous GPS data: implications on present-day relative kinematics. Earth Planet Sci Lett 222:197–208

Fournier RO (1985) The behavior of silica in hydrothermal solutions. Geology and geochemistry of epithermal systems: reviews in economic geology, v 2, Society of Economic Geologists, 45–61

Fournier RO (1987) Conceptual models of brine evolution in magma-hydrothermal systems. U.S. Geol Surv Professional Paper 1350:1487–1508
Fournier RO, Potter RWII (1982) An equation correlating the solubility of quartz in water from 25° to 900°C at pressures up to 10,000 bars. Geochim Cosmochim Acta 46:1969–1973
Fronda AD, Lazaro VS, Halcon RM, Reyes RG (2021) Geothermal energy development: the Philippines country update. Proceedings world geothermal congress 2020+1, Reykjavik, Iceland, 8p
Fukao Y, Obayashi M (2013) Subducted slabs stagnant above, penetrating through, and trapped below the 660 km discontinuity. J Geophys Res. Solid Earth 118:5920–5938. https://doi.org/10.1002/2013JB010466
Garcia J, Walters M, Beal J, Hartline C, Pingol A, Pistone S, Wright M (2012) Overview of the northwest Geysers EGS demonstration project. Stanford University, Stanford, California, January 30–February 1 In: Proceeding of thethirty-seventh workshop on geothermal reservoir engineering, 012, 11p
Genrich JF, Bock Y, McCaffrey R, Prawirodirdjo L, Stevens CW, Puntdewo SSO, Subarya C, Wdowinski S (2000) Distribution of slip at the northern Sumatra fault system. J Geophys Res 105:28327–28341
Geo-Heat Center (2004) State geothermal databases (CD-ROM). Geo-Heat Center, Oregon Institute of Technology.
Geological Survey of Japan (2023) GSJ geochemical reference samples database. http://gbank.gsj.jp/geostandards/welcome.html. Access, July 23, 2023
Global Volcanism Program (2024) Volcanoes of the World. v.5.1.5., Venzke E (ed), Smithsonian Institution. https://doi.org/10.5479/si.GVP.VOTW5-2023.5.1. Access, January 12, 2024
Goyal KP, Conant TT (2010) Performance history of the geysers steam field, California, USA. Geothermics 39:321–328
Gutiérrez-Negrín LCA, Félix IC, Romo-Jones JM, Quijano-León JL (2021) Geothermal energy in Mexico: update and perspectives. Proceedings world geothermal congress 2020+1, Reykjavik, Iceland, 13p
Heming RF, Barnet PR (1986) The petrology and petrochemistry of the Auckland volcanic field. In: Smith IEM (ed) Late cenozoic volcanism in New Zealand, vol 23. The Royal Society of New Zealand, Bulletin, pp 64–75
Henley RH (1984) Chemical structure of geothermal systems. Fluid-mineral equilibria in hydrothermal systems: reviews in economic geology, v 1, Society of Economic Geologists, 9–30
Huttrer GW (2001) The status of world geothermal power generation 1995–2000. Geothermics 30:1–27
Huttrer GW (2021) Geothermal power generation in the world 2015–2020 update report. Proceedings world geothermal congress 2020+1, Reykjavik, Iceland, 17p
Ichikawa H (2017) Seismic tomography globe (an interactive globe on web browsers). http://dagik.org/misc/gst/index.html
Ingebritsen SE, Sorey ML (1988) Vapor-dominated zones within hydrothermal systems: evolution and natural state. J Geophys Res 93:13635–13655
Kanamori H (1978) Chapter 2: structure of the earth. Iwanami Earth Science Course, Volume 1: The Earth, 318p (in Japanese). Iwanami-Shoten, 45–98 (particularly 83)
Kasai K, Sakagawa Y, Komatsu R, Sasaki M, Akaku K, Uchida T (1998) The origin of hypersaline liquid in the Quaternary Kakkonda granite, sampled from WD-1a well, Kakkonda geothermal system, Japan. Geothermics 27:631–645
Kristmannsdóttir H, Arnórsson S, Sveinbjörnsdóttir Á, Ármannsson H (2010) Chemical variety of water in Icelandic heating systems. Proceedings world geothermal congress 2010, Bali, Indonesia, 25–29
Luth WC, Jahns RH, Tuttle OF (1964) The granite system at pressures of 4 to 10 kilobars. J Geophys Res 69:759–773

Mangi PM (2018) Geothermal potential in East Africa—How to realise it. Proceedings of the SDG Short Course III on Exploration and Development of Geothermal Resources, organized by UNU-GTP and KenGen, Lake Bogoria and Lake Naivasha, Kenya, 7–27

Mason B (1966) Principle of geochemistry (3rd ed.). Wiley, New York, 329 p

Matsui T (1996) Chapter 3: differentiation. Iwanami Earth and Planetary Science Course, Volume 1: Introduction to Earth and Planetary Science, 287p (in Japanese). Iwanami-Shoten, pp 101–110

McKibben MA, Elders WA, Raju ASK (2020) Lithium and other geothermal mineral and energy resources beneath the Salton Sea. In: Crisis at the Salton Sea, University of California Riverside Salton Sea Task Force, 123p, 103–119

McCaffrey R (1991) Slip vectors and stretching of the Sumatran fore arc. Geology 19:881–884

McCaffrey R, Zwick P, Bock Y, Prawirodirdjo L, Genrich J, Stevens CW, Puntodewo SSO, Subarya C (2000) Strain partitioning during oblique plate convergence in northern Sumatra: geodetic and seismologic constraints and numerical modeling. J Geophys Res 105:28363–28376

Minissale A, Magro G, Martinelli G, Vaselli O, Tassi GF (2000) Fluid geochemical transect in the Northern Apennines (central-northern Italy): fluid genesis and migration and tectonic implications. Tectonophysics 319:199–222

Mizutani H, Watanabe T (1978) Chapter 4: geothermics. Iwanami Earth Science Course, Volume 1: The Earth, 318p (in Japanese). Iwanami-Shoten, pp 169–223

Muffler LJP, Guffanti M (1975) Introduction. in Muffler, L.J.P. ed., Assessment of geothermal resources of the United States—1978. U.S. Geol Surv Circular 790:1–7

Muraoka H, Gunnlaugsson E, Song Y, Lund J, Bromley CJ, Rybach L (2010a) Host rock controls to thermal water chemistry induced from global comparison. Curr Appl Phys 10:S102–S107. https://doi.org/10.1016/j.cap.2009.11.012

Muraoka H, Kamata H (1983) Displacement distribution along minor fault traces. J Struct Geol 5:483–495

Muraoka H, Nasution A (2004) En echelon volcanic arc as a key to recognize mantle diapirs in the Lesser Sunda arc, eastern Indonesia. J Geothermal Res Soc Japan 26:237–249 (in Japanese with English abstracts)

Muraoka H, Nasution A, Urai M, Takahashi M, Takashima I, Simanjuntak J, Sundhoro H, Aswin D, Nanlohy F, Sitorus K, Takahashi H, Koseki T (2002) Tectonic, volcanic and stratigraphic geology of the Bajawa geothermal field, central Flores, Indonesia. Bull Geol Surv Japan 53:109–138

Muraoka H, Uchida T (2002) Overview of the "Research Cooperation Project on the Exploration of Small-scale Geothermal Resources in the Eastern Part of Indonesia" by the Geological Survey of Japan. Bull Geol Surv Japan 53:63–77

Muraoka H, Ohtani T (2000) Profiling of the Kakkonda geothermal system by bulk rock chemistry of the well WD-1a. Report of Geological Survey of Japan, No 284, 35–55 (in Japanese with English abstracts)

Muraoka H, Sakaguchi K, Tamanyu S, Sasaki M, Shigeno H, Mizugaki K (2007) Atlas of hydrothermal systems in Japan. Geological Survey of Japan, AIST, 110p (in Japanese with English abstract)

Muraoka H, Takahashi M, Sundhoro H, Dwipa S, Soeda Y, Momita M, Shimada K (2010b) Geothermal systems constrained by the Sumatran fault and its pull-apart basins in Sumatra, western Indonesia. Proceedings World Geothermal Congress 2010, Bali, Indonesia, 9p

Muraoka H, Uchida T, Sasada M, Yagi M, Akaku K, Sasaki M, Yasukawa K, Miyazaki S-I, Doi N, Saito S, Sato K, Tanaka S (1998) Deep geothermal resources survey program: igneous, metamorphic and hydrothermal processes in a well encountering 500 °C at 3729 m depth, Kakkonda, Japan. Geothermics 27:507–534

Nakamura H (1964) Exploitation and utilization of geothermal energy. Chigaku-Shuppansha (Earth Science Publisher), 153p (in Japanese)

Nishimura S, Nishida J, Yokoyama T, Hehuwat F (1986) Neo-tectonics of the Strait of Sunda, Indonesia. J SE Asian Earth Sci 1:81–91

NOAA National Centers for Environmental Information (2022) ETOPO 2022 15 arc-second global relief model. NOAA National Centers for Environmental Information. https://doi.org/10.25921/fd45-gt74. Accessed [date]

Obayashi M, Yoshimitsu J, Nolet G, Fukao Y, Shiobara H, Sugioka H, Miyamachi H, Gao Y (2013) Finite frequency whole mantle P-wave tomography: improvement of subducted slab. Geophys Res Lett. https://doi.org/10.1002/2013GL057401

Omenda P, Mangi P, Ofwona C, Mwangi M (2021) Country update report for Kenya 2015–2019. Proceedings World Geothermal Congress 2020+1, Reykjavik, Iceland, 16p

Palano (2015) On the present-day crustal stress, strain-rate fields and mantle anisotropy pattern of Italy. Geophys J Int, 969–985, https://doi.org/10.1093/gji/ggu451

Persoff P, Hulen JB (2001) Hydrologic characterization of reservoir metagraywacke from shallow and deep levels of The Geysers vapor-dominated geothermal system, California, USA. Geothermics 30:169–192

Prawirodirdjo L, Bock Y, Genrich JF, Puntdewo SSO, Rais J, Subarya C, Sutisna S (2000) One century of tectonic deformation along the Sumatran fault from triangulation and Global Positioning System surveys. J Geophys Res 105:28343–28361

Ragnarsson Á, Steingrímsson B., Thorhallsson S (2021) Geothermal development in Iceland 2015–2019. Proceedings world geothermal congress 2020+1, Reykjavik, Iceland, 15p

Romagnoli P, Arias A, Barelli A, Cei M, Casini M (2010) An updated numerical model of the Larderello–travale geothermal system, Italy. Geothermics 39:292–313, https://doi.org/10.1016/j.geothermics.2010.09.010

Rutqvist J, Jeanne P, Dobson PF, Garcia J, Hartline C, Hutchings L, Singha A, Vasco DW, Walters M (2016) The northwest Geysers EGS demonstration project, California—Part 2: Modeling and interpretation. Geothermics 63:120–138

Ryan MP (1994) Chapter 6: Neutral-buoyancy controlled magma transport and storage in mid-ocean ridge magma reservoirs and their sheeted-dyke complex: A summary of basic relationships. In Magmatic Systems (M.P. Ryan, ed.), 401p., Academic Press, San Diego, pp 97–138

Shigeno H (1992) Application of geochemistry in geothermal-field development in Japan. In: D' Amore D (ed) Application of geochemistry in geothermal reservoir development, UNITAR and UNDP, pp 365–382

Sigmundsson F, Einarsson P, Hjartardóttir ÁR, Drouin V, Jónsdóttir K, Árnadóttir T, Geirsson H, Hreinsdóttir S, Li S, Benedikt Gunnar Ófeigsson BG (2020) Geodynamics of Iceland and the signatures of plate spreading. J Volcanol Geothermal Res 391:106436 https://doi.org/10.1016/j.jvolgeores.2018.08.014

Sueyoshi Y, Matsuda K, Shimoike T, Koseki T, Takahashi H, Futagoishi M, Sitorus K, Simanjuntak J (2002) Exploratory well drilling and discharge test of MT-1 and MT-2 in the Mataloko geothermal field, Flores, Indonesia. Bull Geol Surv Japan 53:307–321

Takahashi E (1996) Chapter 3: differentiation. Iwanami Earth and Planetary Science Course, Volume 1: Introduction to Earth and Planetary Science, 287p (in Japanese), Iwanami-Shoten, 111–161

Thompson RC (1991) Structural stratigraphy and intrusive rocks at The Geysers geothermal field. Geothermal Resour Council Trans Monogr Geysers Geothermal Field Special Report 17:59–63

Trčková J, Živor R, Přikryl R (2002) Physical and mechanical properties of selected amphibolite core samples from the Kola Superdeep Borehole KSDB-3. Terra Nova 14(5):379–387. https://doi.org/10.1046/j.1365-3121.2002.00427.x.ISSN0954-4879

Truesdell AH (1984) Introduction to chemical calculations. Fluid-Mineral equilibria in hydrothermal systems: reviews in economic geology, v 1, Society of Economic Geologists, 1–8

Turcotte D, Schubert G (2014) Geodynamics (Third Edition). Cambridge University Press, 623p

Uihlein A (2018) JRC geothermal power plant dataset. European Commission, Joint Research Centre (JRC) PID: http://data.europa.eu/89h/jrc-10128-10001

US Geological Survey (2023) USGS geochemical and microanalytical reference materials distribution by geology, geophysics, and geochemistry science center. https://www.usgs.gov/centers/gggsc/science/usgs-geochemical-and-microanalytical-reference-materials-distribution. Access, July 23, 2023

US Geological Survey (2024) Search earthquake catalog, earthquake hazard program. https://earthquake.usgs.gov/earthquakes/search/. Access, March 9, 2024

Uyeda S (1971) New view of the earth. Iwanami-Shoten, Tokyo, 197p (in Japanese)

Uyeda S, Kanamori H (1978) Subduction of oceanic plates and formation of marginal seas. Kagaku (Science) 48:91–102 (in Japanese)

van Keken PE, Wilson CR (2023) An introductory review of the thermal structure of subduction zones: I—motivation and selected examples. Progress in earth and planetary science, 20p, https://doi.org/10.1186/s40645-023-00573-z

Wessel P, Luis JF, Uieda L, Scharroo R, Wobbe F, Smith WHF, Tian D (2019) The generic mapping tools version 6. Geochem Geophys Geosyst 20:5556–5564. https://doi.org/10.1029/2019GC008515

White DE, Muffler LJP, Truesdell AH (1971) Vapor-dominated hydrothermal systems compared with hot-water systems. Econ Geol 66:75–97

White DE, Williams DL (1975) Summary and conclusions. In: White DE, Williams DL (eds) Assessment of geothermal resources of the United States—1975. U.S. Geological Survey Circular 726, pp 147–155

Wilson CJN, Rogan AM, Smith IEM, Northey DJ, Nairn IA, Houghton BF (1984) Caldera volcanoes of the Taupo Volcanic Zone, New Zealand. J Geophys Res 89:8463–8484

Chapter 2
Tectonic Backgrounds and Advances of Geothermal Developments in Japan

Hirofumi Muraoka and Koichi Tagomori

Abstract Before describing individual geothermal fields in Japan in the following chapters, the first half of this chapter will provide fundamental knowledge of the tectonic background in the Japanese islands, such as tectonics and magma generation, from a nationwide viewpoint. Seismic zones, seismic tomography, and low-velocity zones visualize the Japanese islands' subduction regime and magma generation. The individual roles of the Pacific and the Philippine Sea slabs are described in the slab-overlapped area in central Japan, where the Philippine Sea slab is an insulator from the upper hot mantle wedge to the lower cold Pacific slab. Deep low-frequency earthquakes enable the tracing of replenishment conduits of basaltic magma from the upper mantle to the subvolcanic magma chambers in some volcanoes in Japan. The latter half of this chapter examines Japan's geothermal development, highlighting historical milestones, government policies like the feed-in tariff (FIT), and ongoing R&D efforts. It addresses technical challenges such as supercritical geothermal resources and the need for cooperation with local hot spring communities to achieve Japan's carbon–neutral goals by 2050. The chapter also presents a practical approach to geothermal exploration, emphasizing the importance of accurate reservoir characterization and modeling. It introduces the "Three Key Elements for Geothermal Systems" (3KEYGS) and discusses their role in conceptual models. Additionally, it covers the stages of geothermal power development, focusing on the integration of reservoir simulation for economic viability.

Keywords Japan · Tectonics · Magma generation · Seismic tomography · Slab-overlapped area in central Japan · Deep low-frequency earthquakes · Geothermal development · Government geothermal policies · Geothermal reservoir characterization · Geothermal reservoir modeling

---

H. Muraoka (✉)
Hirosaki University, Hirosaki, Japan
e-mail: hiro@hirosaki-u.ac.jp

K. Tagomori
West Japan Engineering Consultants, Inc., Fukuoka, Japan
e-mail: k-tagomori@wjec.co.jp

J. Kasahara et al. (eds.), *Challenges in Geothermal Development*, Advances in Geological Science, https://doi.org/10.1007/978-981-95-0553-1_2

2.1 Tectonics and Magma Generation in the Japanese Islands

2.1.1 Subduction Regimes of the Japanese Islands

The Japanese islands comprise about 3,700 islands, four of which are relatively large. The largest island, Honshu, has three neighboring large islands, Hokkaido Island to the northeast, Shikoku Island to the south, and Kyushu Island to the southwest (Fig. 2.1). The Japanese islands are subject to two subduction regimes (Fig. 2.1). One is the Pacific Plate, which subducts from the Kuril Trench, Japan Trench, and Izu-Bonin Trench beneath the Northeast Japan arc and the relatively immature Izu-Bonin arc (Fig. 2.1). The other is the Philippine Sea Plate which subducts from the Nankai Trough and Ryukyu Trench beneath the Southwest Japan arc and the relatively immature Ryukyu arc. The former is a global-scale subduction regime, and the latter is a subduction regime of the marginal sea basin formation. Complicated is the latter.

The Philippine Sea Plate is divided into two marginal sea basins bounded by the Kyushu-Palau Ridge: the West Philippine Basin to the west and the Shikoku Basin to the east (Fig. 2.1). The spreading age of the West Philippine Basin was estimated to be 52 to 25 Ma including the Kyushu-Palau Ridge (Ishizuka et al. 2022). It is noted that the age of the oceanic lithosphere sequence consisting of peridotite, cumulate gabbro, sheeted basalt dikes, and basalt pillow lava in ascending order (Fig. 1.3) found at the fore-arc side of the Izu-Bonin-Mariana arc was also estimated to be 52 to 48 Ma (Ishizuka et al. 2011). This suggests that the appearance of the back-arc spreading of the younger Shikoku Basin, caused by the steep Mariana-type subduction, separated the older West Philippine Basin Plate and the older Izu-Bonin-Mariana arc after 25 Ma. The Shikoku Basin's spreading age was coincidentally estimated to be 25 to 15 Ma (Wu et al. 2016; Ishizuka et al. 2022). The activity age of *echelon*-shaped volcanic chains found along the spreading axis of the Shikoku Basin in the NNW-SSE direction was younger than 25 to 15 Ma and was estimated to be 15 to 7 Ma (Ishizuka et al. 2009). Furthermore, the Izu-Bonin-Mariana arc comprises many active Holocene volcanoes and is still growing as an immature island arc.

Thus, the Philippine Sea Plate subduction is characterized by the relatively young slab from 52 Ma to the Recent. The slab is still thin, hot, and therefore buoyant. The northwestward subduction of the buoyant Izu-Bonin arc to the Honshu Island is taking place in typical collision tectonics in the Fossa Magna area (Fig. 2.2). The Itoigawa-Shizuoka Tectonic Line (Fig. 2.2) and its eastern low-altitude valley were called Fossa Magna by Heinrich Edmund Naumann in 1886. The northwestward collision tectonics of the buoyant Izu-Bonin arc are known to have formed the Fossa Magna syntaxis (diagonal) structure, where all the pre-Neogene metamorphic and sedimentary belts were bent northwestward. It was sided by symmetrical northward convex wings on the east and west (Fig. 2.2; Hara et al. 1980; Hirata et al. 2010). Notably, the autochthonous Honshu Island and the allochthonous Izu Peninsula were bounded by Suruga Trough and Sagami Trough in the northern Izu-Bonin arc, where

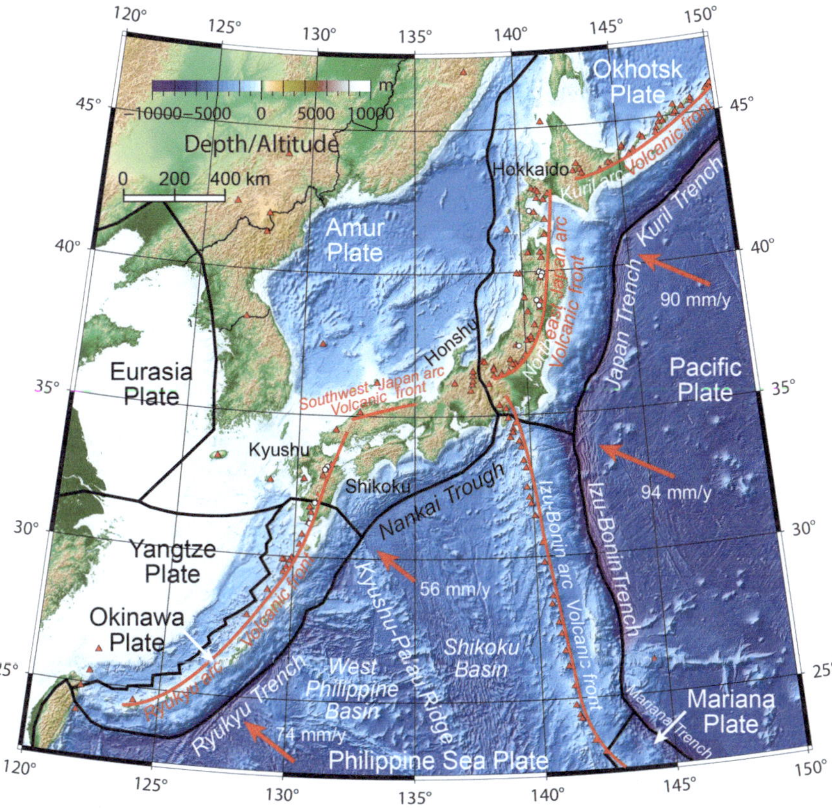

Fig. 2.1 Plate tectonic regimes of the Japanese islands with plate boundaries (thick line, Bird 2003), Holocene volcanoes (red triangle, Global Volcanism Program 2024), and geothermal power plants (open circle, Uihlein 2018). The Lambert projection map was drawn by the ETOPO1 1 arc-minute global relief model (Amante and Eakins 2009) and GMT 6.4.0 (Wessel et al. 2019)

large volcanoes are present, such as Fuji volcano and Hakone caldera (Fig. 2.3; Nakamura and Shimazaki 1981; Hirata et al. 2010). Hence, the Pacific Plate subduction is characterized by the relatively old slab from 150 to 100 Ma. Such an old slab is less buoyant because it is thick and cold.

The three-dimensional distribution of earthquake hypocenters will provide valuable information on the subduction regimes. Figure 2.4 shows 16,311 records of earthquake epicenters with the magnitude higher than 4.0 from January 2012 to February 2024 on the Search Earthquake Catalog of the Earthquake Hazard Program (US Geological Survey 2024). The Wadati-Benioff zone is almost masked by the widespread distribution of shallow-depth earthquakes printed in front (Fig. 2.4). However, some plate boundaries drawn in Fig. 2.1 are denoted by the high-density zones of the shallow-depth earthquakes in Fig. 2.4. Figure 2.5 sorts 5,510 records of earthquakes deeper than 50 km from those of Fig. 2.4. The gradual hypocenter deepening along the Wadati-Benioff zone becomes noticeable.

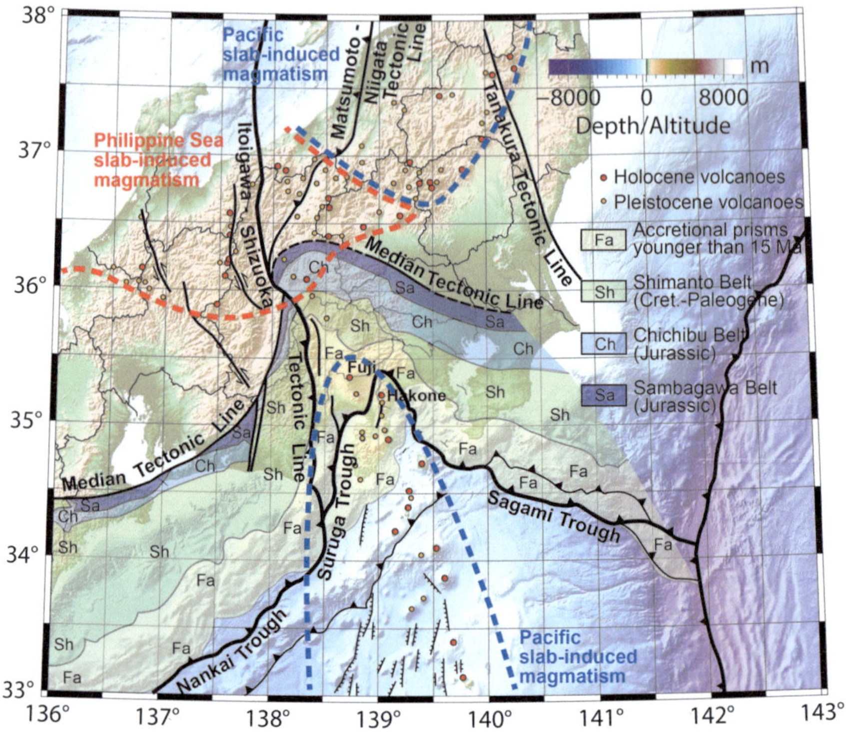

Fig. 2.2 Fossa Magna Syntaxis showing the north-northwestward bending pre-Neogene metamorphic and sedimentary belts (simplified from Hirata et al. 2010). The Lambert projection map was drawn by the ETOPO2022 15 arc-second global relief model (NOAA National Centers for Environmental Information 2022) and GMT 6.4.0 (Wessel et al. 2019)

When we estimate the subduction angle from the Wadati-Benioff zones, the subduction angle of the Pacific Plate is about 40° in the Japan Trench. However, it is increasing to 80° toward the Bonin Trench and the Mariana Trench. The latter subduction regime is called the Mariana-type subduction, where the steep subduction causes counter-upwelling of the back-arc mantle, resulting in the back-arc spreading. This process is proposed as a genetic model of marginal sea basins (Uyeda and Kanamori 1978).

The Northeast Japan arc is composed of two lines of volcanic chains. One is a volcanic chain along the volcanic front. The other is back-arc side volcanic chains (Fig. 2.1). The former chains are situated at a depth of 110 km along the Wadati-Benioff zones. The latter chains are at a depth of 170 km (Fig. 2.5). Particularly, the tendency of the volcanic fronts at a depth of 110 km along the Wadati-Benioff zone is not only found in the Northeast Japan arc but also worldwide. This tendency is presumably controlled by pressure-dependent phenomena such as chlorite and amphibole dehydration reactions. The back-arc volcanic chains are ascribed to the dehydration reaction of phlogopite (Tatsumi 1989). As a result, the subduction angle

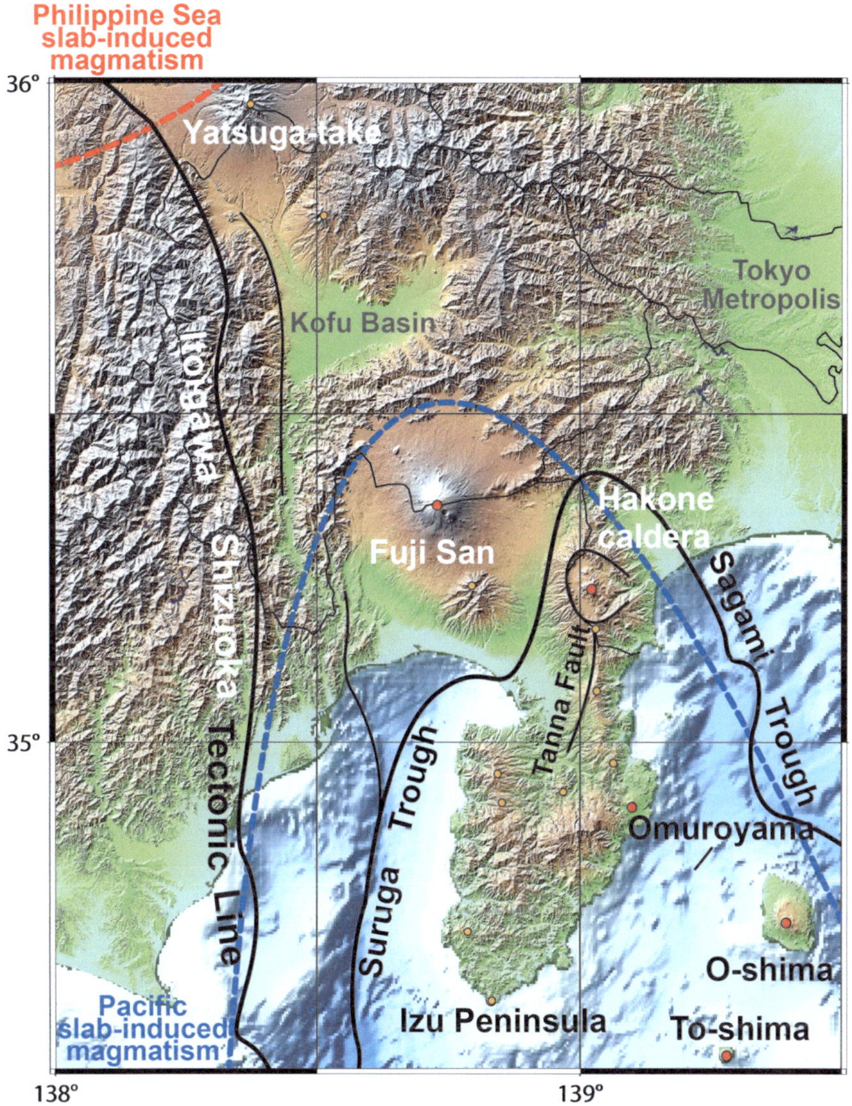

Fig. 2.3 Topography of the northern Izu-Bonin arc, including the Izu Peninsula, Fuji volcano, and Hakone caldera. Tectonic lines were quoted from Hirata et al. (2010). The UTM projection map of land area was drawn by Kasmil 3D using the Basic Geospatial Information with 10 m resolution (Geospatial Information Authority of Japan 2024). The submarine area was drawn by the ETOPO2022 15 arc-second global relief model (NOAA National Centers for Environmental Information 2022) and GMT 6.4.0 (Wessel et al. 2019)

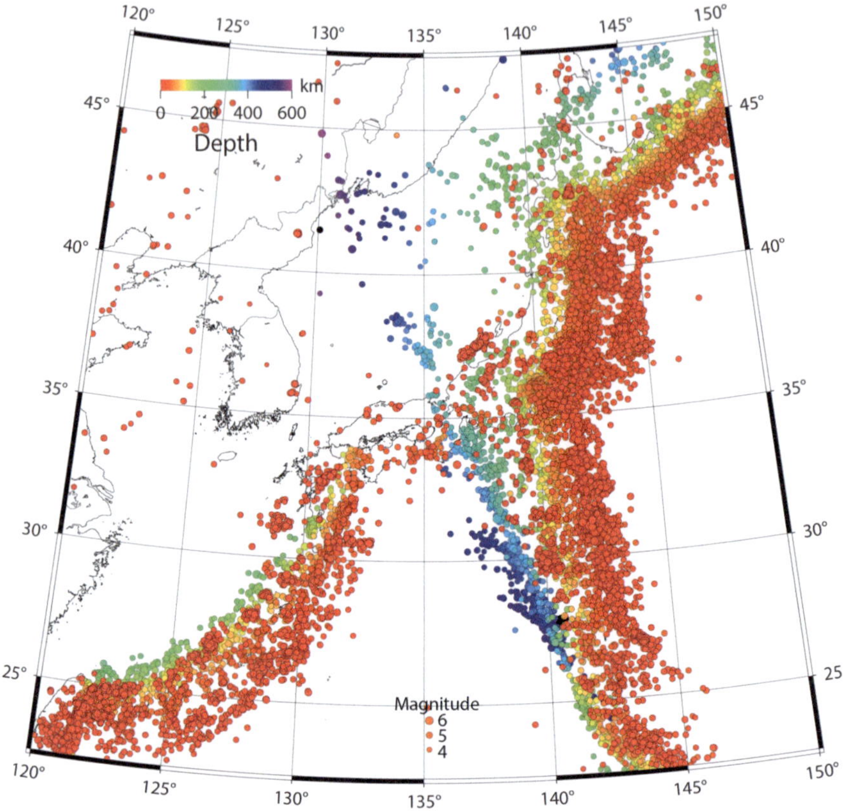

Fig. 2.4 16,311 records of earthquakes from January 2012 to February 2024 on the Search Earthquake Catalog of the Earthquake Hazard Program (US Geological Survey 2024)

negatively correlates with the trench-volcanic front gap. The subduction angle and trench-volcanic front gap are 40° and 300 km, respectively, in the Japan Trench, but are 80° and 200 km, respectively, in the Izu Trench and Bonin Trench because of the steeper Mariana-type subduction (Figs. 2.1 and 2.5). The typical evidence of the back-arc spreading is the Mariana Trough and small back-arc spreading zones are also recognized on the back-arc side of the Izu-Bonin Trench (Uyeda et al. 1983).

The same manner of the Mariana-type subduction regime can also be seen in the Ryukyu Trench, where the Philippine Sea Plate is subducting with an 80° subduction angle and has a 200 kmtrench-volcanic front gap. Following the subduction regime, a back-arc spreading zone called the Okinawa Trough is recognized in the back-arc side of the Ryukyu arc.

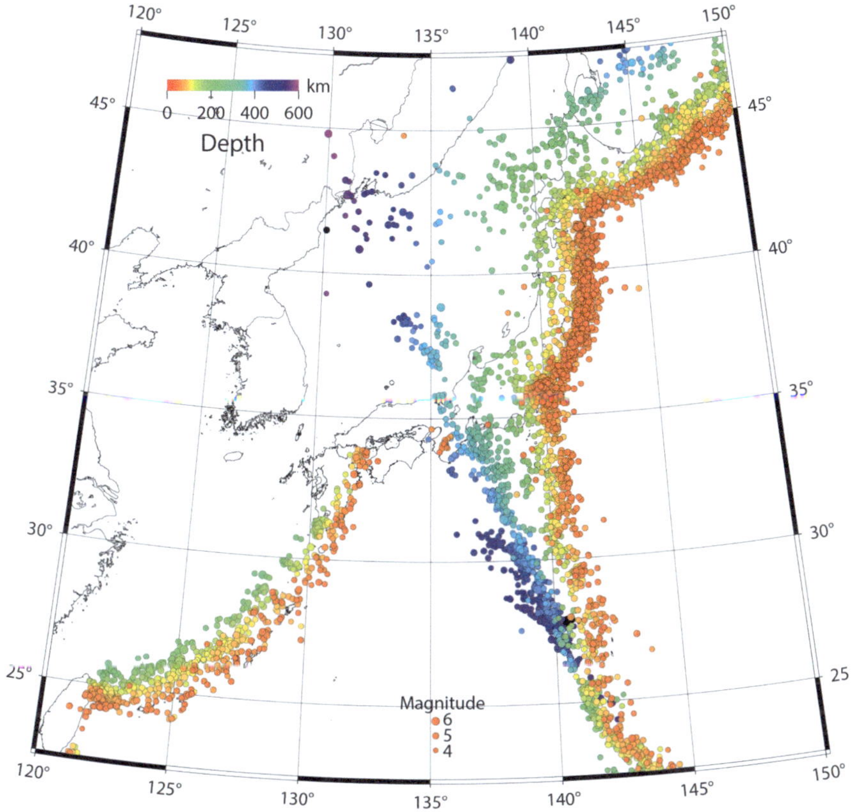

Fig. 2.5 5,510 records of earthquakes deeper than 50 km from January 2012 to February 2024 on the Search Earthquake Catalog of the Earthquake Hazard Program (US Geological Survey 2024)

2.1.2 Seismic Tomography and Magma Generation in the Japanese Islands

Seismic tomography has brought us a powerful imaging tool for magma generation regimes. Until the 1970s, the Wadati-Benioff zone was believed to be nearly equal to the upper surface of the subducting slab, so that the contour map of the depth of the Wadati-Benioff zone was frequently referred to (Fig. 2.6, blue contours; Yoshii 1978). However, through the progress of the earthquake hypocenter resolution and the innovation of seismic tomography, the Wadati-Benioff zone earthquakes were recognized as the double-layered clusters generated within the upper part of the subducting slab in the 1970s (Fig. 2.7; Hasegawa et al. 1978). Particularly, seismic tomography provides the improved depth of the upper surface of the subducting slab by seismic velocity imaging (Fig. 2.7; Nakajima et al. 2001). The refined depth of the upper surface of the subducting slab at the volcanic front by seismic tomography is 10 or 20 km shallower than the previous depth as the Wadati-Benioff zone (Fig. 2.6, red contours; Fig. 2.7). In this context, the depth of the upper surface of the Pacific slab at the volcanic front of the Northeast Japan arc should be modified to be 90 or 100 km (Figs. 2.6 and 2.7).

Current seismic tomography provides beautiful images of the low-velocity anomaly zones related to the subduction zone magmatism. Figure 2.7 shows across-arc cross-sections of S-wave velocity perturbations along the six lines of the Northeast Japan arc (Nakajima et al. 2001; Hasegawa et al. 2008). These cross-sections draw large, partially molten zones in the mantle wedge peridotite, which are almost parallel to the upper surface of the subducting slab and about 50 km above the upper surface of the subducting slab. However, some roots of the partially molten zones seem to be derived from the upper surface of the subducting slab, probably generated by the dehydration reactions from the upper alteration zones of the ocean crust. The largest anomaly of the partially molten zones is observed beneath the volcanic front from a depth slightly deeper than Moho to the inside of the lower crust (Fig. 2.7). This is

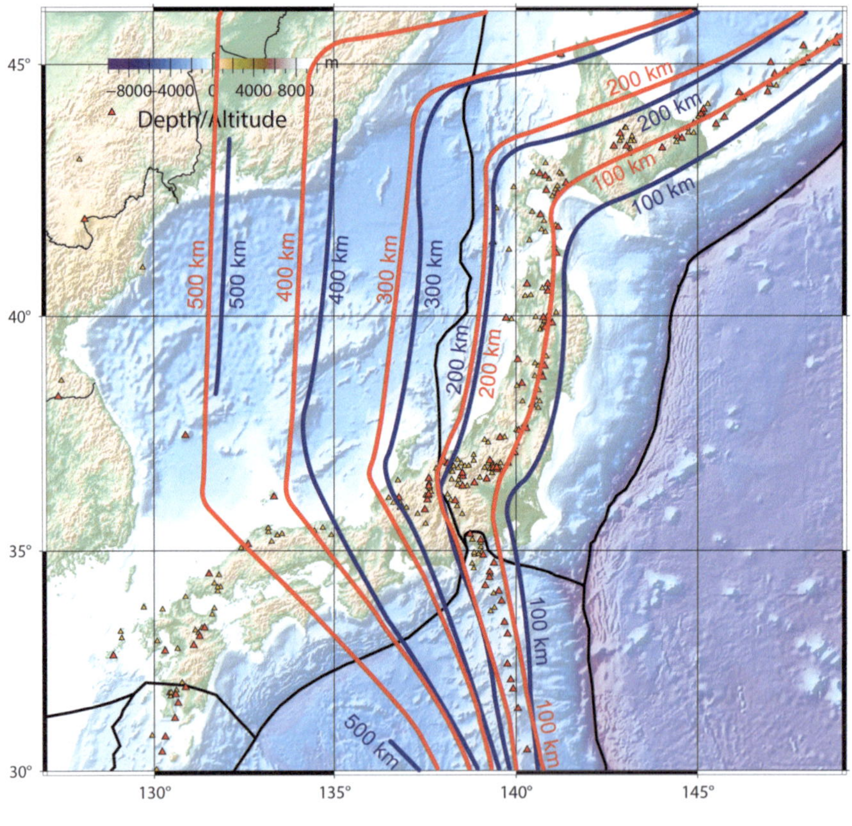

Fig. 2.6 Contour maps of the depths of the Wadati-Benioff zone (blue lines; Yoshii 1978) and the top of the Pacific slab (red lines; Hasegawa et al. 2015). Large red triangle: Holocene volcano, small orange triangle: Pleistocene volcano (Global Volcanism Program 2024)

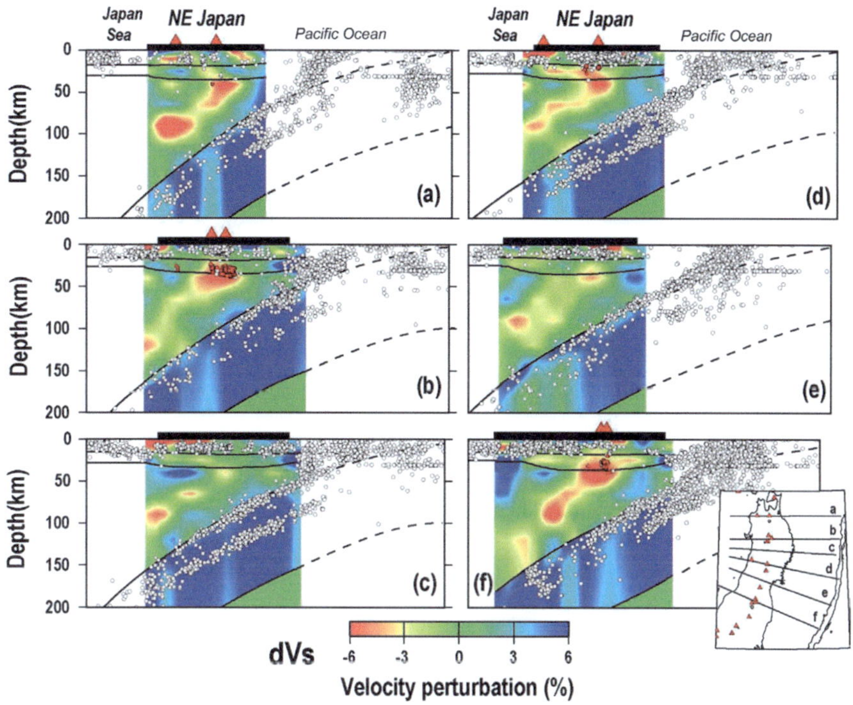

Fig. 2.7 Across-arc cross-sections of S-wave velocity perturbations along lines a to f in the inset map of Northeast Japan (Hasegawa et al. 2008 reproduced from Nakajima et al. 2001). Red triangles and bars at the top represent active volcanoes and the land area, respectively. Dots and red circles show earthquakes and deep, low-frequency microearthquakes, respectively

consistent with the fact that the largest clustered volcanoes are concentrated along the volcanic front.

The same kind of seismic tomography was also performed in Southwest Japan. Figure 2.8 shows across-arc cross-sections of P-wave velocity perturbations along the nine lines of Southwest Japan (Zhao et al. 2012). As going south, the Mariana-type steep subduction of the Philippine Sea slab becomes more obvious. In the same way, the back-arc spreading and volcanism come to be more active toward the south. It should be noted that the depth of the upper surface of the Philippine Sea slab along the volcanic front is 90 km or more in most of the cross-sections, but only the depth of the cross-section (g) at the volcanic front (blue triangle) indicates about 70 km. Many Japanese volcanologists wondered why a caldera is missing between the Aso caldera and Kakuto caldera (Fig. 2.9), away from the regular volcanic spacing of 40 or 50 km. Then, we might obtain the answer from Fig. 2.8 that the depth of the upper surface of the slab (g) was not enough to cause the dehydration reactions of chlorite and amphibole on the given pressure and temperature even along the volcanic front so that the magma generation could not be achieved between the Aso caldera and Kakuto caldera.

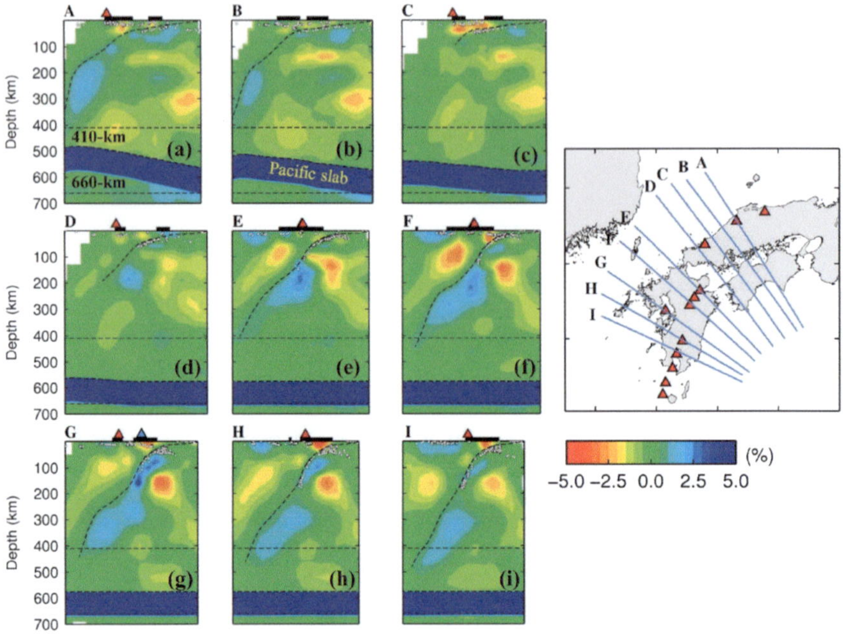

Fig. 2.8 Vertical cross-sections of P-wave velocity perturbation along the profiles shown on the inset map in Southwest Japan (Zhao et al. 2012). The red triangles denote the active arc volcanoes. The black bar at the top of the cross-section denotes the land area. The two dashed lines represent the 410 and 660 km discontinuities. The white dots denote seismicity within a 20 km width of each profile. A dashed line shows the estimated upper boundary of the subducting Philippine Sea Plate slab. The blue triangle on (g) shows the position of the volcanic front estimated from (f) and (h), added in this chapter

Seismic tomography dramatically brings us valuable information. Figure 2.10 summarizes a new contour map of the depth of the upper surface of subducting slabs on the Pacific Plate and the Philippine Sea Plate (Hasegawa et al. 1978, 2015; Nakajima et al. 2001; Hasegawa and Nakajima 2004; Zhao et al. 2012). Particularly, the depth contours of the Philippine Sea Plate were shown in the deeper part by the teleseismic P-wave tomography.

2.1.3 Overlapped Slabs and Magma Generation in Central Japan

The term "hot fingers" was symbolically and conceptually proposed on the clustered volcanoes in Northeast Japan by Tamura et al. (2002), as shown in Fig. 2.11. As has long been known, clustered volcanoes in the Northeast Japan arc tend to be aligned not only in the along-arc direction but also in the across-arc direction. The

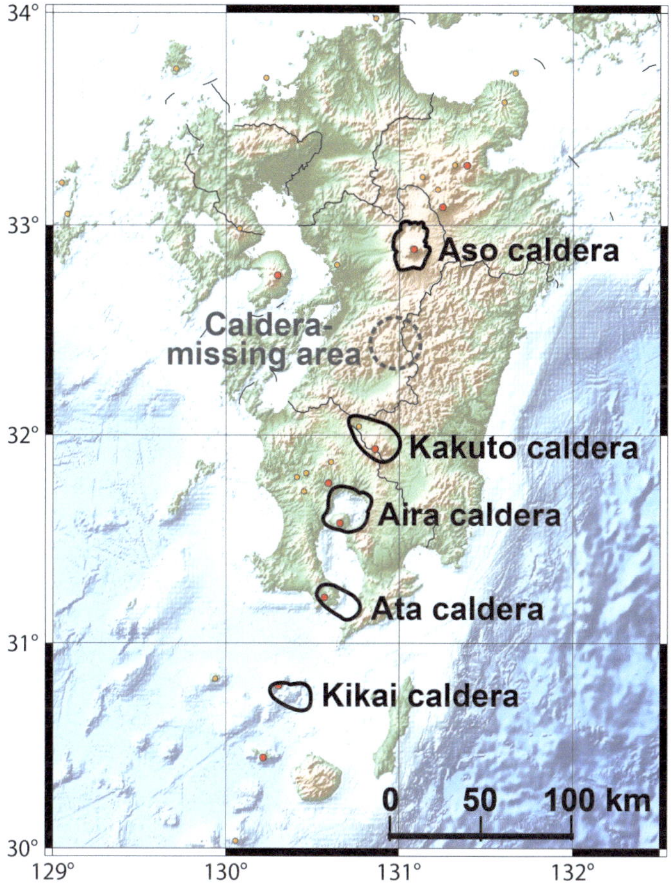

Fig. 2.9 Five collapse calderas aligned with regular spacings except for a missing area in Kyushu

meaning of the across-arc clustering of volcanoes came to be obvious through the recognition of the low-velocity zones in the mantle wedge by seismic tomography (Hasegawa et al. 1978, 2015; Nakajima et al. 2001; Hasegawa and Nakajima 2004; Zhao et al. 2012; Figs. 2.7 and 2.8). The concept of "hot fingers" briefly refined the mutual relationships between these mantle wedge low-velocity zones and clustered volcanoes on the ground surface.

Hot fingers hypothesis classified the Quaternary volcanoes in the Northeast Japan arc into 10 groups of across-arc volcanic clusters; they have an average width of 50 km and are separated by parallel gaps 30–75 km wide (Tamura et al. 2002, 2003; Fig. 2.11). Hot fingers hypothesis suggests that the subducting slab produces fingerlike hot regions in the back-arc mantle, which are rising diagonally toward the fore-arc mantle with time (Tamura et al. 2015; Fig. 2.12).

Recently, central Japan has remarked on the viewpoint of geothermal geology. Almost all Japanese geothermal reservoirs consist of the Miocene, Pliocene, and

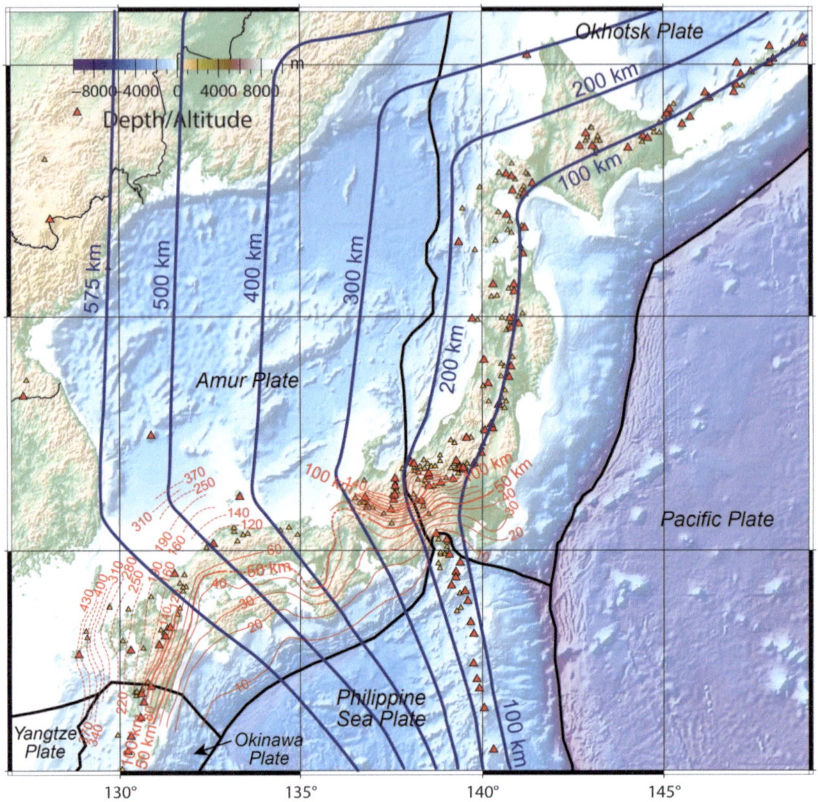

Fig. 2.10 Depth distributions of the upper surface of the Pacific slab and the Philippine Sea slab (Modified from Nakajima et al. 2001; Zhao et al. 2012 and Hasegawa et al. 2015)

Pleistocene strata. Some exceptional cases are the Mesozoic or Paleogene strata. However, several small and relatively high-temperature steam-flash and binary-cycle power plants have recently been developed in the Hida Marginal Belt, central Japan (Figs. 2.13 and 2.14). The Hida Marginal Belt is a 520–270 Ma old serpentinite *mélange*, whose age shows the Cambrian to Permian Era (Kunugiza and Maruyama 2011) and is extraordinarily old compared to known geothermal reservoirs in Japan. This could be ascribed to the unique magmatism in the remarkably active deformation zones near the Itoigawa-Shizuoka Tectonic Line, a plate boundary formed by the colliding subduction tectonics of the Izu-Bonin arc.

Tamura et al. (2002) did not apply the hot finger hypothesis to the clustered volcanoes in central Japan. However, their group 1 volcanic cluster is close to central Japan (Fig. 2.11). Volcanic clusters in central Japan seem complicated and have tight spacings compared to the ten groups of volcanic clusters in Northeast Japan (Figs. 2.11 and 2.13). We wonder if this is due to the unique tectonic setting in central Japan, where the Philippine Sea slab intervenes between the underlying Pacific slab

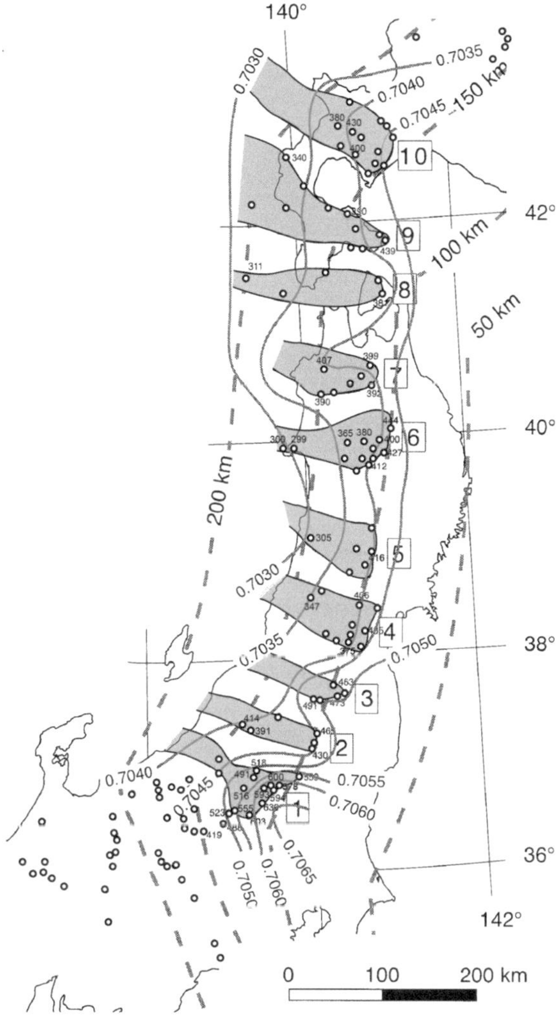

Fig. 2.11 Ten hot fingers estimated beneath across-arc volcano groups in Northeast Japan (Tamura 2003; Tamura et al. 2002). Contour lines: $^{87}Sr/^{86}Sr$ ratio of volcanic rocks. Dashed contour lines: depth to the surface of the seismic zones. Small circle: Quaternary volcano. Numbers labeled volcano: $^{87}Sr/^{86}Sr$ ratio calculated from $(^{87}Sr/^{86}Sr–0.7) \times 10^5$

and the overlying Japan continental crust (Fig. 2.15). Volcanic clusters in northern central Japan seem to be controlled by the Philippine Sea slab, as seen in the results of seismic tomography of Fig. 2.16 (Zhao et al. 2012), where we added the outline shape of the Philippine Sea slab by the thick red lines. Particularly in the cross-sections (b) and (c) of Fig. 2.16, low-velocity zones are obviously situated immediately above the Philippine Sea slab. This suggests that the volcanic clusters in northern central Japan are generated by the dehydration reactions of the Philippine Sea slab rather than by the Pacific slab (Figs. 2.13 and 2.15).

Hasegawa et al. (2007) and Nakajima et al. (2009) drew attention to the existence of "the slab contact zone" between the upper surface of the Pacific slab and the lower surface of the Philippine Sea slab in Tokyo and its vicinity (Fig. 2.13). Normally, the

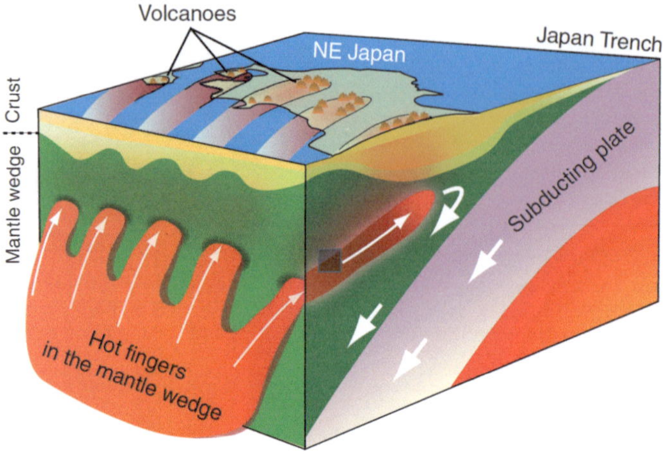

Fig. 2.12 Illustration of three-dimensional hot fingers hypothesis (Tamura et al. 2015)

Pacific slab is heated by the overlying hot mantle wedge (Fig. 1.8). However, the hot upper mantle wedge is thinning out in the slab contact zone (Fig. 2.15). As a result, the seismic belt in the upper plane of the double seismic zones of the Pacific Plate is displaced 150 km westwards by the slab contact zone compared to the Northeast Japan arc (Hasegawa et al. 2007; Nakajima et al. 2009).

Recognizing the slab contact zone is critically important to better understanding the Wadati-Benioff zone's continuity by the Pacific Plate subduction between the Northeast Japan arc and the Izu-Bonin arc. On the small-scale map, like in Fig. 2.5, the Wadati-Benioff zone by the Pacific Plate subduction between the Northeast Japan arc and the Izu-Bonin arc seems entirely continuous with no discontinuity. However, we can find a distinctive discontinuity on the large-scale map, like in Fig. 2.17, where 109,332 earthquakes deeper than 35 km observed in 2020–2021 are plotted. The continuity of the Wadati-Benioff zone is interrupted by the area between the Sagami Trough and the northern edge of the slab contact zone. The reason is quite simple that the shallower part of this area reveals seismic events generated by the shallower Philippine Sea slab subduction rather than by the deeper Pacific slab subduction (Fig. 2.15). Therefore, earthquake clusters in this zone are parallel to the depth contours of the upper surface of the Philippine Sea slab. In other words, the Wadati-Benioff zone by the Pacific Plate subduction is apparently masked by the shallower seismic events by the Philippine Sea slab subduction between the Sagami Trough and the northern edge of the slab contact zone (Fig. 2.17). However, the masking effect exactly occurs in the slab contact zone, indicating that the Philippine Sea slab plays a more important role than the Pacific slab in the shallower part of slab contact zone.

Such a spatial relation is easily confirmed by the interactive use of the three-dimensional hypocenter observation system in Fig. 2.18 (National Research Institute for Earth Science and Disaster Resilience 2024), where the WNW-ESE trending

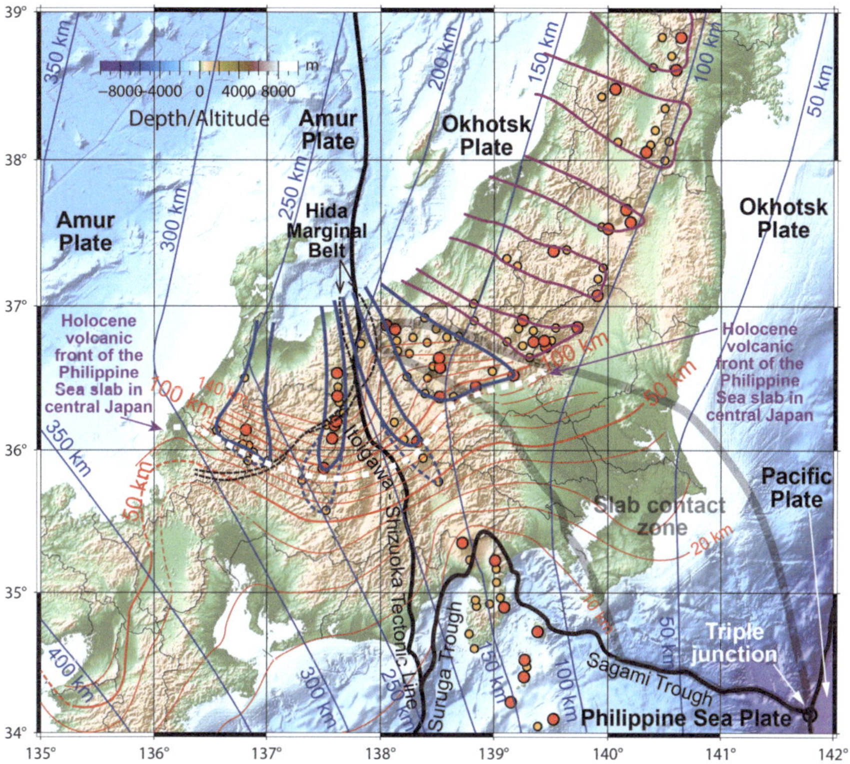

Fig. 2.13 Hot fingers hypothesis applied to central Japan. The thick gray lines: the slab contact zone (Nakajima et al. 2009), thick solid lines: plate boundary (Bird 2003), thick purple lines: hot fingers (Tamura et al. 2002), thick blue lines: hot fingers in central Japan (this book), thin bule contours: the upper surface of the Pacific slab (Zhao et al. 2012), thin red contours: the upper surface of the Philippine Sea slab (Zhao et al. 2012), red circles: Holocene volcanoes, thin solid broken lines: the Hida Marginal Belt (Harayama 1990) and yellow circles: Pleistocene volcanoes (Global volcanism program 2024)

earthquake swarms are observed between the Sagami Trough and the northern edge of the slab contact zone above the Wadati-Benioff zone by the Pacific Plate subduction (Figs. 2.18 and 2.17). These earthquake swarms are considered to be an east-wing seismic zone of the colliding subduction of the Izu-Bonin arc. As explained in the later subsection, a zone of the low-frequency earthquakes or slow slips over 600 km from Shikoku to central Japan along the subducting Philippine Sea Plate at a slab surface depth interval of 35–45 km was discovered (Fig. 2.21; Obara 2002). Low-frequency earthquakes have been distinguished by the band-pass filter 2–8 Hz regardless of their mechanism by JMA since 1997 (Aso and Ide 2014; Kosuga et al. 2017; Aso and Tsai 2014). Figure 2.17 excludes low-frequency earthquakes. Nevertheless, this figure shows that swarms of ordinarily high-frequency earthquakes at a depth interval

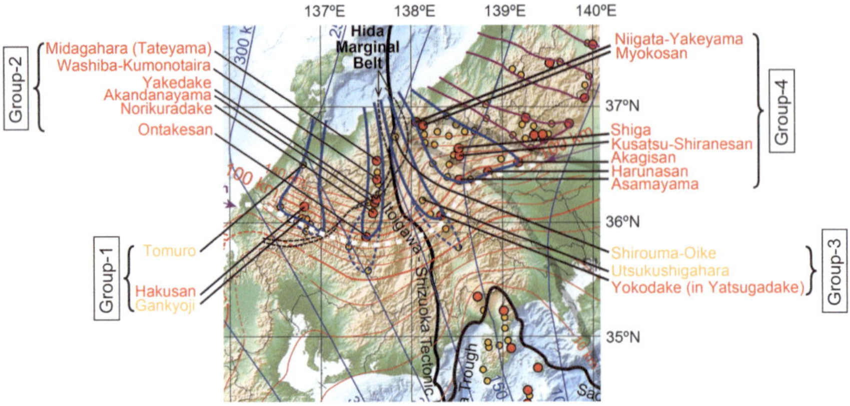

Fig. 2.14 Names of the representative volcanoes of the four hot fingers in Fig. 2.13

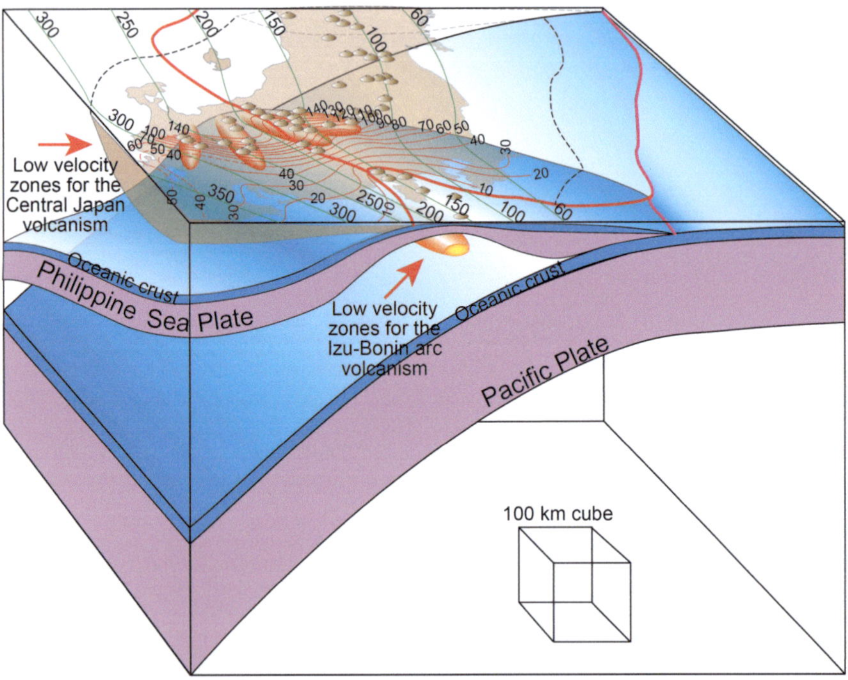

Fig. 2.15 Three-dimensional conceptual illustration of the subduction-magma regime in central Japan. The green and orange contour lines are the depth of the upper surface of the Pacific slab and the Philippine Sea slab projected to the ground surface, respectively

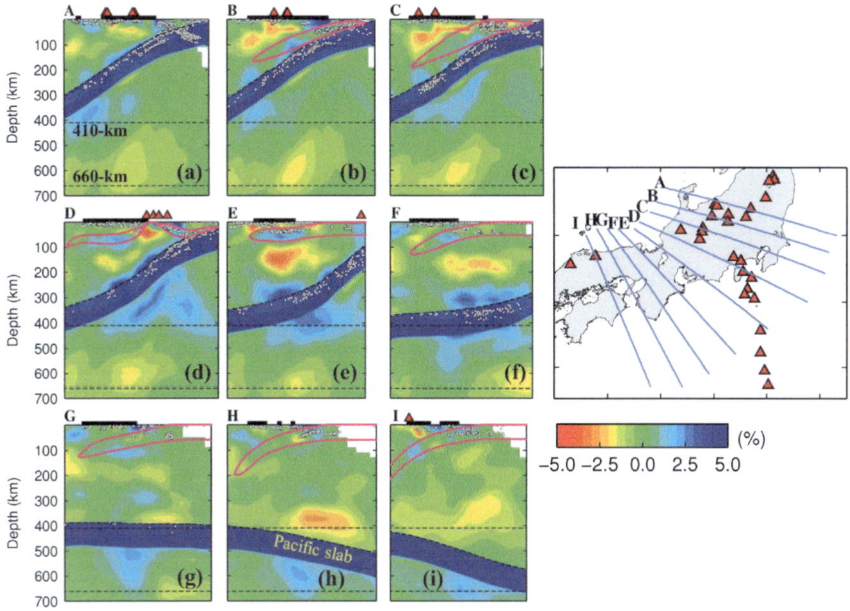

Fig. 2.16 Vertical cross-sections of P-wave velocity perturbation along the profiles shown on the inset map in central Japan (Zhao et al. 2012). The red triangles denote the active arc volcanoes. The black bar at the top of the cross-section denotes the land area. The two dashed lines in each cross-section represent the 410 and 660 km discontinuities. The white dots denote seismicity that occurred within a 20 km width of each profile. The Philippine Sea slab was roughly drawn by the thick red lines in this chapter

of 20–40 km are associated in parallel with the slow-slip zone, in both east and west wings of the Izu-Bonin arc (Figs. 2.17 and 2.18).

Considering from the magmatic point of view, volcanic clusters in northern central Japan must be formed by the low-velocity zones generated by the dehydration reactions of the Philippine Sea slab, because the Pacific slab is constrained to be cold over the 150 km E-W distance by the missing of the overlying hot mantle wedge in the slab contact zone (Hasegawa et al. 2007; Nakajima et al. 2009; Figs. 2.13 and 2.15). Volcanic clusters generated by the low-velocity zones of the Pacific Plate subduction are widespread in the Japanese islands, such as those of the Kuril arc, Northeast Japan arc, and Izu-Bonin arc. However, all the volcanic clusters in Southwest Japan are genetically ascribed to the Philippine Sea Plate subduction. Here, we emphasize that the volcanic clusters in northern central Japan are also genetically ascribed to the Philippine Sea Plate subduction (Figs. 2.15 and 2.16).

We try to apply the hot fingers hypothesis to the volcanic clusters in central Japan, where we think that the hot fingers are virtually the same as the low-velocity zones from the viewpoint of seismic tomography. Four groups of hot fingers are probably defined in central Japan in the slab contact zone and its western extents (Figs. 2.13 and 2.14).

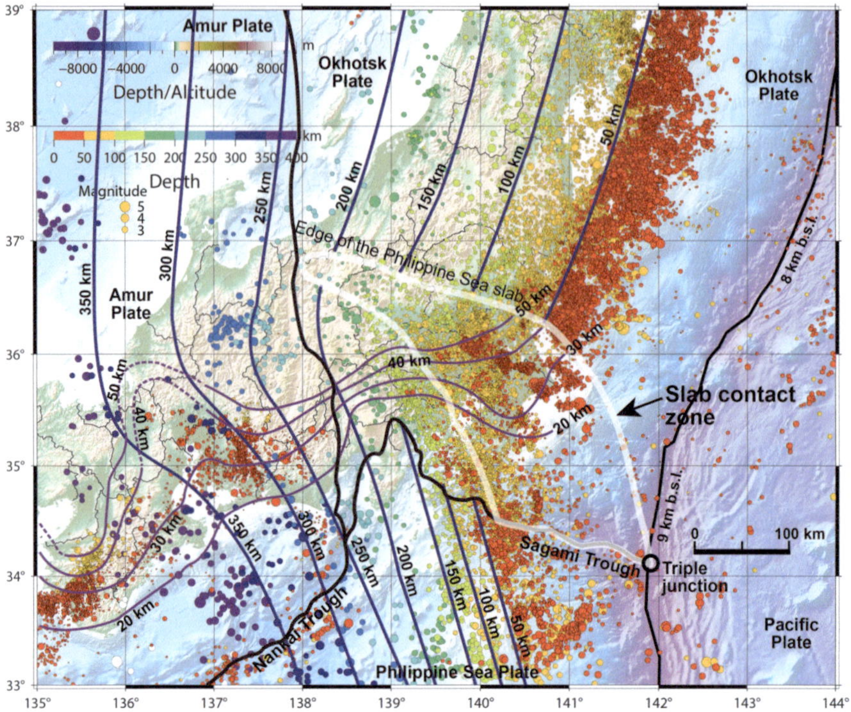

Fig. 2.17 Significance of the slab contact zone on the distribution of the ordinary high-frequency earthquakes in central Japan, extracted from 109,332 earthquakes deeper than 35 km in Japan during 2020 and 2021 (Japan Meteorological Agency 2024). The color of the circle classifies the depths of earthquake hypocenters. Blue lines: the depth contours of the Wadati-Benioff zone drawn from the depth of hypocenters, thick white line: the slab contact zone northern than the Sagami Trough (Nakajima et al. 2009), thick solid lines: plate boundary slightly modified from Bird (2003), purple lines: the depth contours of the upper surface of the Philippine Sea slab (Zhao et al. 2012). The UTM projection map was drawn by the ETOPO2022 15 arc-second global relief model (NOAA National Centers for Environmental Information 2022) and GMT 6.4.0 (Wessel et al. 2019)

Group 1 is called the Tomuro-Hakusan cluster. Group 1 comprises somewhat dissected and relatively older Pleistocene volcanoes (Fig. 2.14). Group 2 is called the Tateyama-Ontakesan cluster. Group 2 is characterized by a narrow alignment on an extraordinarily uplifted zone immediately west of the Itoigawa-Shizuoka Tectonic Line as a plate boundary. The Pleistocene granitic intrusions, such as Takidani Granodiorite and Kurobegawa Granite, are exposed on the surface along this zone by the extraordinary erosion rate due to the rapid uplifting (Harayama 1992; Harayama 2010). High-temperature geothermal manifestations are well known in the wall rock surface of the tunnels through the Pleistocene granitic intrusions (Sekioka 1980; Yuhara et al. 1982). Group 3 is called the Shirouma-Yatsugadake cluster. Group 3 also has a narrow alignment along the subsidence area of the Itoigawa-Shizuoka Tectonic Line as a plate boundary. Group 4 is called the Myokosan-Harunasan cluster

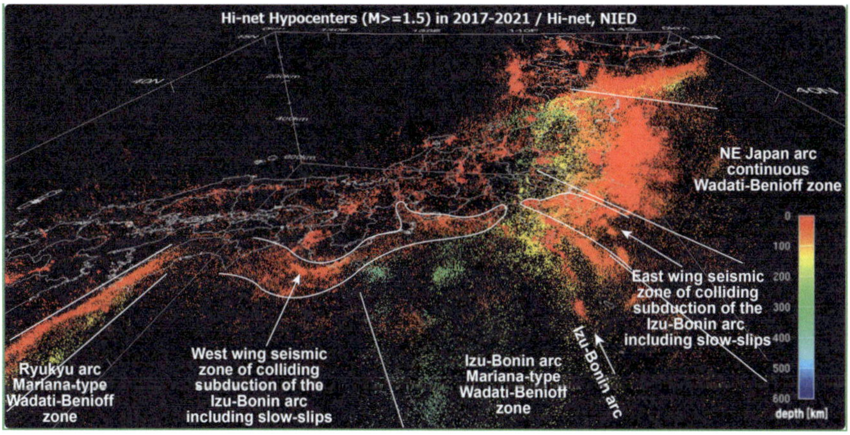

Fig. 2.18 Schematic interpretations added on a snapshot of the interactive three-dimensional hypocenter observation system by the hypocenter data during 2017–2021 in Japan, provided on the website of the MOWLAS (Monitoring of Waves on Land and Seafloor), the National Research Institute for Earth Science and Disaster Resilience (2024)

(Fig. 2.14). Group 4 includes active volcanoes such as Asama Yama, forming a relatively broad zone. Akagisan was previously included in group 1 of Tamura et al. (2002). However, the volcano is here defined as Group 4 due to the inside of the slab contact zone of the Philippine Sea slab in Fig. 2.13.

These four groups of volcanic clusters tend to be perpendicular to the depth contour lines of the upper surface of the Philippine Sea slab (Fig. 2.13). This seems reasonable for the hot fingers' hypothesis, which suggests the diagonally rising counter-flows of the low-velocity zones against the subduction flows.

Figure 2.19 shows part of the contour map of the discharge temperature of Japan's 7,203 hot spring sources (Muraoka et al 2007). Even though most of the recent hot spring sources are developed by drilling, most hot spring wells tend to be shallower than 1.5 km due to the cost-saving demands; then, we could draw the outline distributions of the shallow-depth hydrothermal systems in Japan. The high-temperature anomalies of this map are well consistent with the four volcanic clusters (Fig. 2.13). We also found a hydrothermal cold spot immediately south of the volcanic cluster Group 4, Japan's most conspicuous cold manifestation regarding hydrothermal systems. Japanese hot springs are legally defined as two types: one is spring water or groundwater if the discharge temperature is higher than 25 °C, and the other is spring water or groundwater if the water contains high balneological constituents even below 25 °C. The former is hot springs in a strict sense, and the latter is called "mineral springs." The hydrothermal cold spot in Fig. 2.19 exclusively consists of "mineral springs" less than 25 °C.

Figure 2.20 shows that the cold spot coincides with the $HCO_3^- + CO_3^{2-}$ (mg/kg) dominant hot spring anomaly area. It is reasonable that the magma-induced high-temperature waters are generally dominant in Cl^- and SO_4^{2-} components and vice

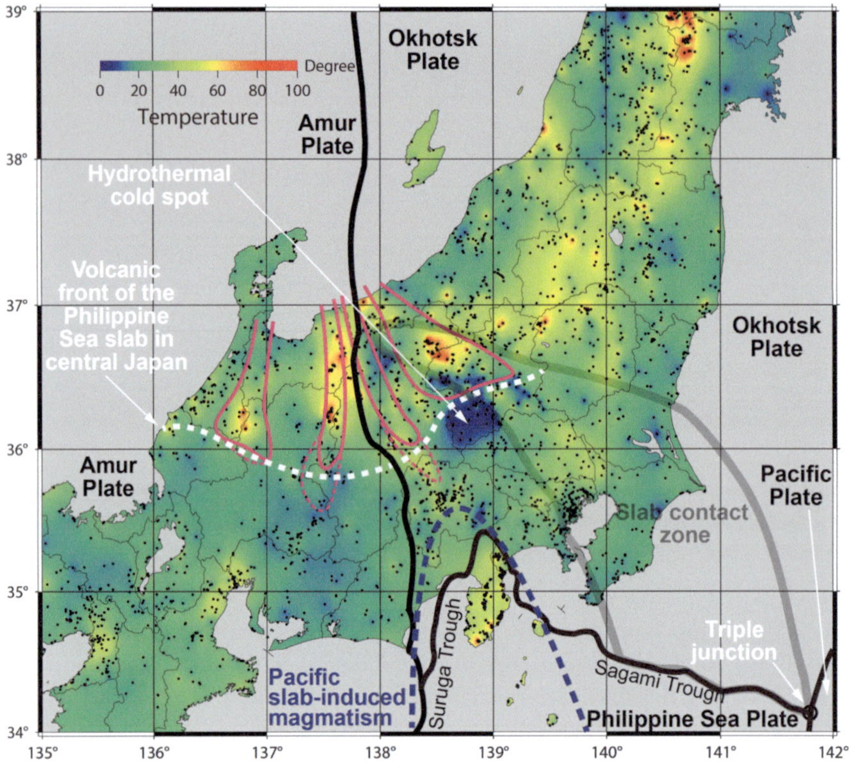

Fig. 2.19 Shallow-depth hydrothermal systems in central Japan contouring the discharge temperature of the 7203 hot springs in Japan (Muraoka et al. 2007)

versa. This area is situated in the Chichibu Belt (Fig. 2.2), which is dominated by limestone. Therefore, these waters have probably interacted with limestone, yielding the $HCO_3^- + CO_3^{2-}$ dominant waters. It is clear that this area is a typical recharge or downflow zone in the hydrothermal convection system and is far from the magmatic fields.

Both anomalies indicate the existence of a hydrothermal recharge and cold spot in the south of Gunma Prefecture, central Japan. The slab contact zone ultimately explains the spot's genesis. The direction of the subduction of the Pacific Plate to the Northeast Japan arc is estimated by the data of the Euler pole (Bird 2003) as shown in Fig. 2.21. The direction is well consistent with the shape of the slab contact zone (Fig. 2.22; Hasegawa et al. 2007). One of the typical dehydration reactions forming the volcanic front is ascribed to the phase change from jadeite-lawsonite-blueschist (JLB) to lawsonite-amphibole-eclogite (LAE) in ocean crust (Hasegawa et al. 2007, 2008). A volcanic front caused by the dehydration of the Pacific slab should originally have been a red broken curve (Fig. 2.22). However, as missing the hot mantle wedge in the slab contact zone, the volcanic front of the Pacific slab must be shifted westward about 150 km in the maximum (Hasegawa et al. 2007). As

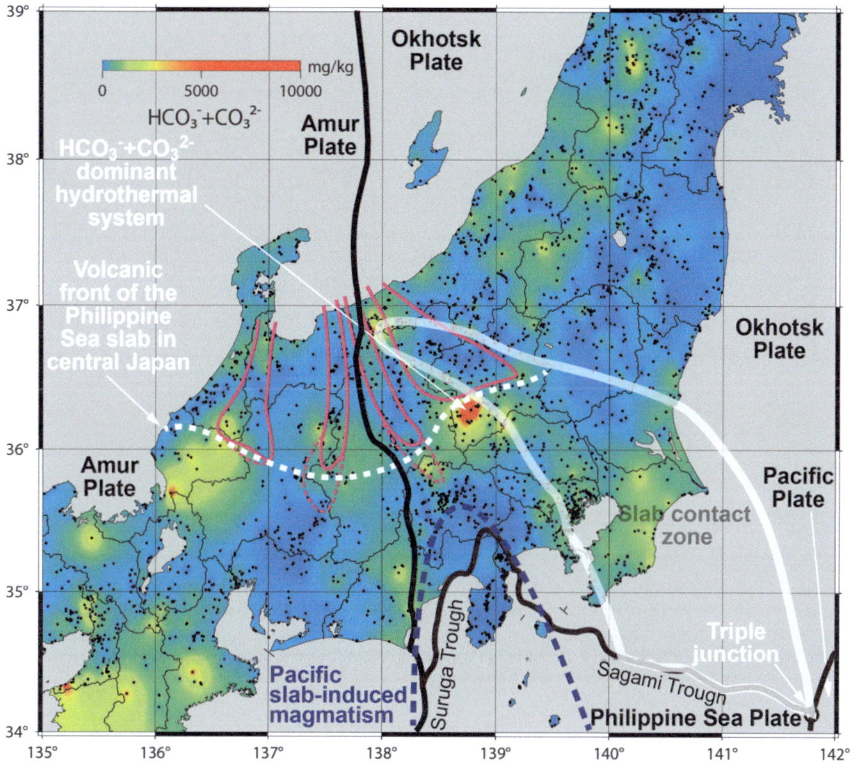

Fig. 2.20 Shallow-depth hydrothermal systems in central Japan contouring the $HCO_3^- + CO_3^{2-}$ contents in mg/kg of the 7203 hot springs in Japan (Muraoka et al. 2007)

Fig. 2.21 Subduction direction of the Pacific Plate estimated from the Euler pole of moving between the Pacific Plate and the North American (Okhotsk) Plate based on the data by Bird (2003)

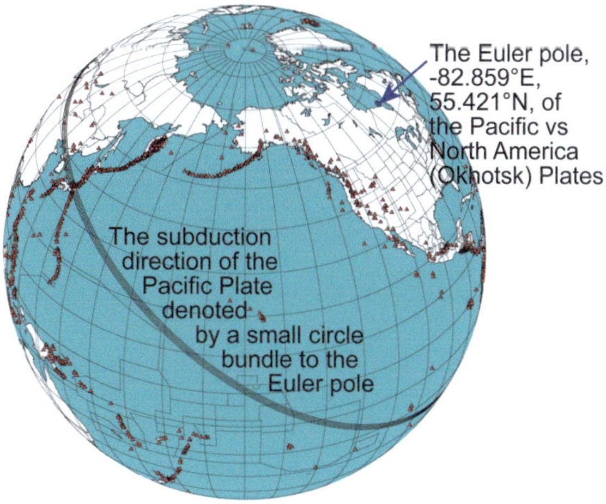

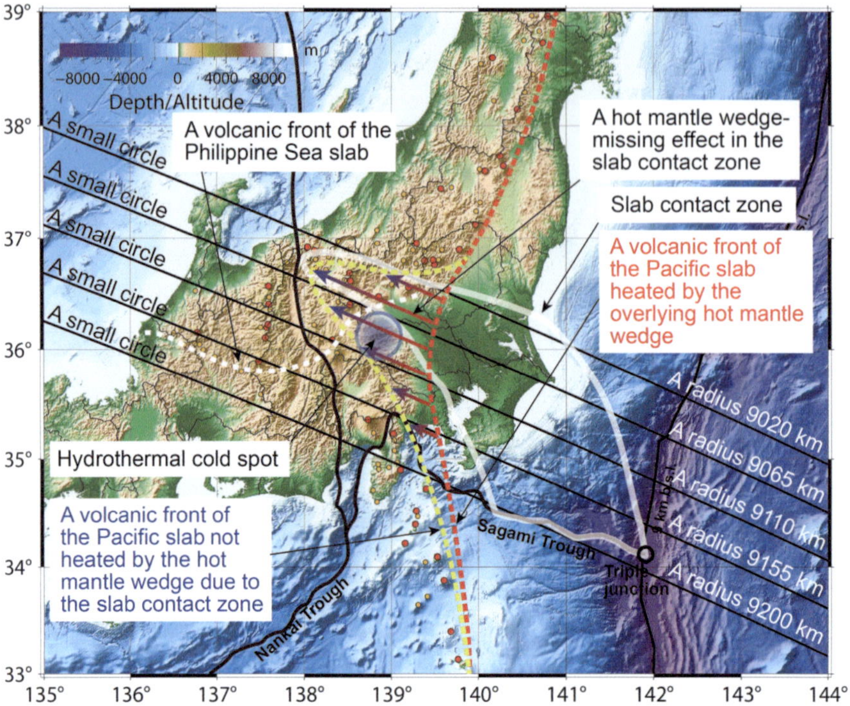

Fig. 2.22 Genesis of the hydrothermal cold spot in the south of Gunma Prefecture, central Japan

shown by a yellow broken curve, this effect could gradually occur in the north and south boundaries (Fig. 2.22).

As a result, Gunma Prefecture may be entirely occupied by a non-magmatic field in terms of the Pacific slab. On the other hand, the Philippine Sea slab-induced magmatism came to form a volcanic front in the middle of Gunma Prefecture, such as Asama Yama, Haruna, and Akagi volcanoes. Then, the non-magmatic area is squeezed into the south of Gunma Prefecture as a hydrothermal cold spot. This area is surrounded by volcanic areas and hydrothermal up-flow zones in the south, west, and north, inevitably becoming cold downflow zones in the regional hydrothermal convection system.

2.1.4 *Deep Low-Frequency Earthquakes and Magma Generation in the Japanese Islands*

This subsection introduces a recent study of the magmatic deep low-frequency earthquakes in Japan by Suzuki et al. (2021, the corresponding author is Hirofumi Muraoka), with slight modifications to draw magmatic conduits from the upper mantle to upper crustal subvolcanic magma chambers.

(1) *Introduction*

A current high-sensitivity seismograph network enables us to observe deep, low-amplitude, and low-frequency earthquakes in Japan (Obara 2002; Aso and Ide 2014; Kosuga et al. 2017). Ordinary high-frequency earthquakes (HFEs) are derived from the brittle and seismogenic layers. The deep low-frequency earthquakes (LFEs) are derived from a so-called aseismic layer beneath the seismogenic layer. The LFEs defined here are 50–1 km deep, have a magnitude of <2.5, and a frequency of 2–8 Hz (Aso and Tsai 2014). These LFEs are distinguished from extremely shallow volcanic tremors (<1 km) that might be caused by the two-phase flash of shallow groundwater, as observed in fumaroles. A non-volcanic LFE zone over 600 km along the subducting Philippine Sea Plate at a slab surface depth interval of 35–45 km was discovered as a dehydration phenomenon from the subducting hydrous slab in Fig. 2.23 (Obara 2002). These LFEs are categorized as non-volcanic LFEs (Obara 2002) and tectonic LFEs (Aso and Ide 2014). The distinctive nature of the tectonic LFE cluster is that it is laterally almost continuous, similar to a curtain, indicating an isothermal and/or isobaric phenomenon controlled by the subduction depth interval.

Other LFE clusters in Japan are associated with Quaternary volcanoes, as shown in Fig. 2.23 (Aso and Ide 2014; Kosuga et al. 2017). They are categorized as volcanic LFEs (Aso and Ide 2014), but in this chapter, they are referred to as "magmatic LFEs" because of their deep origins in the upper mantle. Magmatic LFEs provide key information on the ascent and replenishment of basaltic magma from the upper mantle to upper crustal subvolcanic magma chambers. This aspect was already identified in Hawaii 44 years ago (Aki and Koyanagi 1981). Most of the previous LFE studies, however, focused on the focal mechanism of LFEs themselves. The spatial and temporal relationships of LFEs along with their overhead ordinary HFEs have seldom been investigated. The distinctive nature of the magmatic LFE cluster is that it is laterally discontinuous and sporadic, similar to a diapir, implying a magma rising phenomenon.

The next generation of deep geothermal resources is being investigated. For example, the New Energy and Industrial Technology Development Organization (NEDO) in Japan is promoting the use of supercritical fluid for geothermal power generation under the research project "Research and Development of Supercritical Geothermal Resources." In this context, deep geothermal exploration techniques that overcome depths greater than the conventional geothermal reservoir depths, such as 2 km or 3 km, are anticipated. From this perspective, this study tries to plot hypocenters of the magmatic LFEs and their overhead HFEs beneath six volcanic

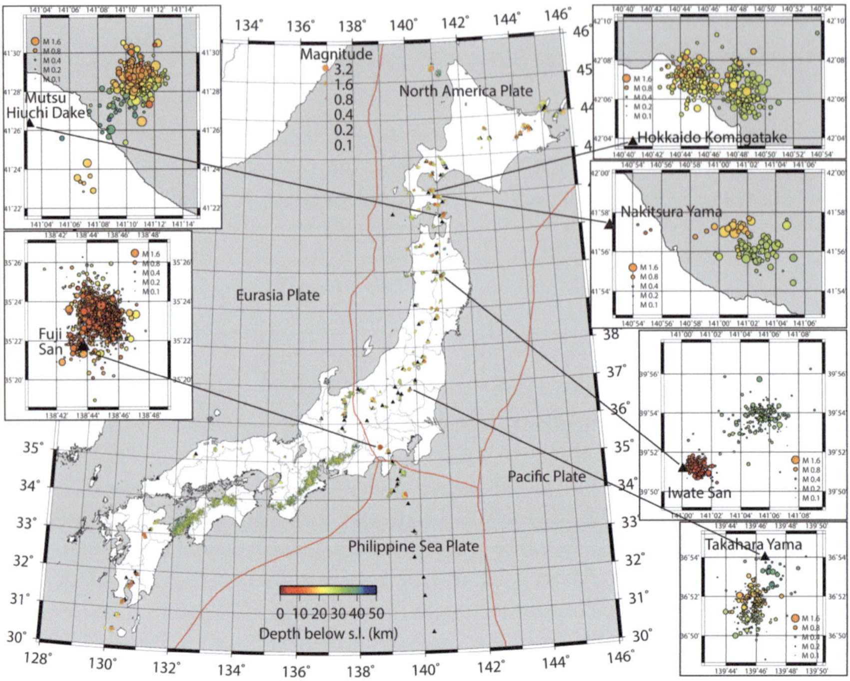

Fig. 2.23 Distribution of 46,657 epicenters of deep low-frequency earthquakes (LFEs) on the plan views in the main part of Japan quoted from the earthquake catalog of the Japan Meteorological Agency (JMA) from the year 2000 to 2016 (Suzuki et al. 2021). Solid triangles show the Quaternary volcanoes. Plan views of the epicenters of LFEs of six volcanic fields are enlarged in insets

fields in northeastern Japan to identify deep magmatic conduits and monitor thermal conditions of the lithosphere (Suzuki et al. 2021).

(2) *Study Areas and Magmatic LFEs Swarm*

Figure 2.23 shows the distribution of 46,657 epicenters of LFEs in a plan view in the main part of Japan. The hypocenter data were taken from the earthquake catalog of the Japan Meteorological Agency (JMA) during the 17 years from the year 2000 to 2016. The LFEs have been distinguished by the band-pass filter 2–8 Hz regardless of their mechanism by JMA since 1997 (Aso and Ide 2014; Kosuga et al. 2017; Aso and Tsai 2014). The earthquake catalog of JMA provides 96 columns of information for each earthquake event in a hypocenter record format. The LFEs can be distinguished from HFEs in the 61st column since 1997, but this procedure has been improved since 2000 (Kosuga et al. 2017). Fumarole-type LFEs that were shallower than 1 km were excluded from this study. Magmatic LFEs are generally associated with Quaternary volcanoes as swarms. Swarms of magmatic LFEs in six volcanic fields—Hokkaido Komagatake, Nakitsura Yama, Mutsu Hiuchi Dake, Iwate San, Takahara Yama, and Fuji San—are shown as insets in Fig. 2.23. Some magmatic LFE clusters consist of only a few to several events, and the six volcanic fields were chosen from the magmatic LFE clusters with relatively large numbers of events. These magmatic LFEs generally form isolated clusters with diameters ranging from 7 to 15 km (Fig. 2.23). It should be noted that most of the LFEs tend to appear toward the east or northeast of the nearest Quaternary volcanoes' summit at a distance of 2 km to 15 km away (Fig. 2.23). The only exception is the LFE cluster in the Takahara Yama volcanic field, but if this cluster is connected to the next nearest volcano, the Nyoho-Akanagi volcano in the southwestern area, then the LFE clusters are situated in the northeastern direction from a Quaternary volcano, which is the same as the other five volcanic fields.

The Moho discontinuity beneath the Japanese islands was recently investigated by three-dimensional seismic tomography (Matsubara et al. 2017). Consequently, the depth of the Moho discontinuity is estimated to be more than 30 km in most of northeastern Japan, and partly more than 35 km along the volcanic front.

(3) *Depth Profiling*

Figure 2.24 shows the distribution of hypocenters of LFEs and HFEs on the cross-sections of the six volcanic fields. It is important to note that the HFEs are distributed in the two depth intervals in the six volcanic fields, namely, 0.00–29.49 km and 113.10–148.47 km in Hokkaido Komagatake; 0.00–29.49 km and 98.69–147.81 km in Nakitsura Yama; 0.81–25.89 km and 90.98–157.43 km in Mutsu Hiuchi Dake; 0.07–11.37 km and 92.83–108.71 km in Iwate San; 1.08–30.65 km and 101.07–137.89 km in Takahara Yama; and 0.00–28.81 and 154.20–207.81 km in Fuji San. These deeper intervals of the HFEs indicate subduction zone earthquakes in a broad sense, including intraslab earthquakes, but only shallower intervals of the HFEs are treated in this study for geothermal purposes (Fig. 2.24). The frequency of HFEs is higher than 8 Hz, and the largest magnitude of the HFEs among the six volcanic fields was 5.3 in Nakitsura Yama. Since the magnitude of the LFEs is 2.5 or less, the average HFEs tend to be slightly larger (Fig. 2.24).

The depths of the Moho discontinuity in the six volcanic fields in Fig. 2.24 were given from the depth contour map by seismic tomography (Matsubara et al. 2017). The LFEs vertically tend to form three clusters on the cross-sections, showing the three depth intervals, 45–31 km in the uppermost of the upper mantle, 39–22 km in the lower crust, and 22–5 km in the upper crust. Compared to the Moho discontinuity, the depth of the Conrad discontinuity is not well demonstrated. For this reason, the depths of the Conrad discontinuity were speculated by the cluster boundaries of the LFEs in the six volcanic fields in Fig. 2.24. The upper mantle clusters can only be recognized in the three volcanic fields—Mutsu Hiuchi Dake (Fig. 2.24c), Iwate San (Fig. 2.24d), and Takahara Yama (Fig. 2.24e). Lower crust clusters and upper crust clusters can be observed in all six volcanic fields (Fig. 2.24). The LFE clusters of the upper crust in Iwate San and Fuji San form vertically elongated ellipsoids, suggesting subvolcanic magma chambers and their peripheral plastic host rock envelopes. Steep walls of the HFE clusters surround the ellipsoidal LFE clusters. The three-level clustering is possibly ascribed to the rheological difference of the lithological composition, such as the peridotite upper mantle, gabbroic lower crust, and granitic upper crust, including quartz-feldspar dominant sedimentary rocks. It is noted that few HFE

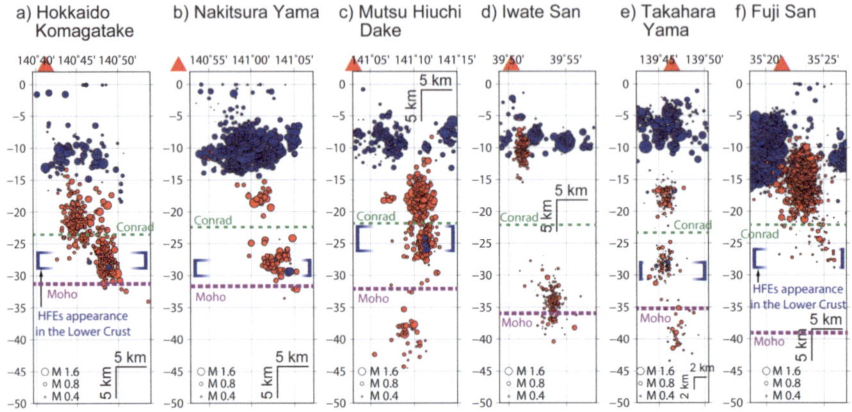

Fig. 2.24 Hypocenter distribution of deep low-frequency earthquakes (LFEs, red circle) and ordinary high-frequency earthquakes (HFEs, blue circle) on the four E-W cross sections (Hokkaido Komagatake, Nakitsura Yama, Mutsu Hiuchi Dake, and Takahara Yama) and on the two N-S cross sections (Iwate San and Fuji San) in the same areas of insets as in Fig. 2.23. A few HFEs are observed in the depth of 31–22 km of the lower crust in the five volcanic fields (Suzuki et al. 2021)

events are almost ubiquitously detected in the depth interval 31–22 km in the five volcanic fields except in Iwate San (Fig. 2.24). The depth interval corresponds to the upper-middle of the lower crust. There are only four events of HFEs in Hokkaido Komagatake (Fig. 2.24a), four events in Nakitsura Yama (Fig. 2.24b), six events in Mutsu Hiuchi Dake (Fig. 2.24c), seven events in Takahara Yama (Fig. 2.24e), and two events in Fuji San (Fig. 2.24f) in the lower crust. With decreasing depth of LFE hypocenters, two or three vertical LFE clusters tend to move laterally from the ENE to the WSW direction by 2 km to 15 km toward the summit of the nearest Quaternary volcanoes. Aseismic gaps are consistently recognized between the magmatic LFE cluster and its overhead HFE cluster throughout the six volcanic fields. However, the aseismic gap is almost vertical in Iwate San and Fuji San because of the high-angle relationships between the LFE and HFE clusters. When the top of the magmatic LFE cluster is wedge-shaped, as observed in Iwate San and Fuji San, its overhead HFE cluster interactively shifts further to the shallower sides, forming an umbrella-shaped envelope.

(4) *Chronological Depth Variations*

Figure 2.25 illustrates the chronological depth variations of the LFE and HFE hypocenters in all six volcanic fields shown in the inset areas of Fig. 2.23, from 2000 to 2016 (Fig. 2.25a–f). Most vertical LFE clusters are stable and active at all depth intervals from the upper mantle to the upper crust throughout the 17-year period (Fig. 2.25a–f). The aseismic gaps between the LFE and HFE clusters are recognized as scarce depth overlaps in Fig. 2.25a–c, e, except for the high-angle contacts of the HFE clusters. During the 17 years, two large-scale earthquakes occurred in the region—one was the 2003 Tokachi-Oki Earthquake M8.0 on September 26, 2003, and the other was the 2011 Tohoku-Oki Earthquake M9.0 on March 11, 2011. The latter is also known as the Great East Japan Earthquake, the name of a historical natural calamity. The timings of the two earthquakes are shown in Fig. 2.25.

(5) *Discussion: Perspective on the Deep Geothermal Energy Developments*

One of the evident historical trends in geothermal energy development is earnestly extending the objective fields to greater depth of the Earth's crust. For example, deeper geothermal wells in Japan were drilled to a depth of 1.5 km by 1970 but rapidly extended to 2.5 km by 1980 and to 3.0 km by 1985 (Saito 1993; Muraoka et al. 1998). It is reasonable that geothermal development is seeking high-enthalpy energy

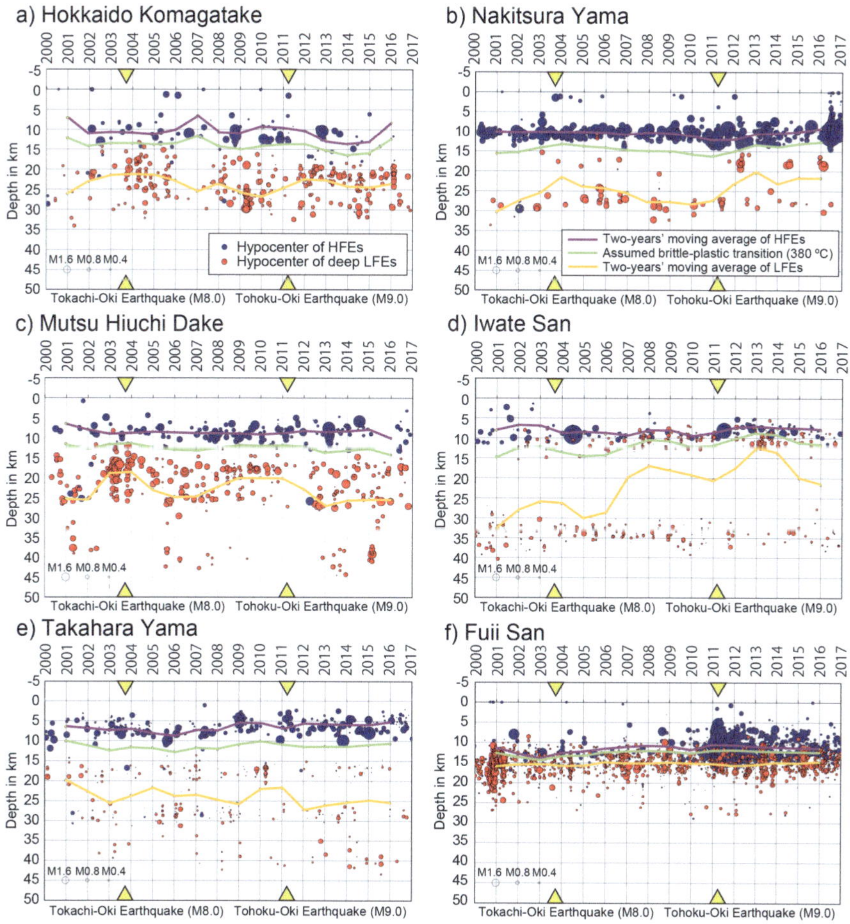

Fig. 2.25 Chronological variations of the hypocenter depth of the deep low-frequency earthquakes (LFEs) and ordinary high-frequency earthquakes (HFEs) in the six volcanic fields, **a** Hokkaido Komagatake, **b** Nakitsura Yama, **c** Mutsu Hiuchi Dake, **d** Iwate San, **e** Takahara Yama, and **f** Fuji San (Suzuki et al. 2021)

resources induced by the geothermal gradient. However, permeability conversely decreases with depth in general. Then, the engineered geothermal system (EGS), including the artificial hydrofracturing technique, has come to be of great interest to geothermal engineers and researchers in the last two decades.

When we consider geothermal development that addresses greater depths of the Earth's crust, exploration technology becomes increasingly challenging. Deep exploratory wells can rarely be attained due to the significantly high costs. Active geophysical exploration methods are higher in costs and close to the limit of the observed depth. Then, passive geophysical exploration methods such as natural earthquake observation are efficient techniques for greater depths. Most active tectonic regions have stationary seismograph networks, and the seismic data are usually open to the public to mitigate earthquake hazards. Therefore, most seismic data is distributed at no cost. This study is one of the efforts in this direction for future deep geothermal energy developments.

As the application of magmatic LFEs to the magma conduits in the greater depths was achieved in Hawaii 44 years ago (Aki and Koyanagi 1981), this method can be applied worldwide, even if a dense seismograph network is necessary.

(6) *Discussion: Spatial Characteristics of Clusters of the Magmatic LFEs*

Based on the above observations, four aspects are discussed. The LFEs tend to form clusters in three depth intervals—the upper mantle, lower crust, and upper crust (Fig. 2.26). These lithological units consist of peridotite rocks, gabbroic rocks, and granitic-sedimentary rocks divided by the Moho discontinuity at about 35 km depth and by the Conrad discontinuity at about 22 km depth (Fig. 2.26a). Clustering can be reasonably explained by the strength envelope model of the continental lithosphere (Fig. 2.26b). The power creep law of wet dunite in the upper mantle was drawn by Chopra and Paterson (1984) and Hirth and Kohlstedt, (1996), the power creep law of wet granulite in the lower crust was drawn by Wilks and Carter (1990) and Burov (2011), and the Byerlee's law of the brittle layers in the lower and upper crusts was drawn by Brace and Kohlstedt (1980) under the strain rate condition of 10^{-14} s^{-1}. The temperature assumption was 800 °C at a depth of 35 km and 380 °C at a depth of 15 km.

The three vertical LFE clusters seem to generate at the maximum strength depth in the upper crust, lower crust, and upper mantle, respectively (Fig. 2.26b). The stress is generally concentrated in the stronger layers in the multi-layer systems, and then the fault rupture tends to disconnect the stronger layers rather than the weaker layers (Muraoka and Kamata 1983). Consequently, the HFE ruptures prefer the maximum strength layers in the multi-layer systems. Whether or not the genesis of the magmatic LFEs also prefers the maximum strength layers is still open to question. The magmatic LFEs occur in a plastic layer, and their focal mechanisms are estimated to be the compensated linear vector dipole type by studying the 38 events in western Japan (Aso and Ide 2014; Julian 1994). The regional stress field systematically controls the dipoles' directions (Aso and Ide 2014). Therefore, even if the magmatic LFEs occur in a plastic layer, it seems possible that they are also preferably clustered in the maximum strength depth of the three lithologic units—the

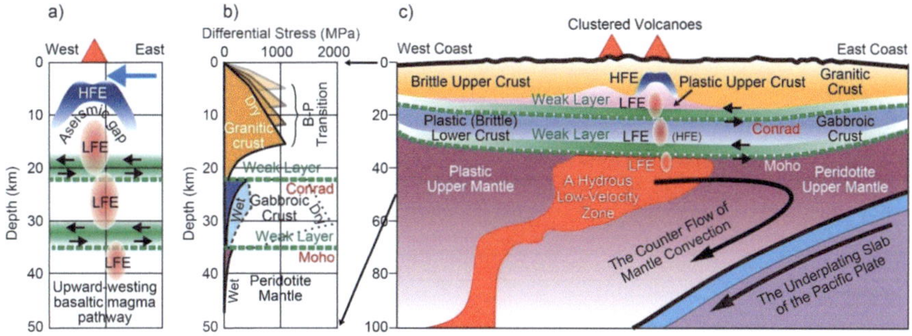

Fig. 2.26 Model of the distribution of the LFEs and HFEs in volcanic fields, northeastern Japan (Suzuki et al. 2021). **a** Simplified spatial distributions of the LFEs and HFEs in volcanic fields in northeastern Japan. **b** Vertical three-LFE clustering is explained by the strength envelope models in the three lithological layers. **c** A model of the basaltic magma replenishment systems to the subvolcanic magma chambers beneath the volcanic fields, where the counter-flow of the mantle convection to the underplating Pacific Plate results in the slight lateral deviation of the basaltic magma conduits from the ENE to WSW directions as the LFEs move to shallower depths

upper mantle, lower crust, and upper crust. A few but ubiquitous numbers of HFEs are detected in the depth interval of 31–22 km. They appear exactly in the lower crust LFE clusters. This fact strongly supports the fact that the LFEs preferably appear at the maximum strength depth in a similar manner to the HFEs.

The LFE clusters tend to appear in the ENE direction from the volcanic summits at 2 km to 15 km. In other words, lateral deviations of the LFE clusters are detected from the ENE to WSW directions at 2 km to 15 km as the LFEs move to shallower depths. This phenomenon could be ascribed to the effect of the counter-flow in the wedge mantle to the Pacific Plate subduction (Fig. 2.26c). Notably, this movement can efficiently occur at the bottom of the lower crust and upper crust because these horizons correspond to the weakest strength envelopes (Fig. 2.26b). The total lateral deviation of the LFEs with depths of 2 km to 15 km can easily be detected in Fig. 2.23. However, when we observe Figs. 2.23 and 2.24, the vertical deviation or tilting has been stable in depth from 45 to 5 km throughout the 17 years. Therefore, although basaltic magma intrusions were quickly ascending, their deviated conduits are controlled under the dynamic equilibrium by the velocity gradient of the counter-flow of the wedge mantle to the Pacific Plate subduction.

(7) *Discussion: The Magmatic LFEs as Rising Basaltic Magma Conduits*

The tectonic LFE cluster in the subducting Philippine Sea Plate at a slab surface depth interval 35–45 km laterally forms an almost continuous zone over 600 km in the along-arc direction of southwestern Japan (Fig. 2.23). This suggests that the tectonic LFE cluster is derived from isothermal and/or isobaric dehydration reactions controlled by the subduction depth interval (Obara 2002). This is also consistent with the fact that the tectonic LFE cluster is situated far more on the fore-arc side rather than the volcanic front of southwestern Japan, showing non-magmatic events. On the contrary, the magmatic LFE clusters in northeastern Japan are associated with the Quaternary volcanoes. They are laterally discontinuous and sporadic. Seismic tomography detected a distinctive low-velocity zone parallel to the subducting deeper slab of the Pacific Plate beneath northeastern Japan (Hasegawa and Nakajima 2004; Fig. 2.27). The low-velocity zone itself is laterally continuous even in the less volcanic across-arc section of northeastern Japan (Nakajima et al. 2001). The zone consists of hydrous and partially molten peridotite receiving water from the progressive dehydration reactions of the subducting hydrous slab, which might be continuous in the along-arc direction (Schmidt and Poli 1998). The most common primary magma from the upper mantle in the island arc region is basaltic magma (Schmidt and Poli 1998) in which water contents of the hydrous mantle should be exclusively partitioned into the rising basaltic magma (Fig. 2.26c). However, primary basaltic magma cannot rise from all over the low-velocity zone (Fig. 2.27) because the dynamics of diapir are subject to the Rayleigh–Taylor-type instability processes, for example, Ramberg (1967) and Whitehead and Helfrich (1990). Rising basaltic magma is sporadically clustered with some spacing (Fig. 2.27). An entity of the magmatic LFEs can thus be explained by the rising basaltic magma from the low-velocity zone (Fig. 2.27). Therefore, the magmatic LFEs provide us with a tool to observe basaltic magma conduits from the upper mantle to the lower and upper crust.

(8) *Discussion: The Brittle-Plastic Transition Depicted by the LFEs and HFEs Boundary*

As the rock deformation phase transitions from the brittle, semi-brittle, and plastic regions with increasing pressure and temperature (Scholz 1988), the temperature threshold of the brittle–plastic transition cannot be uniquely determined in general. However, it was proposed that the temperature of the brittle–plastic transition for silicic rocks under active tectonic regions can be practically restricted to the temperature range of 370–400 °C, as proposed by Fournier (1991). Immediately after Fournier's insight, the well WD-1a was drilled to a depth of 3,729 m, penetrating

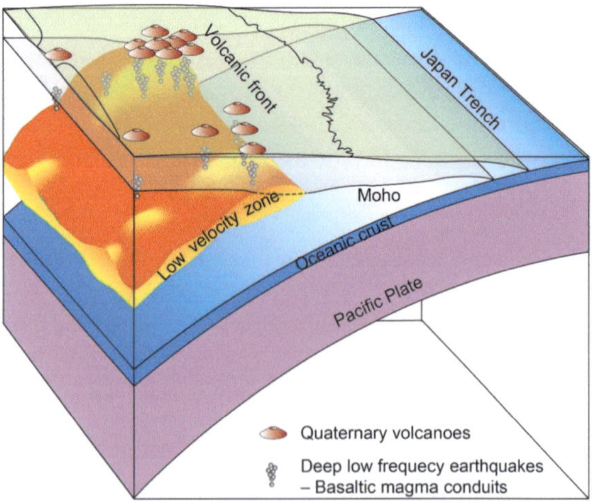

Fig. 2.27 Conceptual model of the sporadic LFEs clusters as rising basaltic magma conduits from the continuous low-velocity zone in northeastern Japan (Suzuki et al. 2021). The low-velocity zone was roughly sketched from Hasegawa and Nakajima (2004)

extremely young granodiorite and tonalite intrusion below 2,860 m in the Kakkonda geothermal field in northeastern Japan in 1995 (Doi et al. 1998). The bottom hole temperature exceeded 500 °C, and the temperature logging curve had an inflection point at a depth of 3,100 m and a temperature of 380 °C (Ikeuchi et al. 1998). This inflection point was ascribed to the bottom of the permeable hydrothermal convection and the brittle-plastic transition (Muraoka et al. 1998). The observed temperature of 380 °C was consistent with the estimated temperature range of 370–400 °C, as proposed by Fournier (1991). The bottom of the seismogenic layer in several volcanic fields in northeastern Japan is also estimated to be 380 °C (Suzuki et al. 2014, 2020). Consequently, the temperature threshold of 380 °C may be reasonable for the brittle–plastic transition on the silicic upper crust in northeastern Japan.

The brittle–plastic transition can then be easily determined by the boundary between the two types of earthquakes providing an intra-crustal isotherm at 380 °C. The aseismic gaps between the magmatic LFEs and their overhead HFEs reveal the brittle-plastic transition and the isotherm at 380 °C.

(9) *Discussion: Possibility of Thermal Monitoring of the Lithosphere by the LFEs and HFEs Boundary*

Based on the above discussion, the chronological depth variation of the LFE and HFE boundary represented by the assumed brittle-plastic transition curve can be used for the dynamic thermal monitoring of the lithosphere (Fig. 2.25). For example, when the chronological depth variations of the assumed brittle-plastic transition in six volcanic fields are compared, the variations throughout the 17-year period are classified into three types (Fig. 2.25). One is a slightly increasing type as seen in Mutsu Hiuchi Dake, which exhibits a cooling trend (Fig. 2.25c). The second is an almost constant type, as observed in Hokkaido Komagatake and Takahara Yama, implying constant thermal trends (Fig. 2.25a,e). The third is a slightly decreasing type as observed in Nakitsura Yama, Iwate San, and Fuji San, which suggests heating trends (Fig. 2.25b,d,f). Steeply facing relationships between the LFEs and HFEs in Iwate San and Fuji San are also consistent with the heating trends as a result of the current active magmatism. Therefore, although more data for the year can be expected, these chronological depth variations of the LFE and HFE boundaries provide a possible tool for the thermal monitoring of the lithosphere. Three volcanic fields have historic eruption records. Magmatic eruptions occurred in Hokkaido Komagatake in 1640, 1694, 1856, 1929, and 1942, in Iwate San at 1686 and 1732, and in Fuji San in 1707 (Kudo and Takarada 2021). Unfortunately, the 17-year observation period of the magmatic LFEs is too short to compare to these eruption records.

(10) *Conclusions*

Some volcanoes have active deep conduits of basaltic magma replenishment from the low-velocity zone to their subvolcanic magma chambers, while others do not comprise replenishment conduits, probably having been lost at some point in their lives. This critically important difference between volcanoes can be detected by signals from magmatic deep low-frequency earthquakes. Therefore, magmatic deep low-frequency earthquakes and their interactive overhead high-frequency earthquakes provide a tool to monitor the thermal evolution of the lithosphere and to explore sustainable geothermal resources with basaltic magma replenishment systems.

2.2 Advances in the Geothermal Exploration and Development in Japan

The March 2011 Great Tohoku Earthquake and Tsunami inflicted catastrophic damage on the Fukushima nuclear power plants, drastically shifting the Japanese public's consensus on the roles and necessity of different types of electric power plants. This led to a unanimous push toward renewable energy sources, including geothermal energy. Japan has a geothermal potential of about 23,500MWe, ranking third in the world.

Despite significant financial, legislative, and institutional support, which has led to numerous new geothermal projects over the past thirteen years, only some have reached the operational stage. Still, many have faced unique barriers inherent to geothermal development. Consequently, as of 2019, Japan's total installed geothermal power capacity is 540 MW, representing a mere 2.3% of its resource potential and serving about 0.2% of its total electricity demand.

In this section, the author aims to describe the current state of geothermal resource developments in Japan, aligning with the government's carbon–neutral policy to achieve its goal by 2050.

2.2.1 Turning Points in Geothermal History in Japan

Japan's first geothermal power plant, the Matsukawa geothermal field, began commercial operation in 1966 with an installed capacity of 9,500 kW. The following year, Kyushu Electric Power Company (Kyushu EPC) started the Otake geothermal power plant, which has an installed capacity of 11,000 kW. Both power plants have been operational for over 55 years.

The oil shocks of 1973 and 1978 forced Japan to reconsider its reliance on fossil fuels, leading to national attention on developing new and alternative energy sources. The Japanese government launched the "Sunshine Project," establishing NEDO (New Energy and Industrial Technology Development Organization) to promote R&D related to new energy and industrial technology.

Following the 2011 earthquake crisis, a feed-in tariff (FIT) system was introduced in 2012. Before FIT, 19 geothermal power plants were operated with a total installed capacity of 525 MW, as shown in Fig. 2.28. The FIT program accelerated geothermal development, constructing many new, albeit smaller, power plants. These new plants are distributed across 69 geothermal areas, with a combined installed capacity of only 80 MW, as shown in Table 2.1.

The government aims to increase installed geothermal capacity by 1,400–1,550 MW by 2030. However, existing legislative and financial incentives alone cannot achieve this goal. As a result, there has been a reassessment of the necessary policies and measures to accelerate the development of large-scale geothermal power plants. The geothermal resources supporting such plants are believed to be located at depths within volcanic national parks. Consequently, legal and environmental regulations must be reviewed and adjusted to facilitate geothermal exploitation. Additionally, it is crucial to develop technologies for exploiting geothermal resources where fluids are stored under supercritical conditions, typically found beneath volcanic fronts at depths of approximately 4,000 to 5,000 m.

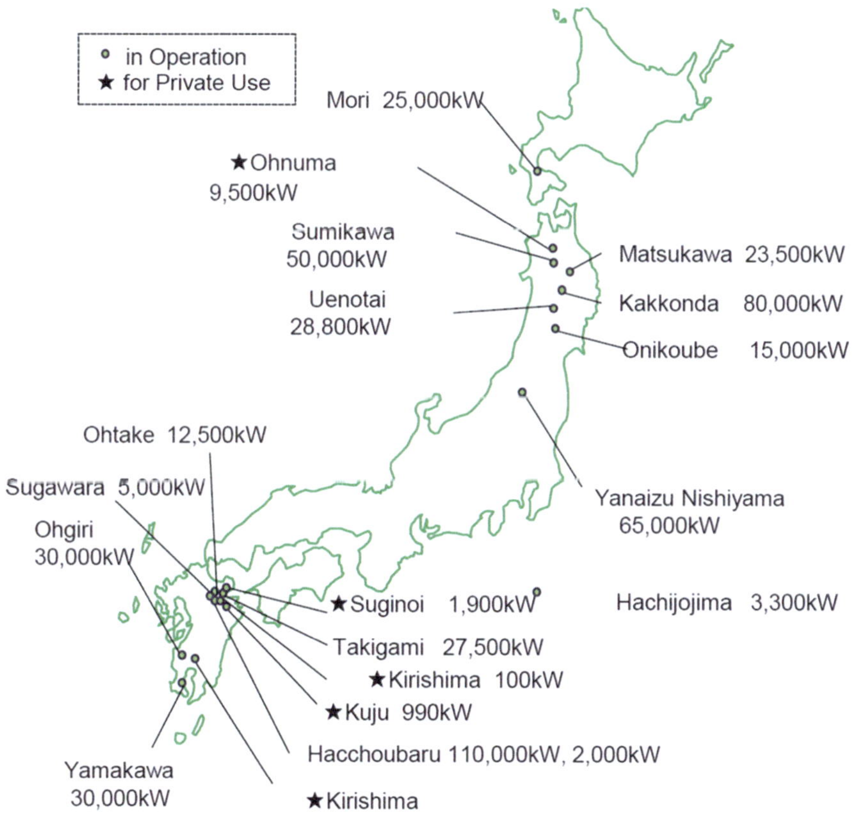

Fig. 2.28 Geothermal power plants before enforcement of the Act of Feed-in Tariff in 2012 (JOGMEC 2017)

Table 2.1 Status of geothermal power plants as of May 2020 following the enforcement of the feed-in tariff ACT in 2012 (JGA 2021)

Category	Newly constructed after FIT Introduction		Under investigation or development	
	Number of sites	Installed capacity (MWe)	Number of sites	Installed capacity (Plan)
Large scale (>10 MWe)	1	46	24	(240 MWe)
Medium scale (1–10 MWe)	6	26	30	(150 MWe)
Small scale (<1 MWe)	62	8	29	(29 MWe)
Total	69	80	83	(419 MWe)

2.2.2 *Historical and Current Geothermal Power Development in Japan*

Figure 2.29 shows Japan's long-term trend of geothermal power Generation from 1966 to 2020, including total installed capacity, annual power generation, and the utilization rate. A sharp increase in both installed capacity and annual power generation was observed from the late 1970s to the mid-1990s, corresponding to the inauguration of plants partially funded by NEDO subsidies.

Although the total installed capacity peaked at 520 MW in 1996, the average utilization rate has gradually decreased since the 1970s, with annual power generation (GWeh) decreasing since 1997. This decline reflects the gradual depletion of the geothermal reservoirs after the commencement of commercial operations. The primary cause of this issue is that the installed capacities of many of the geothermal power plants were too large to sustain long-term commercial operations. Plant operators have attempted several remedial actions to address this problem. Some operators reassessed the optimum capacity that would be financially viable for the plant's lifetime operations and decided to replace the large units with smaller, more sustainable ones. These smaller units began integrating into the grid starting around 2010.

In July 2012, the "Act on Purchase of Renewable Energy Sourced Electricity by Electric Utilities" was enacted, regulating Japan's feed-in tariff (FIT) scheme for renewable energy. The FIT scheme includes two categories based on power plant size, with tariffs ranging from \$0.33 per kW for plants larger than 15 MW to \$0.51 per kW for plants smaller than 15 MW. The purchase price under the FIT scheme is guaranteed for 15 years after the commencement of commercial operations.

As an additional incentive beyond FIT, Japan Oil, Gas and Metals National Corporation (officially renamed to "Japan Organization for Metals and Energy Security" in November 2022, but still abbreviated as "JOGMEC") was entrusted with planning and implementing geothermal exploration and development projects under various governmental financial schemes. These schemes include executing exploration and

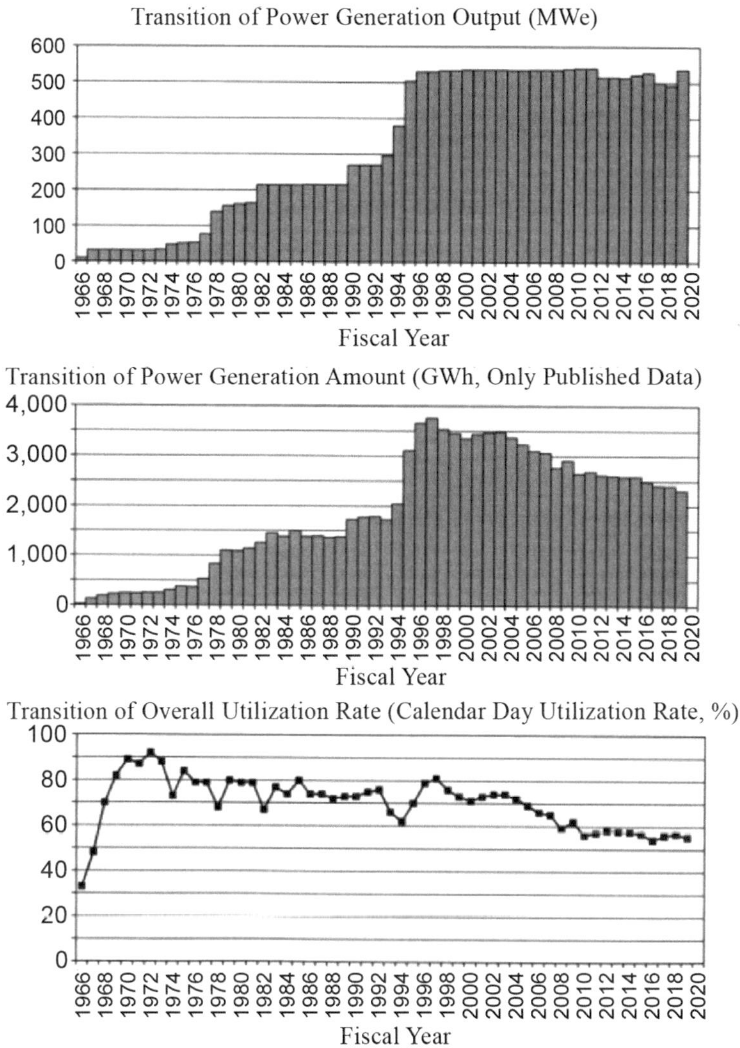

Fig. 2.29 Transition of geothermal power generation from 1966 to 2020 in Japan (TENPES 2021)

drilling projects with full grant subsidies, assisting projects through equity capital investments for private ventures, and mitigating project risk by offering loan liability guarantees.

Figure 2.30 shows Japan's geothermal power plants and ongoing survey projects as of 2024. Among the power plants shown in Fig. 2.30, 53 (each with an installation capacity of less than 1,000 kW) were constructed and operated with the various supports provided by JOGMEC.

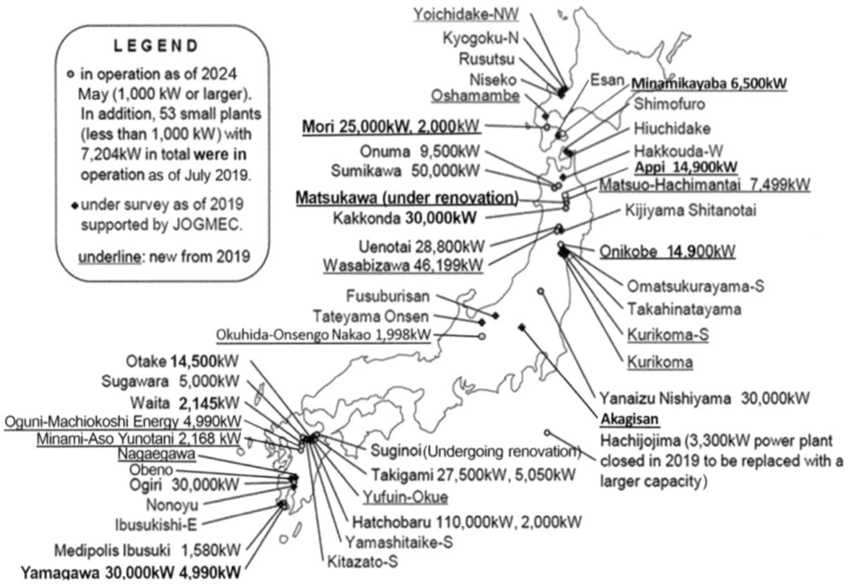

Fig. 2.30 Geothermal power plants and development projects as of April 2024 (JGA 2024 with modification)

Despite the FIT scheme's success in introducing several geothermal power plants, the combined installed capacities of the newly built plants are significantly less than the government's expectations.

2.2.3 Challenges for Accelerating the Geothermal Development

In October 2020, the Japanese government proclaimed a national target of achieving carbon neutrality by 2050, making the shift to clean energy an urgent issue. Regarding geothermal energy, policies must be modified to significantly increase the total installed capacity in megawatts (MW). In other words, large-scale geothermal power plants must be developed. Consequently, new policies and measures for promoting geothermal development have been revised.

1. Eased Regulations in National Parks: The rules regarding geothermal exploration and exploitation within national parks have been eased, as depicted in Fig. 2.31. Geothermal resources in areas classified as Class I to III special zones can now be exploited using directional well drilling from pads located outside the park under certain conditions. However, geothermal exploitation in special protection zones remains prohibited.

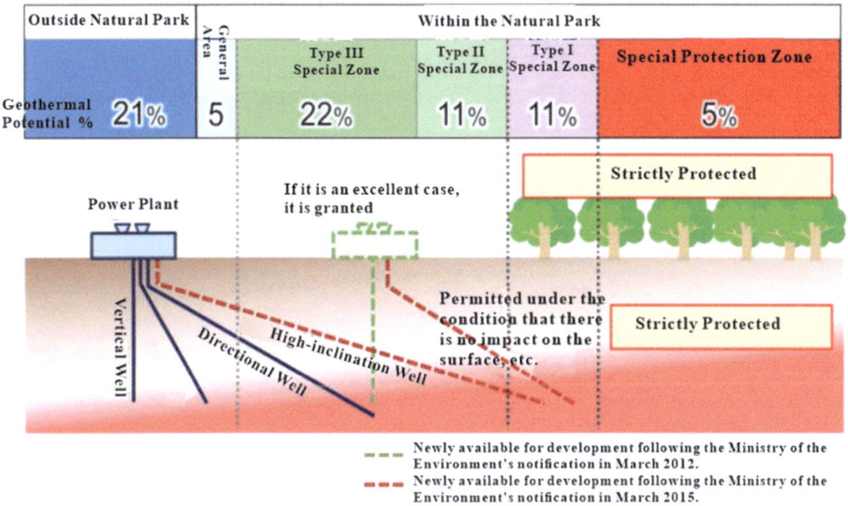

Fig. 2.31 Easing regulations on geothermal development for different types of special zones in national parks and allocation of geothermal potential in each zone (NEDO 2021)

2. Research and Development (R&D) Focus Areas:
   - Enhancing Exploration Accuracy: Using distributed acoustic sensing (DAS) technology to improve the efficiency and precision of geothermal reservoir exploration. Further details on this technology are discussed in Chap. 8.
   - Increasing Well Productivity: Boosting the productivity of geothermal wells through artificial reservoir recharge via water injection.
   - Improving Drilling Efficiency: Utilizing polycrystalline diamond compact (PDC) bits specifically developed for geothermal well environments.
3. Supercritical Geothermal Resources: Ongoing R&D aims to harness supercritical geothermal resources originating from subduction zones. The objective is to develop a pilot power plant and initiate experimental operation by 2040 (Fig. 2.32; Table 2.2).
4. Cost-Effectiveness: Efforts are focused on improving the cost-effectiveness of existing geothermal power plants by optimizing operation and maintenance processes through AI and IoT monitoring technologies.
5. Environmental Assessment Technologies: Efforts are underway to develop new H_2S monitoring devices and simpler computing processes for meteorological phenomena. Additionally, comprehensive studies from scientific, technological, and policy perspectives are in progress to establish cooperative and mutually beneficial relationships with local hot spring communities regarding geothermal development.

Themes 1 and 2 are conducted under the budget of JOGMEC, while themes 3, 4, and 5 fall under the budget of NEDO. These allocations are not exclusive. Both

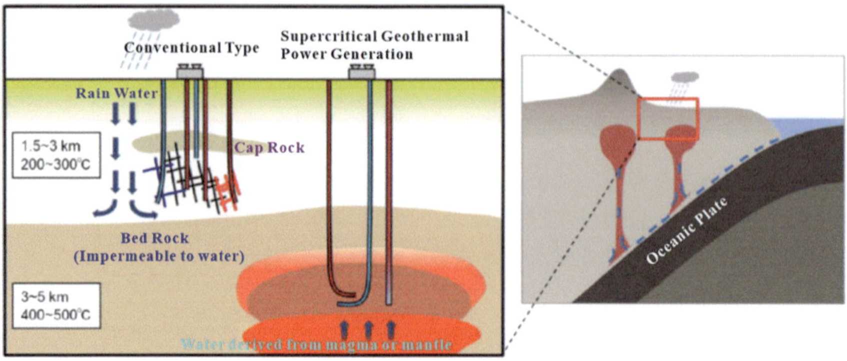

Fig. 2.32 Conventional type geothermal resources and supercritical type geothermal resources under research (NEDO 2021)

Table 2.2 Main features of conventional type geothermal resources and supercritical geothermal resources under research (NEDO 2021)

	Conventional type geothermal resources	Installed capacity
Depth	1,500–3,000 m	3,000–5,000 m
Temperature	200–300 °C	400–500 °C
Fluid origin	Surface rainwater, etc., flows underground through cracks	The origin is seawater that has been drawn underground by plate tectonics
Characteristics	The power generation method varies depending on the temperature of the resource and the steam-water ratio	Higher output is expected compared to conventional power generation

agencies are funded by METI, with NEDO focusing on R&D projects and JOGMEC handling exploration and exploitation projects.

NEDO's R&D projects, especially those under theme 3, are focused on medium- to long-term goals, with the final project expected to conclude by 2050. These initiatives focus on maximizing the sustainable capacity of individual power plants. In contrast, JOGMEC's projects are expected to be completed by 2030, with the specific goal of expanding the nation's geothermal capacity by an additional 1,500 MW.

The technical challenges associated with developing supercritical geothermal resources are more complex and demanding than those for conventional geothermal development. These challenges can be summarized as follows:

1. Reliable Drilling Methods and Equipment: Establish reliable and efficient methods, materials, and equipment for drilling, completion, production, and monitoring of wells reaching depths around 5,000 m under adverse conditions, including elevated temperatures (400–500 °C), corrosive environments (high acidity and salinity), and difficulties in disposing of drilling cuttings under hydraulically underbalanced conditions.

2. Conceptual Reservoir Model Development: Developing a reliable conceptual reservoir model to characterize the behavior of the fluid and rock. This model serves as the basis for establishing exploration methods, the reservoir simulation process with parameters, and ultimately the optimum production plan with its monitoring method.

Addressing these issues requires studying a wide range of technical challenges, including fundamental laboratory tests on rock mechanics, material resistance tests against corrosiveness under high pressure and temperature conditions, and establishing successful numerical reservoir simulations for supercritical fluids. This project, funded by NEDO, is being implemented by a consortium of Japanese institutes, universities, and private companies led by the National Institute of Advanced Industrial Science and Technology (AIST), drawing on experiences and achievements from research projects worldwide (Iceland, USA, New Zealand, etc.).

Hot springs, provided by bathhouses or hotels, have been historically significant in Japan's tourism industry. Hot springs directly use geothermal energy, utilizing natural hot springs or well effluent from shallow, artificially drilled wells. The origin of hot spring water is meteoric, running in a shallow reservoir with a much shorter retention cycle than deep geothermal fluid. Typically, hot spring water temperature ranges between 35 and 50 °C, which is usually insufficient for geothermal power generation. However, many hot spring communities have opposed geothermal power development, fearing that the deep geothermal fluid exploitation may dry up their hot spring resources.

To accelerate geothermal power development, the government recognizes the necessity of gaining the consent and support of hot spring communities. Consequently, this topic has been added to theme 5, aimed at finding ways to cultivate cooperative and mutually beneficial relationships with the local hot spring communities for geothermal power plant development. Thus, hot spring tourism must be addressed in parallel with the R&D projects.

2.3 Practical Approach for Geothermal Exploration and Development

A geothermal reservoir is a subsurface area where geothermal resources are stored. It primarily consists of meteoric-origin water circulating through pores and fractures in hot rock, with its heat source derived from deep-seated magma. Essentially, a geothermal reservoir can be modeled as a naturally formed underground heat exchange network. Its location, geometry, and physical parameters such as temperature and permeability can be estimated through comprehensive analysis of various geoscientific data acquired from the surface and wells.

In geological terms, the lifespan of single or regional volcanic activity is relatively short and follows a specific evolutionary pattern. Consequently, geothermal activity, governed by corresponding volcanic activity, follows suit. Thus, the current state of the geothermal reservoir can be regarded as a transient phenomenon exhibiting complex structural characteristics.

Furthermore, temperature increases the geoscientific complexity in delineating geothermal reservoirs. According to the Arrhenius equation, chemical activity accelerates exponentially as temperature increases. Consequently, not only do the physical parameters of the individual components (e.g., the water viscosity, which is approximately 1,000 μPa·S at ambient conditions decreasing to 86 μPa·S at 300 °C and 86 bar) vary, but so do the chemical reaction characteristics, including acidity, corrosiveness, and solubility involving combinations of minerals, brine, and gas. Due to these reasons and numerous other engineering challenges, geothermal engineering has not yet been mastered. In other words, regardless of the extent of geoscientific studies conducted for a given field, questions about the reliability of estimated results remain. Thus, resource risk cannot be eliminated.

Even if exploration studies successfully identify the reservoir, the project's financial viability remains at risk, especially in cases where development costs, particularly for well drilling, face unexpected overruns. For prospective investors, costs incurred during the exploration and development of subsurface resources should be accountable and fall within the plan. Given the imperfection associated with geothermal engineering, certain resource risks never disappear and must always be considered at any stage of the geothermal development process.

Geothermal resource risk must always be an important factor when making an investment plan as a practical approach to successful geothermal exploration and development. This approach might include starting the construction of the first power plant with a smaller installation capacity than what the resource assessment predicted. Once successful operation is confirmed, during which additional data and experience are acquired both on the subsurface and plant operation, subsequent expansion investment, if any, can be made with much higher confidence.

Geothermal power development is typically divided into two stages: the upstream stage and the downstream stage.

Upstream Stage

The upstream stage involves initial exploration and assessment activities to determine the presence and viability of geothermal resources, laying the groundwork for subsequent development. This stage usually begins with a reconnaissance survey aimed at understanding the geothermal potential of the region and identifying areas with high geothermal potential. The upstream stage may conclude with a detailed resource potential assessment, resulting in a conclusion and recommendation for the succeeding stage. During the upstream stage, several assessments are conducted, including:

- **Environmental Impact Assessment**: Evaluates the potential environmental effects of geothermal development and proposes mitigation measures.
- **Sociological Impact Assessment**: Assesses the social implications of the project, including effects on local communities.

- **Economic Assessment**: Analyzes the project's economic viability, considering local market infrastructure, power demand, transport infrastructure, and power grid connection.

These assessments provide critical data necessary for making decisions about progressing to the downstream stage.

Downstream Stage

The downstream stage involves the actual development and operation of the geothermal power plant. It encompasses engineering, construction, and operational activities to utilize the geothermal resource and start commercial power generation.

As depicted in Fig. 2.33, the upstream stage consists of a series of studies. Once these studies confirm the resource potential, the project can move to the downstream stage, where the actual infrastructure for power generation is built and operationalized. This systematic approach ensures that geothermal projects are both environmentally sustainable and economically viable (Tagomori 2010; Tagomori et al. 2012).

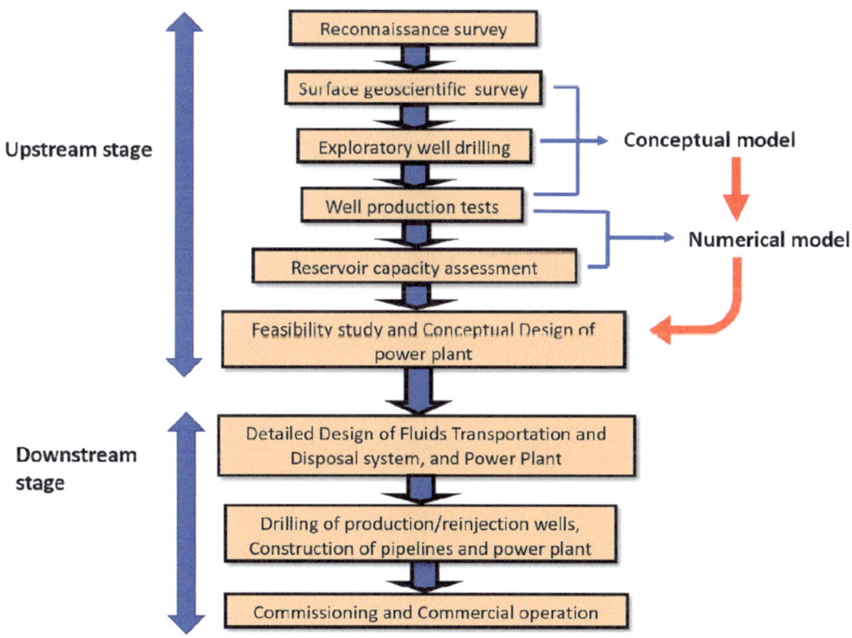

Fig. 2.33 General development process of geothermal power plant construction projects

2.3.1 The 3 Key Elements for Geothermal Systems (3KEYGS)

In the initial reconnaissance survey, the actual work mainly consists of desktop activities such as reviewing existing studies, analyzing satellite images and aerial photos, and examining other available data, including those found online. As the work progresses through the upstream stage, the selection of prospective areas is narrowed down to conduct a more detailed geoscientific survey if necessary. Information acquired from surface geoscientific studies (geological, geochemical, and geophysical methods) and exploratory well testing is used to construct a geothermal conceptual model.

Figure 2.34 illustrates a simplified conceptual model of the Hatchobaru geothermal field in Japan, highlighting the key components of its geothermal system. The model encompasses the geological framework, the hydrological processes—such as meteoric water infiltration, heating, and convection—and the heating mechanism driven by deep-seated magma as the primary heat source. Developing a comprehensive and well-detailed conceptual model is crucial for accurate reservoir simulation analysis, which can be further refined using well-test data for calibration. Reservoir simulation serves as a powerful tool in determining the optimum development scale for power generation.

The following three (3) key elements are generally regarded as essential to implement this process practically and efficiently. Hereinafter, these "Three Key Elements for Geothermal Systems" are referred to as "**3KEYGS**." The concept can be depicted well as "a kettle over a fire," shown in Fig. 2.35.

3KEYGS

1. Geothermal Reservoir Structure and Cap Rock: A geothermal reservoir typically consists of a permeable formation overlain by an impermeable cap rock composed of clay minerals. The reservoir's permeability is a crucial property determining its potential for energy extraction, as it facilitates the movement of geothermal fluids. This permeability arises through porosity, a network of micropores within the rock, or from open fracture formed due to uneven tectonic stress distribution. Faults created under similar conditions indicate the presence of extensive underground fracture systems that enhance permeability. Conversely, various types of clay minerals are produced in geothermal fields through hydrothermal alteration processes influenced primarily by temperature and pressure conditions. Rocks containing these clay minerals exhibit low permeability, acting as sealing caps that confine the convective flow of geothermal fluids, thereby sustaining the reservoir's temperature.
2. High Temperature (Heat Source) at Depth: Reservoir temperatures exceeding approximately 220 °C are required for conventional flash-type geothermal power generation. For binary-type power generation systems, reservoirs with temperatures above 110 °C can be used for commercial operation. Such significant heat typically originates from deep-seated magna. Therefore, geothermal resources

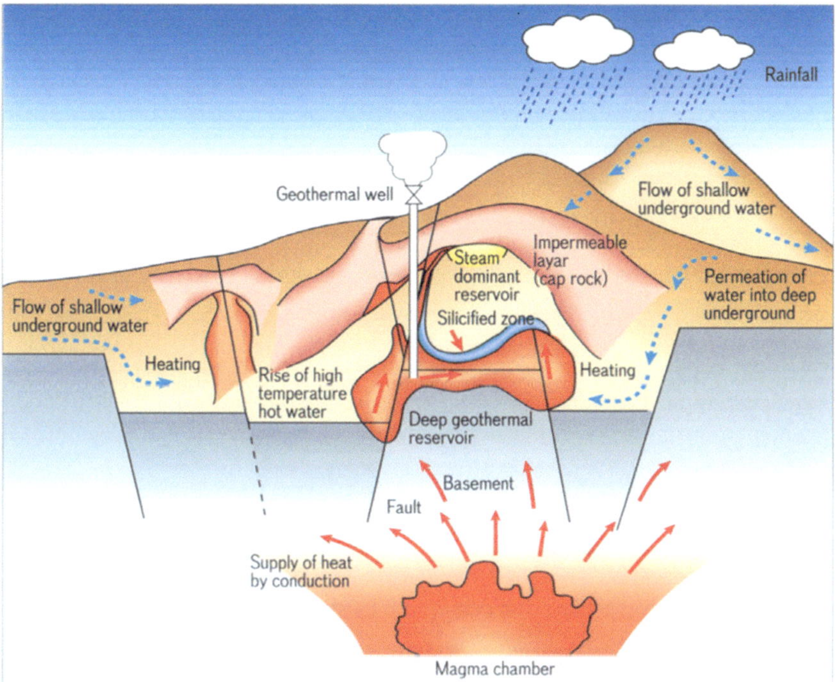

Fig. 2.34 Conceptual model of the Hatchobaru geothermal field (Kyuden Mirai Energy Co., Inc. 2024)

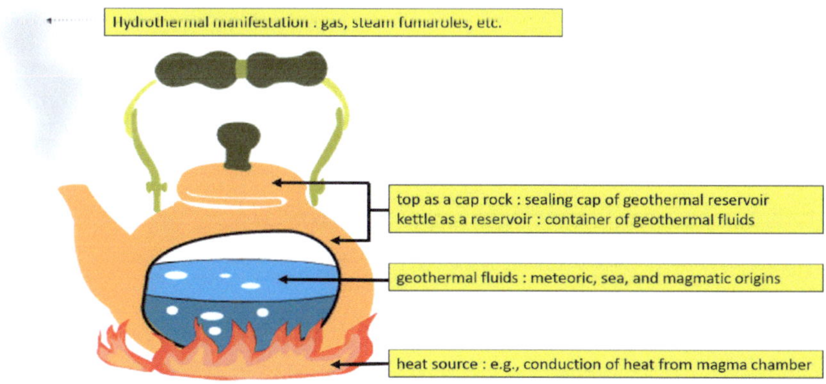

Fig. 2.35 Conceptual illustration comparing "3KEYSGS" to "A kettle over a fire" (Courtesy of Toshiyuki TOSHA)

capable of supporting electric power generation are usually found in regions with current volcanic activity, where this level of heat is prevalent.

3. Geothermal Fluids at Depth: If a well does not encounter abundant fluid at depth, it is called a "dry hole", and geothermal energy cannot be extracted, even if the reservoir temperature is sufficient. In some fields, this lack of permeability can be widespread. Even with abundant fluid initially present, depletion will occur once power plant operation starts. The balance between fluid extraction for power generation and fluid recharge, either naturally or through re-injection, needs careful monitoring and assessment, such as through reservoir simulation. In some cases, repositioning the re-injection area may be necessary to maintain the optimum balance between pressure depletion and temperature cooling. Research and development (R&D) are ongoing to utilize hot but dry geothermal resources by creating artificial fracture systems or drilling wells to form loops, through which water or other heat-transferring medium is circulated to extract geothermal energy.

The comparative effectiveness of various geoscientific methods for addressing the 3 KEYGS issues is shown in Fig. 2.36. While any geoscientific methods can be applied to analyze these issues, it is crucial to understand which methods are most effective for specific problems or combinations of problems related to geothermal reservoirs. For example:

Identifying the location and extent of a prospective geothermal reservoir, which may consist of a permeable and convective reservoir overlain by a hydrothermally altered zone (cap rock), requires integrated analysis based on geological and geophysical methods. This includes fault mapping, hydrothermal alteration zone mapping,

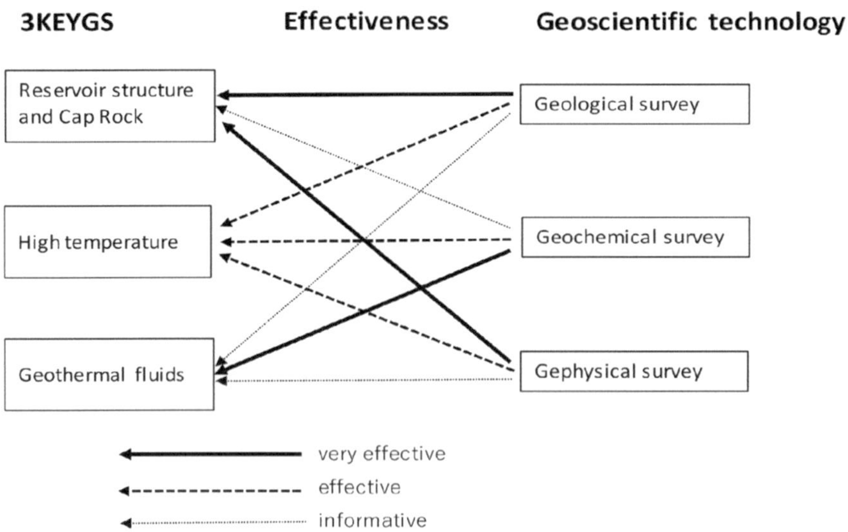

Fig. 2.36 Correlation of effectiveness between 3KEYGS and geoscientific technology

3D resistivity structure analysis, lineament mapping based on resistivity and aerial photos, microseismic analysis, gravity analysis, and more.

- Temperature information for the prospective reservoir at depth can be inferred from geothermometers sampled from surface hydrothermal manifestations. This analysis can be supplemented with geological information, such as the distribution of clay mineral types (which vary depending on the formation temperature when they were formed) and the eruption history of nearby volcanoes, providing additional insights for temperature analysis.
- In some cases, thermal exploration techniques can be effective for locating prospective geothermal reservoirs by analyzing temperature distribution at very shallow depths. Geothermal gradient distribution data obtained from shallow holes can be more economically feasible than drilling deep holes. The effectiveness of this technique varies depending on several factors, including disturbances in temperature distribution caused by shallow groundwater circulation, which can lead to cooling or heating effects.
- Chemical properties of reservoir fluids can be inferred from geochemical analysis on hot springs and steam fumaroles in the vicinity of the geothermal field. Geological and geophysical studies can provide supplementary information, suggesting the existence of hot geothermal fluids at depth. However, the actual reservoir temperature and properties of these fluids must be confirmed by drilling exploratory wells to the target depth.

Drilling exploratory wells is typically conducted during the final stage of exploration, with targets selected based on a comprehensive conceptual model. This model integrates key elements of the 3KEYGS framework, including the reservoir structure (faults, fractured zones and cap rock), expected reservoir temperature distribution, reservoir geometry, and hydrological state. The hydrological analysis encompasses meteoric water sources, circulation pathways, phase transition (such as boiling or condensing), volcanic gas mixing, evaporation, and mineral precipitation. To validate and refine the conceptual model and 3KEYGS data, drilling is essential for obtaining the necessary downhole information (Tagomori et al. 2012; Tagomori and Soeda 2013).

2.3.2 Transforming the Conceptual Model to 3D Numerical Model for Assessing Project Feasibility

To understand the reservoir characteristics of the conceptual model, let us examine the example shown in Fig. 2.34. The geothermal fluids are primarily meteoric in origin, with a retention time of at least 30 years based on isotopic analysis. Initially, cold meteoric water, denser than the surrounding water, permeates and migrates to deep formations through gravity and convection. Most fractures are associated with vertical faults and partially along geological boundaries, serving as the main conduits for geothermal fluids. The hydrothermal alteration zone spreads extensively above the reservoir, acting as an impermeable cap rock. The heat source is thought to be a magma chamber beneath the reservoir. The convective circulation cycle results in reservoir temperatures ranging between 240 and 280 °C, as observed in the Hatchobaru field.

Once the satisfactory conceptual model is complete, a three-dimensional numerical model (hereinafter referred to as "3D-model") is constructed based on the reservoir structure and properties incorporated in the conceptual model. Reservoir simulation programs for geothermal reservoir assessment have been developed since the 1970s. The simulation code usually uses the finite difference method. A simple block-type grid for the 3D model is shown in Fig. 2.37. Recently, non-square type grids have also been developed, which are better adapted to complex geothermal reservoir structures while maximizing computing efficiency.

The reservoir simulator is a powerful tool for assessing whether the reservoir capacity is feasible for commercial power generation, assuming the geothermal conceptual model is adequate. It can be regarded as a tool to test the adequacy of the conceptual model. The simulation result is checked (or calibrated) with wellbore data (such as the temperature profile and downhole fluid pressure) and production data. Furthermore, almost all the data obtained and measured during exploratory well drilling are useful for improving the geothermal conceptual model in terms of 3KEYGS. Unlike surface geoscientific exploration methods, which indirectly infer

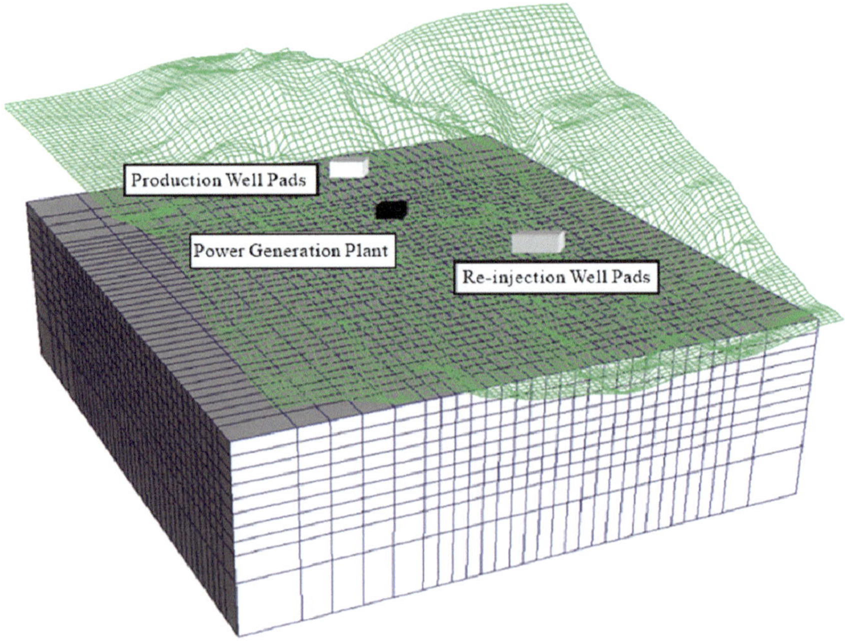

Fig. 2.37 Example of a 3-dimensional block model for numerical simulation

reservoir properties, the information acquired from exploratory wells and associated operations directly reflects geothermal reservoir properties at depth.

However, during the exploration stage, the number of wells available for downhole and production data is limited. Additionally, due to the complex nature of the geothermal reservoir, the conceptual model is far from perfect, especially at this stage. For these reasons, the simulation process is not robust and is subject to a certain error level. Despite these constraints, it is currently the most rational method for geothermal resource evaluation that can withstand business criteria. Based on the simulation results, the power plant construction plan is determined.

The reservoir simulator is based on the diffusivity equation and incorporates Darcy's law and other relevant relationships to account for the underground fluid's thermodynamic behavior. These are processed numerically, usually by the finite difference method, where differential equations are solved by approximating derivatives within each spatially and temporally discretized block or cell. Parameters in each block are meticulously selected to mimic faults, geological boundaries, permeable structures, impermeable structures, and other features that may affect the reservoir's thermodynamic behavior.

In a 3D model, the block size varies typically in proportion to information density, with smaller blocks placed in critical, high-importance areas and larger blocks in less critical surrounding areas. Despite powerful hardware and software availability, simulation quality depends significantly on the reservoir engineer's competence, experience, and ability to effectively manage grid arrangement, cell parameters, computing code (such as liquid-only or liquid–gas two-phase fluids), and boundary conditions. Ultimately, the success of reservoir simulation largely relies on an engineer with substantial experience and proven results in geothermal exploration and development (Tagomori and Soeda 2013).

So far I have discussed the conceptual principle and process involved in reservoir simulation. For actual implementation, a more detailed discussion is necessary. Let us revisit the general overview of the geothermal system. A geothermal system typically involves the convective flow of hot water heated by an underlying heat source. The flow path of this convective plume is influenced by various geological structures such as faults, formation boundaries, and impermeable cap rocks. Consequently, the underground temperature distribution reflects the system's permeability structure. Conversely, the temperature distribution can infer the permeability structure near the geothermal system. This inversion process, conducted during the pre-exploitation stage, is called "Natural state simulation." In this simulation process, the model's permeability structure is adjusted to match the static downhole temperature profiles acquired at the natural state. Due to the inherent nature of this process, the natural state simulation is always performed initially to test the reservoir model, which is then used for the subsequent "History matching simulation."

In the numerical reservoir simulation, differential equations are discretized in spatial and time domains. This allows the simulation to predict the chronological evolution of reservoir states and corresponding parameters due to various disturbances, such as steam production or brine re-injection. "History matching simulation" involves adjusting key properties of the reservoir model to fit actual field data, identifying weaknesses, improving reservoir descriptions, and forming the basis for future performance predictions.

Results from those two simulations are calibrated with information from various borehole measurements and well tests conducted during the exploration phase. Long-term production tests, ideally lasting several months, are preferred as they cause more significant disturbances to the reservoir system, thereby increasing confidence in the model. The design of production tests can be further refined by exploring various well combinations, such as single-well production or simultaneous production from multiple wells, while acquiring data on pressure, temperature, steam and brine flow rates, chemical composition, and well interference.

The "Natural state simulation" aims to replicate the initial states of the geothermal field before any exploitation. In contrast, the "History matching simulation" focuses on modeling the reservoir's transient behavior during production and re-injection in the exploration/exploitation phase. Parameters and boundary conditions in the 3D model are adjusted until the simulation results closely align with the actual field data within an acceptable margin of error.

Once the 3D model is satisfactorily calibrated, it can predict geothermal reservoir performance under various production and re-injection scenarios throughout the power plant's expected commercial life, typically around 30 years. Multiple simulation scenarios can be generated, adjusting variables such as expected power output, plant lifetime, the number and location of production/re-injection wells, and well capacity decline rates. Based on these simulation, the most technically and financially viable scenario is selected for the construction and development plan.

According to the public domain statistics, well drilling constitutes approximately 70 to 80% of the total cost of resource exploration and development, heavily influencing economic feasibility. In modern geothermal projects, the development scenario derived from the 3D model is critical to financial viability, demanding meticulous attention and simulation accuracy. From reconnaissance surveys to resource assessments, every phase must continuously evaluate financial viability. Specifically, well drilling requires careful consideration to ensure necessity and efficiency. If drilling is deemed necessary, it must be executed with both technical precision and cost-efficiency.

References

Aki K, Koyanagi R (1981) Deep volcanic tremor and magma ascent mechanism. J Geophys Res 86:7095–7109

Aso N, Ide S (2014) Focal mechanisms of deep low-frequency earthquakes in Eastern Shimane in Western Japan. J Geophys Res Solid Earth 119:364–377

Aso N, Tsai VC (2014) Cooling magma model for deep volcanic long-period earthquakes. J Geophys Res Solid Earth 119:8442–8456

Bird P (2003) An updated digital model of plate boundaries. Geochem Geophys Geosys 4(3):1027. https://doi.org/10.1029/2001GC000252

Brace WF, Kohlstedt DL (1980) Limits on lithospheric stress imposed by laboratory experiments. J Geophys Res 85:6248–6252

Burov EB (2011) Rheology and strength of the lithosphere. Mar Pet Geol 28:1402–1443

Chopra PN, Paterson MS (1984) The role of water in the deformation of dunite. J Geophys Res 89:7861–7876

Doi N, Kato O, Ikeuchi K, Komatsu R, Miyazaki S, Akaku K, Uchida T (1998) Genesis of the plutonic-hydrothermal system around Quaternary granite in the Kakkonda geothermal system, Japan. Geothermics 27:663–690

Fournier RO (1991) The transition from hydrostatic to greater than hydrostatic fluid pressure in presently active continental hydrothermal systems in crystalline rock. Geophys Res Lett 18:955–958

Geospatial Information Authority of Japan (2024) Basic geospatial information site. https://www.gsi.go.jp/kiban/. Access, October 20, 2024

Geothermal Research Society of Japan (2020) Brochure on geothermal energy in Japan. 4p. https://grsj.gr.jp/english/gej/?l=en_US

Global Volcanism Program (2024) Volcanoes of the world. v.5.1.5., Venzke, E. (ed.)., Smithsonian Institution. https://doi.org/10.5479/si.GVP.VOTW5-2023.5.1. Access, January 12, 2024

Hara I, Shoji K, Sakurai Y, Yokoyama S, Hide K (1980) Origin of the Median tectonic line and its initial shape. Memoirs of the geological society of Japan 18:27–49

Harayama S (1990) Geology of the Kamikochi district. With Sheet Map at 1: 50,000, Geological Survey of Japan, 175p (in Japanese with English abstract)

Harayama S (1992) Youngest exposed granitoid pluton on Earth: cooling and rapid uplift of the Pliocene-Quaternary Takidani Granodiorite in the Japan Alps, central Japan. Geology 20:657–660

Harayama S, Takahashi M, Shukuwa R, Itaya T, Yagi K (2010) High-temperature hot springs and Quaternary Kurobegawa Granite along the Kurobegawa River. J Geol Soc Jpn 116(Supplement):63–81 (in Japanese with English abstract)

Hasegawa A, Umino N, Takagi A (1978) Double-planed structure of the deep seismic zone in the northeastern Japan arc. Tectonophysics 47:43–58

Hasegawa A, Nakajima J (2004) Geophysical constraints on slab subduction and arc magmatism. In: State of the planet: frontiers and challenges in geophysics. AGU Geophysical Monograph 150, IUGG, vol. 19, (Spark, R.S.J., Hawkesworth, C.J., eds.), 81–94

Hasegawa A, Nakajima J, Kita S, Okada T, Matsuzawa T, Kirby SH (2007) Anomalous deepening of a belt of intraslab earthquakes in the Pacific slab crust under Kanto, central Japan: possible anomalous thermal shielding, dehydration reactions, and seismicity caused by shallower cold slab material. Geophys Res Lett 34:L09305. https://doi.org/10.1029/2007GL029616

Hasegawa A, Nakajima J, Kita S, Tsuju Y, Nii K, Okada T, Matsuzawa T, Zhao D (2008) Transportation of H_2O in the NE Japan subduction zone as inferred from seismic observations: Supply of H_2O from the slab to arc crust. J Geogr (Chigaku Zasshi) 117:59–75 (in Japanese with English abstract)

Hasegawa A, Sato H, Nishimura T (2015) Seismology. Introduction to modern earth science series 6, Kyoritsu Publication C., 472p (in Japanese)

Hirata D, Yamashita H, Suzuki K, Hirata T, Li Y, Kon Y (2010) Collision accretion tectonics of the proto-Izu-Mariana arc. J Geogr (Chigaku Zasshi) 119:1125–1160 (in Japanese with English abstract)

Hirth G, Kohlstedt DL (1996) Water in the oceanic upper mantle: Implication of rheology, melt extraction and the evolution of lithosphere. Earth Planet Sci Lett 144:93–108

Ikeuchi K, Doi N, Sakagawa Y, Kamenosono H, Uchida T (1998) High-temperature measurements in well WD-1a and the thermal structure of the Kakkonda geothermal system, Japan. Geothermics 27:591–607

Ishizuka O, Tani K, Reagan MK, Kanayama K, Umino S, Harigane Y, Sakamoto I, Miyajima Y, Yuasa M, Dunkley DJ (2011) The timescales of subduction initiation and subsequent evolution of an oceanic island arc. Earth Planet Sci Lett 306(3–4):229–240. https://doi.org/10.1016/j.epsl.2011.04.006

Ishizuka O, Tani K, Taylor RN, Umino S, Sakamoto I, Yokoyama Y, Shimoda G, Harigane Y, Ohara Y, Conway CE, Perez A, Sekimoto (2022) Origin and age of magmatism in the northern Philippine Sea basins. Geochem Geophys Geosyst 23:e2021GC010242. https://doi.org/10.1029/2021GC010242

Ishizuka O, Yuasa M, Taylor RN, Sakamoto I (2009) Two contrasting magmatic types coexist after the cessation of back-arc spreading. Chem Geol 266:274–296

JGA (Japan Geothermal Association) (2021) Present Situation and issues of geothermal power development in Japan (in Japanese). 17 p. https://www.enaa.or.jp/?fname=gec_2021_1_14.pdf

JGA (Japan Geothermal Association) (2024) Geothermal power plant in Japan (generation-end output > 1,000kW, as of April 2024) (in Japanese). https://www.chinetsukyokai.com/information/nihon.html

Japan Meteorological Agency (2024) The Seismological Bulletin of Japan. https://www.data.jma.go.jp/eqev/data/bulletin/eqdoc_e.html. Access, April 14, 2024

JOGMEC (Japan Oil, Gas, and Metals National Corporation) (2017) Current situation of geothermal power generation in Japan. 15p. https://geothermal.jogmec.go.jp/event/file/2016/session_160602_01.pdf

Julian BR (1994) Volcanic tremor: nonlinear excitation by fluid flow. J Geophys Res 99:11859–11877

Kosuga M, Noro K, Masukawa K (2017) Characteristics of spatiotemporal variations of hypocenters and diversity of waveforms of deep low-frequency earthquakes in northeastern Japan. Bull Earthq Res Inst Univ Tokyo 92:63–80 (in Japanese with English abstract)

Kudo T, Takarada S (2021) Catalog OF ERUPTIVE EVENTS during the Last 10,000 Years in Japan, Version 2.4.1. Available online: https://gbank.gsj.jp/volcano/eruption/index.html. Accessed on 23 February 2021

Kunugiza K, Maruyama S (2011) Geotectonic evolution of the Hida Marginal Belt, Central Japan: Reconstruction of the oldest Pacific-type orogeny of Japan. J Geogr 120:960–980 (in Japanese with English abstract)

Kyuden Mirai Energy Co,. Inc (2024) Brochure: introduction to geothermal power station of Kyuden Mirai Energy Co., Inc. 28 p. https://www.q-mirai.co.jp>files>uploads

Matsubara M, Sato H, Ishiyama T, Van Horne A (2017) Configuration of the Moho discontinuity beneath the Japanese Islands derived from three-dimensional seismic tomography. Tectonophysics 710–711:97–107

Muraoka H, Kamata H (1983) Displacement distribution along minor fault traces. J Struct Geol 5:483–495

Muraoka H, Sakaguchi K, Tamanyu S, Sasaki M, Shigeno H, Mizugaki K (2007) Atlas of hydrothermal systems in Japan. Geological survey of Japan, AIST, 110p. (in Japanese with English abstract)

Muraoka H, Uchida T, Sasada M, Yagi M, Akaku K, Sasaki M, Yasukawa K, Miyazaki S, Doi N, Saito S, Sato K, Tanaka S (1998) Deep geothermal resources survey program: igneous, metamorphic and hydrothermal processes in a well encountering 500ºC at 3729 m depth, Kakkonda, Japan. Geothermics 27:507–534

Nakajima J, Matsuzawa T, Hasegawa A, Zhao D (2001) Three-dimensional structure of Vp, Vs, and Vp/Vs beneath northeastern Japan: Implications for arc magmatism and fluids. J Geophys Res 106:21843–21857

Nakajima J, Tsuji Y, Hasegawa A (2009) Seismic evidence for thermally-controlled dehydration reaction in subducting oceanic crust. Geophys Res Lett 36:L03303. https://doi.org/10.1029/2008GL036865

Nakamura K, Shimazaki K (1981) Sagami Trough, Suruga Trough and subduction of the plate. Kagaku (Science) 51:490–498 (in Japanese)

National Research Institute for Earth Science and Disaster Resilience (2024) 3-D hypocenter distribution. High Sensitivity Seismograph Network Japan. https://www.hinet.bosai.go.jp/topics/ThreeJS/?LANG=en. Access, April 23, 2024

NEDO (New Energy and Industrial Technology Development Organization) (2021) TSC Foresight, Vol. 106, 30 p. https://www.nedo.go.jp/content/100937780.pdf

NOAA National Centers for Environmental Information (2022) ETOPO 2022 15 arc-second global relief model. NOAA National Centers for Environmental Information. https://doi.org/10.25921/fd45-gt74. Access, April 14, 2024

Obara K (2002) Nonvolcanic deep tremor associated with subduction in southwest Japan. Science 296:165–168

Ramberg H (1967) Gravity deformation and earth's crust as studied by centrifuged models. Academic Press, New York, NY, USA, p 224p

Saito S (1993) A trend of geothermal drillings in Japan. J Geothermal Res Soc Japan 15:104–110 (in Japanese)

Schmidt MW, Poli S (1998) Experimentally based water budgets for dehydrating slabs and consequences for arc magma generation. Earth Planet Sci Lett 163:361–379

Scholz CH (1988) The brittle-plastic transition and the depth of seismic faulting. Geol Rundsch 77:319–328

Sekioka M (1980) Unsystematic estimate of heat stored in hot rocks. J Geothermal Res Soc Japan 2:71–79 (in Japanese with English abstract)

Suzuki Y, Ioka S, Muraoka H (2014) Determining the maximum depth of hydrothermal circulation using geothermal mapping and seismicity to delineate the depth to brittle-plastic transition in northern Honshu, Japan. Energies 7:3503–3511

Suzuki Y, Muraoka H, Asanuma H (2020) Validation and evaluation of an estimation method for deep thermal structures using an activity index in major geothermal fields in northeastern Japan. Energies 13:4684

Suzuki Y, Muraoka H, Asanuma H (2021) Thermal monitoring of the lithosphere by the interaction of deep low-frequency earthquakes in northern Japan. Energies 14:1546. https://doi.org/10.3390/en14061546

Tagomori K (2010) Current status and subjects on geothermal reservoir exploration and assessment. Environmental Solution Technology 9(6):19–25 (in Japanese)

Tagomori K, Saito H, Tokita H, Matsuda K (2012) Practical approach for exploitation and evaluation of geothermal reservoir, and expected subjects for further geothermal resource development. Geothermal and volcanological research report of Kyushu University, 2012, No.20, 46–54 (in Japanese)

Tagomori K, Soeda Y (2013) Geothermal resource survey and its assessment. Energy Resour 34(1):29–33 (in Japanese)

Tamura Y (2003) Genetic relationship between Quaternary NE Japan arc magmas and Miocene Japan Sea back-arc basin basalts : implications for a dynamic model of hot fingers in the mantle wedge. J Geogr 112(5):781–793 (in Japanese with English abstract)

Tamura Y, Tatsumi Y, Zhao D, Kido Y, Shukuno H (2002) Hot fingers in the mantle wedge: new insights into magma genesis in subduction zones. Earth Planet Sci Lett 197:105–116

Tamura Y, Busby CJ, Blum P, and the Expedition 350 Scientists (2015) Expedition 350 summary. Proceedings of the international ocean discovery program (IODP), Volume 350 publications.iodp.org. https://doi.org/10.14379/iodp.proc.350.101.2015

Tatsumi Y (1989) Migration of H_2O and genesis of basalt magmas in subduction zones. J Geophys Res Solid Earth 94:4697–4707

TENPES (Thermal and Nuclear Power Engineering Society) (2021) Current situation and trend of geothermal power generation 2020. 141p (in Japanese)

Uihlein A (2018) JRC geothermal power plant dataset. European Commission, Joint Research Centre (JRC) PID: http://data.europa.eu/89h/jrc-10128-10001.

US Geological Survey (2024) Search earthquake catalog, earthquake hazard program. https://earthquake.usgs.gov/earthquakes/search/. Access, March 17, 2024

Uyeda S, Kanamori H (1978) Subduction of oceanic plates and formation of marginal seas. Kagaku (Science) 48:91–102 (in Japanese)

Uyeda S, Tamaki K, Kushiro I (1983) Tectonics of the eastern margins of the Philippine Sea—the Mariana arc and the Bonin arc. Kagaku (Science) 53:439–447 (in Japanese)

Wessel P, Luis JF, Uieda L, Scharroo R, Wobbe F, Smith WHF, Tian D (2019) The generic mapping tools version 6. Geochem Geophys Geosyst 20:5556–5564. https://doi.org/10.1029/2019GC008515

Whitehead JA, Helfrich KR (1990) Magma waves and diapiric dynamics. In Magma Transport and Storage; Ryan, M.P., Ed.; John Wiley & Sons: Chichester, UK, 420 p, 53–76

Wilks KR, Carter NL (1990) Rheology of some continental lower crustal rocks. Tectonophysics 182:57–77

Wu J, Suppe J, Lu R, Kanda R (2016) Philippine Sea and East Asian plate tectonics since 52 Ma constrained by new subducted slab reconstruction methods. J Geophys Res Solid Earth 121:4670–4741. https://doi.org/10.1002/2016JB012923

Yoshii T (1978) Fundamental geophysical data in the surroundings of the Japanese Islands. Kagaku (Science) 48:489–494

Yuhara K, Sekioka M, Ito Y, Saito T (1982) Emissivity correction for the wall temperatures measured by infrared radiation thermometers in a hot tunnel, Kurobe. J Geothermal Res Soc Japan 4:107–121

Zhao D, Yanada T, Hasegawa A, Umino N, Wei W (2012) Imaging the subducting slabs and mantle upwelling under the Japan Islands. Geophys J Int 190:816–828. https://doi.org/10.1111/j.1365-246X.2012.05550.x

Chapter 3
High-Temperature Hydrothermal Systems Induced from Shallow-Depth Magma Intrusions in Japan

Hirofumi Muraoka

Abstract Recently, many geothermal wells have encountered consolidated magma intrusions worldwide. These intrusions are likely geothermal heat sources. In this chapter, we consider the genetic relationships between the depth of the high-level magma intrusions and high-temperature hydrothermal systems in Japan, particularly from the viewpoint of the magma intrusion's buoyancy.

Keywords Shallow-depth magma intrusions · High-temperature hydrothermal systems · Buoyancy of magma · Wall rock density · Tectonics in Japan

3.1 High-Temperature Hydrothermal Systems in Japan

3.1.1 Trinity of the Geothermal Systems

The trinity of the geothermal systems is composed of the geothermal heat sources, geothermal reservoirs, and geothermal fluids as schematically illustrated in Fig. 3.1. The geothermal heat sources in the high-temperature geothermal fields are likely high-level magma intrusions and their consolidated high-temperature equivalents. Idealistic shapes are suspected by eroded old examples of plutonic intrusions, as shown in Fig. 3.2. Some of them have ring dike complex features at their tops, which represent magmatic local radial extension tectonics induced from the dome-shaped extension of the roofs by magma intrusions themselves. The southernmost elliptical intrusion in Fig. 3.2 has about 19 km diameter and about 284 km^2 in the width of the area. If the area of the intrusion is less than 100 km^2, we call it stock; if more than 100 km^2, we call it batholith. Most of the elliptical intrusions in Fig. 3.2 can thus be called batholith. The area of the Felsite intrusion in The Geysers geothermal field, USA, is estimated to be about 300 km^2 and can be called a batholith. However,

H. Muraoka (✉)
Hirosaki University, Hirosaki, Japan
e-mail: hiro@hirosaki-u.ac.jp

J. Kasahara et al. (eds.), *Challenges in Geothermal Development*, Advances in Geological Science, https://doi.org/10.1007/978-981-95-0553-1_3

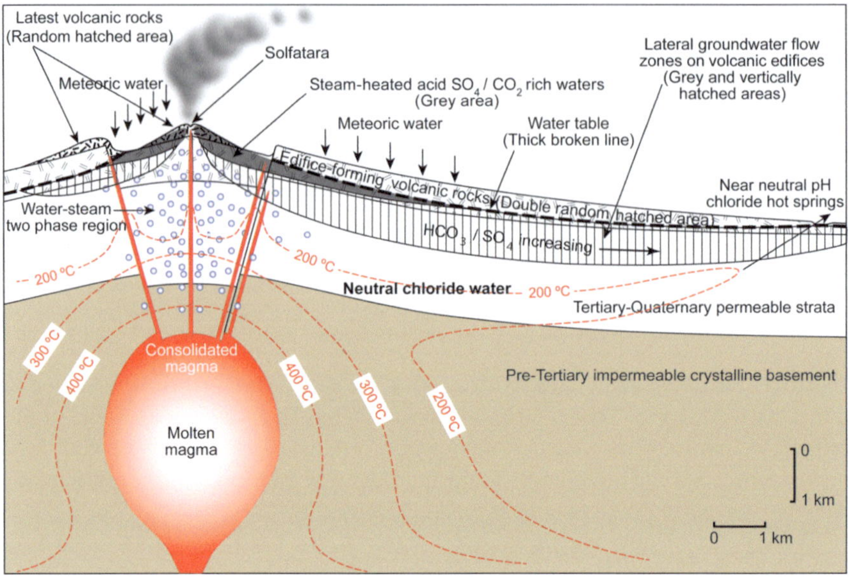

Fig. 3.1 Schematic illustration of the trinity of the geothermal systems. The trinity of the geothermal systems comprises geothermal heat sources such as magma intrusions, geothermal reservoirs such as faults, and geothermal fluids such as brine and steam. Modified from Muraoka (2018) and based on Henely (1984)

a batholith-scale young heat source intrusion has rarely been found in Japan by this time.

Most geothermal reservoirs are known to be faults. Outcrops of faults cannot easily be found in humid climate regions with thick soil deposits and vegetation covers. Nevertheless, it is important to find the reservoir fault leading to the success of geothermal drilling. Figure 3.3 shows an outcrop of the relatively large-scale fault with a 4-m-thick gouge in northernmost Honshu, Shimokita Peninsula, Japan (Suzuki et al. 2017). An arrow shows the location of this outcrop in Fig. 3.4. Figure 3.4 independently provides a map of silicified alteration zones (Tomiyama et al. 2007), before the finding of the fault outcrop. However, the silicified alteration zones are well concentrated on the hanging walls of the Hiuchi Dake East Slope Fault, which tilts 77° to the east. The stream pH mapping indicates that the fault zone is neutral among the surrounding low-pH regions and is prospected to be a reservoir fault (Suzuki et al. 2017). There are few cases, but some geothermal reservoirs consist of intrusions with cooling joints. Cooling joints are penetrative structures over the magma intrusions and give effective permeability to the entire body. Figure 3.5 shows the cooling joints of the 2.7-m-thick andesitic dike, probably serving as a reservoir for the Raiden hot springs in Hokkaido, Japan.

Table 3.1 summarizes the tectonic field, structure, geological formation, and geological age of the geothermal reservoirs in Japan's representative geothermal

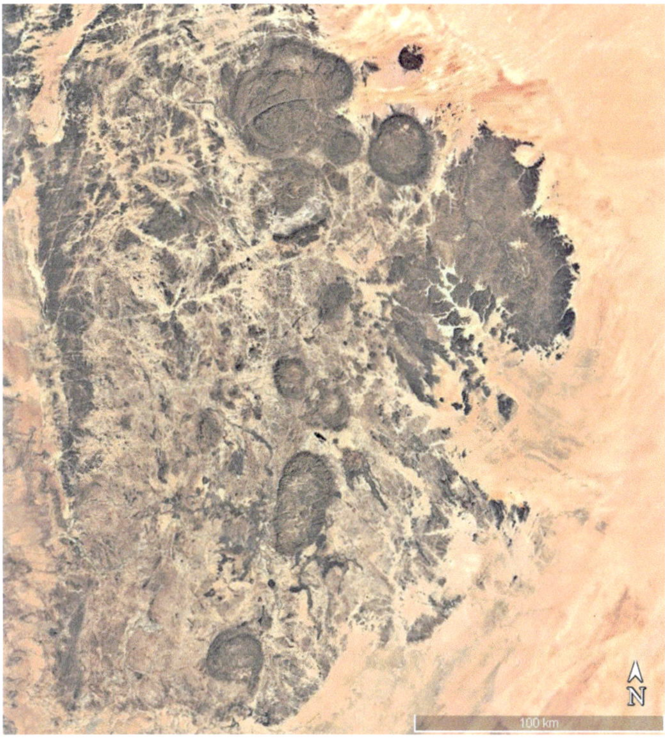

Fig. 3.2 Elliptical per-alkaline granitic intrusions intruded into the Pre-Cambrian strata in Air Plateau, Niger, Africa. The satellite imagery is used by the courtesy of the Google Earth

power plants (Japan Geothermal Energy Association 2000). When we look at reservoir structure, most of the reservoir structure consists of faults, probably up to 90%. However, the intrusions also play an important role, such as the Kakkonda granite in Kakkonda, quartz porphyry in Uenotai, and dacite intrusion in Yamagawa, probably up to 10%. The age of the reservoir is almost entirely the Miocene, Pliocene, and Pleistocene, although the Pre-Tertiary units, the Jurassic-Paleogene accretionary prisms or Cretaceous granitic rocks, make a small contribution as reservoirs at the base, such as in Kakkonda, Uenotai, Onikobe, Hatchobaru, and Ogiri. Mori is a special case where the reservoir is the Jurassic accretionary prism comprising limestone in the wall rocks of the vent of the funnel-shaped Nigorikawa caldera. The permeability was obtained by the vent fracturing along the magma conduit during the Quaternary age.

As for the geothermal fluids in Japan, volcanic gases are intensively rising from the subvolcanic magma chamber to the volcanic eruption crater (Fig. 3.1). These volcanic gases are condensed and oxidized with groundwater near the surface, forming acid sulfate water. They are inadequate for geothermal power generation because of the corrosion of the casing and pipeline. Neutral chloride water is known to be adequate

Fig. 3.3 An outcrop of the Mutsu Hiuchi Dake East Slope Fault, northernmost of Honshu, Shimokita Peninsula, Japan. The thickness of the fault gouge is 4 m (Muraoka 2018; Suzuki et al. 2017)

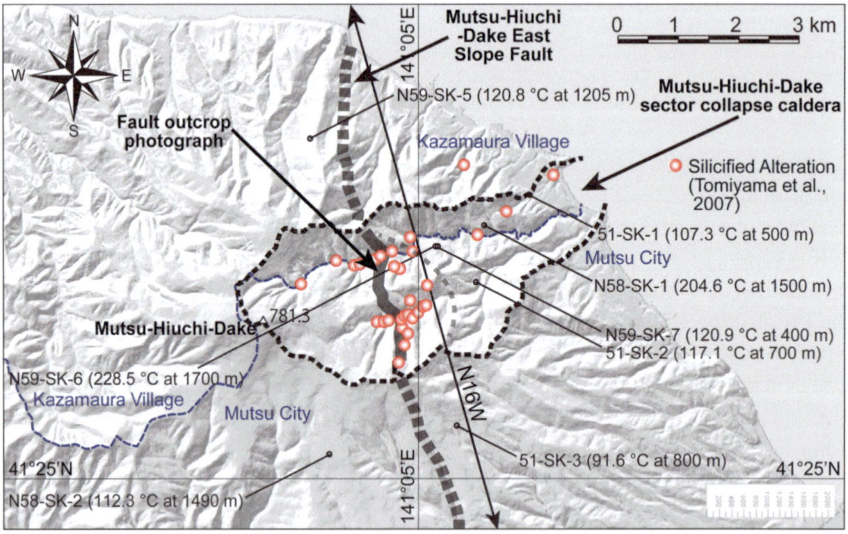

Fig. 3.4 Silicified alteration zones (red circle) mapped by Tomiyama et al. (2007). The Mutsu Hiuchi Dake East Slope Fault and its outcrop are shown as a thick line and an arrow (Muraoka 2018; Suzuki et al. 2017)

Fig. 3.5 A Quaternary hypersthene andesite dike, 2.7 m thick and WNW-ESE trending at the Kasupeno Cape in the west of Raiden volcano, Hokkaido (Photograph by Muraoka on April 4 2016; originally described by Yamagishi et al. 1976). Host rocks consist of 1.5 million years old hypersthene andesite hyaloclastite (New Energy Development Organization 1987). The dike forms a dike swarm, and its columnar joints are discharging at the eastern extension, the Raiden hot springs. The dike was originally formed as one of the magmatic conduits of Raiden volcano (Muraoka 2018)

for geothermal power generation (Henley 1984; Fig. 3.1). Even in the acid sulfate water, if the water fully interacts with host rocks for an extended period, matured water such as neutral chloride water will be yielded. Figure 3.6 shows the pH values versus the known well depths of hot springs in Japan, where a primary cluster in pH ranges from 6 to 10, but a small acid cluster is concentrated from 1 to 4. The well depth of the acid cluster tends to be much shallower than that of the neutral cluster. This suggests that the water–rock interactions at greater depths tend to neutralize the water. This is also supported by the fact that the acidic kaolinite alteration zones are empirically found in shallower depths.

In summary, acid water is usually found in shallow depths through two processes. One is that the rising volcanic gases are condensed and oxidized in the shallow groundwater near the volcanic vent craters, yielding acidic water. The other is that acid water tends to be neutralized through the water–rock interactions at a greater depth by the reducibility of rocks.

Table 3.1 Summary of geothermal reservoirs and their geological age in Japan (Japan Geothermal Energy Association 2000)

Name of plant	Prefecture	Tectonic field	Geological settings	Reservoir depth	Reservoir structure	Reservoir formations	Main reservoir age	Contribution by the Pre-Tertiary units
Mori	Hokkaido	Reverse fault	Nigorikawa caldera	736–3250 m	Vent fracturing of the funnel-shaped caldera	Kamiiso Group (chert, slate, limestone)	Jurassic accretionary prism (the Iwaizumi Terrane)	100% but by the vent fracturing of the funnel-shaped caldera
Ohnuma	Akita	Reverse fault	Hachimantai	1485–1714 m	Hanawa Depression Zone	Takaishizawa Formation	Miocene	Nothing
Sumikawa	Akita	Reverse fault	Hachimantai	1552–2512 m	Propylite Fault, Akagawa Fault	Neogene granodiorite, Altered andesite, Dacitic tuff	Miocene	Slightly
Matsukawa	Iwate	Reverse fault	Hachimantai	1080–1500 m	Sumikawa Fault, Diorite porphyry	Kunimitoge Formation, Takinoue Onsen Fromation, Diorite porphyry, Quartz diorite	Miocene, Pliocene, Pleistocene	Nothing

(continued)

Table 3.1 (continued)

Name of plant	Prefecture	Tectonic field	Geological settings	Reservoir depth	Reservoir structure	Reservoir formations	Main reservoir age	Contribution by the Pre-Tertiary units
Kakkonda	Iwate	Reverse fault	Hachimantai	944–3003 m	Kakkonda Fault, Takinoue Fault, Kakkonda granite	Pre-Tertiary slate, Kunimitoge Formation, Takinoue Onsen Formation, Kakkonda granite	Miocene, Pliocene, Pleistocene, Jurassic accretionary prism (the North Kitakami Terrane)	Slightly contributed to the deeper reservoir but more contributed by the Pleistocene Kakkonda granite
Uenotai	Akita	Reverse fault	Kurikoma	1191–2228 m	Doroyu Fault, Quartz porphyry	Pre-Tertiary crystalline schist and serpentinite, Doroyu Formation, Minasegawa Formation, Quartz porphyry	Miocene, Pliocene, Jurassic accretionary prism (the South Kitakami Terrane)	Definitely
Onikobe	Miyagi	Reverse fault	Kurikoma, Onikobe caldera	228–1500 m	NW–SE faults?	Pre-Tertiary granite, Kanisawa Formation, Sannozawa Formation, Akazawa Formation	Miocene, Pliocene, Pleistocene	Slightly

(continued)

Table 3.1 (continued)

Name of plant	Prefecture	Tectonic field	Geological settings	Reservoir depth	Reservoir structure	Reservoir formations	Main reservoir age	Contribution by the Pre-Tertiary units
Yanaizu-Nishiyama	Fukushima	Reverse fault	Back-arc side of Bandai	1561–2700 m	Chinoikezawa Fault, Sarukurazawa Fault	Takizawagawa Formation	Early Miocene	Nothing
Takigami	Oita	Normal fault	Beppu-Shimabara Graben	1104–2707 m	Noine Fault, Teratoko Fault	Usa Group, Lower Member of Takigami Formation	Miocene, Pliocene, Early Pleistocene	Nothing
Otake	Oita	Normal fault	Beppu-Shimabara Graben	372–1562 m	Yokoo Fault, Otake Fault	Usa Group, Hohi Volcanic Rocks	Miocene, Pliocene, Early Pleistocene	Nothing
Hatchobaru	Oita	Normal fault	Beppu-Shimabara Graben	759–3031 m	Hatchobaru Fault, Komatsunoike Sub-Fault	Pre-Tertiary crystalline schist, Cretaceous granodiorite, Usa Group, Hohi Volcanic Rocks	Miocene, Pliocene, Early Pleistocene	Slightly

(continued)

Table 3.1 (continued)

Name of plant	Prefecture	Tectonic field	Geological settings	Reservoir depth	Reservoir structure	Reservoir formations	Main reservoir age	Contribution by the Pre-Tertiary units
Ogiri	Kagoshima	Normal fault	Kagoshima Graben	1009–1501 m	Ginyu Fault, Ginyu Sub-Fault	Shimanto Supergroup, Kirishima Welded Tuffs, Ebino Group, Iino Lava, Makizono Lava	Early Pleistocene	Slightly
Yamagawa	Kagoshima	Normal fault	Kagoshima Graben	1500–2500 m	Dacite Intrusion, Onkadobira Faullt	Nansatsu Group, Yamagawa Formation, Dacite Intrusions	Miocene, Pliocene, Early Pleistocene	Nothing

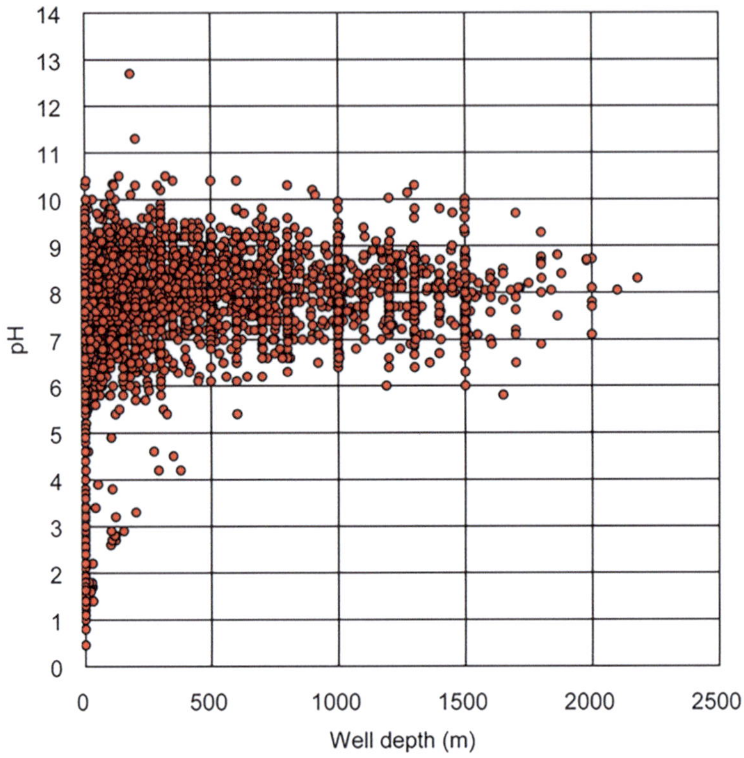

Fig. 3.6 Well depth and pH relationships of hot springs in Japan (Muraoka et al. 2007)

3.1.2 Distribution of High-Temperature Hydrothermal Systems in Japan

One of Japan's most popular leisure activities is staying in hot spring inns. As a result, 27,915 hot spring sources have been developed in Japan as of March 2022 (Ministry of Environment 2022). Japan is probably the largest hot spring utilization country in the world. We tried to collect published data on hot springs in Japan and obtained 7,203 data, as shown in Fig. 3.7 (Muraoka et al. 2007). Most of these hot springs are developed by drilling nowadays. However, many boiling springs are traditionally known as natural springs associated with the Quaternary volcanoes. When we compare these boiling hot springs with the distribution of the Holocene volcanoes (Fig. 3.8), their spatial correspondences are obvious.

Even though the recent hot spring sources are developed by drillings, most of the hot spring wells tend to be shallower than 1.5 km by the cost-saving demands. Then, we could draw the outline distributions of the shallow-depth hydrothermal systems in Japan by contouring the discharge temperatures of 7,203 hot springs in Fig. 3.9 (Muraoka et al. 2007). Most of the high-temperature hydrothermal systems

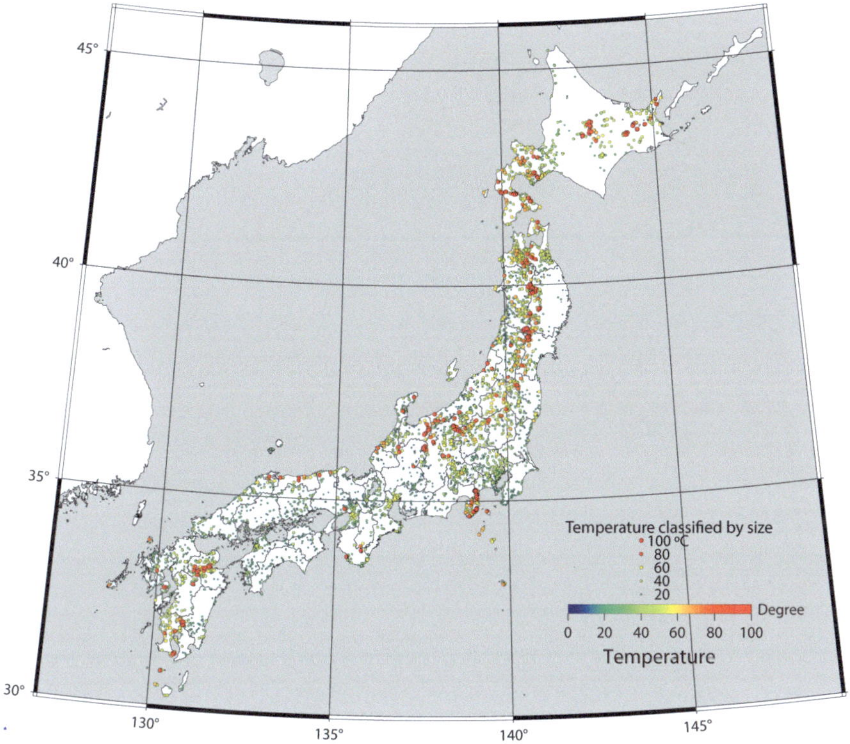

Fig. 3.7 Distribution of 7,203 hot springs in main part of Japan (Muraoka et al. 2007)

are exactly associated with the Holocene volcanic chains in Japan (Fig. 3.8), so their distributions are separated in Northeast Japan and Kyushu Island. Western Japan, such as the Kinki Province, Chugoku Province, and Shikoku Island are mainly composed of the Pre-Jurassic and Jurassic accretional prisms, including two series of paired metamorphic belts (Miyashiro 1973) and are not very active on the Quaternary volcanism. Then, the high-temperature hydrothermal systems are relatively rare. However, two boiling hot springs more than 90 °C are known in the forearc side of the volcanic front of this region: the Arima hot spring near Kobe City and the Yunomine hot spring in Kii Peninsula south of Osaka City (Sano and Wakita 1985). Particularly, the Hongu hot springs, including the Yunomine hot spring, form a wide high-temperature anomaly in the Kii Peninsula in Fig. 3.9. The two factors recently explain genesis of the Hongu hot springs: One is the high heat-flow anomaly associated with the northwestward subduction of the spreading axis of the Philippine Sea Plate, and the other is the high permeability of columnar joints of the 14-Ma old quartz porphyry sheets serving for the efficient hydrothermal convections (Muraoka 2008).

Northern Japan, including northern Honshu and Hokkaido, is a cold and snowy region with high electricity demands and earnest demands for the direct use of

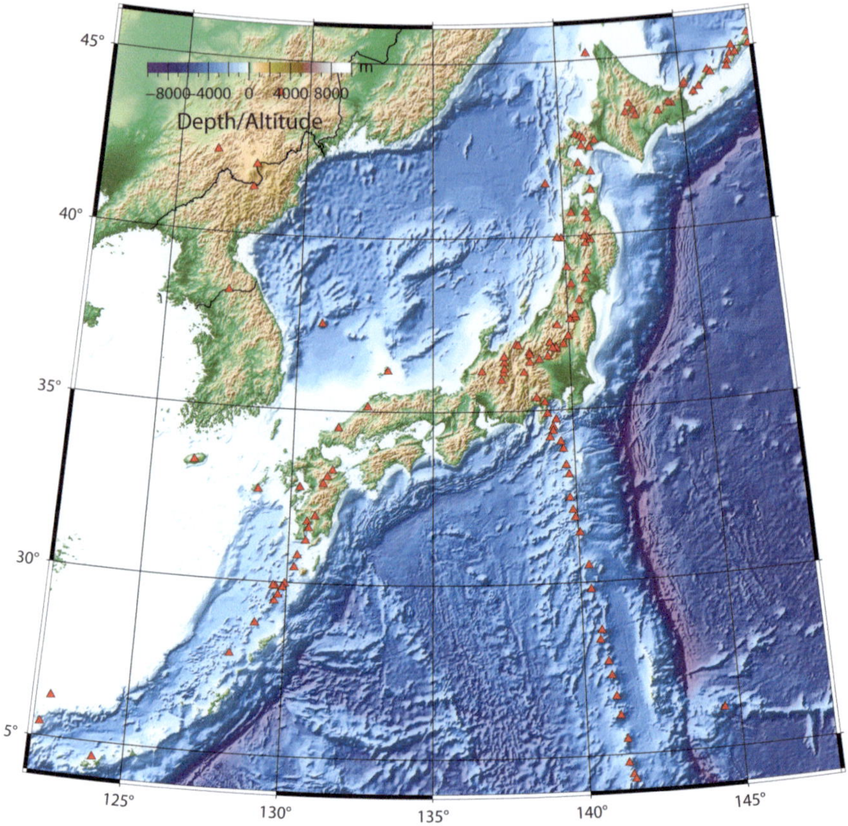

Fig. 3.8 Distribution of the Holocene volcanoes (red triangle) in Japan (Global Volcanism Program 2024), based on the ETOPO1 1 arc-minute global relief model (Amante and Eakins 2009) and GMT 6.4.0 (Wessel et al. 2019)

thermal water during the long winter. Coinciding with these colder regions, numerous hydrothermal systems are recognized (Fig. 3.9). Developments of the cascade geothermal utilization from the steam flash power generation, binary cycle power generation, space heating, greenhouse, and snow melting are much more expected in northern Japan.

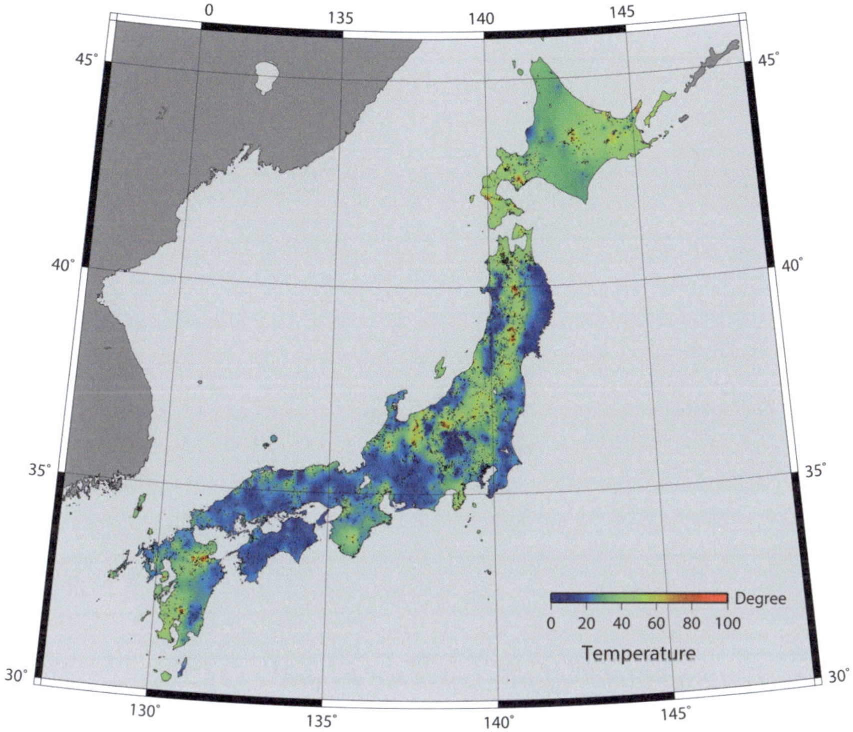

Fig. 3.9 Shallow-depth hydrothermal systems in Japan (Muraoka et al. 2007)

3.2 Emplacement Depth of High-Level Magma Intrusions in Japan

3.2.1 Driving Force of Magma Ascent

The driving force of magma ascent is ascribed to the two mechanisms (Fig. 3.10). One is the buoyancy of the low-density molten magma to the surrounding high-density host rocks. The buoyancy mechanism is always effective in all the tectonic fields but is particularly important in the contraction tectonic fields. A typical example is the rise of the diapiric magma intrusions, as seen in the ellipse-shaped consolidated granitic stocks or batholiths (Fig. 3.2). The other is a pumping mechanism. When the dike-shaped magma intrusions are connected from the magma source regions and the magma pressure exceeds the tectonic stress to the normal to the dike surface in the magma source regions, magma will ascend by the pumping mechanism (Fig. 3.10; Rubin 1990; Reches et al. 1993). This mechanism is essentially important in the extension tectonic fields like the mid-ocean ridges, because of the dike swarms and the weak lateral stresses by the spreading in the axial valley.

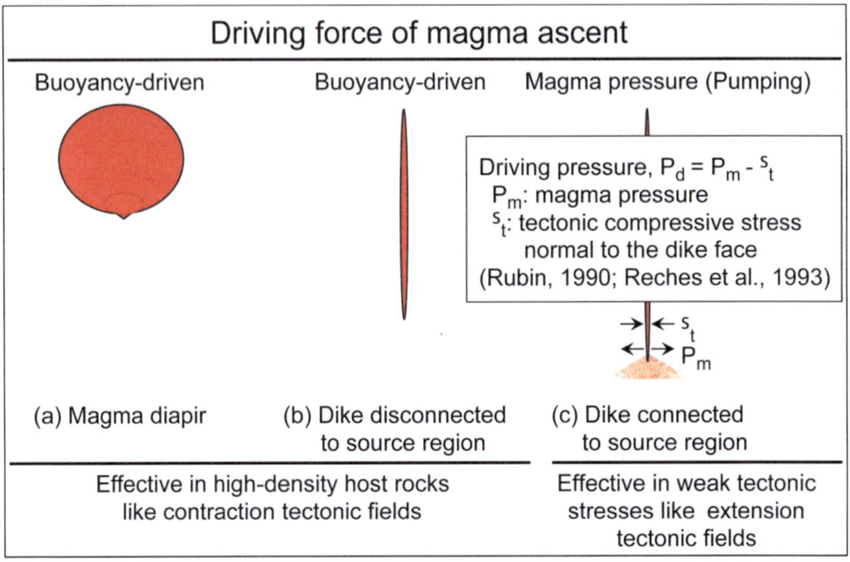

Fig. 3.10 Concept of driving force of magma ascent (Muraoka et al. 2006)

3.2.2 *Tectonic Fields of the Japanese Islands*

Tectonic fields are critically important for better understanding the geothermal system because the type of the geothermal system is completely distinct among the different tectonic fields. The fault configurations can define three types of tectonic fields, as shown in Fig. 3.11.

A vertical overburden stress could be evenly constant over the tectonic fields, and therefore, this stress can be used as an indicator to compare with horizontal stresses. A vertical overburden stress is the maximum stress in the normal fault fields, and the two lateral axes of stresses are smaller than this. As a result, the deformation of the

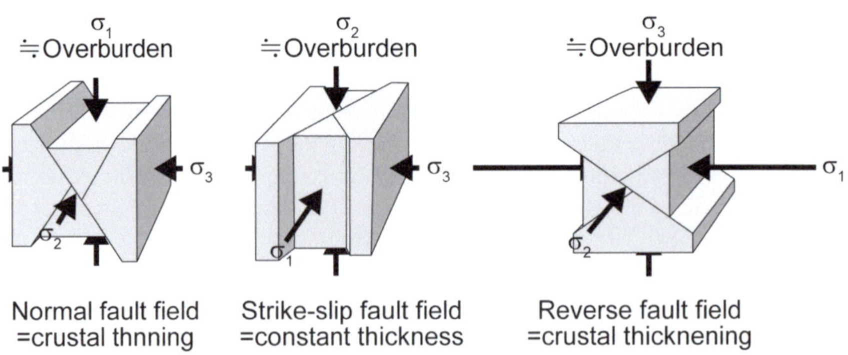

Fig. 3.11 Three types of tectonic fields represented by the fault configurations

field is characterized by crustal thinning and area widening. On the contrary, the two lateral axes of stresses are larger than the vertical overburden stress in the reverse fault fields, so the deformation of the field is characterized by crustal thickening and area narrowing. This is precisely the context of orogeny in the narrow sense. The strike-slip field is neutral between them, and the deformation of the field tends to be constant on the crustal thickness and area width (Fig. 3.11).

If the total area can be assumed to be constant on the surface of the Earth, the normal fault and reverse fault fields would be evenly distributed. The normal fault fields are distributed in the ocean ridges and continental rift zones. The reverse fault fields are distributed in the subduction zones and continental collision zones. The Japanese Islands are typical of the subduction zones, and the reverse fault fields should be predominant over the normal fault fields.

The azimuths of the sigma horizontal maximum of the crustal stress of main part of Japan were estimated from the focal mechanism solutions and are published on the Website at https://www.gsj.jp/publications/pub/openfile/openfile0738.html (Kiguchi et al. 2024). The map is valuable but somehow complicated (Fig. 3.12). Then, we tried to simplify it under some geological aspects (Fig. 3.13).

As the Japanese Islands are situated in the compressive subduction zone, most areas are widely subject to the reverse fault fields in Northeast Japan and central Japan, as expected above (Figs. 3.12 and 3.13). Only the relatively stable Southwest Japan, such as Kyushu, western Honshu (Chugoku Provincc), and part of Shikoku, are recognized as the strike-slip fault fields. The Fossa Magna region is also a strike-slip fault field, probably due to the collision tectonics by the northwestward subduction of the buoyant Izu-Bonin active volcanic arc.

The normal fault fields are relatively restricted in Japan. Two typical normal fault fields in Kyushu, the Beppu-Shimabara Graben and Kagoshima Graben, can be explained as the back-arc spreading zones due to the 80° steep subduction of the Philippine Sea Plate from the Ryukyu Trench (Uyeda and Kanamori 1978). A similar example of the back-arc spreading is well known in the Taupo Volcanic Zone in New Zealand. The Jyoban-Choshi Forearc Extension Zone cannot be explained by the back-arc spreading (Fig. 3.13). When we consider the northeastern continuation of the Median Tectonic Line and the Ryoke metamorphic belt, the Futaba Tectonic Line and the Abukuma metamorphic belt could be possible candidates. Miyashiro (1973) considered the Abukuma metamorphic belt a northeastern continuation of the Ryoke metamorphic belt. Hara et al. (1980) presented a regional structural map of Southwest Japan (Fig. 3.14) where the Ryoke metamorphic belt structurally continued to the Abukuma metamorphic belt, and the Median Tectonic Line nearly continued to the Futaba Tectonic Line. On this structural map, it should be noted that a very tight bending is revealed in the Jyoban-Choshi area (Fig. 3.14). A tight, simple bending fold inevitably gives rise to an extension zone on the outside rim of the fold (Fig. 3.15). Thus, a possible explanation is that the extraordinary tight bending of the Ryoke-Abukuma belt in the Jyoban-Choshi area between the west-northwestward subduction of the Pacific Plate and the north-northwestward colliding subduction of the Izu-Bonin arc could be a cause of the Jyoban-Choshi Forearc Extension Zone (Fig. 3.13).

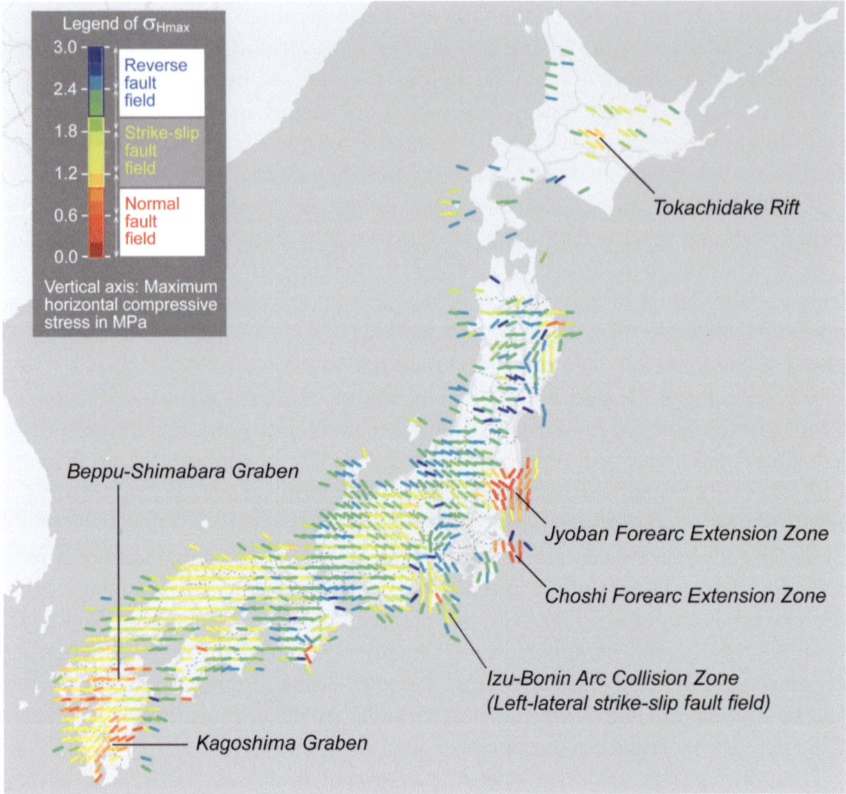

Fig. 3.12 Distribution of the three fault type fields and SH_{max} azimuth (Kiguchi et al. 2024)

3.2.3 Buoyancy of Magma Induced from the Host Rocks in Japan

Throughout the 1980s and 1990s, many geothermal wells penetrated young plutonic stock or batholith intrusions in the world which were consolidated but still kept sub-solidus high-temperature conditions under the cooling process from the magma. They play a role of the heat sources for each geothermal system. In other words, these drilling events encounter an entity of the heat sources in the trinity of the geothermal systems, which was high-level magma intrusions as has long been speculated.

Remarkable is the fact that they were exclusively reported from the contraction tectonic fields (Table 3.2; Muraoka and Yano 1998). When we think the principle of the magma ascent by buoyancy (Fig. 3.10), it is not necessarily strange because the more the contraction tectonics make dense host rocks, the more the magma is able to ascend by buoyancy. This principle could be easily checked using the core sample density from geothermal wells.

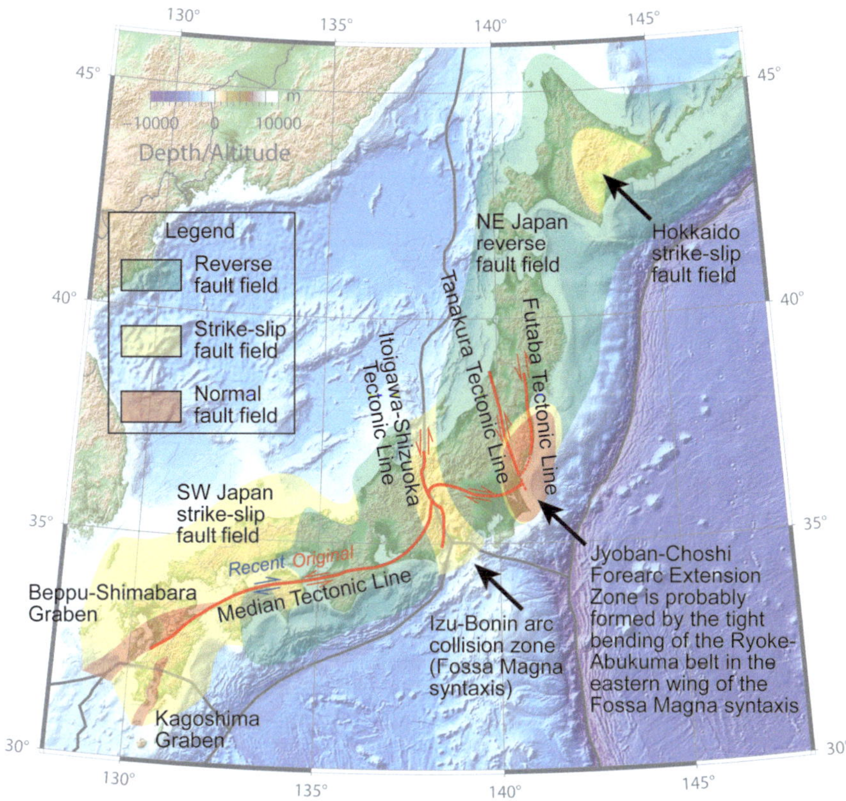

Fig. 3.13 Simplified and geologically interpreted map of three types of fault fields in Japan

The Geological Survey of Japan constructed a geothermal database on geothermal exploratory wells from the 33 fields in Japan (Yano et al. 1989). The fields numbered 1 to 6 are compiled from the normal fault fields such as the Beppu-Shimabara Graben and Kagoshima Graben, Kyushu (Fig. 3.13). The fields numbered 19 to 22 are compiled from the typical reverse fault fields such as the Hachimantai, Hanamaki, and Akinomiya fields, northern Honshu (Fig. 3.13). Therefore, two groups represent contrasting endmembers for the extension tectonic fields and contraction tectonic fields in Japan. We used the core density data measured under the wet-constraint state because all the cores were taken from geothermal fields in more or less hydrothermal conditions.

Figures 3.16 and 3.17 show 1,799 core density data from 27 wells in Kyushu and 1,182 core density data from 44 wells in northern Honshu, respectively. The northern Honshu cluster is biased to the higher-density side with depths than the Kyushu cluster. The superimposed comparison makes the density difference more

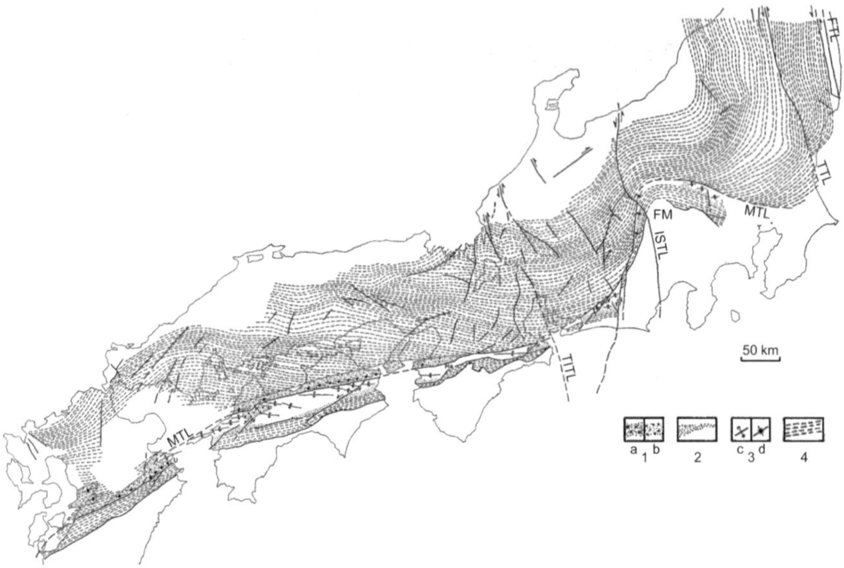

Fig. 3.14 A structural map of Southwest Japan (Hara et al. 1980). 1. Sediments after the Median Tectonic Line (a: Mesozoic, b: Tertiary), 2. Mikabu Greenrocks (ophiolite), 3. a: Crestal traces of upright folds in the Sambagawa Belt, b: Trough traces of upright folds, 4. Axial trends of major structure developed during the phase of upright folding, MTL: Median Tectonic Line, FTL: Futaba Tectonic Line, TTL: Tanakura Tectonic Line, ISTL: Itoigawa-Shizuoka Tectonic Line, TITL: Tsuruga-Isewan Tectonic Line, FM: Fossa Magna. The immediate north of the Median Tectonic Line indicates the high-temperature/low-pressure type Ryoke-Abukuma Metamorphic Belt

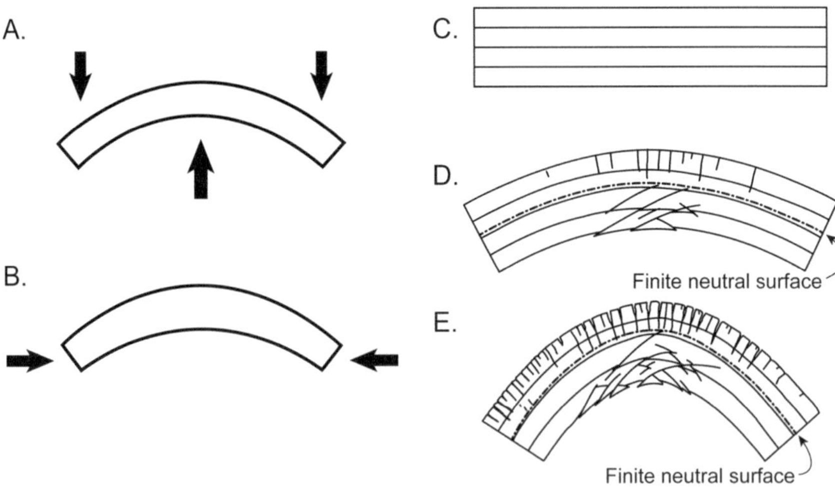

Fig. 3.15 An extension stress field yielded by a tight bending fold. **a** The driving force of a bending fold. **b** The driving force of a buckling fold. **c–e** Progressive development of a bending fold with extension fissures and conjugate shear faults bounded by a finite neutral surface under tangential longitudinal strain (Modified from Ramsey 1967; Kakimi 1978; Ramsay and Huber 1987)

Table 3.2 Comparison of magma intrusions beneath geothermal fields between the extension- and contraction tectonic settings. Modified from Muraoka and Yano (1998)

Regional field	Representative field	Tectonic setting	Major fault type	Magma pluton	Depth of top of pluton	References
Iceland	Nesjavellir	Extension tectonics Speading ridge	Normal fault	Lack (dike swarm)		Franzson (1995)
Salton Trough	Cerro Prieto	Extension tectonics Speading ridge	Normal fault	Lack (dike swarm)		Elders et al. (1997)
Tuscany	Larderello	Extension tectonics Back-arc spreading	Normal fault	Seismic K-horizon	3.50 km?	Romagnoli et al. (2010)
Taupo	Ohaaki	Extension tectonics Back-arc spreading	Normal fault	Magnetized body	4 km?	Soengkono (1995)
Kyushu	Hatchobaru	Extension tectonics Back-arc spreading	Normal fault	Thermal simulation	4 km?	Ehara and Morita (1995)
The Geysers	The Geysers	Contraction tectonics Ridge subduction	Strike-slip fault	The Geysers felsite	1.20 km	Thompson (1989)
Philippines	Tongonan	Contraction tectonics Subduction zone	Reverse fault	Mahiao diorite	1.60 km	Lovelock et al. (1982)
Kamchatka	Mutnovsky	Contraction tectonics Subduction zone	Reverse fault	Mutnovsky diorite	1.00 km	Kiryukhin (1993)
Northeast Japan	Nyuto	Contraction tectonics Subduction zone	Reverse fault	Nyuto diorite	1.34 km	NEDO (1992)
	Kakkonda			Kakkonda granite	1.95 km	Kato and Sato (1995)

evident (Fig. 3.18). When we draw the logarithmic regression curves, the difference is statistically evident (Fig. 3.19). The density of the rocks in such a shallow-depth range is easily changed by the compaction of the porosity so that the density rapidly increases with the depth. If we assume a magma intrusion with the density of 2.5 g/cm^3, the neutral buoyancy depth in northern Honshu is about 1 km but that in

Kyushu is about 4 km (Fig. 3.19). It is evident that the higher density of host rocks causes shallow-depth magma emplacement in contraction tectonic fields. However, the magma density also changes with depth; therefore, the following subsection will provide a more detailed discussion.

3.2.4 Estimates of the Density of Magma from the Major Chemical Compositions

Recently, the density of silicate melts can be roughly estimated from the major element chemical compositions using the coefficients of the volumetric expansion or compressibility of the individual oxide components as a function of pressure and temperature under an assumption of the addition rate basis (Lange and Carmichael 1990; Taniguchi 2001). Taniguchi (2001) introduced the method proposed by Lange and Carmichael (1990) in detail. As easily expected, an error in this kind of method is still significant. For instance, the density of the simplified magma system consisting of 64% diopside and 36% anorthite in the molecular weight percent is experimentally measured to be 2.70 g/cm^3. However, this method gives a density of 2.79 g/cm^3, which is 3.3% larger than the correct answer (Taniguchi 2001). However, if we think of the difficulty of measuring the density of high-temperature and high-pressure silicate melts in the molten magma chamber or laboratory, this order of error can be compromised.

We shall calculate five compositions of typical volcanic rocks from Japan (Table 3.3; Aoki 1978). This chapter's primary concern is the variation of the magma density caused by the depth or pressure. The temperature of the magma must be assumed in the calculation of the density of magma by this method. The temperature of the magma varies in a wide range between the liquidus and solidus temperatures, particularly wide on the mafic magma. It is not easy to find a simple function, but here, the temperature of magma is roughly and empirically given by the following equation as a function of the SiO_2 wt %:

$$T\text{magma} = -20 \cdot W_{SiO_2} + 2473 \quad (3.1)$$

Here, T_{magma} is the temperature of the magma in K, and W_{SiO2} is the SiO_2 content in the weight percent of the magma. For example, if the SiO_2 wt % of the magma is 70 wt % as dacite in Table 3.3, the dacitic magma is given to be 1073 K or 800 °C.

To calculate the magma density, we here use the chemical composition of dacite in Table 3.3 as an example. A calculation procedure is expressed in Table 3.4. The first step is that the chemical composition of oxides in wt % is converted to the molecular ratio divided by the molecular weight of the individual oxide component (Table 3.4). However, some oxides must be ignored because the necessary parameter

Table 3.3 Compositions of representative volcanic rocks from Japan (Aoki 1978). 100 wt % recalculation was made on an anhydrous basis

Series	Tholeiite	Calc-Alkali	Calc-Alkali	Calc-Alkali	Alkali
Rock name	Basalt	Andesite	Andesite	Dacite	Rhyolite
Analysed number	41	70	55	7	8
	wt%	wt%	wt%	wt%	wt%
SiO_2	51.64	58.48	63.29	69.71	73.13
TiO_2	0.87	0.96	0.74	0.70	0.20
Al_2O_3	18.87	16.91	16.13	14.98	13.70
Fe_2O_3	3.26	3.04	3.14	1.87	1.48
FeO	6.82	5.20	3.76	2.15	1.23
MnO	0.14	0.13	0.13	0.10	0.06
MgO	5.16	3.95	2.38	1.09	0.15
CaO	10.74	7.48	5.95	3.96	0.58
Na_2O	2.11	2.68	3.27	4.12	4.31
K_2O	0.29	1.02	1.05	1.14	5.11
P_2O_5	0.10	0.15	0.16	0.18	0.05
Total	100	100	100	100	100

is not given in Table 3.5, and/or the amount is negligible. Then, the molecular fraction is recalculated by the other oxides.

Equation (3.1) can be used to calculate magma's temperature, and the molecular weight of magma can be calculated from the following Eq. (3.2) using the molecular fraction of the individual oxides.

$$W_{liquid} = \left(X_{SiO_2}/1\right)\cdot M_{SiO_2} + \left(X_{Al_2O_3}/1\right)\cdot M_{Al_2O_3} + \left(X_{Fe_2O_3}/1\right)\cdot M_{Fe_2O_3} + (X_{FeO}/1)M_{FeO} + \left(X_{MgO}/1\right)\cdot M_{MgO} + (X_{CaO}/1)\cdot M_{CaO} + \left(X_{Na_2O}/1\right)\cdot M_{Na_2O} + \left(X_{K_2O}/1\right)\cdot M_{K_2O} \quad (3.2)$$

Here, W_{liquid} is the molecular weight of magma, X_i is the mole fraction of the component i, and M_i is the molecular weight of the component i. When we input real numbers to the Eq. (3.2), we obtain 65.0045 g/mole by the following formula (3.3).

$$W_{\text{liquid}} = (0.761/1)\cdot 60.082 + (0.096/1)\cdot 101.959 + (0.008/1)\cdot 159.687 + (0.02/1)\cdot 71.844 + (0.018/1)\cdot 40.303 + (0.047/1)\cdot 56.077 + (0.043/1)\cdot 61.977 + (0.008/1)\cdot 94.196 \quad (3.3)$$

The next step is to acquire the molecular volume of the individual oxide from the following Eq. (3.4):

$$V_{liquid} = \sum\left\{X_i\cdot\left[V_{1,1673\text{K}} + dV_i/dT\cdot(T - 1673\text{K}) + dV_i/dP\cdot(P - 1\text{b})\right]\right\} + X_{\text{Al}_2\text{O}_3}\cdot X_{\text{Na}_2\text{O}}\cdot\left(dV_{\text{Na}_2\text{O}\cdot\text{Al}_2\text{O}_3}/dP\right)_{1673K} \quad (3.4)$$

Table 3.4 Calculation procedure for the density of magma (Lange and Carmichael 1990; Taniguchi 2001)

	Chemical composition	Molcular weight	Molcular ratio	Ignore treatment	Molcular fraction	Temp	Molcular weight of magma	V_i at 10 MPa	D_{liquid} at 10 MPa	V_i at 100 MPa	D_{liquid} at 100 MPa
Expression	W_i	M_i	X_i		X_i	T	W_{liquid}		D_{liquid}		D_{liquid}
Unit	wt %	g/mole				K	g/mole	cm^3/mole	g/cm^3	cm^3/mole	g/cm^3
SiO_2	69.71	60.082	1.160	1.160	0.761			20.457		20.327	
TiO_2	0.7	80.865	0.009	–	–			–		–	
Al_2O_3	14.98	101.959	0.147	0.147	0.096			3.410		3.390	
Fe_2O_3	1.87	159.687	0.012	0.012	0.008			0.293		0.291	
FeO	2.15	71.844	0.030	0.030	0.020			0.238		0.237	
MnO	0.1	70.937	0.001	–	–			–		–	
MgO	1.09	40.303	0.027	0.027	0.018			0.178		0.178	
CaO	3.96	56.077	0.071	0.071	0.046			0.697		0.698	
Na_2O	4.12	61.977	0.066	0.066	0.044			1.045		1.036	
K_2O	1.14	94.196	0.012	0.012	0.008			0.309		0.304	
P_2O_5	0.18	141.941	0.001	–	–			–		–	
Total	100		1.536	1.525	1.000			26.626		26.462	
Result						1073	65.0045		2.441		2.456
Equation						Equation (3.1)	Equation (3.2)	Equation (3.4)	Equation (3.5)	Equation (3.4)	Equation (3.5)

Table 3.5 Parameters for the magma density estimation under high-pressure and high-temperature conditions (Taniguchi 2001; Lange and Carmichael 1990)

Oxide component	Mole volume	Coefficient of thermal expansion to mole volume	Coefficient of pressure compressibility to mole volume	Coefficient of pressure and temperature to mole volume
Expression	$V_{1,1673\ K}$	$(dV_1/dT)_{1b}$	$(dV_1/dP)_{1673\ K}$	$[(dV_1/dP)]/dT$
Unit	cm^3/mole	$10^{-3}cm^3$/mole · K	$10^{-4}cm^3$/mole · b	$10^{-7}cm^3$/mole · b · K
SiO_2	26.90	0.00	−1.89	1.3
TiO_2	23.16	7.24	−2.31	–
Al_2O_3	37.11	2.62	−2.26	2.7
Fe_2O_3	42.13	9.09	−2.53	3.1
FeO	13.65	2.92	−0.45	−1.8
MgO	11.45	2.62	0.27	−1.3
CaO	16.57	2.92	0.34	−2.9
Na_2O	28.78	7.41	−2.40	−6.6
K_2O	45.84	11.91	−6.75	−14.5
$Na_2O \cdot Al_2O_3$	–	–	10.18	–

Here, most of the parameters are given in Table 3.5. This equation consists of two parts: One is the term enclosed by sigma, which adds the partial molecular volumes of the individual oxides, and the other is the correction term; when Na_2O and Al_2O_3 are dominant, the addition rate basis does not work and the correction factor is required. However, the latter term is so small in many cases that it can be virtually neglected. Then, the D_{liquid} is obtained by the following Eq. (3.5).

$$D_{\text{liquid}} = W_{\text{liquid}}/V_{\text{liquid}} \tag{3.5}$$

The calculation of magma density is only shown for two pressure conditions in Table 3.4, but we actually calculated for many pressure conditions, as discussed below.

3.2.5 Buoyancy of Magma Induced from the Magma Compositions in Japan

Density curves estimated for the five magmas in Table 3.3 are shown in Fig. 3.20. The density curves of the upper crust are drawn from the regression curves in Fig. 3.19. These density estimates can only be obtained from the pressure rather than the depth. Therefore, we need a conversion between the pressure and depth. For example, a depth of 35 km to the Moho in continental regions corresponds to a pressure of about

1 GPa (1000 MPa or 10 kbar) by Wyllie (1971). This indicates 28.6 MPa/km as a conversion factor. This might be true in the deeper range of the crust. However, our concerns are about shallow-depth magma intrusions. Brace and Kohlstedt (1980) showed a pressure range from 24.0 MPa/km to 27.5 MPa/km in the shallow-depth crust. We have core density data (Figs. 3.16, 3.17, 3.18 and 3.19), and then, 25.0 MPa/km seems reasonable in Japan. Therefore, Fig. 3.20 shows the two vertical axes: a pressure in MPa and a depth in km with a conversion factor of 25.0 MPa/km.

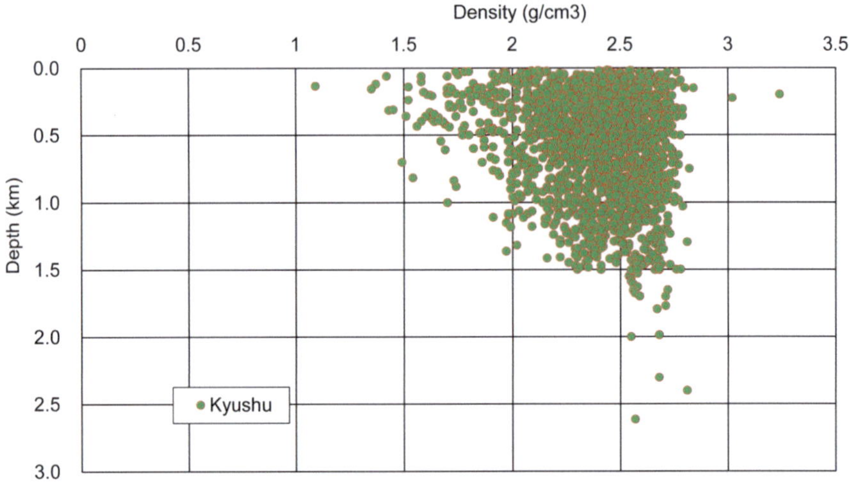

Fig. 3.16 1,799 core density data from 27 wells in Kyushu

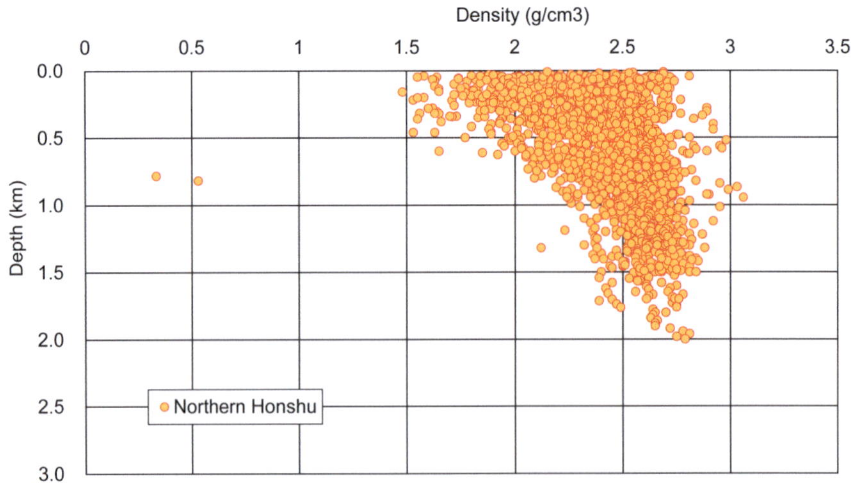

Fig. 3.17 1,182 core density data from 44 wells in northern Honshu

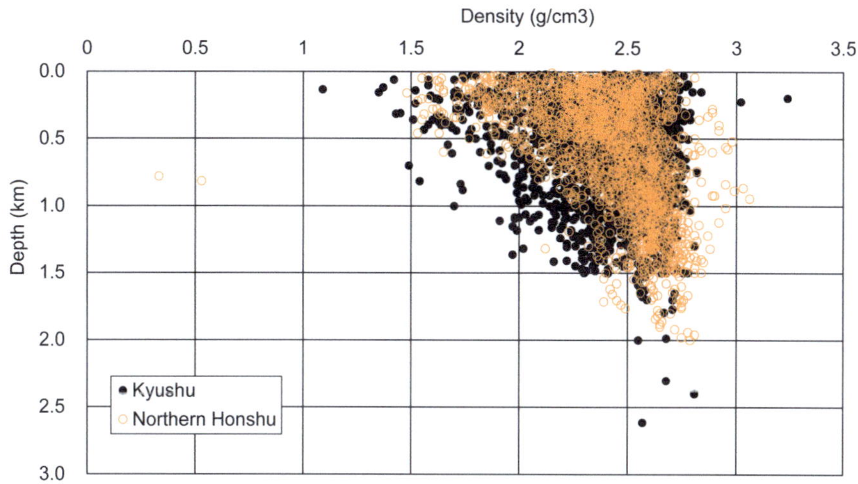

Fig. 3.18 Superimposed comparison of core density data between Kyushu and northern Honshu (modified from Muraoka and Yano 1998)

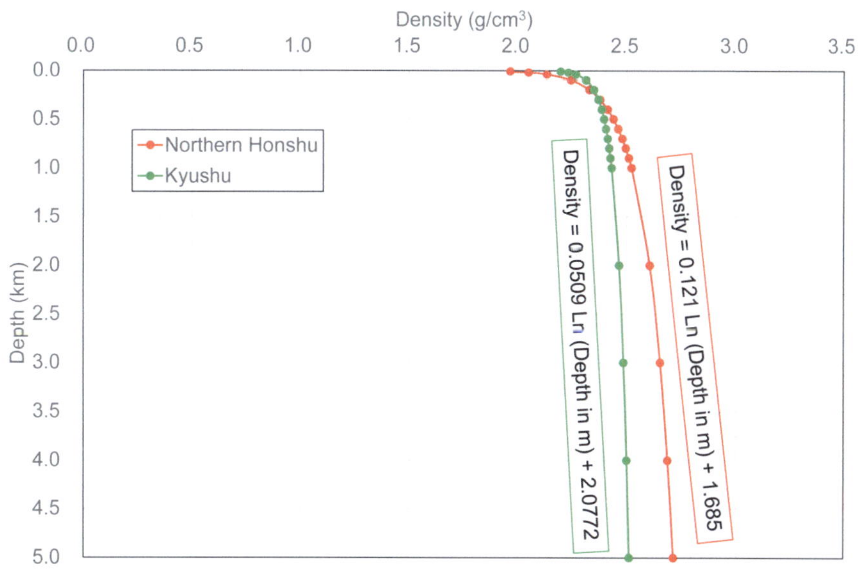

Fig. 3.19 Logarithmic regression curves of core density data between Kyushu and northern Honshu (modified from Muraoka and Yano 1998)

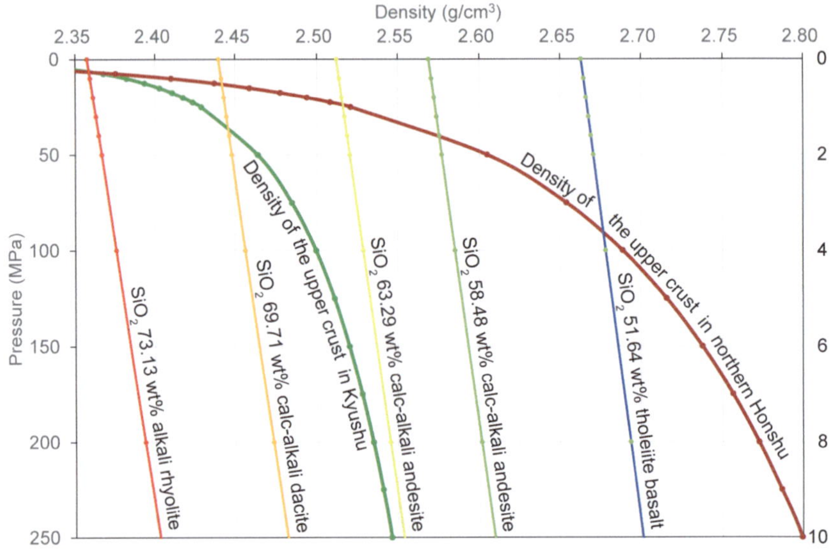

Fig. 3.20 Neutral buoyancy depths of high-level magma intrusions in Japan. Density curves of five magmas are calculated from the compositions in Table 3.3. The density curves of the upper crust are calculated from the regression curves in Fig. 3.19

The magma density systematically decreases with the SiO_2 contents of magmas (Fig. 3.20). When we carefully observe the five magma density curves, we find that the more silicic the magma, the more the curves show a low gradient in a strict sense. This is because the silicic magmas with high silica polymerization degrees have large compressibility with pressure rather than the mafic magmas. The neutral buoyancy depth of magma varies with the density of the host rocks and the density of magmas themselves (Fig. 3.20). The neutral buoyancy depth of magma is dramatically different between the low-density host rocks in the normal fault field like Kyushu and the high-density host rocks in the reverse fault field like northern Honshu. A neutral buoyancy depth is only observed on the rhyolite or dacite magmas at 0.4 km and 1.5 km in Kyushu on this diagram. On the contrary, the neutral buoyancy depth is observed on the silicic to mafic magmas at 0.4 km, 0.5 km, 1.0 km, 1.6 km, and 3.6 km in northern Honshu.

The meaning of the neutral buoyancy depth of magma is only a limit for the buoyancy and ascending. Actual emplacement depths could be, more or less, deeper than that due to a variety of factors such as cooling, crystallization, degassing, and encountering groundwater. A simplified broad interpretation of the neutral buoyancy depth of magma might be the depth of the top of the magma intrusions. Actually, the neutral buoyancy depths of 0.5 km, 1.0 km, and 1.6 km for dacitic to andesitic magmas in northern Honshu (Fig. 3.20) are close to the known depths of the top of the magma intrusions in the contraction tectonic fields (Table 3.2). We know that active volcanisms occur in Kyushu. The neutral buoyancy depths of magma in Kyushu seem slightly paradoxical (Fig. 3.20). However, andesite in Kyushu tends to be slightly silicic, such as 56–67 wt % in Sakurajima volcano and 51–67 wt % in Kuju volcano in the SiO_2 content (Ono et al. 1981). Therefore, the neutral buoyancy depths of the silicic end members of andesite are possibly close to the depth of 4 km in Fig. 3.20. In addition, this paradox could probably be filled with the pumping mechanism by the back-arc spreading or extension tectonics instead of the buoyancy mechanism (Fig. 3.10).

A remaining issue is that, even though the tholeiitic basalt magma has a high density, the magma is inevitably less dense than the consolidated equivalents. This principle is so important in such immature arcs as the Izu-Bonin arc and the southern Ryukyu arc where host rocks are always composed of consolidated equivalents such as tholeiitic gabbro, diabase, or basalt. In this situation, the tholeiitic basalt magma is still buoyant enough to form shallow-depth magma intrusions. The Hachijo-Jima geothermal power plant with the installed capacity of 3.3 MWe could have been constructed at the southern foot of the basaltic Higashi-Yama volcano, Hachijo-Jima Island in the Izu-Bonin arc in 1999 (This plant is now under replacing). The successful reason was probably ascribed to the fact that the tholeiitic basaltic magma has a high density; nevertheless, the host rocks are further denser because of the consolidated equivalents and the magma could have formed a shallow-depth intrusion.

3.3 Emplacement Depth of Magma Intrusions Versus Thermal Structure in Japan

To generate high-temperature hydrothermal reservoirs with 180–370 °C, enough for the steam flash type power generation, the high-temperature magmatic heat sources, which are magma intrusions, providing the high-temperature thermal fields with 650–1,200 °C at the shallower crust or their hot consolidated equivalents are inevitably required.

Figure 3.21 shows cooling temperature profiles of magma by thermal conduction. The analytical solution for the rectangular magma bodies (Lachenbruch et al. 1976) was used in the calculations. The rectangular magma bodies were stacked vertically close to the spheroid (Muraoka and Matsubayashi 1994). This is volumetrically comparable with a vertically axially symmetric spheroid with a 5 km lateral radius and a 2 km vertical radius. We assume the initial temperature of the magma is 850 °C, the latent heat of crystallization is 200 °C, and then the resultant initial temperature is 1,050 °C. The cooling temperature profiles were calculated every 20,000 years, incrementing from 20,000 to 200,000 years after the intrusion.

Commercial-based geothermal wells commonly do not exceed 3 km for cost-saving demands; therefore, we focus on the magma cooling temperature at a depth of 3 km. If the magma is emplaced at a depth of 10 km, as shown in the left figure of Fig. 3.21, this magma can raise the groundwater temperature at a depth of 3 km by only 30 °C throughout the entire cooling history of the magma. Hence, if the magma is emplaced at a depth of 5 km, as shown in the right figure of Fig. 3.21, this

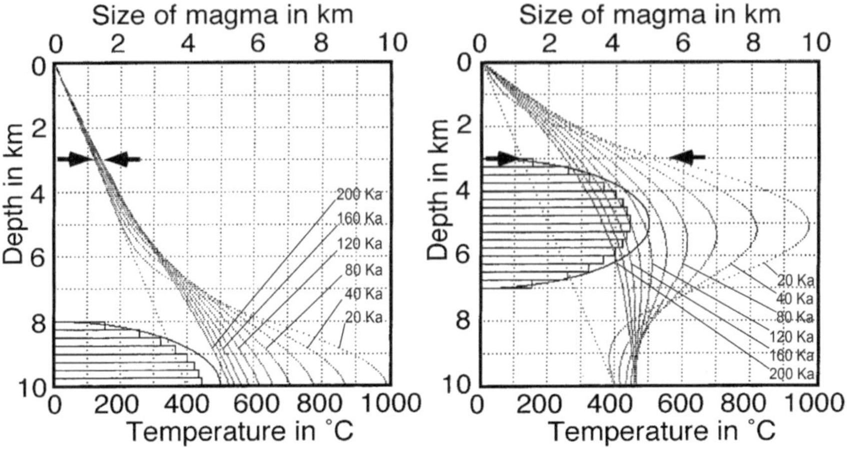

Fig. 3.21 Depth effect of geothermal heat sources (Muraoka 2018; Muraoka and Matsubayashi 1994). Curves labeled with age in Ka show cooling temperature profiles at the time after the intrusion of magma

magma can raise the groundwater temperature at a depth of 3 km by 400 °C or more throughout the entire cooling history of the magma.

Thus, the emplacement depth of the heat source magma intrusions is critically important for generating high-temperature geothermal systems. Deep magma intrusions can never form high-temperature geothermal reservoirs in the shallow-depth crust; only shallow-depth magma intrusions can generate high-temperature geothermal reservoirs in the shallow-depth crust.

3.4 Emplacement Depth of Magma Intrusions Versus High-Temperature Hydrothermal Systems in Japan

Section 3.1 described that almost all the high-temperature hydrothermal systems in Japan are associated with the Holocene or late-Pleistocene volcanoes. This suggests that subvolcanic magmas generate the high-temperature hydrothermal systems in Japan. Then, Sect. 3.2 mentioned magma ascent mechanisms, tectonic regimes in Japan, and young plutonic intrusions often penetrated by geothermal wells in the world during the 1980s and 1990s. However, they are biased to find in the contraction tectonic fields. In addition, the geneses of the shallow-depth magma intrusions by the buoyancy mechanism in Japan were discussed. Section 3.3 emphasized that only shallow-depth magma intrusions can generate high-temperature hydrothermal systems. We mention the mutual relationships between the high-temperature hydrothermal systems and shallow-depth magma intrusions in Sect. 3.4 for closing Chap. 3.

Figure 3.22 shows the temperature logging curves in the Hohi area, Kyushu, Japan (Tamanyu 1994). The Hohi area is the eastern part of the Beppu-Shimabara Graben in Kyushu, Japan, the typical normal fault field. Figure 3.23 shows the temperature logging curves in the Hachimantai area, northern Honshu, Japan (Tamanyu 1994). This is the typical reverse fault field.

The depth-temperature curves are quite different between the Hohi and Hachimantai areas (Figs. 3.22 and 3.23). Not any wells exceed 250 °C at a depth of 2,000 m in Kyushu, while about half of the wells exceed 250 ºC at a depth of 2,000 m in northern Honshu. This contrast indicates that the heat source magma intrusions are shallower in northern Honshu and deeper in Kyushu, controlled by the host rock density between the contraction and the extension tectonic fields.

Figure 3.24 shows a geothermal gradient map of Japan (Yano et al. 1999). The same tendency is observed from this map. Despite the dominant volcano-geothermal activity in Kyushu, geothermal gradients do not necessarily seem to be high in Kyushu. Hence, geothermal gradients are explicitly high in Northeast Japan. This is consistent with the hypothesis that the emplacement depths of the heat source magma intrusions are deeper in the extension tectonic fields and shallower in the contraction tectonic fields.

However, it only concerns the thermal structure and does not directly mean the total geothermal potential because we have another important parameter, permeability. The extension tectonic fields like Kyushu are blessed with high pcrmcability, and therefore, the total geothermal potential might be unexpectedly well balanced between the contraction and extension tectonic fields.

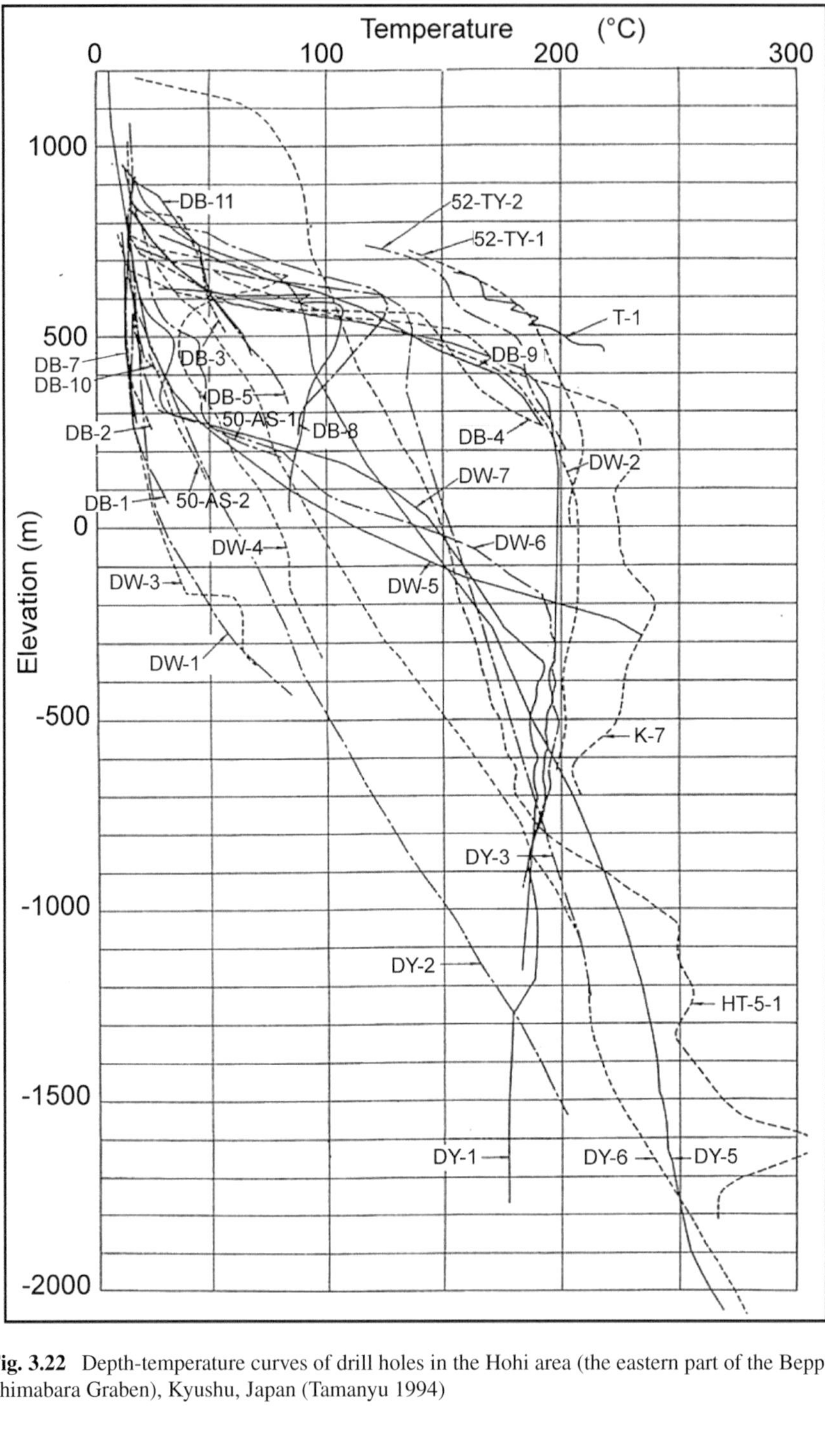

Fig. 3.22 Depth-temperature curves of drill holes in the Hohi area (the eastern part of the Beppu-Shimabara Graben), Kyushu, Japan (Tamanyu 1994)

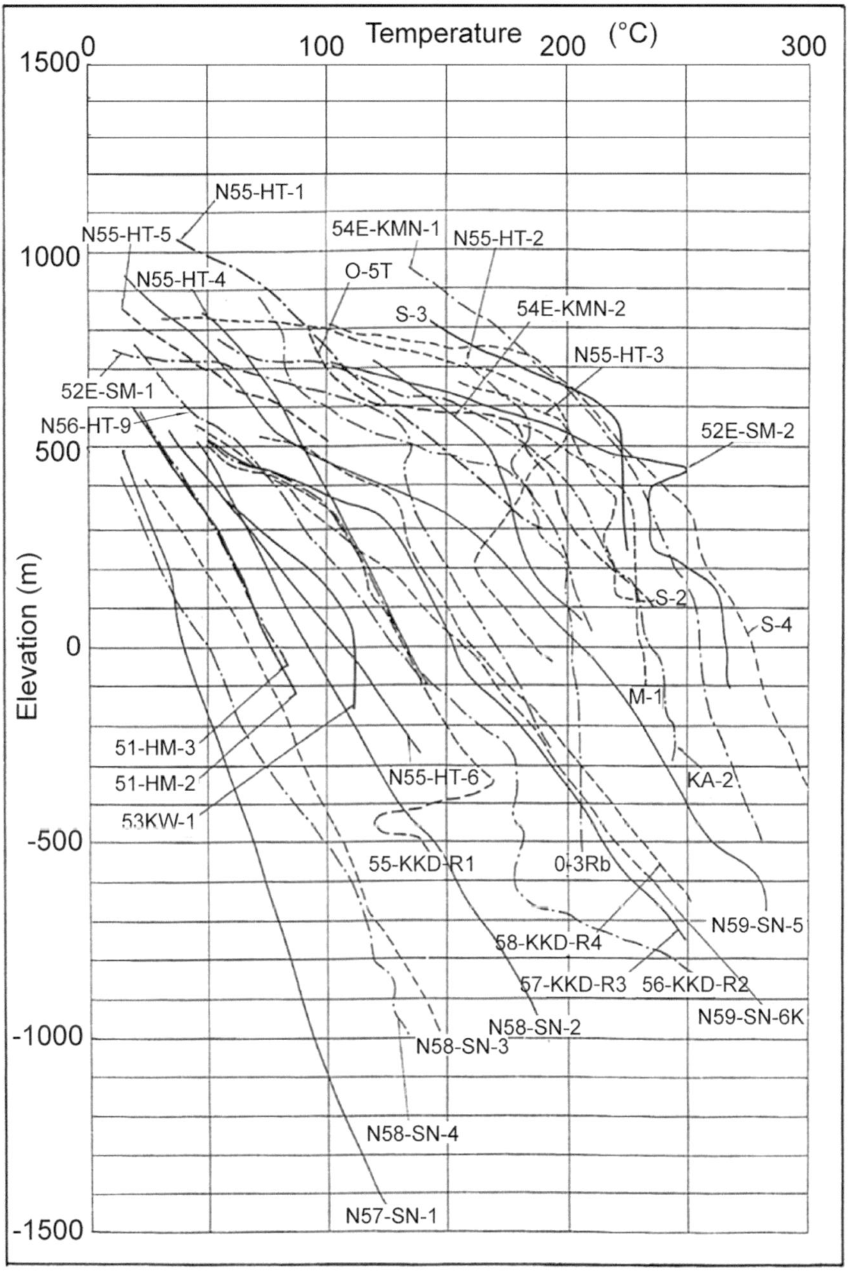

Fig. 3.23 Depth-temperature curves of drill holes in the Hachimantai area, northern Honshu, Japan (Tamanyu 1994)

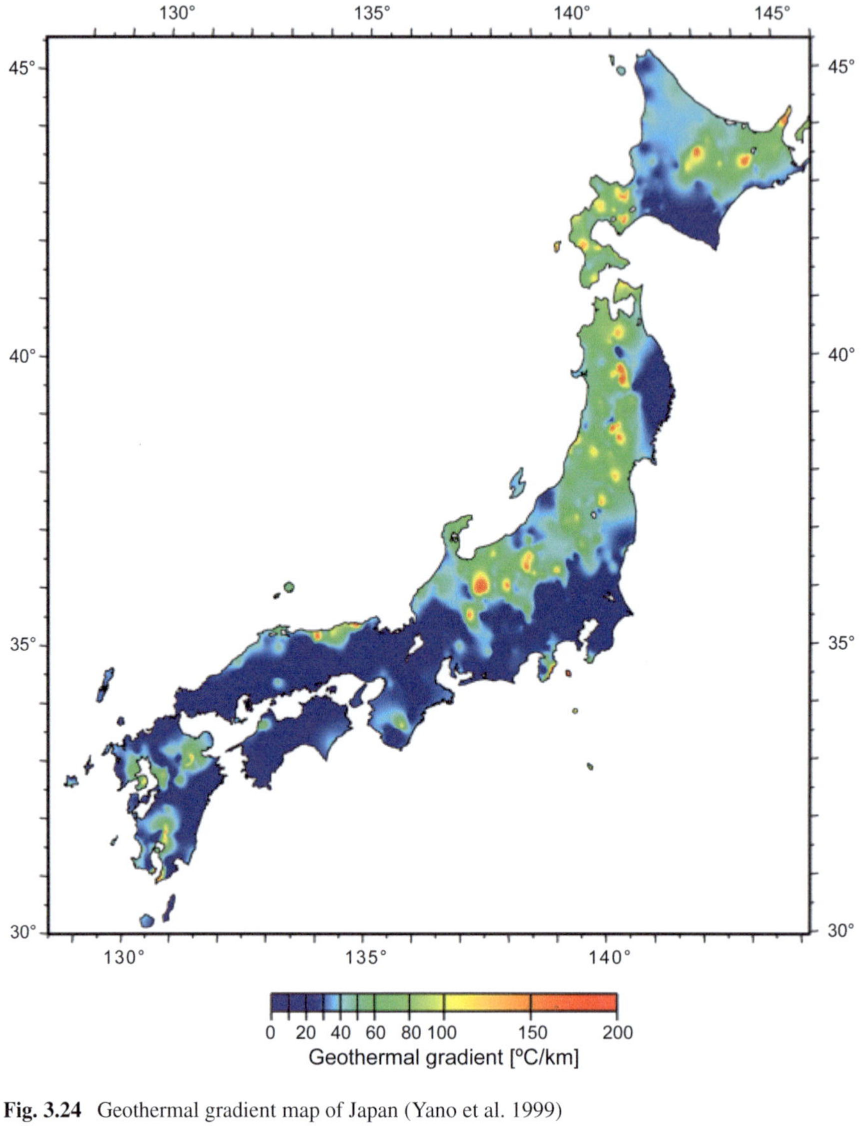

Fig. 3.24 Geothermal gradient map of Japan (Yano et al. 1999)

References

Amante C, Eakins BW (2009) ETOPO1 1 arc-minute global relief model: procedures, data sources and analysis. NOAA Technical Memorandum NESDIS NGDC-24. National Geophysical Data Center, NOAA. https://doi.org/10.7289/V5C8276M. Accessed 3 Jan 2023

Aoki K (1978) Chapter 5: origin of volcanic rocks in Japan. Iwanami Earth Science Course, Volume 3: Material Science of the Earth II, 260p (in Japanese). *Iwanami-Shoten*, 153–170

Brace WF, Kohlstedt DL (1980) Limits on lithospheric stress imposed by laboratory experiments. J Geophys Res 85:6248–6252

Ehara S, Morita K (1995) Thermal structure of Kuju Volcano, Central Kyushu, Japan and an estimation of the heat extraction rate from the volcanic geothermal reservoir. Proceedings of World Geothermal Congress ’95, Florence, Italy, 2719–2721
Elders WA, Williams AE, Biehler S (1997) What lies beneath the Cerro Prieto geothermal field? Geothermal Resources Council Transactions 21:171–179
Franzson H (1995) Nesjavellir high-temperature field. Guidebook for the World Geothermal Congress ’95, Excursion to Iceland, 25–32
Global Volcanism Program (2024) Volcanoes of the World. v.5.1.5., Venzke, E. (ed.)., Smithsonian Institution. Access, January 12, 2024
Hara I, Shoji K, Sakurai Y, Yokoyama S, Hide K (1980) Origin of the median tectonic line and its initial shape. Memoirs of the Geological Society of Japan, No.18, 27–49
Henley RH (1984) Chemical structure of geothermal systems. Fluid-Mineral equilibria in hydrothermal systems: reviews in economic geology, v 1, Society of Economic Geologists, 9–30
Japan Geothermal Energy Association (2000) Directory of facilities of geothermal power plants in Japan, 254 p (in Japanese)
Kakimi T (1978) Analyses of geologic structure. The Association for the Geological Collaboration in Japan, 240 p
Kato O, Sato K (1995) Development of deep-seated geothermal reservoir bringing the Quaternary granite into focus in the Kakkonda geothermal field, Northeast Japan. Resour Geol 45:131–144 (in Japanese with English abstract)
Kiguchi T, Imanishi K, Nakai M (2024) SH_{max} azimuth map (Japan, 0.2-degree mesh). Website on the crustal stress database. https://gbank.gsj.jp/crstress/english/
Kiryukhin AV (1993) High temperature fluid flow in the Dachny field of the Mutnovsky hydrothermal system, Russia. Geothermics 22:49 64
Lachenbruch AH, Sass JH, Munroe RJ, Moses TH Jr (1976) Geothermal setting and simple heat conduction models for the Long Valley caldera. J Geophys Res 81:769–784
Lange RL, Carmichael ISE (1990) Chapter 2. Thermodynamic properties of silicate liquids with emphasis on density, thermal expansion and compressibility. Rev Mineralogy 24:25–64
Lovelock BG, Cope DM, Balstar AJ (1982) A hydrothermal model of the Tongonan geothermal field. Proceedings pacific geothermal conference, Auckland, 259–264
Ministry of Environment (2022) Report of the utilization state of hot springs in Japan. https://www.env.go.jp/nature/onsen/pdf/3-5_p_1.pdf (in Japanese)
Miyashiro A (1973) Metamorphism and metamorphic belts. George Allen & Unwin, 492 p
Muraoka H (2008) K-Ar ages and chemistry of the Miocene intrusive rocks in the Hongu hot spring area, Wakayama Prefecture, Japan. Bull Geol Surv Japan 59(1/2):27–43 (in Japanese with English abstract)
Muraoka H (2018) Expectation for hydrology in the field of geothermal resource exploration and exploitation. J Groundw Hydrol 60:463–471 (in Japanese with English abstract)
Muraoka H, Ishizuka O, Chen J, Suojia, Miyazaki S (2006) A hypothesis on concealed magmatism in southern Tibet. Proceedings of the 7th Asian geothermal symposium, July 25–26, Qingdao, People’s Republic of China, 135–143
Muraoka H, Matsubayashi O (1994) Estimate of magma chamber size from the thickness of contact metamorphic aureoles. The memoirs of the geological society of Japan, No 43, 73–86 (in Japanese with English abstract)
Muraoka H, Sakaguchi K, Tamanyu S, Sasaki M, Shigeno H, Mizugaki K (2007) Atlas of hydrothermal systems in Japan. Survey of Japan, AIST, 110p. (in Japanese with English abstract)
Muraoka H, Yano Y (1998) Why neo-plutons are deeper in extension tectonic fields and shallower in contraction tectonic fields? Proceedings of the 20th New Zealand Geothermal Workshop, pp 109–114
New Energy and Industrial Technology Development Organization (1992) Report on the geothermal development promotion survey, No 27, “Eastern Tazawako”, 1021 p (in Japanese)

New Energy Development Organization (1987) Atlas of the results of the nationwide geothermal exploration survey project (2nd stage): volcano-related hydrothermal convection system type-1 Niseko Area, New Energy Development Organization, 92 p
Ono K, Soya T, Mimura K (1981) Volcano of Japan (Scale 1: 2,000,000, Map Series No.11, Second Edition). Geological Survey of Japan
Ramsey JG (1967) Folding and fracturing of rocks. McGraw-Hill, New York, 568 p
Ramsey JG, Huber MI (1987) The techniques of modern structural geology: Vol 2, Folds and Fractures. Academic Press Inc. (London) Ltd. 309–700 p
Reche Z, Fink JH, Agnon A (1993) Control of dike emplacement by vertically-varying tectonic stresses. EOS Transactions., AGU, Spring Meet. Suppl., 291
Romagnoli P, Arias A, Barelli A, Cei M, Casini M (2010) An updated numerical model of the Larderello–Travale geothermal system, Italy. Geothermics 39:292–313, https://doi.org/10.1016/j.geothermics.2010.09.010
Rubin AM (1990) A comparison of rift-zone tectonics in Iceland and Hawaii. Bull Volcanol 52:302–319
Sano O, Wakita H (1985) Geographical distribution of $^{3}He/^{4}He$ ratios in Japan: implications for arc tectonics and incipient magmatism. J Geophys Res 88:8729–8741
Soengkono S (1995) A magnetic model for deep plutonic bodies beneath the central Taupo Volcanic Zone, North Island, New Zealand. J Volcanol Geoth Res 68:193–207
Suzuki Y, Ioka S, Muraoka H (2017) Geothermal resource exploration by stream pH mapping in Mutsu Hiuchi Dake volcano, Japan. Energies 10:1009. https://doi.org/10.3390/en10071009
Tamanyu S (1994) Magma reservoirs from the viewpoint of geothermal modelling-Examples from the Hohi, Sengan and Kurikoma geothermal areas in Japan. Memoirs of the geological society of Japan, No 43, 141–155 (in Japanese with English abstract)
Taniguchi H (2001) Invitation to magma science. Shokabo 179 p (in Japanese)
Thompson RC (1989) Structural stratigraphy and intrusive rocks at The Geysers geothermal field. Geothermal Resour Council Trans 13:481–485
Tomiyama S, Umeda K, Hanamuro T, Takashima I, Hayashi S, Negishi Y, Masudome Y (2007) Alteration type and thermoluminescence ages of altered rocks in the Mutsu-Hiuchidake volcanic field, Shimokita peninsula, NE Japan. Japanese Mag Mineral Petrol Sci 36:111–121 (in Japanese with English abstract)
Uyeda S, Kanamori H (1978) Subduction of oceanic plates and formation of marginal seas. Kagaku (Science) 48:91–102 (in Japanese)
Wessel P, Luis JF, Uieda L, Scharroo R, Wobbe F, Smith WHF, Tian D (2019) The generic mapping tools version 6. Geochem Geophys Geosyst 20:5556–5564. https://doi.org/10.1029/2019GC008515
Wyllie PJ (1971) The dynamic earth: textbook in geosciences. Wiley, 416 p
Yamagishi H, Kohnoya M, Andoh S (1976) Explanatory text of the geological map of Japan, scale 1: 50,000 "Shimakotan". Geological Survey of Hokkaido, 24p (in Japanese with English abstract)
Yano Y, Suda Y, Tamanyu S (1989) Well data compiled from Japanese nation-wide geothermal surveys, Part 1 Core sample data. Report of Geological Survey of Japan, No. 271, 832 p
Yano Y, Tanaka A, Takahashi M, Okubo Y, Sasada M, Umeda K, Nakatsuka N (1999) Geothermal Gradient Map of Japan (Scale 1:3,000,000). Geological Survey of Japan, AIST, 6p (in Japanese with English abstract)

Chapter 4
Fundamentals of the Brittle–Ductile Transition in the Crust

Toshiaki Masuda and Yasutomo Omori

Abstract In this chapter we briefly review the results of HT–HP rock deformation experiments. Given that rocks in geothermal fields are of sufficiently high temperature to give rise to plastic deformation, we focus on the transition between brittle and plastic deformation. By evaluating the shape change of rock bodies that results from changes in temperature and pressure, we show that it is difficult to undertake in situ observations of ongoing plastic deformation in geothermal fields because of the negligibly small shape change that results from plastic deformation at expected natural strain rates of ~10^{-14}/s.

Keywords Brittle–ductile transition · Flow law · HT–HP rock deformation experiments · Plastic deformation · Stress

4.1 Brittle Deformation at Shallow Depths in the Crust

Brittle fracturing and frictional sliding occur in rocks at shallow depths in the crust. Our knowledge of these deformation mechanisms stems mainly from high-temperature and high-pressure deformation experiments (e.g., Griggs and Handin 1960; Heard et al. 1972; Karato and Wenk 2002; for details of the experimental techniques, see Sect. 4.2.1). Figure 4.1 shows the stress differences (= differential stresses) during experimental fracturing of the Westerly granite (data from Scholz 2002) and the level of frictional sliding in terms of confining pressure, as determined by Byerlee (1978). At confining pressures up to 600 MPa, the stress differences for fracturing are higher than those for frictional sliding. Although rocks in the shallow crust contain many fractures, the figure shows that frictional sliding is the main deformation mechanism in the shallow crust, because frictional sliding occurs at a

T. Masuda (✉) · Y. Omori
Shizuoka University, Shizuoka, Japan
e-mail: masuda.toshiaki@shizuoka.ac.jp

Y. Omori
e-mail: ohmori.yasutomo@shizuoka.ac.jp

J. Kasahara et al. (eds.), *Challenges in Geothermal Development*, Advances in Geological Science, https://doi.org/10.1007/978-981-95-0553-1_4

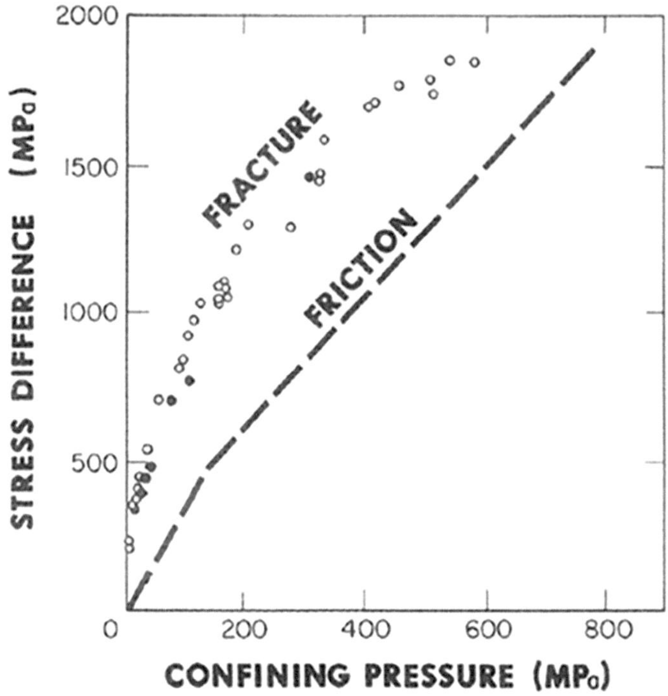

Fig. 4.1 Influence of pressure on the fracture strength of the Westerly granite and the frictional strength for an optimally oriented plane, as compiled by Scholz (2002). The vertical axis represents the stress difference ($\sigma_1 - \sigma_3 =$ differential stress), and the horizontal axis represents the confining pressure in MPa. For data sources, see Scholz (2002, p. 24)

lower stress difference than the does fracture formation. We note that the experiments of Byerlee (1978) regarding frictional sliding involved various rock types, including limestone, sandstone, granite, and gneiss, and the differential stress required to induce frictional sliding appears to be independent of rock type. For many more details on the brittle deformation of rocks, see Paterson and Wong (2005).

4.2 Ductile Deformation in the Deep Crust

At deeper levels in the crust (>10–15 km), rocks undergo ductile deformation rather than fracturing or frictional sliding. Here, the term "ductile" is used to denote a substantial change in shape of a given volume of rock without fracturing. This definition (e.g., Paterson and Wong 2005) is an essentially macroscopic or phenomenological one, and it takes no account of the microscopic mechanisms by which the deformation occurs. If ductile deformation is accommodated mainly by the microscopic movement of dislocations in crystal lattices, we prefer the term "plastic" flow.

4.2.1 Experimental Techniques

High-temperature (HT) and high-pressure (HP) deformation experiments started in the 1950s and became popular in the 1960s, thanks to new experimental techniques developed especially by D. T. Griggs and coworkers (e.g., Griggs and Handin 1960). Such HT–HP experiments can produce plastic flow in rocks. Figure 4.2 shows the specimen assembly for the solid-confining-media type apparatus (the Griggs type apparatus) used by Gleason and Tullis (1995). The cylindrical specimen is surrounded by various confining media (e.g., KCl/NaCl salt and pyrophyllite) that ensure an environment of deformation that allows the specimen to flow at a constant and uniform temperature and under uniform stress conditions, thus producing a homogeneous strain in the specimen. The specimen assembly is placed in the pressure vessel, and the ambient confining pressure is applied by pushing the outer piston using a hand pump. The axial force, applied by the inner piston, is driven by an electric motor. Thus, the axial deformation rate (~strain rate) is controlled by the gear system of the apparatus. The temperature is controlled by a graphite heater using electrical currents and is monitored by a thermocouple located next to the jacketed sample.

Figure 4.3 shows an example of a specimen before and after a modern HP–HT experiment (Stipp et al. 2006). The specimen after deformation is still cylindrical, thus demonstrating the occurrence of homogeneous deformation. In the earlier HP–HT experiments, the specimens after deformation always became barrel-shaped (see symbols of the specimens in Fig. 4.7), indicating inhomogeneous deformation. Figure 4.3 shows how experimental techniques have advanced over time.

4.2.2 Microstructures Produced by Plastic Deformation of Quartz

The most representative microstructure produced by plastic deformation of quartz aggregates is that of dynamically recrystallized grains. In deformation experiments of quartz rocks (Fig. 4.4), small newly recrystallized quartz grains appear mostly at the boundaries of pre-existing host quartz grains that show undulatory extinction as a result of the accumulation of dislocations during the plastic deformation.

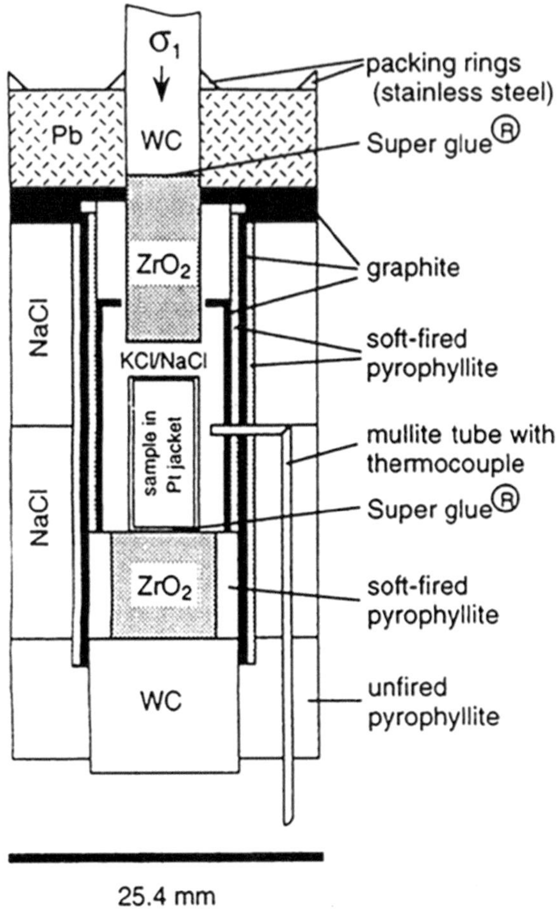

Fig. 4.2 Cross section of the specimen assembly (the molten salt cell) used by Gleason and Tullis (1995). WC indicates tungsten carbide. The sample at the center is jacketed by Pt and Ni and surrounded by various materials. Temperature is monitored by a thermocouple placed inside the inner graphite heater

4.2.3 Flow Law

A flow law can be represented by an equation that formulates the relationship between differential stress (σ) and strain rate ($\dot{\varepsilon}$). A typical flow law for rocks and minerals is given by the power law, expressed as follows:

$$\dot{\varepsilon} = A\sigma^n \exp\left(-\frac{Q + PV}{RT}\right)$$

Here, P is the pressure, T is the absolute temperature, R is the gas constant, Q is the activation energy, V is the activation volume, n is the stress exponent, and A is a constant (e.g., Karato 2008). P is usually ignored and deleted from the flow law when one considers crustal deformation, given the small activation volume of minerals and the small variations in pressure in the continental crust. Thus, the flow law becomes:

Fig. 4.3 Photograph of the Pt/Ni-jacketed specimen before (left) and after (right) the deformation experiment of Stipp et al. (2006). The cylindrical shape of the deformed specimen demonstrates that the deformation was homogeneous during the experiment. The bulge on the top and bottom of the deformed specimen is due to the plastic deformation of the upper and lower disk of the Pt/Ni jacket

$$\dot{\varepsilon} = A\sigma^n \exp\left(-\frac{Q}{RT}\right)$$

The flow law has three empirical parameters that should be determined experimentally: A, Q, and n. Q describes the influence of temperature, while n represents the influence of differential stress on the strain rate. If $n = 1$, the material is Newtonian, in which case the strain rate is proportional to the magnitude of differential stress.

As quartz is a common mineral in the crust, numerous researchers have performed deformation experiments on this mineral and proposed various parameters for the applicable flow law. Table 4.1 gives the values of A, Q, and n as determined in eight experimental studies. The values vary, and Fig. 4.5 shows how they differ in temperature–differential-stress space. The six panels in Fig. 4.5 show strain rates from $\dot{\varepsilon} = 10^{-16}$/s to $\dot{\varepsilon} = 10^{-8}$/s. Among the eight flow laws, the most popular may be that of Gleason and Tullis (1995). The stress exponent of Gleason and Tullis (1995) is 4. This means that a 10 × increase in differential stress gives rise to a 10,000 × increase in the strain rate. Thus, the strain rate appears to be highly sensitive to the magnitude of differential stress.

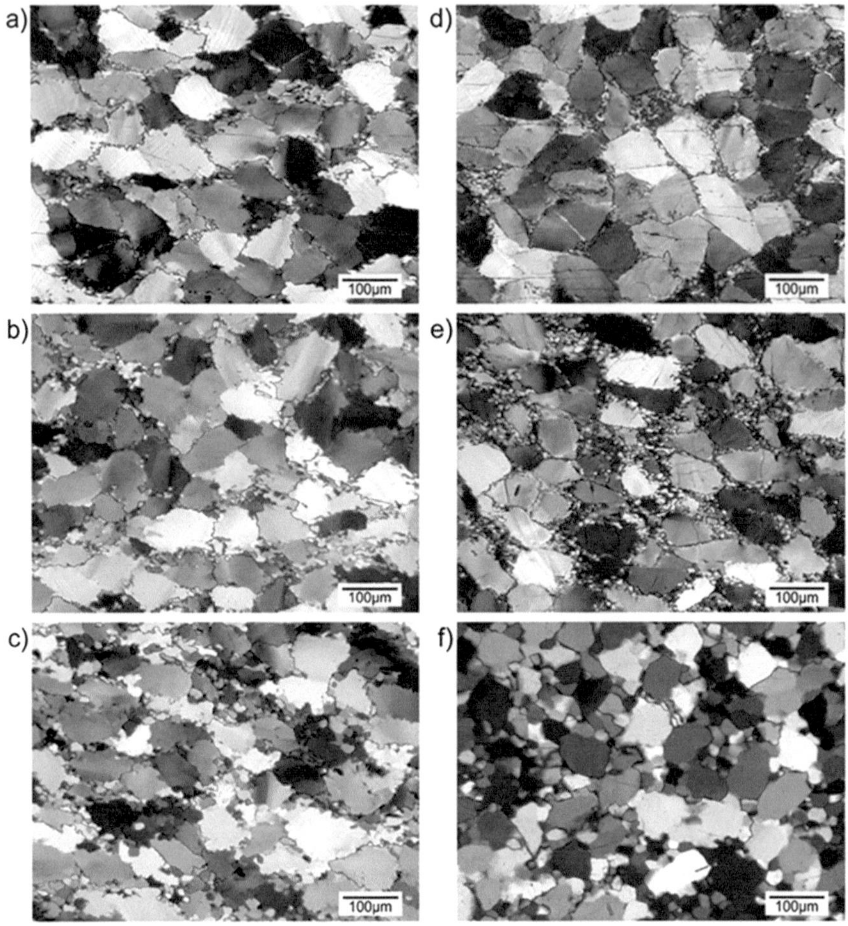

Fig. 4.4 Photomicrographs of experimentally deformed Black Hills quartzite taken under the optical microscope with crossed polars (Stipp et al. 2006). The confining pressure was 1.5 GPa for all experiments. **a–c** Water-added samples. **d–f** Vacuum-dried samples. The temperatures were **a** 800 °C, **b** 900 °C, **c** 950 °C, **d** 1000 °C, **e** 1100 °C, and **f** 1100 °C. The strain rate for **a–e** was ~1.9–2.4 × 10^{-6}/s and for **f** 2.2 × 10^{-7}/s. The axial compressive strain was **a** 41%, **b** 33%, **c** 29%, **d** 25%, **e** 28%, and **f** 7%. Small equant grains are dynamically recrystallized new grains produced during the experiments

Table 4.1 Values of *A*, *Q*, and *n* from eight previous studies. H & C = Hansen and Carter, J et al. = Jaoul et al., K & T = Kronenberg and Tullis, K et al. = Koch et al., P & L = Paterson and Luan, G & T = Gleason and Tullis, H et al. = Hirth et al., R & B = Rutter and Brodie

	Hansen and Carter (1982)	Jaoul et al. (1984)	Kronenberg and Tullis (1984)	Koch et al. (1989)	Paterson and Luan (1990)	Gleason and Tullis (1995)	Hirth et al. (2001)	Rutter and Brodie (2004)
log A (MPân*s)	−1.7	−2.54	−5.66	−5.94	−7.18	−4	−11.2	−4.93
Q (KJ/mol)	137	151	120	134	135	223	135	242
n	1.8	1.8	2.7	2.7	3.1	4	4	2.97
					*dry		* dry	

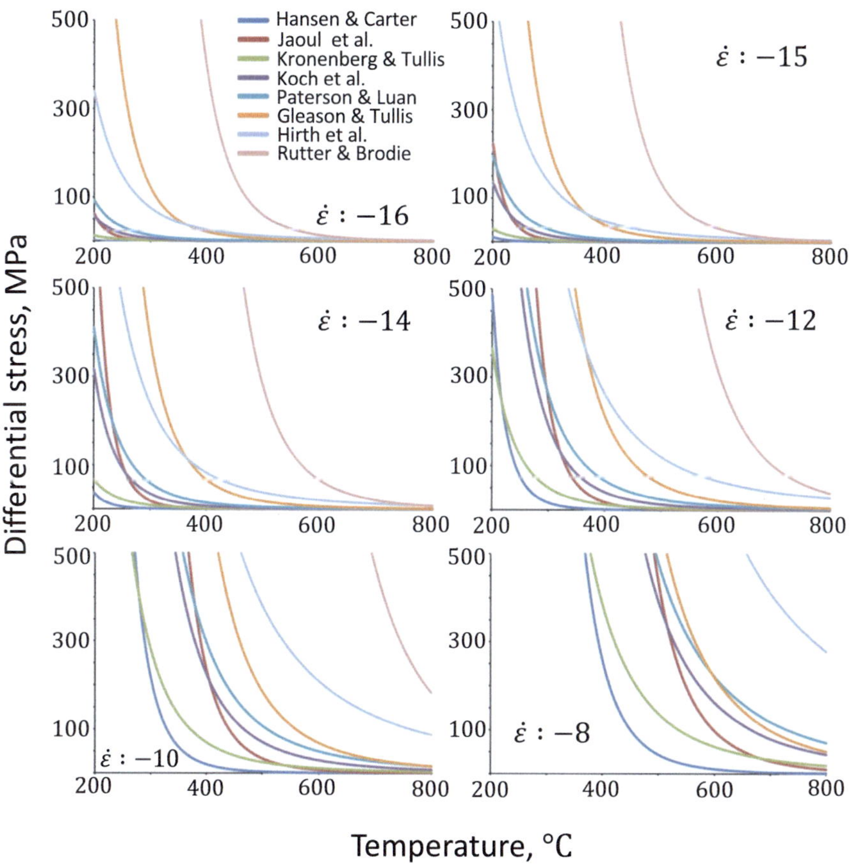

Fig. 4.5 Relationship between temperature (horizontal axis) and differential stress (vertical axis) at different strain rates from 10^{-16} to 10^{-8}/s (e.g., $\dot{\varepsilon} = -10$ corresponds to a strain rate of 10^{-10}/s)

4.3 Brittle–Ductile Transition

4.3.1 *Experimental Demonstration of the Brittle–Ductile Transition*

Heard (1960) demonstrated experimentally that at a constant strain rate (10^{-4}/s), a given rock (Solnhofen limestone in this case) can behave differently when pressure and temperature conditions differ, but with the ductile deformation always taking place on the higher P–T side of the brittle–ductile transition (Fig. 4.6). The data in Fig. 4.6 further show that the brittle–ductile transition in the uniaxial compression experiment occurred at a much lower pressure than in the case of a uniaxial extension experiment.

Tullis and Yund (1977) performed deformation experiments on the Westerly granite at various pressures and temperatures with a strain rate of 10^{-6}/s. The results are shown schematically in Fig. 4.7. In the figure, two dark vertical bars indicate

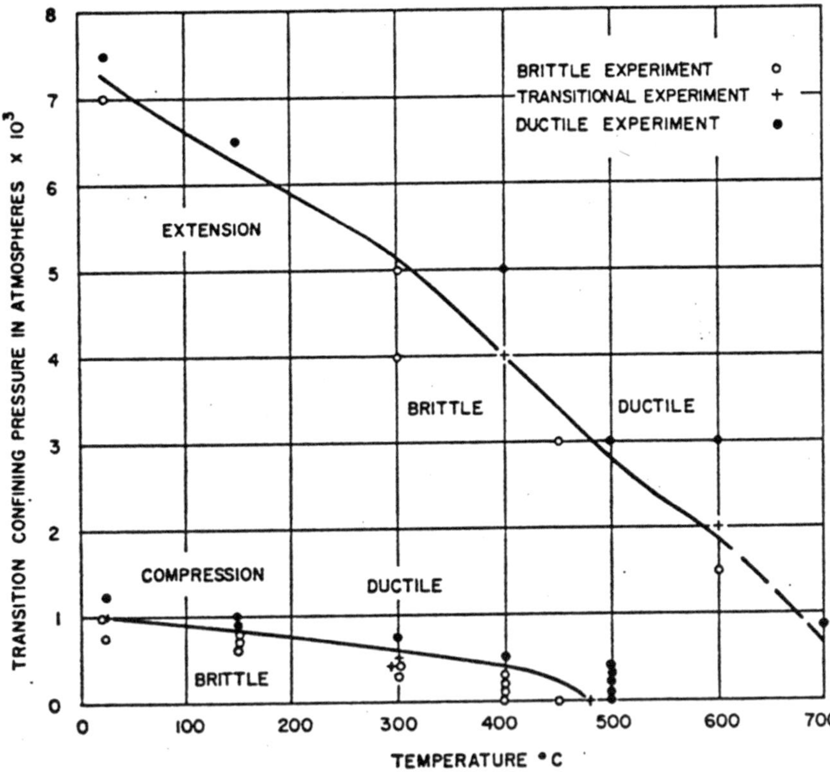

Fig. 4.6 Temperature–pressure conditions of brittle and ductile deformation of the Solnhofen limestone (after Heard 1960). Data for extension and compression are plotted

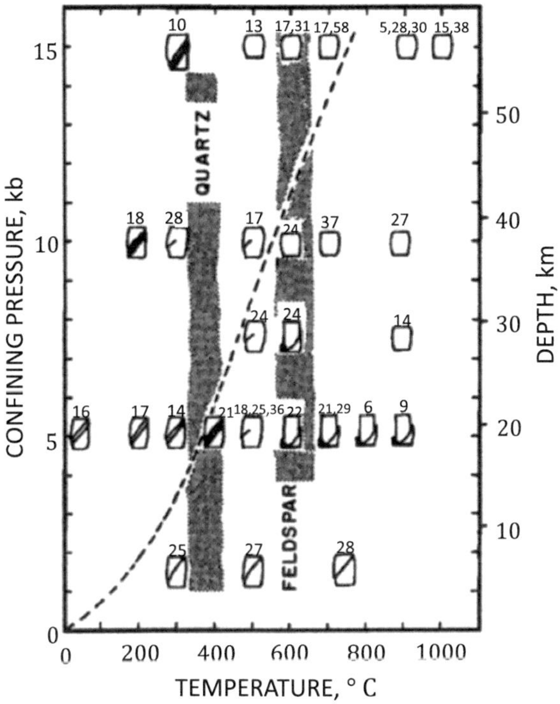

Fig. 4.7 Summary of experimental conditions and microscopic observations of the Westerly granite (after Tullis and Yund 1977). The pressure and temperature conditions of the experiments are indicated by the symbols. The numbers next to the symbols refer to the percentage shortening of the sample. All experiments were performed under dry conditions at a constant strain rate of 10^{-6}/s. The symbols indicate the types of macroscopic yield behavior for a single major fault at 1.5 kbar, multiple faults with a small offset, a single wide fault zone, a sharp fault extending only part way through the sample, a single very sharp fault with increasing temperature at 5 kbar, and relatively homogeneous shortening at higher temperatures and pressures. The transition region from dominantly microcracks to dominantly dislocation glide and climb is shown for quartz and feldspars. The dashed line is the calculated geotherm (model B) for San Andreas, from Lachenbruch and Sass (1973)

a change in the deformation mode of quartz (left bar) and feldspar (right bar). On the lower-temperature side of each bar, the minerals develop microcracks, while on the higher-temperature side, they undergo ductile deformation. The figure shows that granite is still brittle at temperatures above 800 °C at a confining pressure of 5 kb (500 MPa). The dashed line is the geotherm. This diagram clearly demonstrates that the two modes of deformation (brittle and ductile) can be assigned to different pressure–temperature conditions in the continental crust.

Paterson and Wong (2005) reviewed the nature and conditions of the brittle–ductile transition, provided examples, and assessed the transition in detail.

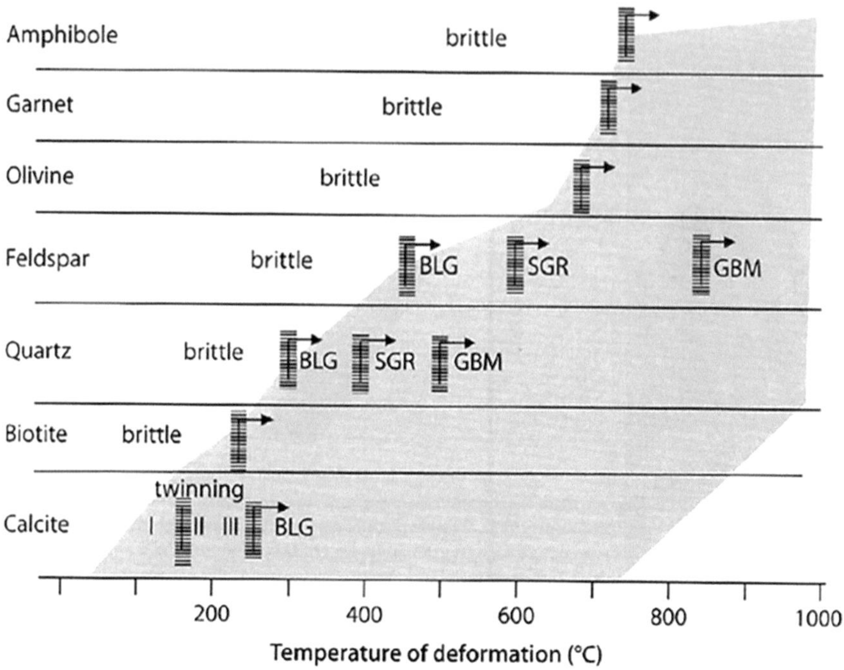

Fig. 4.8 Temperature dependence of deformation mechanisms for different minerals (after Passchier and Trouw 2005). Bars indicate transition zones. Arrows indicate the effect of strain rate. BLG (bulging), SGR (subgrain rotation), and GBM (grain boundary migration) are the main types of recrystallization. The shaded area is the domain of crystal-plastic deformation

4.3.2 Brittle–Plastic Transition for Various Minerals

The temperature of the brittle–plastic transition is variable, depending on the mineral. Passchier and Trouw (2005) provided the transition temperature for seven minerals (Fig. 4.8), without considering pressure because the influence of pressure appears to be negligible in the case of crustal deformation. Of the seven minerals considered, the temperature of the brittle–plastic transition is lowest for calcite (<200 °C) and highest for amphibole (~700 °C).

4.3.3 Stress Profiles of Faulting and Plastic Flow in Typical Continental Crust

Figure 4.9 shows schematic profiles of differential stress with respect to depth in typical continental crust (Passchier and Trouw 2005). The three straight lines indicate three different tectonic fields for faulting. T is the stress profile for the thrusting field,

S is for the strike-slip faulting field, and N for the normal-faulting field. They are all modes of brittle deformation. Such straight lines imply that the strength of rocks in the brittle field is controlled mostly by pressure and that it is largely independent of temperature in the crust. The curves in Fig. 4.9 represent stress profiles for plastic deformation at strain rates of 10^{-15}/s and 10^{-12}/s. As the typical strain rate of plastic deformation in orogenic belts is thought to be on the order of 10^{-14}/s (e.g., Pfiffner and Ramsay 1982), the differential stress for natural deformations occurs in the area between the two curves.

The shear resistance for iso-strain rate curves at deep levels in the crust is < 100 MPa, but the shear resistance increases rapidly with decreasing depth (Fig. 4.9). Although the influence of temperature is not shown in Fig. 4.9, it is appropriately incorporated by using a constant geothermal gradient. The important aspect of Fig. 4.9 is the depth at which the lines representing brittle fracturing intersect the curves representing dislocation creep. In general, at shallower depths, brittle deformation is prevalent at lower differential stresses, whereas plastic deformation is prevalent at greater depths.

Shimamoto et al. (1989) proposed a fault-zone model that includes semibrittle and semiductile regimes between the brittle and ductile regimes (Fig. 4.10). Given that rock is a mixture of various minerals, the brittle–ductile transition cannot be simply described. At a given temperature, some minerals deform plastically while

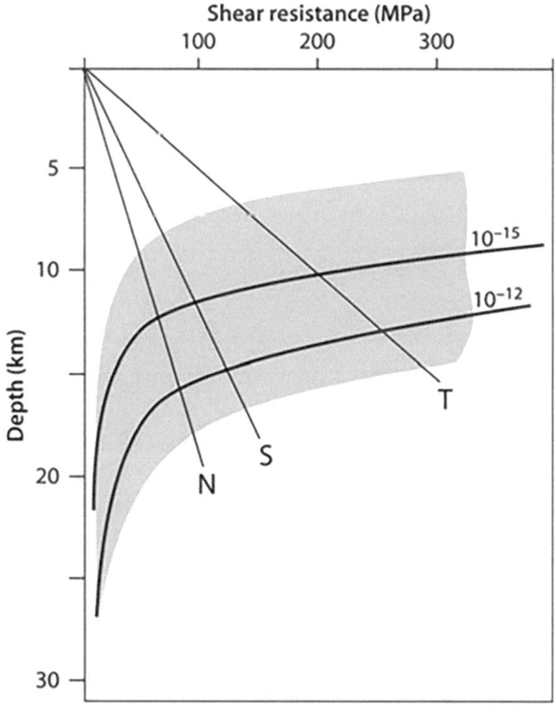

Fig. 4.9 Depth–strength diagram from Passchier and Trouw (2005) for quartz for a geothermal gradient of 30 °C/km and with hydrostatic fluid pressure on brittle faults (based on Sibson 1983, and Gleason and Tullis 1995). Straight lines show the strength of brittle fracturing for thrusts (T), strike-slip (S), and normal faults (N). The curved lines are for dislocation creep at strain rates of 10^{-15} and 10^{-12}/s. Shading indicates the estimated error for the dislocation creep curves due to uncertainties in the flow laws to be used and estimates of rheological parameters (Gleason and Tullis 1995)

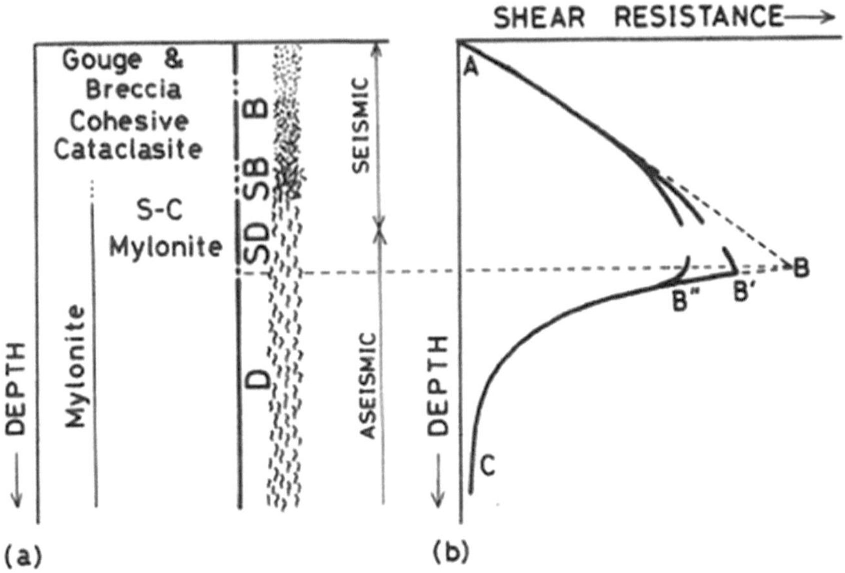

Fig. 4.10 Fault-zone model of Shimamoto (1989). **a** Depth relationships among the brittle (B), semibrittle (SB), semiductile (SD), and ductile (D) regimes together with their corresponding fault rocks. **b** Presumed depth profile of fault strength

other coexisting minerals behave in a brittle manner (see Fig. 4.8). Figure 4.10 shows the various levels of depth in the crust where mylonites and ultramylonites might be produced. Because these rock types usually contain both brittle and plastic minerals in various proportions, their strength (i.e., resistance to deformation) does not change sharply with depth.

4.3.4 Seismicity Beneath a Japanese Island

The Tohoku district in the northern part of Honshu, Japan, is a region of intense seismicity related to subduction of the Pacific Plate beneath the North American Plate (Fig. 4.11). The spatial distribution of hypocenters in the region shows two parallel arrays of seismicity that reflect the location of the plate boundary (Fig. 11b). The domain below these arrays is the Pacific Plate, which is thought to be rigid. Another region of seismicity exists in the interior of the continental crust, with the lower limit of this region being sub-parallel to the surface of the crust. A wedge-shaped aseismic domain occurs between this region and the two sloping arrays of hypocenters that mark the plate boundary. The rocks in this domain are considered to be ductile rather than rigid. Therefore, Fig. 4.11 shows an example of the brittle–ductile transition.

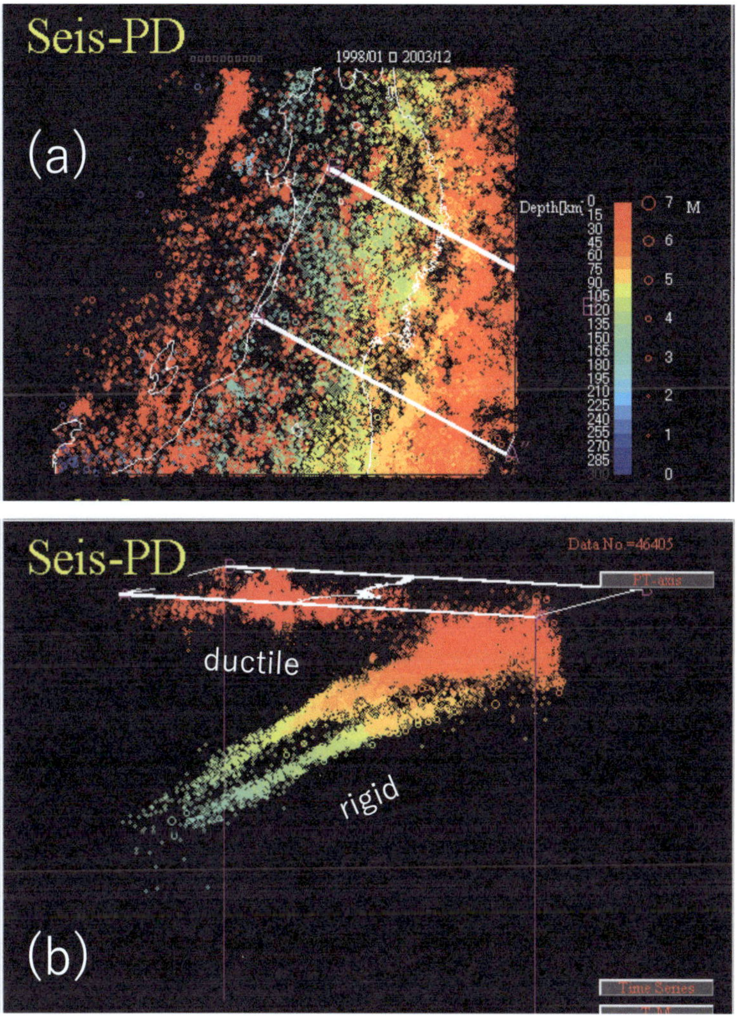

Fig. 4.11 Epicenters and three-dimensional hypocenters of northeastern Japan viewed with Seis-PD (a MATLAB program composed by R. Ikuta of Shizuoka University (we thank him for providing this figure)) using the data supplied by the Meteorological Agent of Japan. **a** Locations of epicenters. The thin white line shows the boundary between land and sea. The color and radius of each circle indicate the depth and magnitude of the earthquake, respectively (see the scales at right). The white rectangle indicates the data shown in (**b**). **b** Hypocenters of earthquakes in a three-dimensional cross section of the area marked by the white box in (**a**). "ductile" and "rigid" are attached by the authors. See text for details

4.4 Problems in Extrapolating Experimental Data to Natural Deformation

There are problems in applying the results of experimental deformation at a strain rate of 10^{-4} to 10^{-7}/s to the interpretation of natural deformation with a strain rate of 10^{-13} to 10^{-15}/s (Paterson 1987).

4.4.1 Influence of Water Content on the Flow Laws

Paterson (1987) placed special emphasis on the uncertainties surrounding the influence of water on the plastic deformation of quartz (i.e., hydrolytic weakening). After the discovery of hydrolytic weakening by Griggs and Blacic (1964, 1965), the mechanism by which quartz weakens has been of central interest to many researchers.

Hirth et al. (2001) proposed the following flow law incorporating the influence of water:

$$\dot{\varepsilon} = A \bullet f_{H_2O}^{m} \bullet \sigma^{n} \bullet \exp\left(-\frac{Q}{RT}\right)$$

Here, $f_{H_2O}^{m}$ is the fugacity of water (f_{H_2O}) and its exponent (m), and other symbols are as above. Despite this advance, the influence of water on the flow law remains poorly understood.

4.4.2 Influence of σ_2 on Rock Deformation

The fracture strength of rocks has been studied mostly using cylindrical specimens in uniaxial compression experiments, where σ_2 cannot be controlled independently of σ_3 because $\sigma_2 = \sigma_3$ during the deformation. Mogi (1967, 1971) addressed this problem by using a rectangular specimen set in an apparatus designed specifically for the application of stress in three orthogonal orientations, and it was clearly shown that σ_2 has an influence on the fracture strength of rocks. Kern (1971) investigated the triaxial plastic flow of Solnhofen limestone by using a system with three orthogonal pistons and a rectangle-shaped specimen. However, a lack of experimental data means the influence of σ_2 on brittle and plastic deformation remains poorly understood.

4.4.3 Three-Dimensional Flow Law

Natural deformation in orogenic belts occurs in three dimensions. As most HP–HT deformation experiments have axial symmetry, they are not directly applicable to investigations of natural deformation, and the flow laws developed for one-dimensional deformation, as described above, are inadequate for three-dimensional analysis. The three-dimensional power law (the von Mises–Levy equation) is given as follows:

$$\dot{\varepsilon}_{ij} = B \bullet II_{\sigma'}^{(n-1)/2} \bullet \sigma_{ij}' \bullet \exp\left(-\frac{Q}{RT}\right)$$

Here, $\dot{\varepsilon}_{ij}$ is the strain rate tensor, σ_{ij}' is the deviatric stress tensor, B is a constant, $II_{\sigma'}$ is the second invariant of the deviatric stresses, and other symbols are as for one-dimensional flow laws (e.g., Ranalli 1995; Karato 2008). It should be emphasized that the three-dimensional flow law is formulated based on the relationship between the strain rate tensor and the deviatoric stress tensor. As mechanical properties are usually described in terms of principal stresses ($\sigma_1, \sigma_2, \sigma_3$), we are not familiar with expressions using deviatoric stresses (σ_1', σ_2', σ_3'). It is possible that the magnitude of Q depends on the stress rather than being constant. These uncertainties show that our understanding of rock deformation is incomplete.

4.5 Comments on Deformation in the Kakkonda Geothermal Field

4.5.1 Seismicity in the Kakkonda Geothermal Field

According to the seismic analysis of Tosha et al. (1998), microearthquakes in the Kakkonda geothermal field (northern Honshu, Japan) are concentrated in a narrow zone around two intrusions of dacite at depths of <2 km. Such a nonuniform occurrence of earthquakes indicates that the crustal structure in the geothermal field close to magmatic and volcanic rocks is complicated and that the three-dimensional stress field is not uniform. Thus, in this case, it is difficult to apply any general scheme for the deformation of rocks in typical crust. The stress distribution in a geothermal field might be strongly inhomogeneous, and hypocenters are likely to be localized in narrow zones or close to intrusive rocks.

4.5.2 Transformation Superplasticity

In the branch of materials science that deals with ceramics, superplasticity is characterized by two behaviors (e.g., Johnson 1970): the ability to be plastically deformed without fracturing up to higher strains (>>100% strain) and the ability to be plastically deformed at very low differential stresses. Consequently, transformation superplasticity can be defined as superplasticity that takes place during the transformation of materials and chemical reactions. During such a transformation, the atoms in the crystal lattice become mobile, meaning that plastic deformation occurs easily due to the enhanced diffusivity of atoms and active dislocation movements in the crystal lattice. Poirier (1982) used the term "transformation plasticity" and Ruff and Kanamori (1983) used the term "transformation superplasticity" for the same behavior of rocks and minerals.

Transformation superplasticity may take place in geothermal fields, where high-temperature plutonic bodies cause ongoing metamorphism in the surrounding host rocks (e.g., Watanabe et al. 2017). However, as shown in Sect. 4.5.3, it might be difficult to find evidence for the occurrence of superplasticity.

4.5.3 Plastic Deformation vs Thermal Expansion

As the plastic deformation of rocks occurs at a low rate, it is difficult to directly observe such deformation in deep rocks. Here, we present an example of change in shape that might advance our understanding of plastic deformation.

We consider a 10-km-long cylinder that is elongated by 1 cm per year. As the average strain rate ($\dot{\varepsilon}$) is calculated as $\dot{\varepsilon} = \varepsilon/t$, we obtain:

$$\dot{\varepsilon} = \frac{\varepsilon}{t} = \frac{\ln\left(\frac{10000.01}{10000}\right)}{3 \times 10^7 (s)} \sim \frac{10 \times 10^{-7}}{3 \times 10^7} \sim 3 \times 10^{-14}/s$$

Here, we use the natural logarithm of strain and a time unit of 1 s. The strain rate of $3 \times 10^{-14}/s$ is almost equivalent to the estimated natural strain rate of orogenic belts (e.g., Pfiffner and Ramsay 1982). This strain rate corresponds to an elongation of 0.01 μ m per hour for a 1-m-long cylinder. Such a deformation would be impossible to detect by eye, and the 1-m-long rock cylinder would appear to be a rigid body.

We consider the thermal expansion of the 1-m-long rock cylinder as the temperature changes. The coefficient of thermal expansion for minerals is usually on the order of 10^{-6} to 10^{-5} /°C (e.g., Morimoto et al. 1975; Fei 1995), and we use this value for our rock cylinder. When the temperature increases by 1 °C, the 1-m-long rock cylinder becomes elongated by 1–10 μ m. Therefore, if the temperature change occurs over a short time (e.g., 1 h), the apparent change in shape corresponds to a strain rate on the order of 3×10^{-9}/s. Thus, the apparent change in shape of a rock body in a geothermal field due to a change in temperature is much faster than that due to plastic deformation. We note that the temperature-induced change in shape is not a plastic deformation, because dislocations are not principally responsible for accommodating the change in shape.

4.5.4 Expansion vs Compression

Minerals shrink with decreasing temperature and expand with decreasing pressure. Here, we consider a volume change in rocks exhumed from deeper levels to the Earth's surface. The compressibility of rocks is on the order of 10–100 GPa (e.g., Morimoto et al. 1975; Knittle 1996), and the coefficient of thermal expansion is on the order of 10^{-5}to10^{-6}/°C (e.g., Morimoto et al. 1975; Fei 1995).

As an example of change in shape, we consider a body of rock at a temperature of 500 °C and a pressure of 400 MPa (~10 km depth) that is exhumed to the surface of the Earth (20 °C and 0.1 MPa). In such a case, the magnitude of thermal shrinkage (on the order of 10^{-3}to10^{-4}) is larger than the magnitude of expansion due to the decrease in pressure (on the order of 10^{-4}to10^{-5}), resulting in shrinkage of the rock body. The rocks at the surface of Earth are expected to have many open cracks or other cavities, leading to higher permeability and lower strength than rocks at deeper levels.

4.5.5 Metamorphic Reactions Under Stress

In the Kakkonda geothermal field, the temperature (>250 °C) is high enough to drive metamorphism in the rocks around plutons (Muraoka et al. 1998). If the differential stress is sufficiently high, plastic deformation is possible even at depths of < 10–15 km, shallower than the brittle–plastic transition for normal crust. However, as the strain rate is presumably very low, the plastic deformation is unlikely to be measurable. As chemical reactions may take place in such rocks to produce new metamorphic minerals, the existence of differential stress may be discernible if a preferred orientation of platy metamorphic minerals (e.g., muscovite and chlorite) is detected as the minerals nucleate and grow (e.g., Masuda and Omori 2021). Such a preferred orientation is probably produced under the influence of differential stress.

4.5.6 Difficulties in Measuring the in Situ Stress in the Crust

Finally, we consider stress in the crust. If the in situ stress can be appropriately measured in rocks, we might advance our understanding of the brittle deformation in an area of intense seismic activity or the metamorphic processes that occur in geothermal fields such as Kakkonda and directly evaluate the ongoing brittle and ductile deformation. Although several methods have been proposed to measure in situ stresses directly (e.g., Zoback 2007; Zang and Stephansson 2010), it seems to be difficult to perform such precise stress measurements in a geothermal field.

References

Byerlee J (1978) Friction of rocks. Pure Appl Geophys 116:615–626

Fei Y (1995) Thermal expansion. In: Ahrens AJ (ed), Mineral physics & crystallography, a handbook of physical constants. AGU Reference Shelf 2, 29–44

Gleason GC, Tullis J (1995) A flow law for dislocation creep of quartz aggregates determined with the molten salt cell. Tectonophysics 247:1–23

Griggs DT, Blacic JD (1964) The strength of quartz in the ductile regime (abstr.) EOS. Trans American Geophys Union 45:102–103

Griggs DT, Blacic JD (1965) Quartz—anomalous weakness of synthetic crystals. Science 147:292–295

Griggs DT, Handin J (1960) Rock deformation. Geol Soc America Memoir 79, 382 p

Hansen FD, Carter NL (1982) Creep of selected crustal rocks at 1000 MPa. EOS Trans Amer Geophys Union 63:437

Heard HC (1960) Transition from brittle fracture to ductile flow in Solnhofen limestone as a function of temperature, confining pressure, and interstitial fluid pressure. In: Griggs D, Handin J (eds) Rock deformation. Geological Society of America Memoir, 79, pp 193–226

Heard HC, Borg IY, Carter NL, Raleigh CB (1972) Flow and fracture of rocks. Geophysical Monograph, 16, American Geophysical Union, Washington D. C., 352 p

Hirth G, Teyssier C, Dunlap WJ (2001) An evaluation of quartzite flow laws based on comparisons between experimentally and naturally deformed rocks. Int J Earth Sci 90:77–87

Jaoul O, Tullis J, Kronenberg A (1984) The effect of varying water content on the creep behavior of Heavitree quartzite. J Geophys Res 89:4298–4312

Johnson RH (1970) Superplasticity. Metall Rev 146:115–134

Karato S (2008) Deformation of earth materials—an introduction to the rheology of solid earth. Cambridge University Press, Cambridge, p 463

Karato S, Wenk H-R (2002) Deformation of minerals and rocks. Reviews in mineralogy & geochemistry, 51, Mineralogical Society of America, Washington D.C., 420 p

Kern H (1971) Dreiaxiale Verformungen an Solnhofener Kalkstein im Temperaturbereich von 20–50 °C. Röntgenographische Gefügeuntersuchungen mit dem Texturgoniometer. Contribution Mineral Petrol 31:39–66

Knittle E (1996) Static compression measurements of equations of state. In: Ahrens AJ (ed), Mineral physics & crystallography, a Hnsdbook of physical constants. AGU Reference Shelf 2, pp 98–142

Koch PS, Christie JM, Ord A, George Jr.RP (1989) Effect of water on the rheology of experimentally deformed quartzite. J Geophys Res 94:13975–13996

Kronenberg A, Tullis J (1984) Flow strength of quartz aggregates: grain size and pressure effects due to hydrolytic weakening. J Geophys Res 89:4281–4297

Lachenbruch AH, Sass JH (1973) Thermo-mechanical aspects of the San Andreas fault system. In: Proceedings of the conferences on tectonic problems of the san Andreas fault system, Publ. Geol. Sci., 13, edited by Kovack, R.L., Nur, A., pp. 192–203, Stanford University, Stanford, California

Masuda T, Omori Y (2021) A kinetic model of oriented nucleation under nonhydrostatic stress: implications for the preferred orientation of columnar and platy minerals in metamorphic rocks. J Struct Geol 176:104957

Mogi K (1967) Effect of intermediate principal stress on rock failure. J Geophys Res 72:5117–5131

Mogi K (1971) Effect of the triaxial stress system on the failure of dolomite and limestone. Tectonophysics 11:111–127
Morimoto N, Sunagawa I, Miyashiro A (1975) Kobutsugaku. Iwanami, Tokyo, 640 p. (Textbook of mineralogy written in Japanese)
Muraoka H, Uchida T, Sasada M, Yagi M, Akaku K, Sasaki M, Yasukawa K, Miyazaki S, Doi N, Saito S, Sato K, Tanaka S (1998) Deep geothermal resources survey program: igneous, metmorphic and hydrothermal processes in a well encountering 500 °C at 3729 m depth, Kakkonda, Japan. Geothermics 27:507–534
Passchier CW, Trouw RAJ (2005) Micro-tectonics. Springer, Berlin Heidelberg, p 289
Paterson MS (1987) Problems in the extrapolation of laboratory rheological data. Tectonophysics 133:33–43
Paterson MS, Luan FC (1990) Quartzite rheology under geological conditions. In: Knipe RJ, Rutter EH (eds), Deformation mechanisms, rheology and tectonics. Geological Society Special Publication, 54, pp 299–307
Paterson MS, Wong T-F (2005) Experimental rock deformation—the brittle field, 2nd edn. Springer, Berlin Heidelberg, p 347
Pfiffner OA, Ramsay JG (1982) Constraints on geological strain rates: arguments from finite strain states of naturally deformed rocks. J Geophys Res 87:311–321
Poirier JP (1982) On transformation plasticity. J Geophys Res 87:6791–6797
Ranalli G (1995) Rheology of the earth, 2ed. Chapman & Hall, London, p 413
Ruff L, Kanamori H (1983) Seismic coupling and uncoupling at subduction zones. Tectonophysics 99:99–117
Rutter EH, Brodie KH (2004) Experimental intracrystalline plastic flow in hot-pressed synthetic quartzite prepared from Brazilian quartz crystals. J Struct Geol 26:259–270
Scholz CH (2002) The mechanics of earthquakes and faulting. Cambridge University Press, Cambridge, p 496
Shimamoto T (1989) The origin of S-C mylonites and a new fault-zone model. J Struct Geol 11:51–64
Sibson RH (1983) Continental fault structure and the shallow earthquake source. J Geol Soc London 140:741–769
Stipp M, Tullis J, Behrens H (2006) Effect of water on the dislocation creep microstructure and flow stress of quartz and implications for the recrystallized grain size piezometer. J Geophys Res 111:B04201. https://doi.org/10.1029/2005JB003852
Tosha T, Sugihara M, Nishi Y (1998) Revised hypocenter solutions for microearthquakes in the Kakkonda geothermal field, Japan. Geothermics 27:553–571
Tullis J, Yund RA (1977) Experimental deformation of dry Westerly granite. J Geophys Res 82:5705–5718
Watanabe N, Numakura T, Sakaguchi K, Saishu H, Okamoto A, Ingebritsen SE, Tsuchiya N (2017) Potentially exploitable supercritical geothermal resources in the ductile crust. Nat Geosci 10:140–144
Zang A, Stephansson O (2010) Stress field of the earth's crust. Springer, Dordrecht, p 322
Zobach MD (2007) Reservoir geodynamics. Cambridge University Press, Cambridge, p 449

Chapter 5
General Exploration Methods Applied for Geothermal Development

Koichi Tagomori and Hirofumi Muraoka

Abstract This chapter systematically explores the general methodologies utilized in geothermal exploration, focusing on geologic, geochemical, and geophysical approaches. Geothermal development progresses from identifying subsurface geothermal reservoirs to installing energy production infrastructure. Initial exploration relies on analyzing unknown subsurface structures and estimating reservoir potential. Geologic exploration employs remote sensing, structural analysis, and alteration zone mapping techniques to identify heat sources and reservoir characteristics. Geochemical exploration emphasizes fluid analysis, geothermometers, and mixing models to estimate reservoir conditions. Geophysical exploration, including gravity and resistivity methods, provides critical insights into subsurface structures by analyzing physical anomalies. The magnetotelluric (MT) method is particularly emphasized for its efficiency in detecting low resistivity zones associated with geothermal reservoirs. Case studies highlight the importance of integrating multidisciplinary approaches for successful geothermal resource identification, reducing exploration risks, and enhancing drilling accuracy.

Keywords Geothermal exploration · Geologic exploration · Geochemical exploration · Geophysical exploration · Magnetotelluric (MT) method · Gravity survey · Resistivity anomalies · Microgravity monitoring · 3D inversion modeling

K. Tagomori (✉)
West Japan Engineering Consultants, Inc., Fukuoka, Japan
e-mail: k-tagomori@wjec.co.jp

H. Muraoka
Hirosaki University, Hirosaki, Japan
e-mail: hiro@hirosaki-u.ac.jp

J. Kasahara et al. (eds.), *Challenges in Geothermal Development*, Advances in Geological Science, https://doi.org/10.1007/978-981-95-0553-1_5

5.1 Geologic and Geochemical Exploration for Geothermal Development

5.1.1 Geologic Exploration for Geothermal Development

The final objective of geothermal exploitation for geothermal power development is to find high-quality geothermal fluids that work efficiently in the turbine. We must explore high-temperature geothermal heat sources to acquire high-enthalpy geothermal fluids and high-pressure geothermal reservoirs to acquire high-pressure geothermal fluids. Geologic explorations can contribute to these demands in the early stage of geothermal exploration. Representative geologic exploration techniques will be briefly described in the following paragraphs.

(1) *Geologic Remote Sensing*

Recent progress in satellite remote sensing enables the prompt evaluation of a given area in a reconnaissance survey even before the field survey. We can investigate the fundamental information on volcanic and structural geology from multispectral imagery, synthetic aperture radar (SAR) imagery, and digital elevation model (DEM) maps with high resolution. Using the thermal infrared imagery acquired at night, we can directly extract high-temperature anomaly areas discharging hot water, as shown in Fig. 5.1 (Urai et al. 2002). The color gradation mainly reflects the elevation-induced cooling ground surface temperature. The only exception is a bright spot found at the Nage hot spring area in the Nage caldera with a 5 km diameter (Muraoka et al. 2002). The bright spot at the Nage area indicates a large steaming ground extending 200 m by 500 m, discharging 80–72 °C hot water and mixing with cold stream water (Takahashi et al. 2002).

(2) *Volcanic Geology*

Most high-temperature geothermal reservoirs are associated with shallow-depth magma chambers and their consolidated equivalents, young plutonic intrusions as heat sources. These magmatic intrusions commonly form volcanic expressions on the surface because the volatilization of shallow-depth magma chamber causes the magma pressure to rise. When the magma pressure exceeds the overburden pressure, a volcanic eruption occurs. Therefore, heat sources can be explored by the geomorphological evaluation of volcanic edifices, such as calderas and lava domes, and volcanic ejecta, such as lava-flow plateaus and pyroclastic-flow plateaus.

(3) *Structural Geology*

Most geothermal reservoirs consist of faults, which are essential targets for geothermal drilling. Normal faults are characteristic in extension tectonic fields such as mid-ocean ridges, continental rift zones, and back-arc spreading zones. They often form the horst–graben (basin and range) topography. High permeability is expected in the extension tectonic fields. Reverse faults are characteristic in contraction tectonic fields such as subduction zones, and they often form the pile-nappe

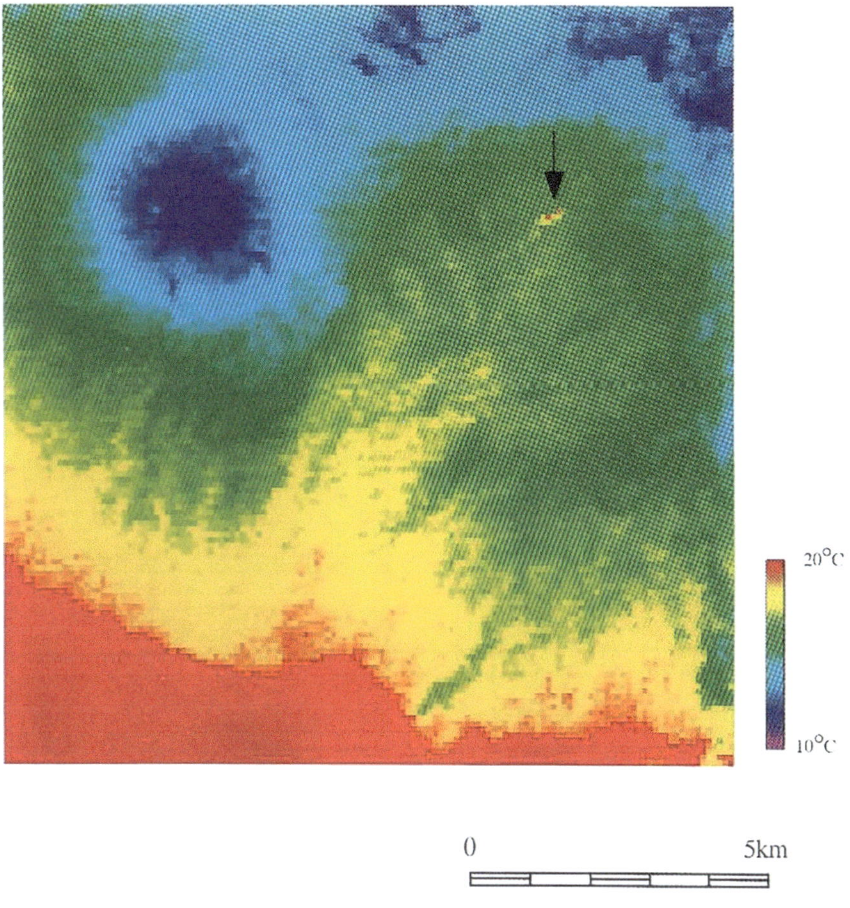

Fig. 5.1 Brightness temperature distribution of the southern Ngada District, Flores Island, eastern Indonesia, derived from ASTER TIR (thermal infrared) band 13 observed on June 23 (22:41, local time), 2001. An arrow indicates the Nage geothermal field, central Flores Island, eastern Indonesia (Urai et al. 2002)

structure. Permeability tends to be restricted in the contraction tectonic fields by the compaction due to the tectonic overstresses. Strike-slip faults are found in the neutral tectonic fields. Findings of faults and their evaluations are one of the most critical subjects of structural geology for geothermal exploration.

(4) *Alteration Zoning*

Steaming grounds and high-temperature hot spring areas often form hydrothermal alteration zones. Particularly, acidic alteration zones are generally observed near the volcanic centers of andesite volcanoes. Hydrothermal alteration zones are also observed along the main reservoir faults. Zonal mapping of hydrothermal alteration minerals is one of the most important surveys for geothermal geology. It provides us with the extent, temperature, and pH conditions of the hydrothermal systems concerned (Izawa 1996). Representative alteration minerals and their stability regions in the temperature and pH conditions are beneficial (Izawa 1996; See Section 12.1 in detail).

(5) *Age Dating*

Radiometric dating methods such as K–Ar and Ar–Ar dating can commonly estimate the ages of young volcanic rocks and plutonic rocks. The thermoluminescence dating can also estimate the ages of hydrothermal alterations (Takashima et al. 2002).

One of the typical applications of volcanology to geothermal exploration is to use the Smith–Shaw diagram (Fig. 5.2), which is composed of two parameters: One is the volume of magma chambers, and the other is the age of volcanic rocks (Smith and Shaw 1979). A series of slant lines indicates theoretical cooling models. The Smith–Shaw diagram can evaluate the heat source potential. If a given magma system is young and voluminous, the system could be active as a heat source. If a given magma system is old and less voluminous, the system could be extinct as a heat source.

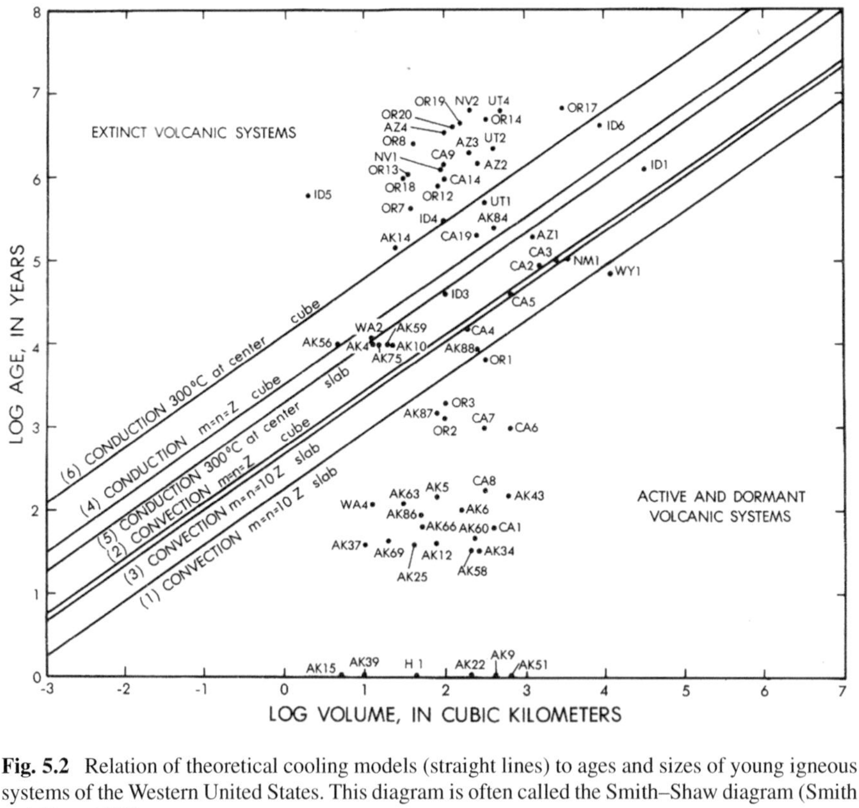

Fig. 5.2 Relation of theoretical cooling models (straight lines) to ages and sizes of young igneous systems of the Western United States. This diagram is often called the Smith–Shaw diagram (Smith and Shaw 1979)

WY1, NW1, and CA3 in Fig. 5.2 indicate the Yellowstone caldera, Long Valley caldera, and Valles caldera, respectively. The Yellowstone caldera is 70 km by 45 km and is one of the largest calderas in the world, along with the Toba caldera in Indonesia.

5.1.2 Geochemical Exploration for Geothermal Development

Geochemical explorations treat geothermal fluids as the most important resource for geothermal power generation. Hot water, steam, and gas are collected from the steaming grounds and hot spring fields. Their major components, minor components, and isotope ratios are analyzed for various geochemical modeling. Geochemical explorations can contribute to these demands in the early stage of geothermal exploration. Representative geochemical exploration techniques will be briefly described in the following paragraphs.

(1) *Fluid Characterization*

Geothermal fluids are a result of the interactions with a variety of source materials. The typical source materials are upper-crustal rocks, meteoric waters, sea waters, magma, and magmatic gases. Estimating source materials of geothermal fluids is critically important to construct evolutionary models of the given geothermal systems. Then, the estimation of source materials is commonly tried in the first step of geochemical exploration. One of the powerful tools for this estimation is the stable isotope analyses, particularly δD and $\delta^{18}O$ analyses.

(2) *Solute Geothermometer*

Solute geothermometers are one of the most frequently used geochemical exploration techniques to estimate reservoir temperatures (Fournier and Truesdell 1973; Fournier 1977; Fournier and Potter 1979; Giggenbach 1988). See representative geothermometers in Table 10.1.1 in detail. Solute geothermometer is divided into two types: One is based on the solubility equilibria, and the other is based on the cation exchange equilibria. The quartz geothermometer is the former case, and most alkali geothermometers are the latter. The method assumes that the chemical equilibria in the reservoirs are probably maintained during the rising of fluids to the surface by hydrothermal convection.

Unfortunately, solute geothermometers do not provide any depth information about the reservoirs. In addition, when we use alkali solute geothermometers in the hot springs in the marine sedimentary basins, most methods give false reservoir temperatures due to the effect of the interstitial relic sea waters of marine sediments (Suzuki et al. 2016). Using the ratio, the Na-K geothermometer relatively avoids this effect (Suzuki et al. 2016).

(3) *Mixing Model*

Mixing of various sources among groundwaters, seawaters, steams, or gases almost always occurs in hydrothermal convection. The mass balance of end-member components quickly estimates their mixing effects. From a geothermal energy point of view, enthalpy is one of the most important parameters. In this context, the enthalpy–silica concentration mixing model and enthalpy–chloride concentration mixing model are frequently applied in the geochemical explorations (see Figs. 10.1 and 10.2 in detail).

5.2 Geophysical Exploration for Geothermal Development

5.2.1 Role of Geophysical Exploration in Geothermal Development

Similar to geologic and/or geochemical surveys, geophysical exploration in geothermal development aims to:

a. Delineate Geologic Structures: Identify geologic structures associated with geothermal systems and determine the position and extent of prospective geothermal reservoirs by analyzing anomalies in physical properties at various depths and profiles.
b. Reduce Exploration Risk: Minimize the risk of geothermal resource exploitation by increasing the success rate of well drilling. This is achieved by improving the accuracy of geothermal reservoir models by incorporating geophysical data into a conceptual geothermal model.
c. Monitor Reservoir Behavior (Time Lapse)**:** Track changes in geothermal reservoir behavior by regularly measuring changes in physical parameters before and after the commencement of power plant operations. Techniques such as microgravity and microseismicity measurements at stationary observation points are effectively used for this monitoring.

Geothermal fields are not limited to areas with visible thermal manifestations at the surface. Fields lacking surface evidence have also been exploited for power generation and thermal utilization. Geophysical exploration is valuable for investigating potential geothermal areas, regardless of visible thermal manifestations, as it can detect the distribution of physical properties of rocks at depth by measuring natural or artificial signals transmitted through the underground. Physical properties such as resistivity, rock density, magnetic susceptibility, seismic hypocenter distribution, and heat flow are measured to identify geologic structures and features directly or indirectly associated with geothermal systems. Geophysical exploration can be considered an indirect method of investigating underground geologic structures as an alternative to drilling.

Figure 5.3 illustrates commonly applied geophysical methods for geothermal exploration, their expected outcomes, and their relation to the "3KEYGS" (refer

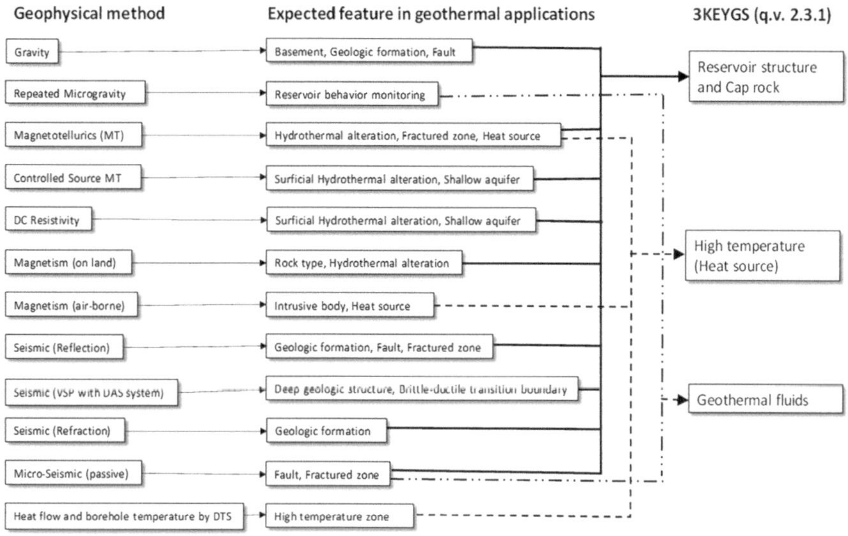

Fig. 5.3 Geophysical methods commonly applied for geothermal exploration and their expected outcomes

to Sect. 2.3.1). During the reconnaissance stage (see Fig. 2.33), various preferential geoscientific methods are combined to identify a potential area worth detailed investigation in the subsequent surface geoscientific survey. Regarding economy and efficiency in executing a geothermal power development project, geophysical methods are more costly than geologic and geochemical methods. In particular, these methods require extensive fieldwork to transport sophisticated instruments and equipment to each measuring station in mountainous areas with rugged terrain and difficult access. Therefore, geophysical methods are typically applied in focused areas based on results from previous geoscientific studies conducted during the reconnaissance stage.

To ensure the financial viability of a geothermal power development project, it is crucial to select the most suitable geophysical methods that can be conducted within the planned budget for the exploration stage, which is part of the overall project budget. Based on case histories from the 1980s to the present, gravity and resistivity methods have been applied in most geothermal fields. These two methods are considered standard and essential techniques for geothermal exploration.

In Japan, gravity and resistivity techniques have been used as standard methods, yielding significant results in identifying geologic structures associated with geothermal systems and pinpointing drilling targets. This is further detailed in the following section.

Even when geothermal systems are delineated using various geoscientific methods, the final exploration stage invariably involves drilling several exploratory wells to gather detailed information about the geothermal reservoir characteristics.

Other geophysical techniques can be added as needed, considering the local geomorphic characteristics and geologic background, but they should remain within the project's economic constraints. Recently, the seismic method using DAS (fiber-optic distributed acoustic sensor) systems has been experimentally applied to detect the location of geothermal reservoirs (Kasahara et al. 2022, 2024a, b), as discussed in Chap. 8 of Part I. The temperature measurements in the borehole have been carried out using the DTS technology (see Chap. 7 of Part I). Recently, the gravity gradient using the full tensor gradiometer (FTG) has been used. Some examples of measurements and the principle of the gravity gradient are also shown in Chap. 8. In addition to the resistivity measurements by MT method, resistivity log in borehole has been used. The resistivity log is well fitted to the Vp profile in Takigami, in Kyushu, Japan (Kasahara et al. 2024 and internal information of Idemitsu Kosan Co., Ltd.).

5.2.2 *Gravity Method*

A gravimeter, as shown in Fig. 5.4, is an instrument used for gravity surveys and is comparatively lightweight and portable compared to those used in other geophysical methods. Since gravity surveys can be conducted using existing roads and mountain trails, extensive vegetation clearing in thick bush areas is unnecessary, minimizing damage to the natural environment. The resolution of gravity measurements is a few μgal. The fluid flow near surface seems affect to a few μ gal. To obtain the time-lapse of gravity, the absolute gravity meter can be used.

Fig. 5.4 Gravity measurement in a field

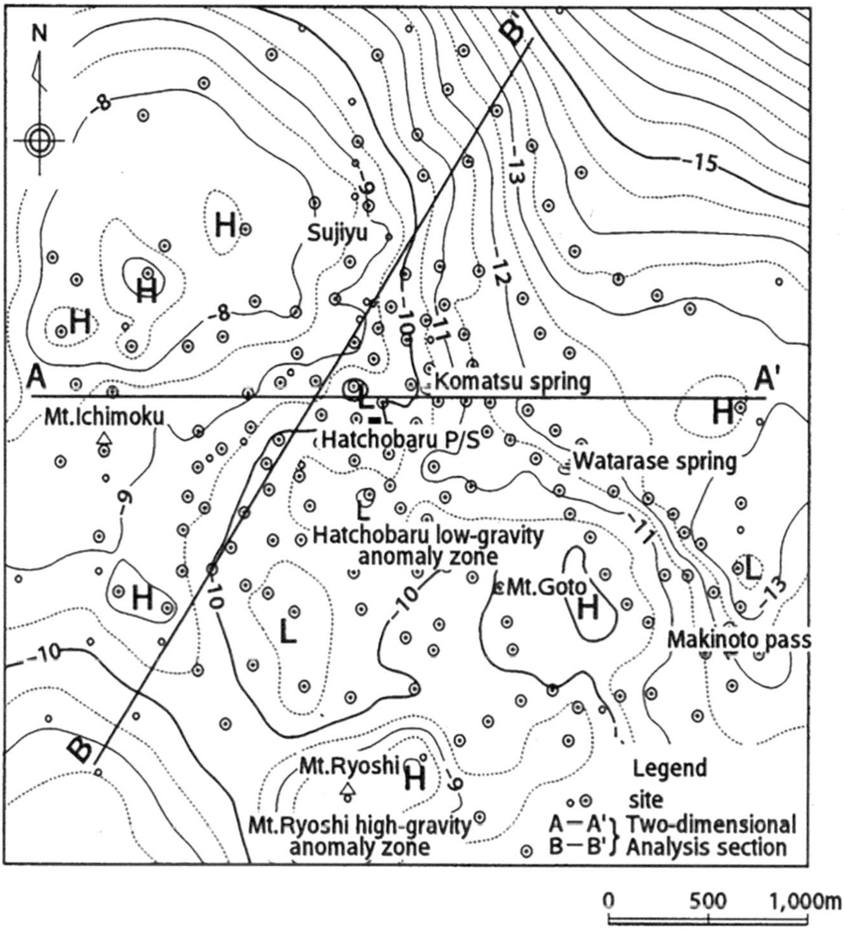

Fig. 5.5 Bouguer anomaly in the Hatchobaru geothermal field, ($\rho = 2.25$ g/cm^3), contour interval = 0.5mgal (Tagomori et al. 1997a, b)

The gravity method is frequently applied for both regional and detailed surveys to detect prospective locations where geothermal reservoirs are expected at depth.

For regional surveys, the gravity method aims to identify extensive geologic structures within and around the expected geothermal development area. These structures include the geometry of the geologic basement, intrusive bodies, large-scale fault systems, and other regional structural features. Following the reconnaissance survey, more promising areas are identified through integrated interpretation based on previous gravity surveys and other geoscientific data. A detailed gravity survey can then be planned to extract gravity lineaments obtained from filtering analysis, focusing on these prospective areas.

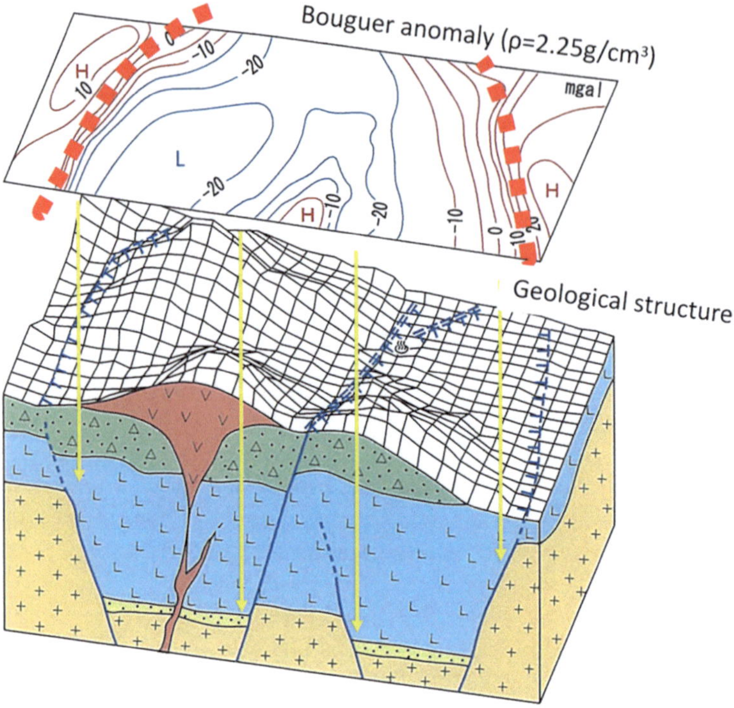

Fig. 5.6 Correlation between Bouguer anomalies and geologic structure in the Hatchobaru geothermal system

The gravity lineaments extracted through the filtering process at various depths are thought to reflect faults or linear fractures, which may be closely related to geothermal systems. Figures 5.5 and 5.6 show the Bouguer anomaly map and an illustration of Bouguer anomalies versus geologic structures in the Hatchobaru geothermal field in Japan, respectively. In Fig. 5.6, a comparison of Bouguer anomalies with geologic structures reveals that the geometry of the geologic basement correlates with Bouguer high and low anomalies. Additionally, the portions of discontinuous geologic layers with faulting systems correspond to gravity lineaments extracted from gravity changes with steep gradients. The deep zones of these gravity lineaments are potential candidates for well-drilling targets, as they may yield productive geothermal steam and hot water.

During the filtering analysis process, residual gravities using several spatial ranges are applied to eliminate the regional trend of the Bouguer anomaly. This helps delineate clear images of lineaments associated with fractured zones related to geothermal systems at depth.

Regarding item "c (Monitor Reservoir Behavior)" in the previous Sect. 5.2.1, the first practical measurements of precise gravimetry in geothermal research at the Wairakei field were established by Hunt (1970, 1983, 1988) and Allis and Hunt (1986). In Japan, repeated measurements of microgravity changes began in 1990 to monitor reservoir mass changes associated with the production and reinjection of

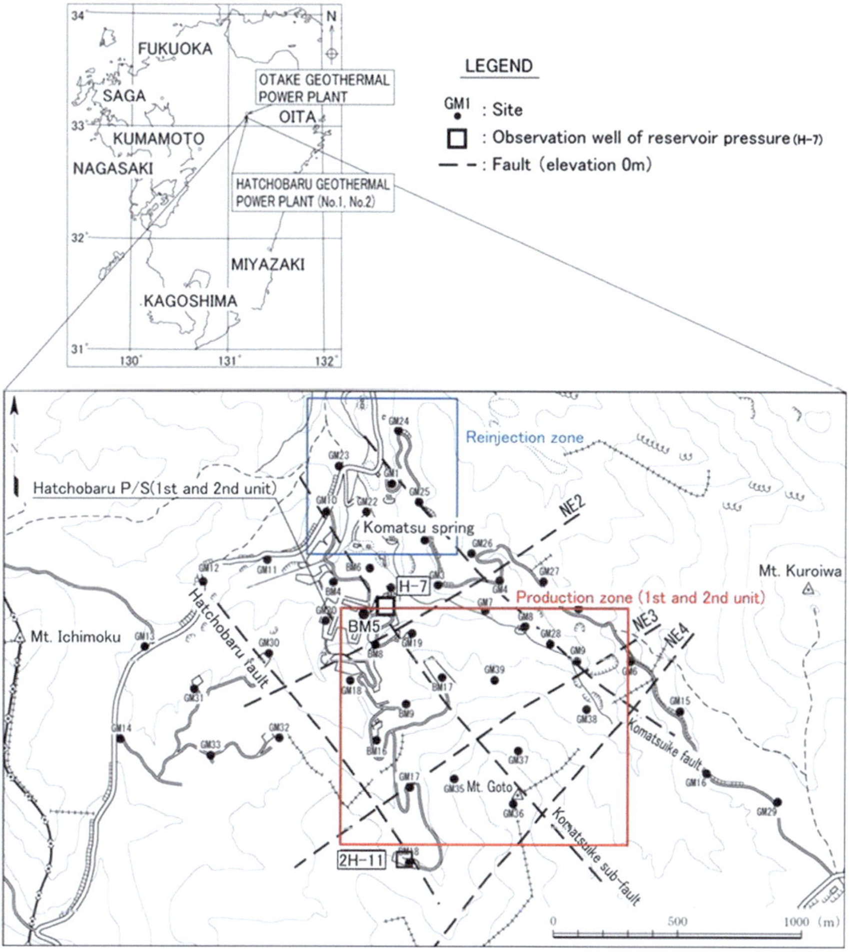

Fig. 5.7 Locations of gravity measurement stations and reservoir pressure observation wells in the Hatchobaru geothermal field (Saito et al. 2006)

geothermal fluids. This was prompted by an increase in output from 55 to 110 MWe at the Hatchobaru geothermal power plant. The locations of gravity measurement stations (Benchmarks: BM) and reservoir pressure observation wells are shown in Fig. 5.7 (Saito et al. 2006).

Since maintaining many observation wells to cover an entire reservoir is costly, combining a few observation wells and strategically placed gravity monitoring stations might be an effective alternative. Figure 5.8 shows correlation curves between pressure changes in the H-7 observation well and gravity changes at BM5 near the H-7 well. Except for a brief period when helium gas was not charged in the capillary tube, both curves showed a strong correlation, with a coefficient of 0.82 (Tagomori et al. 1997a, b).

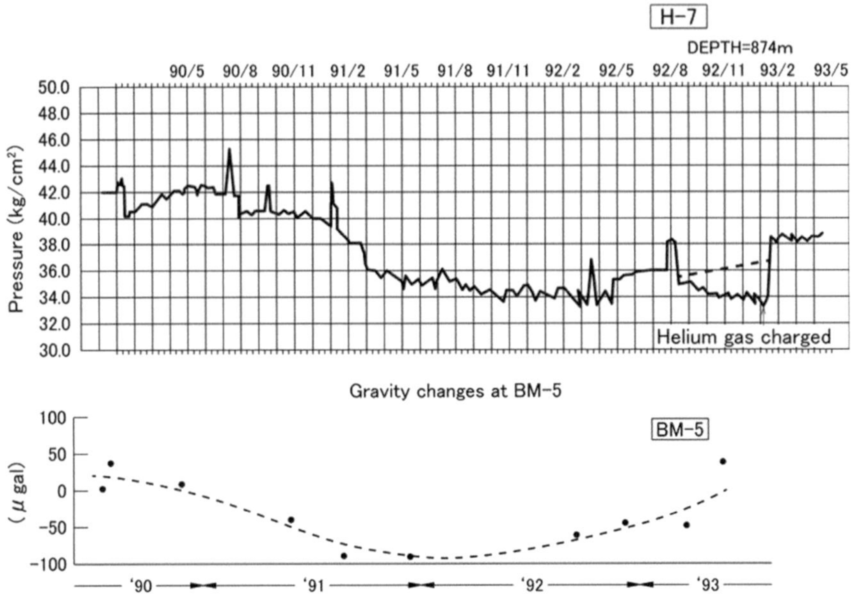

Fig. 5.8 Comparison between pressure changes of H-7 observation well and gravity changes at BM5 gravity monitoring station (Tagomori et al. 1996)

Thus, in some geothermal power plants, monitoring microgravity changes is expected to be a valuable method for monitoring geothermal reservoir behavior due to the production and reinjection of geothermal fluids, potentially reducing the need for extensive observation wells.

5.2.3 Resistivity Method

The resistivity technique is widely used in geothermal exploration due to its effectiveness in identifying geologic features and defining target zones for deep well drilling.

Geothermal reservoirs typically exist in volcanic sediments with highly permeable fractured zones saturated with hot fluids. These zones are often overlain by an impermeable hydrothermal alteration zone, or "cap rock," which acts as a sealing layer for the geothermal system's convective heat flow.

Hydrothermally altered formations containing clay minerals are common in the shallow formations above geothermal reservoirs. These rocks show low resistivities, contrasting sharply with surrounding high-resistivity fresh volcanic rocks, which helps identify the unique resistivity characteristics of geothermal reservoirs.

However, hydrothermally altered formations are known to exhibit very low permeability, so the presence of a low resistivity altered formation does not necessarily indicate a productive zone.

Before the 1980s, DC resistivity techniques (such as Schlumberger and Wenner methods) were commonly used to delineate layered resistivity structures. However, these techniques often face challenges in mountainous areas due to the need for straight cable setups extending several kilometers. In the early 1970s, geothermal development companies experimentally introduced magnetotelluric (MT) techniques by using instruments initially developed for oil exploration. These techniques solved the limitations of DC resistivity methods, particularly in difficult terrains. Over time, several improvements were made to MT instruments, making them lighter and more compact, thus facilitating efficient field measurements in mountainous regions. Today, the MT method has become the mainstream technology for geothermal exploration.

In addition to instrument modifications, the rise of real-time data processing technologies using microcomputers has drastically improved both data acquisition and processing quality in the field. The MT technique measures natural fluctuations of electromagnetic fields caused by solar activity (see Fig. 5.9). For MT measurements, two sets of electrodes (Ex and Ey) and three magnetic coil magnetometers (Hx, Hy, and Hz) are placed on the ground to measure time-domain variations of electric and magnetic fields across a wide frequency range, typically from several hundred Hz to 0.001 Hz. The relationship between exploration depth and frequency in MT measurement is expressed by the following equation (Cagniard 1953):

$$\delta = 503\sqrt{\frac{\rho}{f}} \tag{5.1}$$

Here,

δ:skin depth (meter)
ρ: resistivity (ohm-m) of the conductor (i.e., underground)
f: frequency (Hz).

The skin depth (δ) corresponds to the MT penetration depth based on the theory of the skin effect in electromagnetic fields, representing the distance below the surface of a conductor where the current density has diminished to 1/e of its surface value. In a homogeneous medium, resistivity can be calculated from the measured electromagnetic field using the following equation:

$$\rho_a = \frac{1}{5f}\left|\frac{E}{H}\right|^2 \tag{5.2}$$

Here,

ρ_a:apparent resistivity (ohm-m)
f: frequency (H_z)
E: electric field (mV/km) (electric field in 50–100 meter distance)
H: orthogonal magnetic field (nT) (E_x/H_y, Ey/H_x).

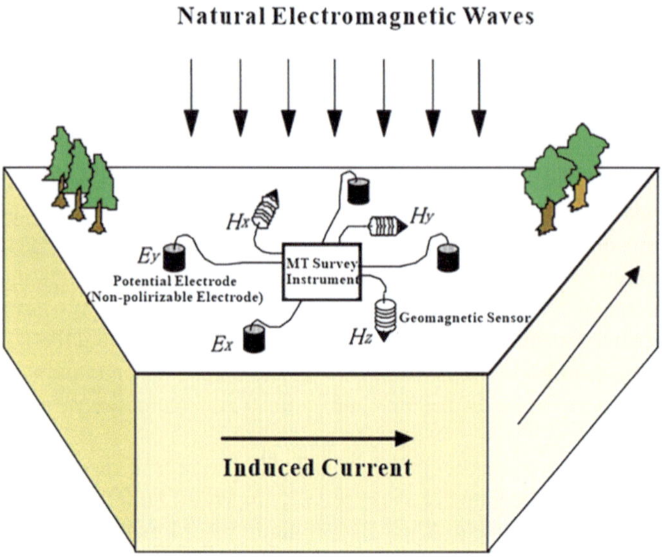

Fig. 5.9 Field measurement layout for a MT station (AIST, 2024a, b)

The earth consists of various geologic layers and rock types. Therefore, the calculated resistivity represents an average resistivity from shallow (high frequency) to deep (low frequency) depths. This averaged resistivity is expressed as "apparent resistivity" (ρ_a). By measuring a wide range of electromagnetic field frequencies, resistivities from shallow to deep zones can be calculated, providing a comprehensive view of subsurface structures.

Despite its widespread use, the MT method has some limitations. Firstly, because MT is a passive method that measures natural signals, data quality can be significantly affected by solar activities and cultural noise. To mitigate this issue, a simultaneous measurement using a remote reference station (noise reduction by stacking procedures) must be incorporated. This approach helps to reduce the impact of noise on the data quality. Secondly, the MT method is not well suited for measuring shallow depths. To address this limitation, other geophysical methods need to be employed to complement the MT data. The results obtained from these additional methods are then combined and jointly inverted for coherent interpretation. This process helps to avoid the effect known as "static shift," which is a potential distortion caused by local resistivity bodies near the surface, resulting in an incorrect resistivity distribution at depth. Techniques such as time-domain electromagnetic (TEM) and DC resistivity are often applied to offset the static shift effect.

Although the presence of fractures cannot be detected directly, low resistivity anomalies associated with several factors–such as clay minerals, temperature, porosity, water saturation, and chemical composition (Cl^-, $SO4^{2-}$, $CO3^{2-}$, Na^+, Ca^{2+}, etc.) of the geothermal fluids–provide essential insights into geothermal systems. These anomalies can be associated with the geothermal reservoir or the cap rock, which acts as a sealing layer composed of clay minerals. Accordingly, detecting low resistivity anomalies is crucial in the context of geothermal resource exploration, whether those anomalies reflect the geothermal reservoir itself or the overlying cap rock zones. These anomalies offer crucial information for understanding the geothermal reservoir structure.

Figure 5.10 illustrates the geologic and resistivity sections in the Ogiri geothermal area in Kyushu Island, Japan, located in southern Japan. In these figures, it is evident that the low resistivity zone correlates well with the clay cap distribution formed by smectite alteration. The geothermal reservoir is located beneath this smectite cap rock, extending along fractures associated with vertical faulting systems. This correlation between the low resistivity zones and the smectite cap rock highlights the effectiveness of resistivity methods in identifying key geologic features associated with geothermal systems. The presence of the clay cap rock above the reservoir is crucial as it acts as a sealing layer, trapping the geothermal fluids within the reservoir. The fractures and fault systems play a significant role in providing pathways for the geothermal fluids, contributing to the permeability of the reservoir. By using resistivity techniques, such as the MT method, it is possible to map these subsurface features accurately, facilitating the identification of potential drilling targets and enhancing the understanding of the geothermal system's structure.

Figure 5.11 presents a schematic diagram of a resistivity model related to a fault-controlled geothermal system, based on case studies developed in Japan. To simplify the explanation, the resistivity layers in the diagram are composed of three layers: the "resistive overburden (ρ_1)" near the surface, the "low resistivity layer (ρ_2)" in the middle, and the "electrical basement (ρ_3)" at the bottom. The low resistivity layer is further divided into two zones: the "very low anomaly zone (ρ_α)" and the "low anomaly zone (ρ_β)" surrounding it. Initially, the "very low anomaly zone (ρ_α)" in

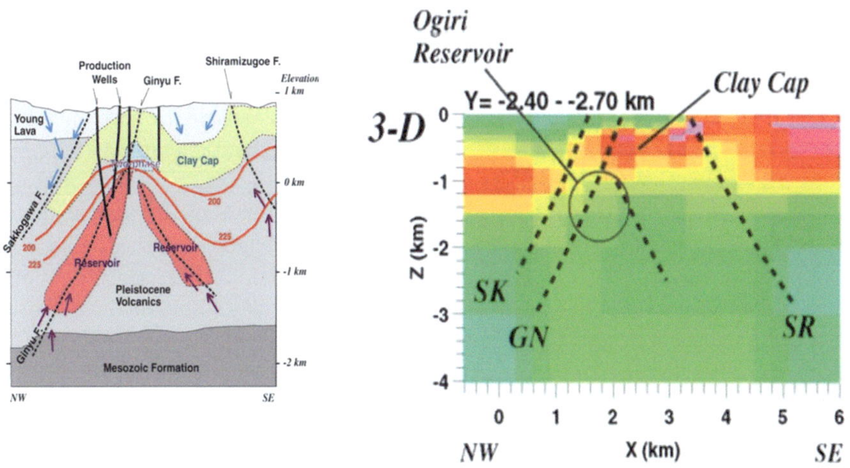

Fig. 5.10 Geothermal geologic section (left) and the corresponding resistivity section analyzed by 3D inversion (right) in the Ogiri geothermal area (Uchida 2005)

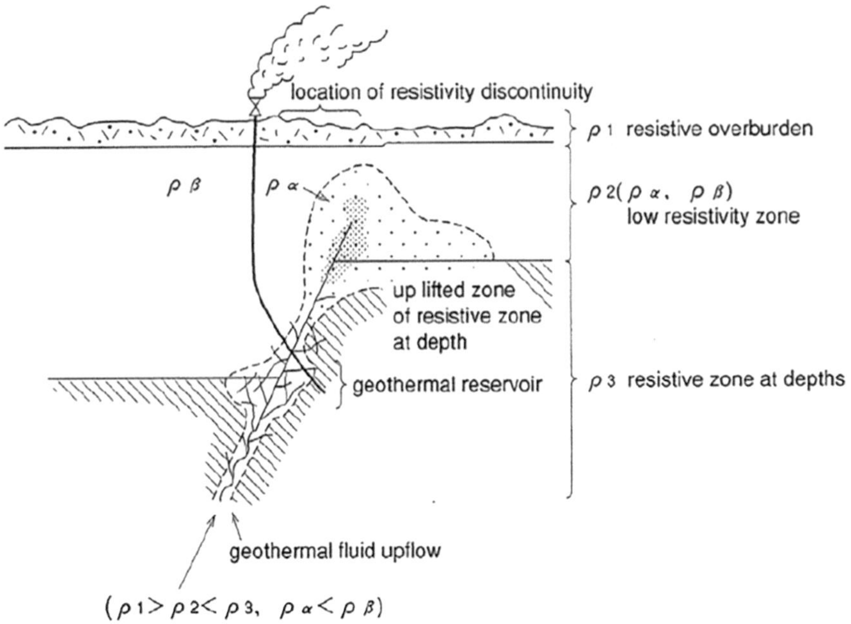

Fig. 5.11 Schematic diagram of a distinctive resistivity model associated with vertical fault systems in the Hatchobaru geothermal reservoir, Japan (Tagomori et al. 1990)

the low resistivity layers was interpreted as an indication of the geothermal reservoir itself. However, drilling results revealed that the reservoirs were located much deeper, beneath the "very low anomaly zone (ρ_α)."

Through the development of many case histories, it was understood that the "very low anomaly zones (ρ_α)" mostly reflect the extent of hydrothermal alteration composed of clay minerals, particularly smectite, which can mask the signals associated with the geothermal reservoir at greater depths. Consequently, in the context of resistivity structures within fault-controlled geothermal systems, areas where the "very low anomaly zone"—indicative of hydrothermal alteration cap rocks—coincides with discontinuities caused by depth variations in the electrical basement are considered prime candidates for geothermal resource exploitation. This insight aids in refining the interpretation of resistivity data, leading to more precise identification of geothermal reservoirs. By recognizing the role of hydrothermal alteration zones and their relationship with fault systems, geoscientists can better target drilling efforts, improving the success rate of geothermal resource exploration.

In Fig. 5.12, resistivity sections based on 3D inversion analysis at the Otake-Hatchobaru geothermal field, located in the central part of Kyushu Island, Japan, are depicted. The dashed lines in these sections represent resistivity discontinuities beneath the very low anomaly zone, corresponding to the locations of geologic faults expected to be geothermal reservoirs (Tagomori et al. 2012). Production wells

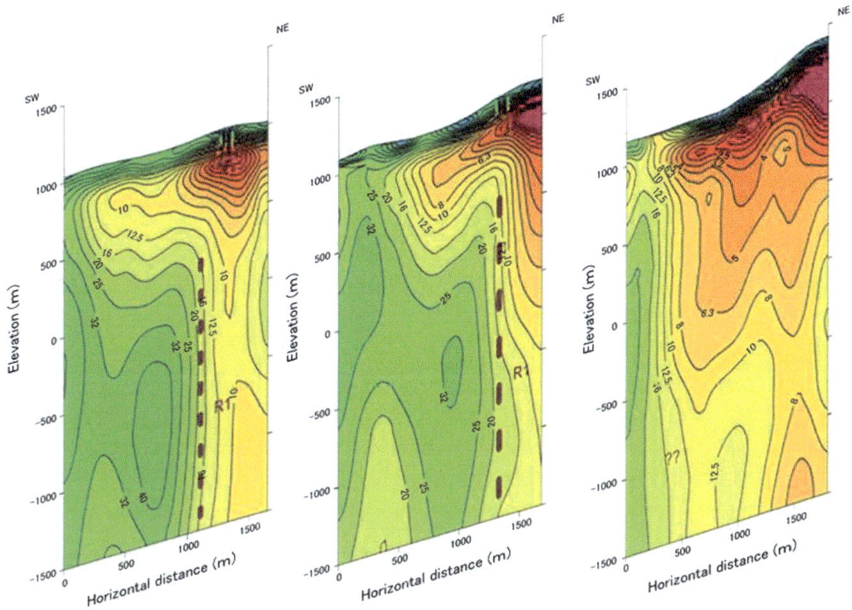

Fig. 5.12 Resistivity sections by 3D inversion results of MT survey in the Otake-Hatchobaru geothermal reservoir in Kyushu Island, Japan (Saito et al. 2011). Contours are electrical resistivity in Ohm-meter. The vertical axis is depth in meters

have been directionally drilled into the deeper parts of these resistivity discontinuities, successfully producing steam and hot water for geothermal power generation. It is important to note that the resistivity structures of geothermal fields can vary significantly from one field to another and are often much more complex than the schematic diagram presented in Fig. 5.11. Therefore, the resistivity model described above should be regarded as a basic concept to aid in interpreting geothermal systems where low resistivity zones are widely distributed. Each geothermal field requires careful, site-specific analysis to map its resistivity structure accurately and identify potential drilling targets.

References

AIST (2024a) AIST exploration geophysics research group website. https://unit.aist.go.jp/georesenv/explogeo/casestudy/casestudy01.html

Allis RG, Hunt TM (1986) Analysis of exploitation induced gravity changes at Wairakei geothermal field. Geophysics 51:1647–1660

Cagniard L (1953) Basic theory of the magneto-telluric method of geophysical prospecting. Geophysics 18:605–635

Exploration Geophysics Research Group, National Institute of Advanced Industrial Technology (AIST) (2024b). https://unit.aist.go.jp/georesenv/explogeo/casestudy/casestudy01.html

Fournier RO (1977) Chemical geothermometers and mixing models for geothermal systems. Geothermics 5:41–50
Fournier RO, Potter RW (1979) Magnesium correction to the Na-K-Ca chemical geothermometer. Geochim Cosmochim Acta 43:1543–1550
Fournier RO, Truesdell AH (1973) An empirical Na-K-Ca geothermometer for natural waters. Geochim Cosmochim Acta 37:1255–1275
Giggenbach WF (1988) Geothermal solute equilibria. Derivation of Na-K-Mg-Ca geoindicators. Geochim Cosmochim Acta 52:2749–2765
Hunt TM (1970) Net mass loss from the Wairakei Geothermal Field, New Zealand. Geothermics Special Issue 2:487–492
Hunt TM (1983) Recharge in Wairakei geothermal field for 1974–1983. In: Proceedings of the 5th NZ geothermal workshop, pp 49–54
Hunt TM (1988) Update on gravity changes in the main production borefield at Wairakei. In Proceedings 10th New Zealand geothermal workshop, pp 129–132
Izawa E (1996) Chapter 8: exploration of resources. In: Iwanami Earth and Planetary Science Course, vol 4: Observation of the Earth, p 330 (in Japanese). Iwanami-Shoten, pp 291–323
Kasahara J, Hasada Y, Kuzume H, Mikada H, Fujise Y (2022) The DAS geothermal study in the geothermal borehole of the Ohnuma geothermal power plant in Hachimantai, Honshu, Japan. J Geotherm Res Soc Jpn 44(2):73–84
Kasahara J, HasadaY, Mikada H (2024a) Geothermal reservoir characterization using distributed acoustic sensing from vertical seismic profiling in six geothermal fields in Japan. In: Distributed acoustic sensing in borehole geophysics. AGU monograph
Kasahara J, Hasada Y, Furuya S, Shimazaki M, Mikada H, Ohnuma H, Fujise Y (2024b) DAS and DTS geophysical study in the Takigami geothermal field in northeastern Kyushu, Japan. In: Distributed acoustic sensing in borehole geophysics. AGU monograph
Muraoka H, Nasution A, Urai M, Takahashi M, Takashima I, Simanjuntak J, Sundhoro H, Aswin D, Nanlohy F, Sitorus K, Takahashi H, Koseki T (2002) Tectonic, volcanic and stratigraphic geology of the Bajawa geothermal field, central Flores, Indonesia. Bull Geol Surv Jpn 53:109–138
Saito H, Tagomori K, Honda M, Kasagi J, Fujise Y, Ehara S (2006) Reservoir behavior inferred from Quasi-3D analysis of gravity changes in the Hatchobaru geothermal field. J Geotherm Res Soc Jpn 45(3):181–197
Saito H, Honda M, Senju T (2011) Geothermal structure in the South-Eastern Area of the Otake Geothermal field derived from magnetotelluric survey. In: Proceedings of annual meeting of geothermal research society of Japan, A30
Smith RL, Shaw HR (1979) Igneous-related geothermal systems. In: Muffler LJP (ed) Assessment of geothermal resources of the United States-1978, U.S. Geological Survey Circular, vol 726, pp 12–17
Suzuki Y, Ioka S, Muraoka H (2016) Comparative study of the subsurface thermal structure in northern Honshu, Japan, based on normalized temperature data and solute geothermometers. Energies 9:382. https://doi.org/10.3390/en9050382
Tagomori K, Honda M, Shimoike T, Matsumoto T, Mimura T, Akiyoshi M, Ushijima K (1990) Geoelectric structure of the Hatchobaru geothermal field and the adaptability of electrical exploration methods for its understandings. CHIMETSU 27(4):49–68
Tagomori K, Ehara S, Nagano H, Oishi K (1996) Study on reservoir behavior based on gravity changes in the Hatchobaru geothermal field. J Geotherm Res Soc Jpn 18(2):91–105
Tagomori K, Honda M, Nagano H, Akiyoshi M, Haruguchi K, Ushijima K (1997a) Geothermal structure in the Hatchobaru geothermal field derived from detailed gravity and resistivity structure. CHINETSU 34(4):1–17
Tagomori K, Saito H, Honda M (1997) Study on reservoir behavior from micro-gravity changes monitored in the Hatchobaru geothermal field, Japan. In: Proceedings of the 18th PNOC-EDC geothermal conference, pp 280–287

Tagomori K, Saito H, Tokita H, Matsuda K (2012) Practical approach for exploitation and evaluation of geothermal reservoir, and expected subjects for further geothermal resource development, geothermal and volcanological research report of Kyushu University, No. 20, pp 46–54
Takahashi M, Urai M, Yasukawa K, Muraoka H, Masuda K, Akasaka H, Koseki T, Hisatani K, Kusnadi D, Sulaeman B, Nsution A (2002) Geochemical characteristics of hot spring water in Bajawa area, central Flores, Indonesia. Bull Geol Surv Jpn 53:183–199
Takashima I, Nasution A, Muraoka H (2002) Thermoluminescence dating of volcanic and altered rocks in the Bajawa geothermal area, central Flores, Indonesia. Bull Geol Surv Jpn 53:139–146
Uchida T (2005) Three-dimensional magnetotelluric investigation in geothermal fields in Japan and Indonesia. In: Proceedings, world geothermal congress, pp 24–29
Urai M, Muraoka H, Nasution A (2002) Satellite remote sensing data and their interpretations for geothermal applications: a case study on the Ngada District, central Flores, Indonesia. Bull Geol Surv Jpn 53:99–108

Chapter 6
Case Studies I: High- and Ultra-High-Temperature Geothermal Fields in Japan and the World

Koichi Tagomori, Naotsugu Ikeda, Ko Sato, Hirofumi Muraoka, and Junzo Kasahara

Abstract This chapter provides a comprehensive overview of high- and ultra-high-temperature geothermal fields in Japan and globally. Section 6.1 focuses on major fracture-type geothermal reservoirs in Kyushu, Japan, including Hatchobaru, Takigami, Ogiri, and Yamagawa. These fields are situated within tectonic grabens and volcanic zones, where geothermal fluids circulate through fault-controlled reservoirs. Technologies such as double flash and binary systems are employed, with geophysical methods like magnetotellurics (MT), gravity surveys, and distributed sensing used for reservoir characterization. Section 6.2 explores ultra-high-temperature geothermal challenges, emphasizing the Kakkonda field, where drilling to 3,729 m revealed >500 °C superheated zones within young plutonic rocks. The chapter introduces the brittle-plastic transition and engineered geothermal systems (EGS) in plastic zones. The high-temperature challenges in Kilauea lava lake drilling in the USA, the DESCRAMBLE project in Italy, and the IDDP project in Iceland are briefly reviewed. These case studies underscore the need for integrated geoscience, engineering innovation, and long-term strategies for future geothermal energy development.

K. Tagomori (✉)
West Japan Engineering Consultants, Inc., Fukuoka, Japan
e-mail: k-tagomori@wjec.co.jp

N. Ikeda
Welma Co. Ltd., Fukuoka, Japan
e-mail: n-ikeda@welma.co.jp

K. Sato
Socio Engineering Co., Ltd., Tokyo, Japan
e-mail: satoukou6450@gmail.com

H. Muraoka
Hirosaki University, Hirosaki, Japan
e-mail: hiro@hirosaki-u.ac.jp

J. Kasahara
Shizuoka University, Shizuoka, Japan
e-mail: kasahara.junzo@shizuoka.ac.jp

J. Kasahara et al. (eds.), *Challenges in Geothermal Development*, Advances in Geological Science, https://doi.org/10.1007/978-981-95-0553-1_6

Keywords Graben system · Double flash system · Magnetotelluric (MT) survey · Clay cap · Binary power plant · Brittle-plastic transition · Supercritical geothermal · Ultra-high-temperature geothermal · Engineered Geothermal System (EGS) · Kyushu · Hatchobaru · Otake · Takigami · Ogiri · Yamagawa · Kakkonda · Kilauea · Larderello · Reykjanes

6.1 Geothermal Fields in Kyushu, Japan

In this section, we describe the details of power plant in Kyushu island, Japan. There are five major geothermal power plans in Kyushu, i.e., Hatchobaru, Otake, Takigami, Ogiri, and Yamagawa.

6.1.1 Development Status of the Major Geothermal Fields in Kyushu, Japan

Kyushu Island is tectonically underlain by the Philippine Sea Plate (PHS). The PHS is divided by the NW–SE trending Palau-Kyushu Ridge into the Miocene's hot Shikoku Basin segment and the Paleogene's cold West Philippine Basin segment. The subduction of the Philippine Sea Plate beneath the southwestern Japan arc (Eurasian plate) commenced 6–4 million years ago (Ma) at the convergent plate boundaries: the Nankai Trough in the eastern part and the Ryukyu (Nansei-Shoto) Trench in the western part (Fig. 6.1 left). This subduction led to the development of large graben systems and the activity of arc-type magma in the frontal arc region (Kamata 1989).

Geothermal activities such as fumaroles and hot springs are widely distributed in and around these volcanoes and volcanic fronts. These manifestations have been well studied from geoscientific perspectives, and numerous sites have been exploited for tourism, contributing to the traditional "Onsen" (Hot spring or thermal bath) culture, which has been popular since ancient times.

Another tectonic feature characterizing the geothermal system in Kyushu is the graben formation, which occurs under extensional tectonic fields. These underground formations under extensional stress tend to exhibit higher permeability, enhancing geothermal potential.

Figure 6.1 and Table 6.1 summarize 17 active volcanoes in Kyushu Island and the Satsunan Islands (a chain of volcanic islands south of Kyushu consisting of many insular volcanoes). The location of the Beppu-Shimabara graben and Kagoshima graben (indicated in blue scripts) are also indicated. Most of Kyushu's major geothermal power plants are located within these grabens.

The history of geothermal resource exploration and exploitation in Kyushu is rich and dates back to the early 20th century. The first geothermal well drilling intended for electric power generation was carried out in Beppu, Oita, prefecture in

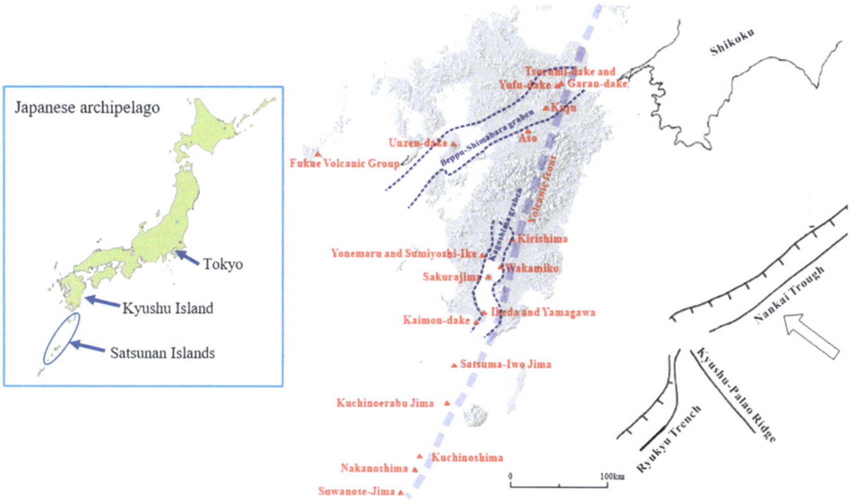

Fig. 6.1 Tectonic setting around Kyushu and the location of active volcanoes, grabens, and volcanic fronts. The thick, broken line shows the volcanic front. The arrow shows the direction of subduction of the Philippine Sea Plate

Table 6.1 Location and eruption ages of active volcanoes in Kyushu Island and its vicinity (also refer to Fig. 6.1) (JMA 2013)

Name of Active Volcano	Location (degrees)		Altitude (m)	Mountain or Spot Erupted	Eruption Monitoring	Latest Eruption Age (A.D.)
	North Latitude	East Longitude				
Tsurumidake	33.29	131.43	1,375	Tsurumidake	✓	867
Garandake	33.32	131.43	1,045	Garandake		1000 and several hundred years ago
Yufudake	33.28	131.39	1,583	Yufudake	-	about 2000 years ago
Kuju	33.09	131.25	1,791	Nakadake	✓	1996
	33.09	131.23	1,762	Hossho		
Aso	32.88	131.10	1,592	Takadake	✓	2020
Unzendake	32.76	130.30	1,483	Heisei Shinzan	✓	1995
Fukue volcanic group	32.66	128.85	315	Onidake	-	2000 - 3000 years ago
Kirishima	31.93	130.86	1,700	Karakunidake	✓	2018
Yonemaru	31.78	130.57	15	center of Yonemaru	-	7000 - 8000 years ago
Sumiyoshiike	31.77	130.59	40	center of Sumiyoshiike		
Wakamiko	31.66	130.80	-77	central cone	-	-
Sakurajima	31.59	130.66	1,117	Ontake	✓	2021
Ikeda	31.21	130.57	256	Nabeshimadake	-	about 4300 years ago
Yamagawa	31.21	130.64	3	Banshohana		
Kaimondake	31.18	130.53	924	Kaimonndake	-	885
Satsuma-Iwojima	30.79	130.31	704	Iwodake	✓	2020
Kuchinoerabujima	30.44	130.22	657	Furudake	✓	2020
Kuchinoshima	29.97	129.93	628	Maedake	-	thousands of years ago
	29.96	129.93	425	Moedake		
Nakanoshima	29.85	129.86	979	Ontake	-	1914
Suwanosejima	29.64	129.71	796	Ontake	✓	2021

1919. This early effort led to the successful generation of 1.12 kW of electricity in 1925. However, more coordinated geothermal resource exploration involving governmental research institutions, university laboratories, and private companies began after World War II.

The first commercial geothermal power plant, the Otake Geothermal Power Plant (GPP) began operating in 1967 with an installed capacity of 12.5 MWe. This milestone came just one year after the nation's first geothermal power plant started operation in 1966 at the Matsukawa GPP.

Following this, significant developments included:

- Hatchobaru Unit 1: Started operation in 1977 with a capacity of 55 MWe.
- Hatchobaru Unit 2: Commenced operation in 1990, also with a capacity of 55 MWe.
- Other Major Geothermal Power Plants: Constructed during the 1990s.

After the Takigami geothermal power plant was completed in 1996, the construction of new geothermal power plants in Kyushu stalled for nearly two decades. The situation changed with the introduction of a new Feed-In-Tariff (FIT) program and government deregulations regarding national parks. These changes attracted existing and new private developers to invest in geothermal projects, some of which have already succeeded.

Figure 6.2 and Table 6.2 illustrate the location and summary of geothermal power plants in Kyushu with more than 1 MWe installed capacity. Most of Kyushu's major geothermal power plants are located around areas of current volcanic activity and the graben. Conversely, minor geothermal manifestations are ubiquitous throughout the territory and are utilized for Onsen (hot spring) tourism.

The following section briefly explains some of the major geothermal power plants in Kyushu. It is important to note that the grid electric power in western Japan operates at 60 Hz, which imposes additional constraints on turbine unit capacity compared to the eastern part of Japan, where 50 Hz is adopted.

6.1.2 Hatchobaru Geothermal Field, Central Kyushu

6.1.2.1 Introduction

The Hatchobaru geothermal power plant is located around the western edge of the Kuju volcanic mountain range, located within the central part of the Beppu-Shimabara Graben, which runs east–west through central Kyushu and at about 20 km to the southwest of the power plant, the famous Aso Caldera is located (Figs. 6.1, 6.2 and 6.3 and Tables 6.1 and 6.2). The area occupies the eastern part of Aso-Kuju National Park, which was designated by the Ministry of Environment of Japan in 1934 (light brown area in Fig. 6.3). Several geothermal plants, including Takigami GPP, Otake GPP, and Sugawara GPP, are located along the rim of the buried Shishimuta caldera.

The power plant, developed and operated by Kyushu Electric Power Company (Kyushu EPC), consists of two units, each with a capacity of 55 MWe, utilizing a double flash system where primary and secondary steam are separated at the steam

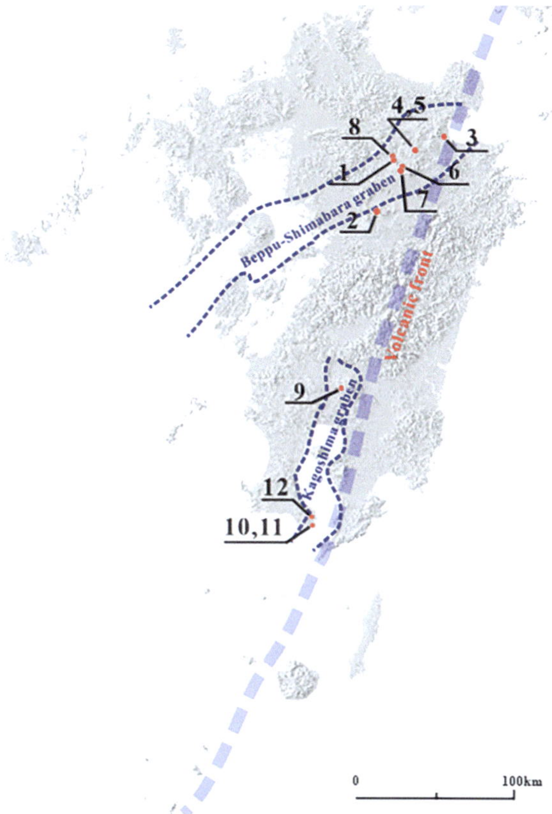

Fig. 6.2 Location of the major geothermal power plants in Kyushu. The description of the donated number is shown in Table 6.2

separator and flasher, respectively. This double flash system, jointly developed by Kyushu EPC and Mitsubishi Heavy Industries, Ltd., was first commercially introduced in the world with the Hatchobaru No.1 unit in June 1977. The unit-2 started commercial operation in June 1990.

Additionally, a 2 MWe binary plant was added, utilizing steam from waned production well. Approximately 2 km to the north, the Otake geothermal power plant has been operational since 1967 and was renovated and uprated from 12.5 MWe to 14.5 MWe in October 2020. Both Hatchobaru and Otake power plants use two-phase fluid transport piping system that connects the wellhead to the separator located close to the power plant.

6.1.2.2 Geothermal Systems of the Hatchobaru Geothermal Field

According to data published by the Thermal and Nuclear Electric Power Engineering Society (2021) in 2020, about 90 production and reinjection wells have been drilled

Table 6.2 Geothermal power plants in Kyushu with more than 1 MW installed capacity (TENPES 2021 with updates)

No	Plant name	Location	Installed capacity (MWe)	Technology	COD
1	Waita GPP	Kumamoto Pref.	1.995	SF	2015.6.16
2	Minami-Aso Yunotani GPP		2.168	SF	2023.3.3
3	Suginoi Hotel	Oita Pref.	1.9	SF	2006.4.1
4	Takigami GPP		27.5	SF	1996.11.1
5	Takigami Binary GPP		5.05	B	2017.3.1
6	Otake GPP (renovated with uprating)		13.7	SF	2020.10.5 Originally 1967.8.11
7	Hatchobaru GPP		55	DF	1977.6.24
			55	DF	1990.6.22
			2	B	20006.4.1
8	Sugawara Binary GPP		5	B	2015.6.29
9	Ogiri GPP	Kagoshima Pref.	30	SF	1996.3.1
10	Yamagawa GPP		30	SF	1995.3.1
11	Yamagawa Binary GPP		4.99	B	2018.2.23
12	Medipolice Ibusuki GPP		1.58	B	2015.2.18

* SF: single flash, DF: double flash, B: binary

in the Hatchobaru field. Due to topographic constraints, the extension area reached by the Hatchobaru wells is rather small, considering the plant's size.

Based on well data and other geoscientific studies, a geothermal reservoir conceptual model was proposed (Figs. 6.4 and 6.5 while referring to Fig. 6.3 as well).

The main features of the Hatchobaru geothermal systems include:

- Reservoir temperatures: Ranging from approximately 240–300 °C along fractured zones associated with fault systems.
- Reservoir type: Water-dominated, with production zones at depths of 1,000–2,300 m along mostly vertical fault zones within andesitic volcanic rocks.
- Heat source: Related to the late Quaternary volcanic activity of Kuju Volcano, with geothermal fluids originating from meteoric water based on isotope analysis.
- Five fault systems: Fault systems (Komatsuike, Komatsuike-sub, NE3, NE4, and Hatchobaru fault) contribute to the productive geothermal reservoirs, sustaining a 110 MW power generation capacity,

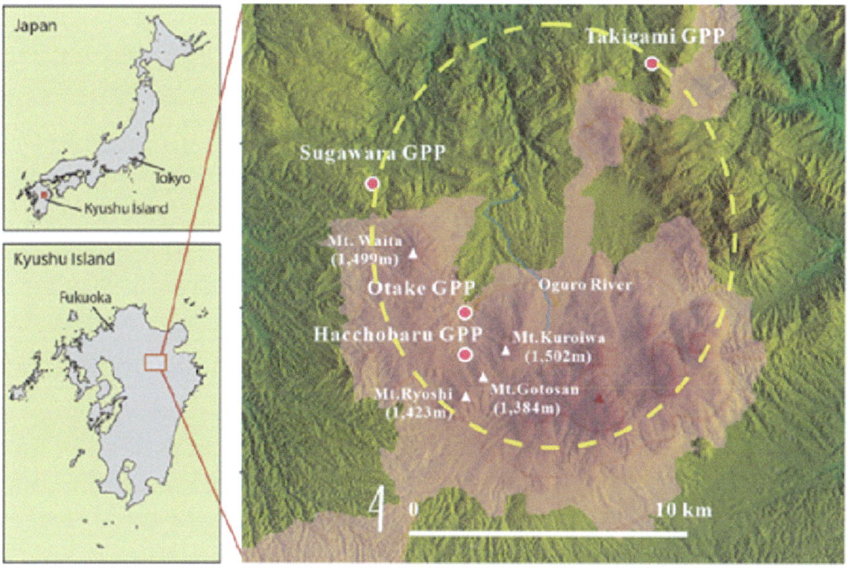

Fig. 6.3 Location map of Hatchobaru geothermal power plant and the nearby geothermal power plants with the surrounding volcanic mountains and approximate boundary of the buried Shishimuta Caldera structure inferred from gravity anomaly (GSI 1997 with modification)

Fig. 6.4 Overview of the Hatchobaru GPP, looking toward west-northwest from a production well pad (Kyuden Mirai Energy 2024)

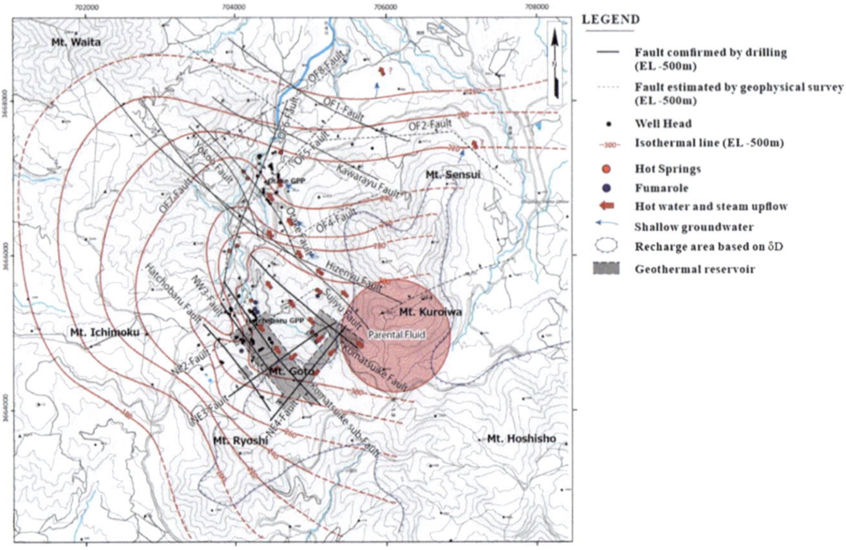

Fig. 6.5 Plan view of the geothermal conceptual model of the Hatchobaru geothermal field (Tsutsumi et al. 2021)

- Alteration minerals: Cap rock containing clay minerals such as smectite, alunite, and kaolinite developed near the surface to depths around 500 m above sea level, extending mostly along NW-trending faults.

6.1.2.3 Applied Geophysical Exploration in the Hatchobaru Geothermal Field

As Japan's largest geothermal power plant, the Hatchobaru field has been the site for testing and applying various geoscientific technologies, including those at the R&D stage. In other sections of this book, significant findings made using the fiber optic DTS (distributed temperature sensor) logging system (Chapter Part I-7) and the fiber optic DAS (distributed acoustic sensor) logging system are introduced (Chapter Part I-8) (Kasahara et al. 2024, 2025).

The subsurface resistivity structure is crucial in constructing the Hatchobaru geothermal reservoir model. The low resistivity distribution of clay minerals in the cap rock is key to identifying the reservoir structure (Fig. 6.6). Consequently, resistivity methods are highly effective for imaging subsurface structures. In the Hatchobaru field, MT (Magnetotelluric) data acquisition campaigns were conducted in 1986, 1994, and 2010, covering nearly 100 survey stations with less than 200-m spacing in the relatively small area around the plant (Fig. 6.7).

Data processing technology has evolved from 2D forward modelling (Tagomori et al. 1997) to 3D inversion analysis with topography compensation. Inagaki et al.

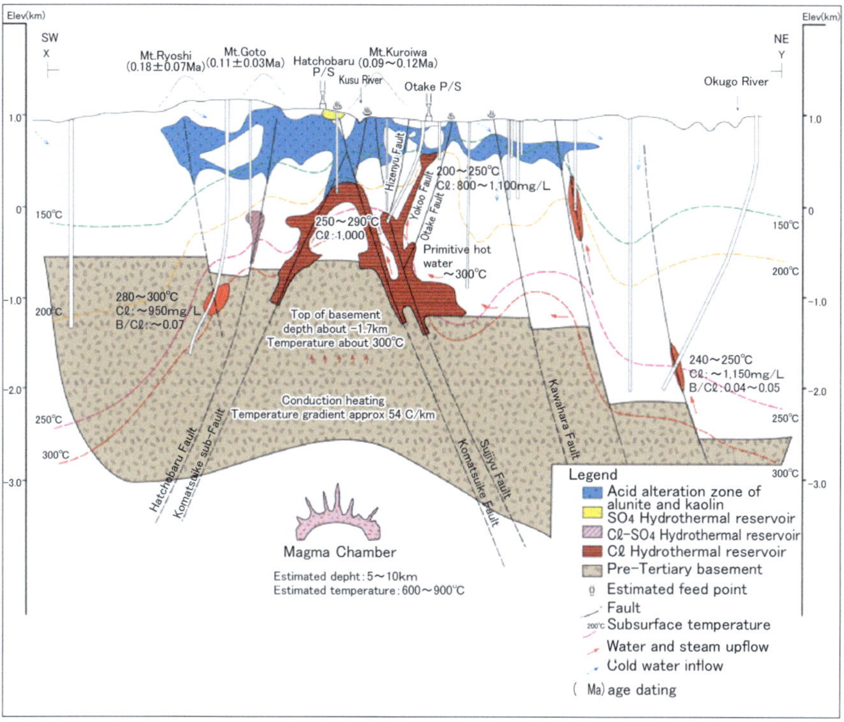

Fig. 6.6 Geothermal reservoir model along NW–SE cross section in the Hatchobaru geothermal field (Tsutsumi et al. 2021)

(2022) analyzed the quality between 2D MT forward modeling and 3D MT inversion models based on well data. They concluded that the resistivity distributions from the 3D MT inversion better matched the hydrothermal mineralogy and subsurface temperature distribution observed in the wells (Figs. 6.8 and 6.9). Therefore, the 3D MT inversion analysis provides a more reliable basis for building a geothermal conceptual model for Hatchobaru field. Using DAS technology, Kasahara et al. (2024) studied the geothermal characteristics in the Hatchobaru-Otake geothermal field. The results show high seismic reflective zone are distributing below Otake geothermal field at the depth below 2 km depth. They observed more than 30 microearthquakes during two-week survey period. The hypocenters are distributed below the Hatchobaru-Otake geothermal filed mostly below 2 km to 4 km depth. The seismic reflections and microearthquake hypocenter distribution supports the geothermal model proposed by Momita et al. (2000).

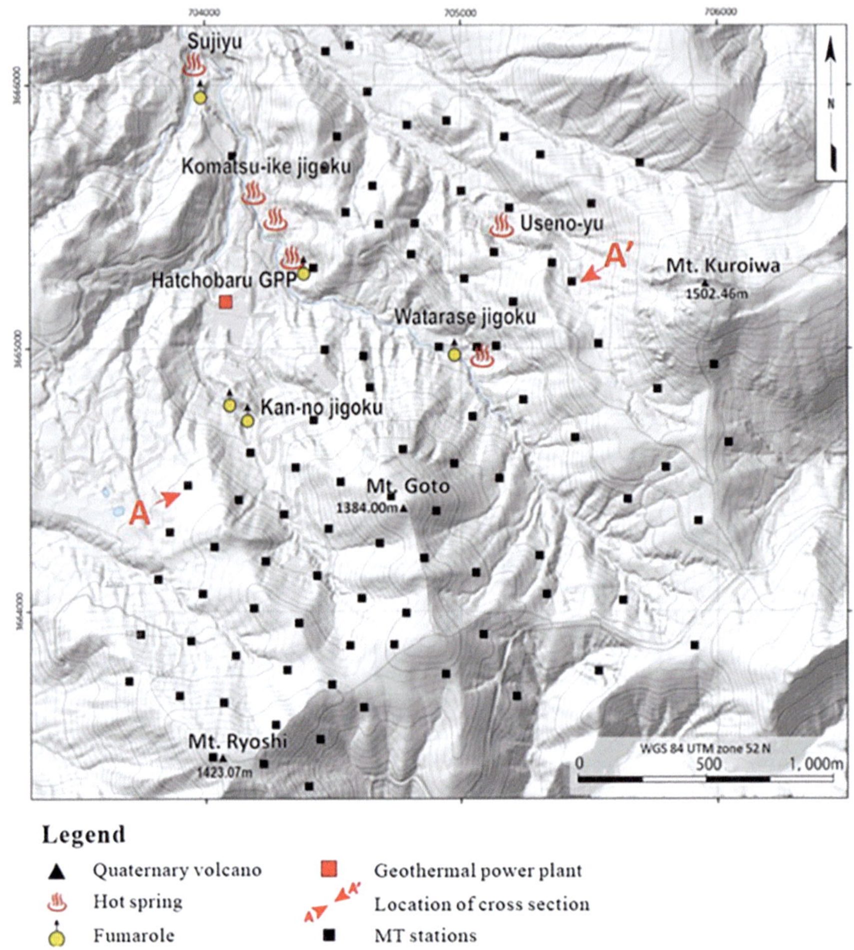

Fig. 6.7 Map showing the placement of MT stations, data from which were processed for the 3D Inversion analysis for the Hatchobaru geothermal field (Inagaki et al. 2022)

6.1.3 Takigami Geothermal Field, Central Kyushu

6.1.3.1 Introduction

The Takigami geothermal power plant is approximately 15 km north-northeast of the Hatchobaru geothermal field, along the ridge of Shishimuta caldera as shown in Fig. 6.3. It commenced commercial operations in November 1996 with a rated output of 27.5 MWe. While the development of geothermal resources was carried out by Idemitsu Kosan (Idemitsu), the plant's operation is managed jointly by Idemitsu and Kyushu EPC. Idemitsu is responsible for geothermal field management and

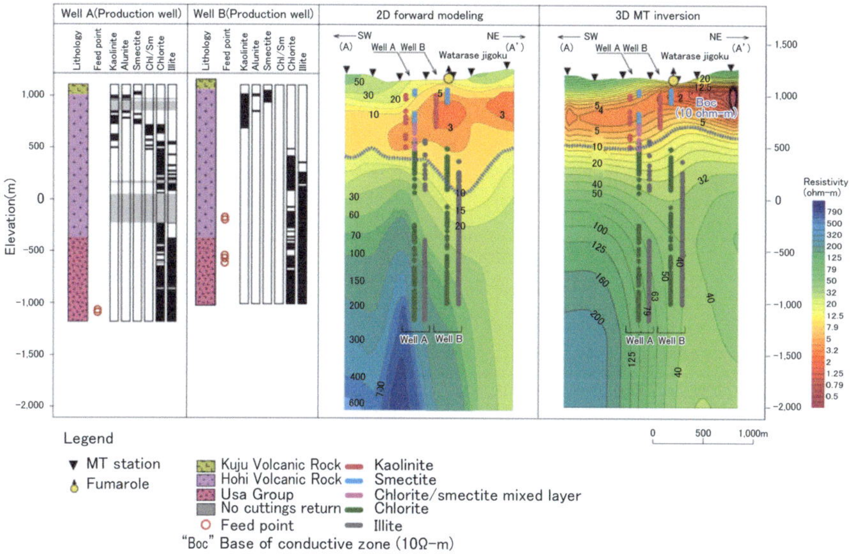

Fig. 6.8 Comparison between 2D MT forward analysis and 3D MT Inversion analysis along A-A' (Fig. 6.43), feed points, and occurrence of alteration minerals in well A and B. The dashed line highlights the base of the conductive zone (Boc: 10 Ω-m) (Inagaki et al. 2022)

steam supply, while Kyushu EPC handles the energy conversion. The power plant operations are controlled remotely from the Otake power plant (Fig. 6.10).

The re-injected brine, with ample extractable energy (~1,000 t/hr at 130 °C in Fig. 6.11), indicated the potential for further utilization. Consequently, a 5 MWe binary power plant was constructed in March 2017 on an existing production well pad owned and operated by Idemitsu.

6.1.3.2 Geothermal Systems of the Takigami Geothermal Field

The initial layout of drilling pads, well traces, pipelines, and the power plant, along with the geological setting of the Takigami geothermal field, is depicted in Fig. 6.12.

The geologic cross-section of the Takigami field and the subsurface isothermal contour is shown in Fig. 6.13. Based on drilling findings and other surface and subsurface studies, Takenaka et al. (1991) proposed a geothermal reservoir model (Fig. 6.14). Key features of the field include:

1. Geothermal fluid origin: Derived from meteoric water, recharged in mountainous areas over 880 m in altitude south of the field.
2. Reservoir division: The reservoir is divided into two parts:
   - Southwestern: Deeper, higher temperature (~250 °C) and salinity (Cl: ~600 ppm)

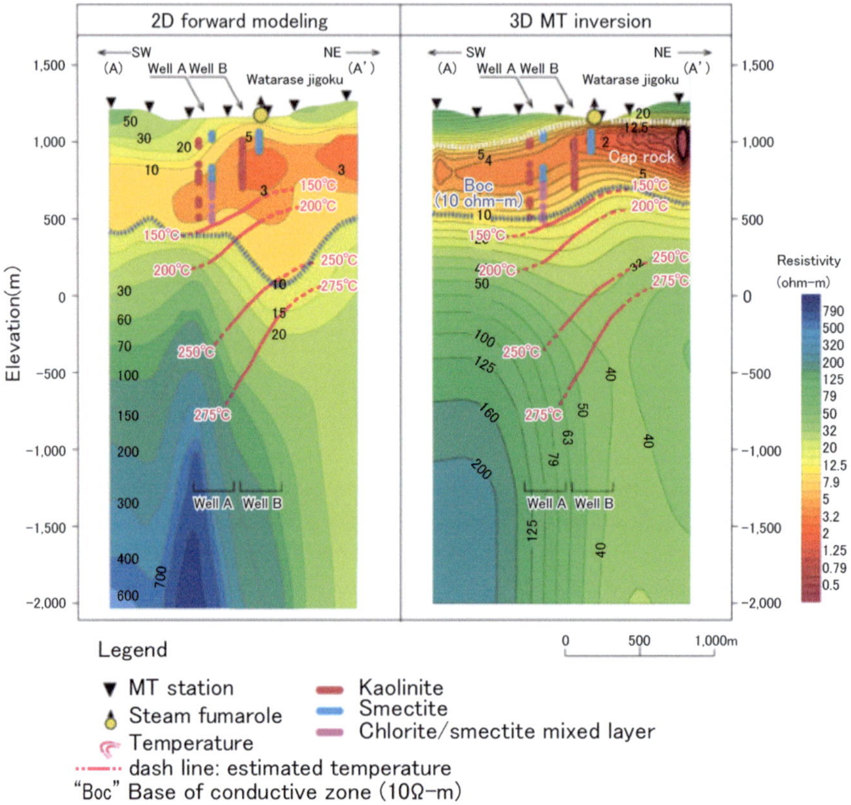

Fig. 6.9 Comparison between 2D MT forward analysis and 3D MT Inversion analysis along A-A' (Fig. 6.43). Similar to Fig. 6.8, with addition of isothermal temperature contours inferred from the well bore data are omission of chlorite and illite distribution in the wells (Inagaki et al. 2022)

Fig. 6.10 Picture of Takigami geothermal power plant (left) and the Takigami binary geothermal power plant (right) (Idemitsu Kosan 2024)

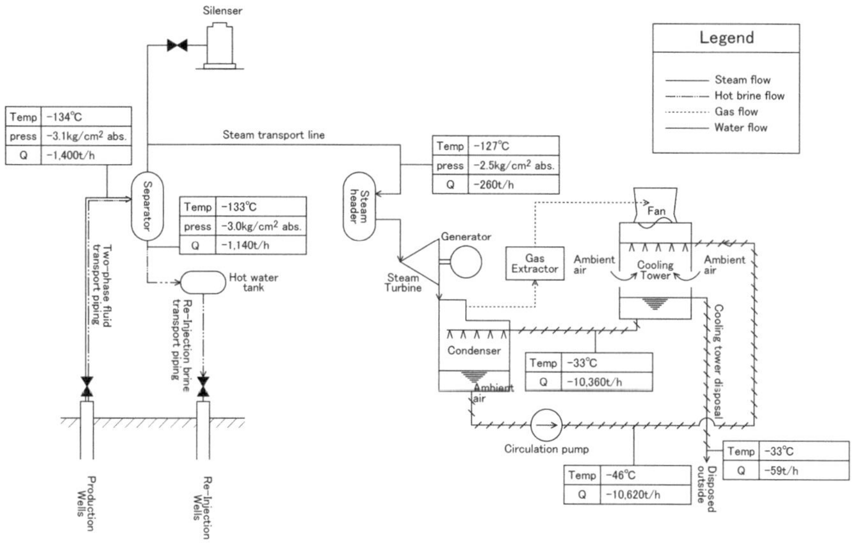

Fig. 6.11 The original flow design of the Takigami geothermal power plant (Takenaka et al. 1995)

- Northwestern: Shallower, cooler, more fractured, with moderate temperature (16–0210 °C) and salinity (Cl: ~ 450 ppm), diluted by deep formation brine.

3. Southwestern reservoir: Exhibits higher temperature salinity, ascending laterally to the northeast.
4. Northeastern fluid: This fluid shows varying sulfate concentrations controlled by the solubility of anhydrite, depending on fluid temperature.

The deployment of a carbonate-scale abatement system in the Takigami field marks a significant achievement. Early findings from nearby exploratory wells indicated carbonate-scale precipitation in the Takigami production wells. After evaluating various mitigation options, the injection of a chemical inhibitor (poly-acrylic acid) into the borehole was chosen. This scale abatement is necessary for wells in the eastern part of the field but not for those in other areas. The conceptual design of the scale abatement system is depicted in Fig. 6.15.

6.1.3.3 Applied Geophysical Exploration in the Takigami Geothermal Field

The area around the Takigami geothermal field is associated with a negative Bouguer anomaly, possibly due to the buried Shishimuta caldera. The field is situated on the eastern margin of this low-gravity zone. The northeastern part of the geothermal area extends to a local gravitational high, while the western portion shows a low gravity

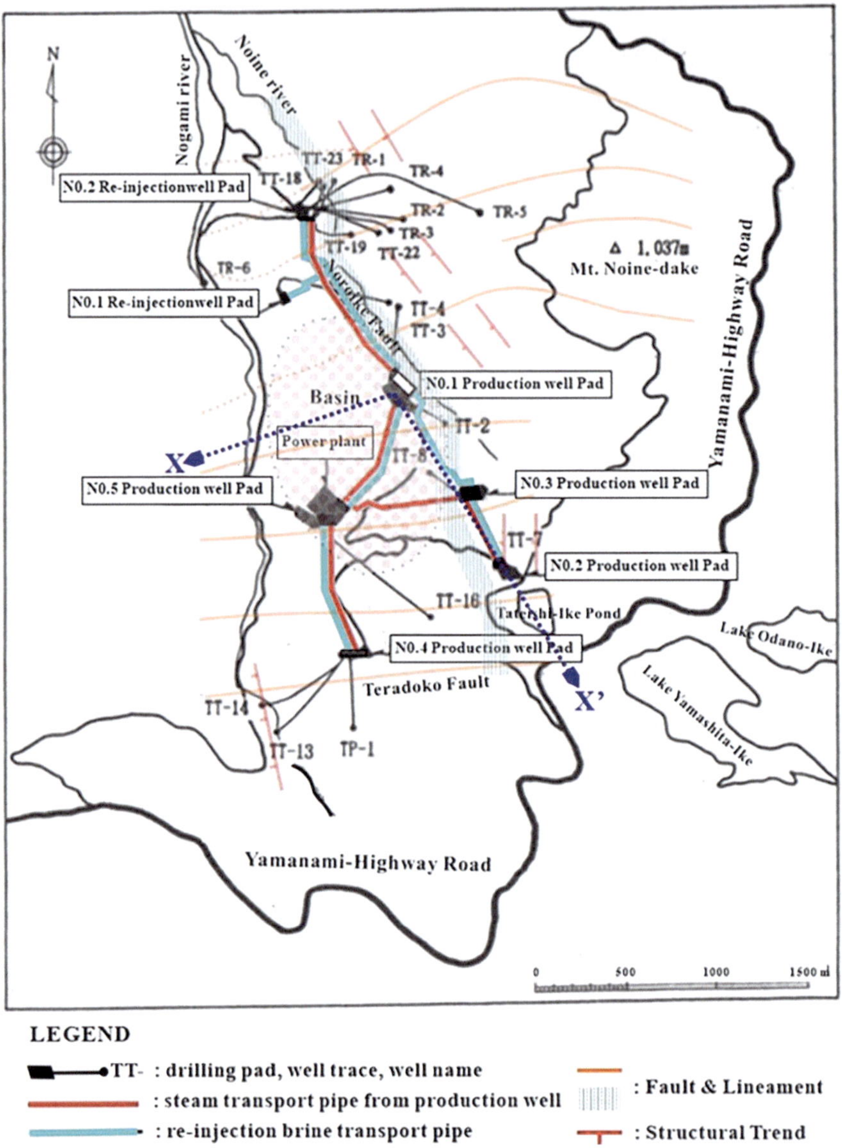

Fig. 6.12 Power plant lay-out and notable geological feature of the Takigami geothermal field (Takenaka et al. 1991 with modification)

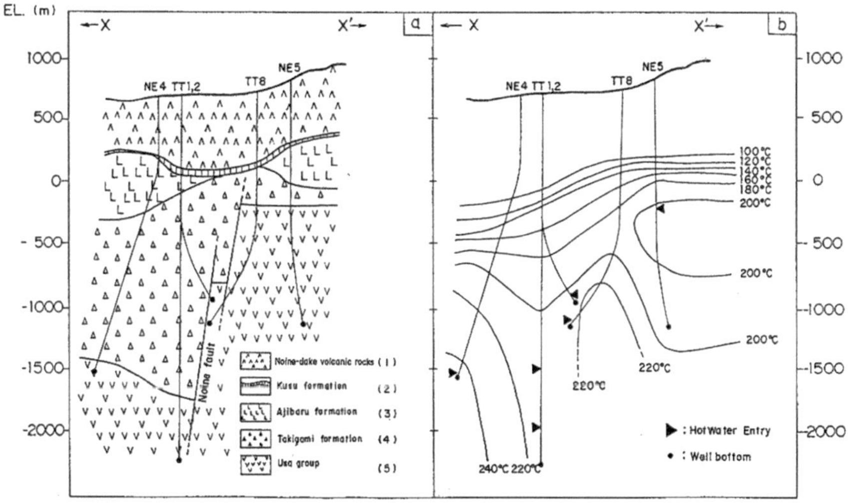

Fig. 6.13 **a** Generalized geologic cross-section of the Takigami field (X–X' in Fig. 6.12), **b** Measured downhole temperatures, deduced isotherms, and location of the hot water entries (Takenaka et al. 1991)

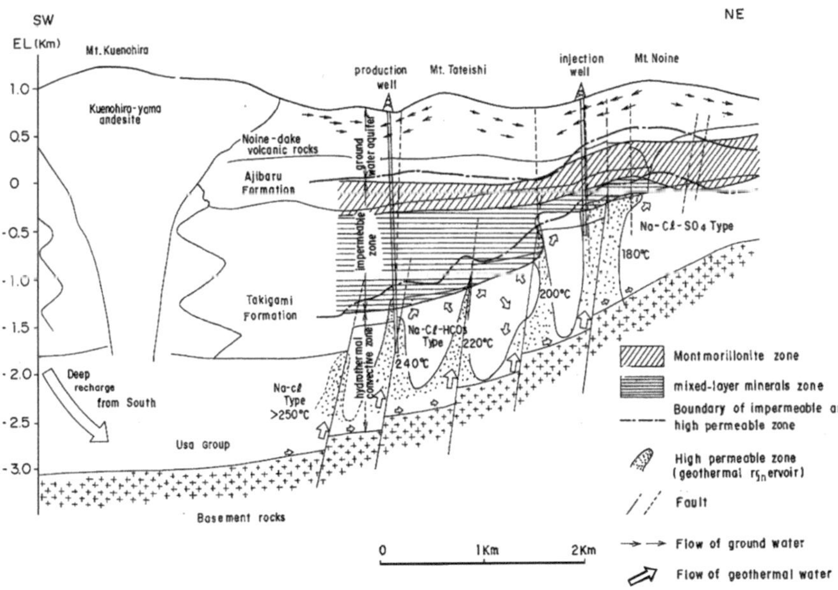

Fig. 6.14 Geothermal reservoir model of the Takigami field (Takenaka et al. 1991)

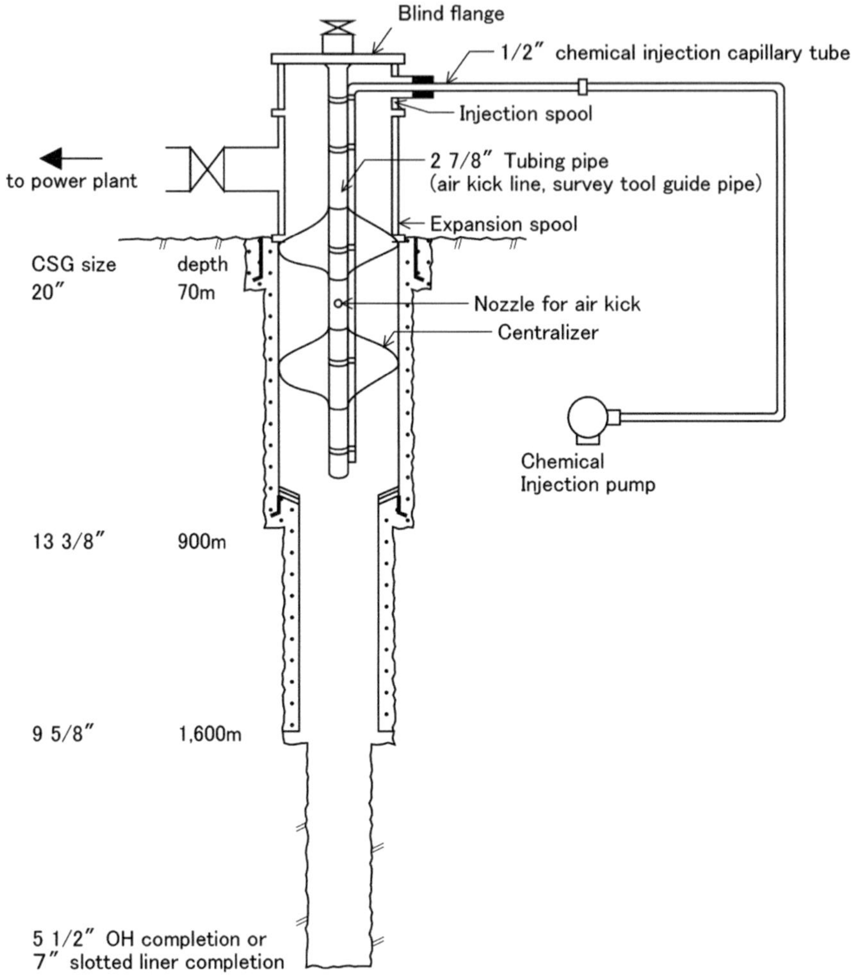

Fig. 6.15 Conceptual diagram of the chemical injection system adopted in some of the production wells in the Takigami field (Takenaka et al. 1995)

anomaly (−35 mgal). A north–south striking Noine fault is believed to exist in the area, related to a large-scale volcano-tectonic depression (Kamata 1989) (Fig. 6.16).

Geophysical technologies, such as MT (Magnetotelluric) and Remote Sensing, played a crucial role in siting geothermal wells, especially during the early exploration stages, when few surface geothermal manifestations were found in the vicinity of the field. The author references a paper discussing MT inversion in the Takigami field (Ushijima et al. 2005). Although the inversion was two-dimensional, the results showed good agreement with the well data, including electrical resistivity logs, lost circulation zones during drilling, and temperature distributions (Fig. 6.17).

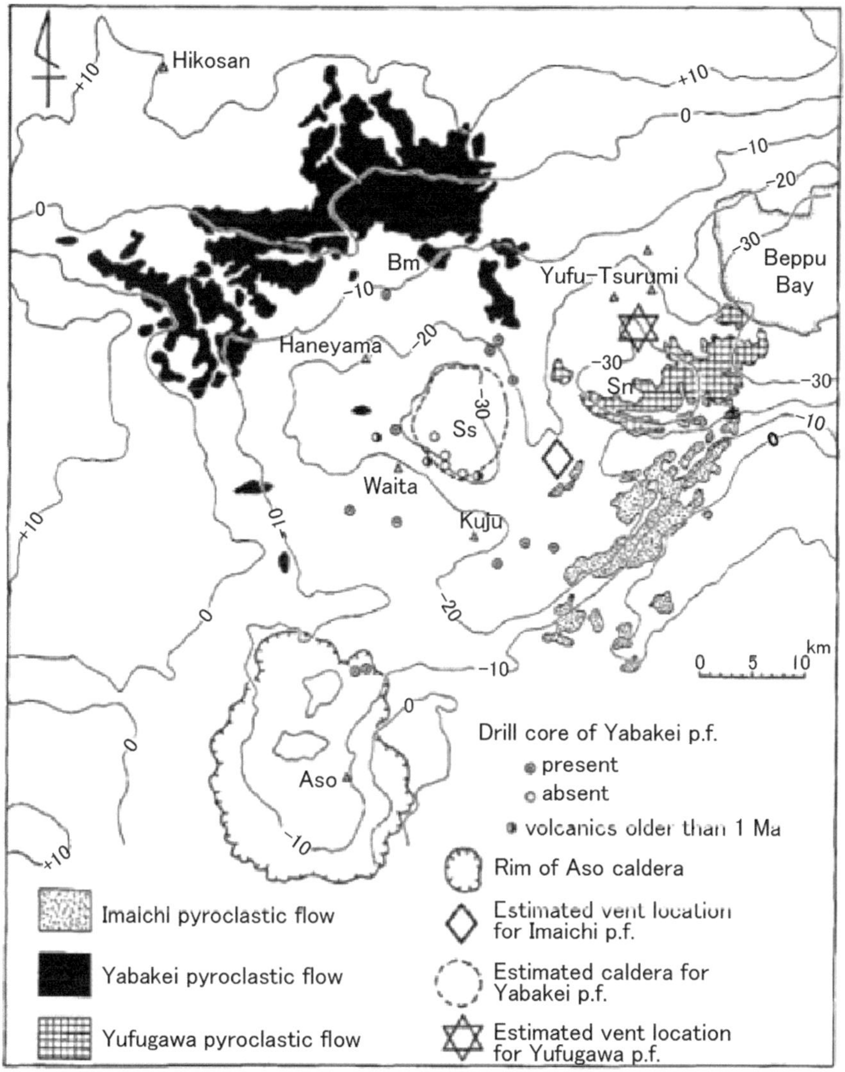

Fig. 6.16 Distribution and vents of three large-scale Pleistocene pyroclastic-flow deposits including Yabakei pyroclastic flow (Shishimuta caldera) in the eastern edge of Beppu-Shimabara graben. Contours show Bouguer anomaly in mgal (Kamata 1989)

Recently, Kasahara et al. (2025) surveyed the Takigami geothermal filed using DAS technology. The EW Vp profile across the TP-02 wellhead is shown in Fig. 6.18. The vertical fault is 600 m east of TP-02 wellhead. They made seismic migration image of seismic reflection obtained by DAS records (Figs. 6.19 and 6.20). Higher seismic refection images distribute from1 km to 4 km depth. These could be caused by geothermal reservoir filing the fractures.

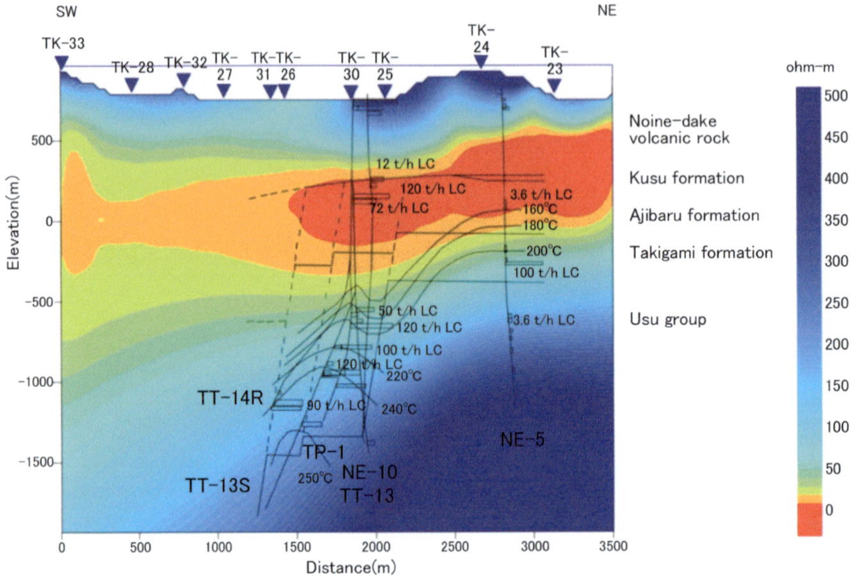

Fig. 6.17 Resistivity section derived from 2-D inversion of MT data together with data (Ushijima et al. 2005)

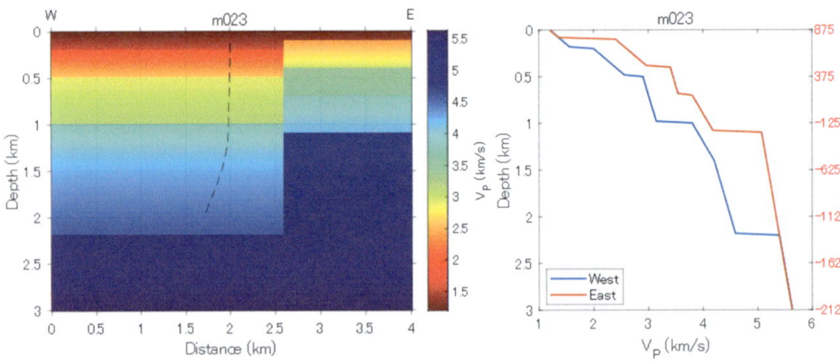

Fig. 6.18 Vp structure of EW depth section obtained by the DAS seismic survey (Kasahara et al. 2025). The right diagram shows the Vp depth profile of the east and western side of the vertical fault at 2.5 km distance. The broke line shows the TP-02 geothermal well

6.1.4 Ogiri Geothermal Field, Southern Kyushu

6.1.4.1 Introduction

The Ogiri geothermal power plant is located in the southern part of Kyushu Island, on the western slope of the Kirishima volcanic region, one of Japan's largest Quaternary

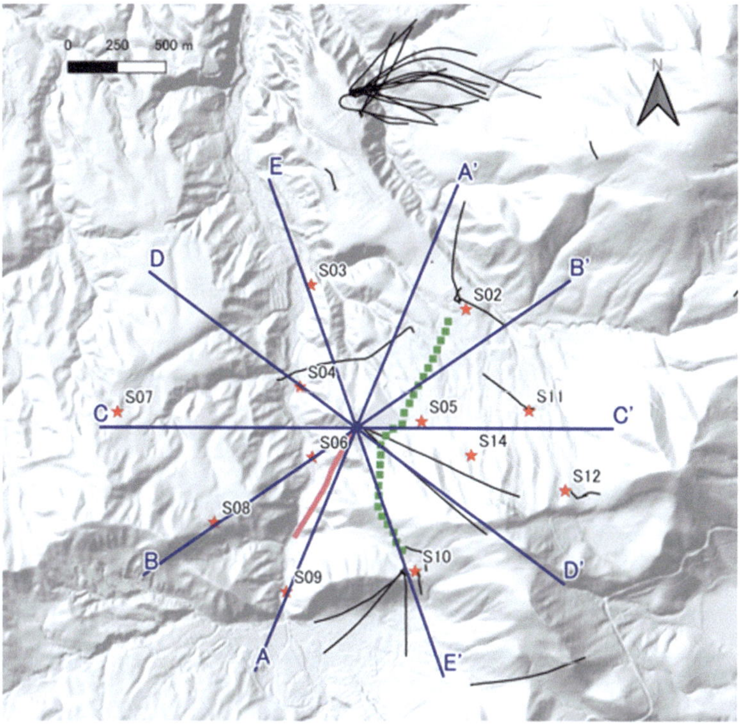

Fig. 6.19 Location map of migration imaging of seismic reflection sections (shown by blue lines) (Kasahara et al. 2025). The refection images are shown in Fig. 6.20. The pink line shows projection of the TP—02 geothermal well

volcanoes. The plant is situated within the west Kirishima geothermal field, with an elevation ranging between 800 and 1000 m (Fig. 6.21). Geothermal exploration for electric power development began in 1973 by private companies, followed by public institutions in 1975.

Commercial operation of the plant started in 1996 with an installed capacity of 30MWe. The development model of the Ogiri plant is similar to the Takigami plant. For Oogiri, Nittetsu Kagoshima Geothermal Co. Ltd (Nittestu) managed the geothermal resource development, while the plant operation was a joint effort between Nittestu and Kyushu EPC. Nittetsu managed the geothermal field and steam supply, whereas Kyushu EPC handled energy conversion. According to Takayama et al. (2014), the plant had an average utilization factor of over 90% from its inception until 2014, covering 18 years. By 2017, due to a decline in reservoir productivity, the nominal output was changed to 28MWe, with a utilization factor of around 75% (TENPES 2021), which is still relatively high compared to other geothermal power plants in Japan.

The plant facilities, including well pads and piping, are laid out over a gentle mountain slope, covering an area of approximately 500 m in width and 1,500 m in

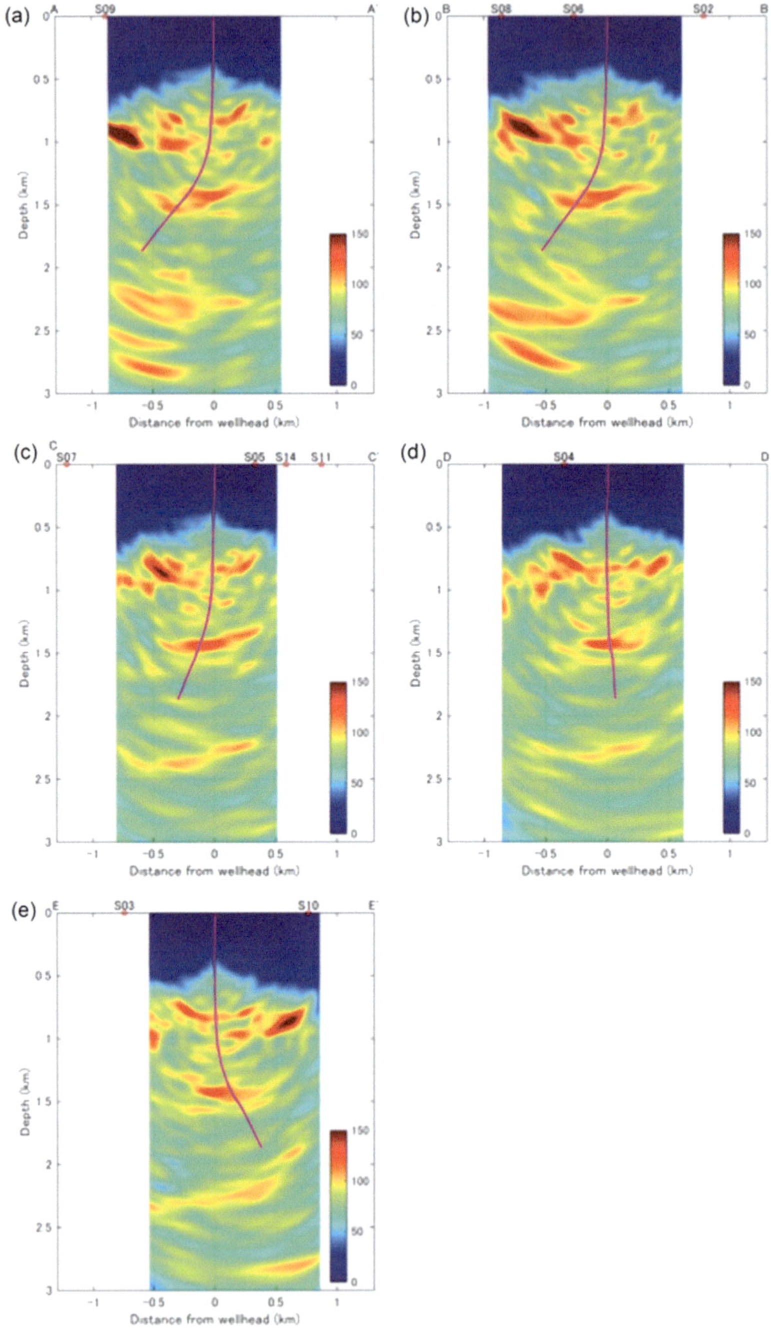

Fig. 6.20 Seismic migration images of Takigami geothermal field obtained by DAS seismic survey (Kasahara et al. 2025). The location of each cross-section is shown in Fig. 6.19. The color shows the intensity of seismic reflection. The reddish color is higher seismic reflection

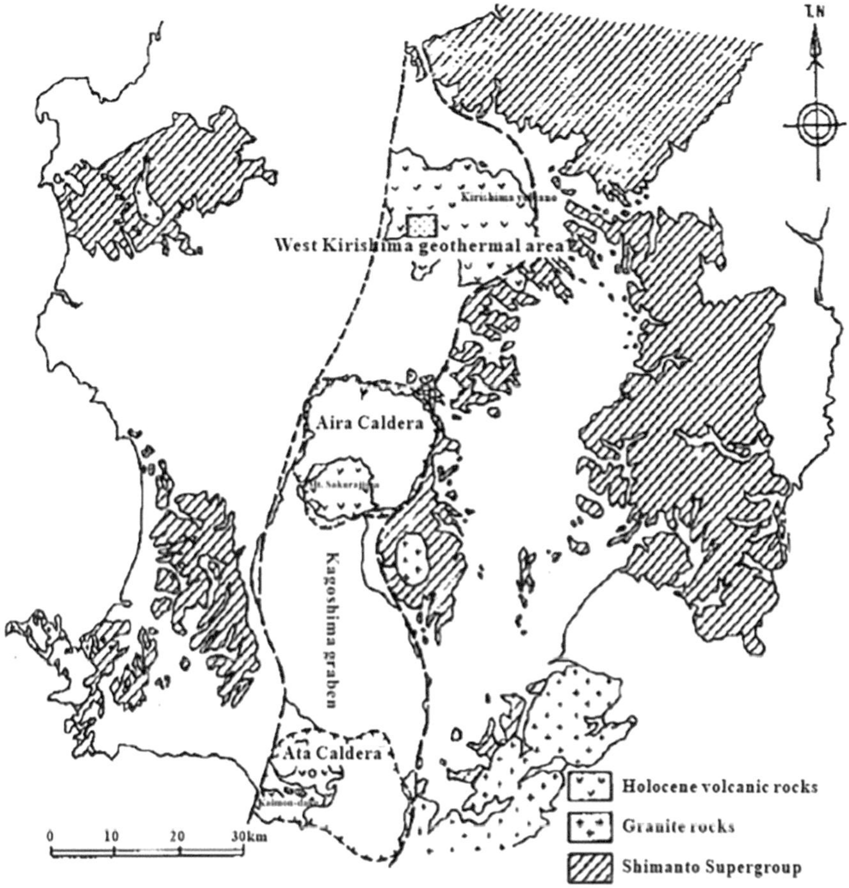

Fig. 6.21 Regional geological setting in and around the West Kirishima geothermal area (Goko 1995)

length (Fig. 6.22). Separated brine from the separators is re-injected by gravity. There are three drilling pads (A, B, and C) for production wells located in the eastern part of the plant premises. These pads are 130 m higher in elevation than the re-injection well pads (E, D, and F) in the western part. As of March 2020, there were 13 production wells and 5 reinjection wells in service, compared to 10 and 7, respectively, when the commercial operation began (Fig. 6.23).

6.1.4.2 Geothermal Systems of the Ogiri Geothermal Field

The geothermal structure and hydrological features of the Ogiri geothermal systems are extensively discussed by Goko (2000), Horikoshi et al. (2005), Itoi et al. (2010), and Takayama et al. (2014). According to these studies, the productive geothermal

Fig. 6.22 Picture of the Ogiri geothermal power plant (Kyuden Mirai Energy 2024)

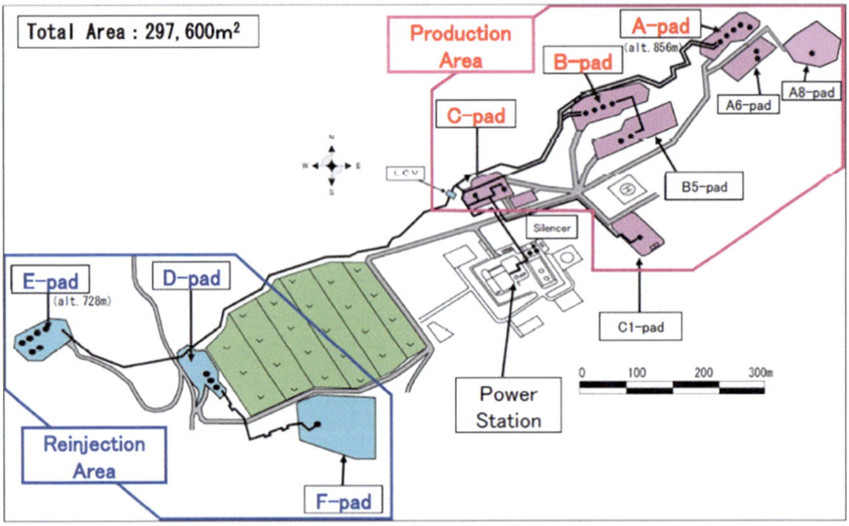

Fig. 6.23 Layout of the surface facilities including the drilling pads (A-F) and pipe lines of the Ogiri geothermal power plant (Takayama et al. 2014)

zone is mainly controlled by two vertical faulting systems trending NW-SE and ENE-WSW (Figs. 6.24 and 6.25). The Ginyu fault, trending ENE-SWS, and associated extensional fractures in an echelon pattern are major steam productive systems and permeable zones for reinjecting hot water.

Figure 6.26 presents a simplified geothermal model near the Oogiri field (Uchida 2005). In this model, the Mesozoic basement layer is situated at a depth of about 2.5 km, overlain by the Quaternary volcanic layers. The Ogiri geothermal reservoir is associated with the Ginyu Fault and is located in the Quaternary volcanic layers. High-temperature anomalies beneath the Ginyu and Shiramizugoe Faults are indicated by temperature contours of 200 and 225 °C; it can be deduced that there are high-temperature anomalies beneath the Ginyu and Shiramizugoe Faults. The clay

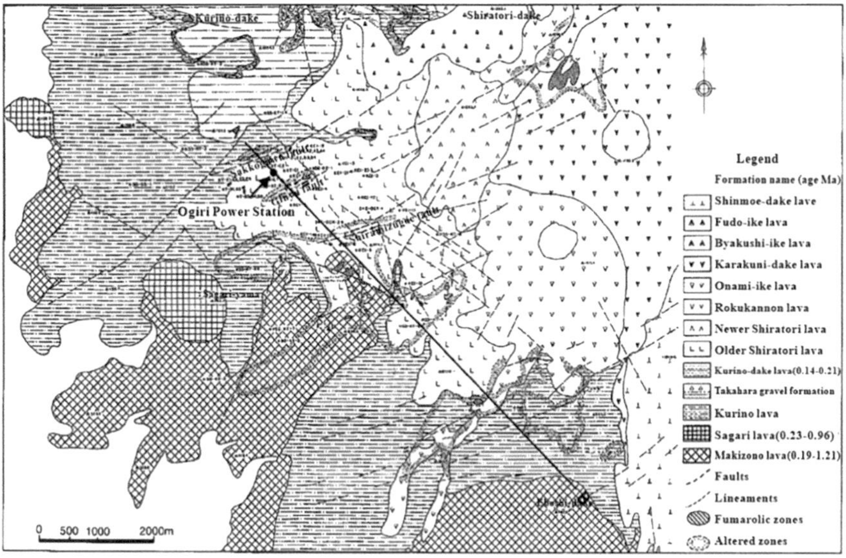

Fig. 6.24 Geologic map of the West Kirishima geothermal area. A-B represents line of the cross-section for Fig. 6.25 (Horikoshi et al. 2005)

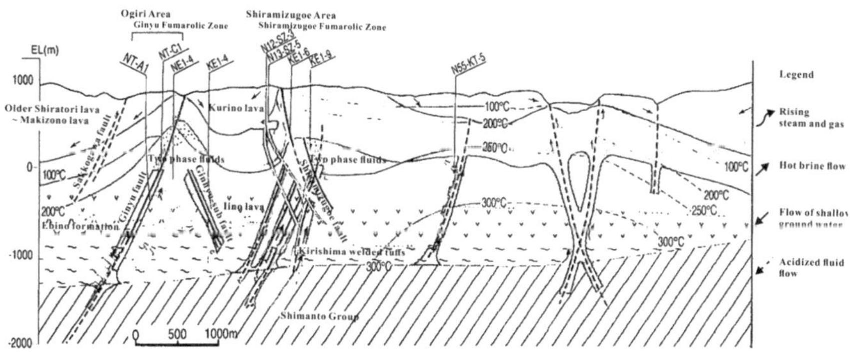

Fig. 6.25 Cross-section of line A-B, showing geology in the West Kirishima geothermal area. Isotherms are in °C (Horikoshi et al. 2005)

cap, dominated by low-temperature clay minerals such as smectite, is distributed in zones below 200 °C.

The main features of the Ogiri geothermal systems are:

- Reservoir temperatures: Mostly constant ranging around 230–232 °C along fractured zones associated with the Ginyu fault and auxiliary fault systems.

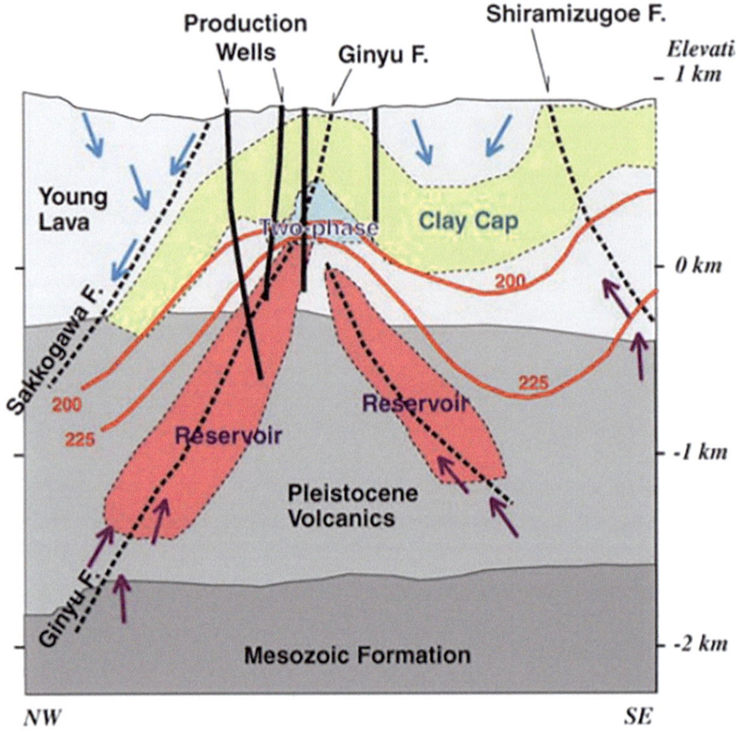

Fig. 6.26 Cross-section depicting a simplified geothermal model along a NE-SW line of the Ogiri geothermal field (Uchida 2005)

- Reservoir type: Water-dominated with production zones ranging about 600–1,300 m, primarily along the Ginyu fault, auxiliary faults, and associated fractures within andesitic volcanic rocks. A partially recognized steam-dominated reservoir.
- Subsurface temperatures gradient: Increases from the northeastern to the south-western region, indicating a heat source associated with the post-igneous activity of the Kirishima volcanic region. The geothermal fluid of meteoric origin recharges into the Ginyu fault from deeper zones in the east and northeast.
- Ginyu fault: Major steam productive reservoir with near-planar fracture parallel to the fault plane, ensuring power generation output equivalent to 30 MWe for about 18 years since commercial operation began in 1996.
- Clay cap: Over the entire Ogiri geothermal field, smectite and other altered minerals are well observed. These minerals seal pores and void spaces, forming impermeable layers that prevent cold groundwater infiltration and retain heat from ascending hot convective flow from the deep hot reservoir.

6.1.4.3 Applied Geophysical Exploration in the Ogiri Geothermal Field

As with many other geothermal fields in Japan, various geophysical technologies have been implemented in the Ogiri field, providing useful information for each survey's objectives. Among these technologies, MT shows excellent cost performance for geothermal field applications.

This book introduces an achievement in MT analysis based on the work of Uchida (2005), who developed a 3-D inversion code for MT data. One key element of this code is the successful adaptation of the Bayesian criterion ABIC for optimizing the trade-off between minimizing data misfit, model roughness, and static shifts.

The MT field data is derived from surveys conducted by New Energy Development Organization (NEDO), Nittetsu, and AIST (National Institute of Advanced Industrial Science and Technology) since 1996 over the Ogiri and Shiramizugoe geothermal fields. Survey lines were set in the NW-SE direction, perpendicular to the geologic strike of the faults, with spacing of 250–500 m and MT stations arranged at intervals of 150–300 m. Approximately 170 MT stations were used, as shown in Fig. 6.27. Remote reference analysis was applied to all the data using a reference site located about 40 km south of the survey area.

In Fig. 6.27, the 3D inversion model is compared with the 2D inversion results of TM-mode data along Lines A and B. Generally, the 2D and 3D models render consistent results. However, for Line B, the 2D model failed to identify the lower boundary of the low-resistivity anomaly, making the 3D model more realistic.

The resistivity distribution in the shallow layers (less than 300 m deep) shows a clear contrast between the high resistivity zone in the northern half and the low-resistivity zone in the southern half, aligning with surface manifestations shown in Figs. 6.24 and 6.25. At mid-depth, the entire area shows low resistivity. At deeper depths, high-resistivity anomalies are recognized at the center of the region, while the subsurface has high resistivity at depths greater than 2 km. The low-resistivity anomalies are associated with the Ginyu and Shiramizugoe faults, likely due to intensively altered clay minerals. The shallow high-resistivity zone on the northern side of the Sakkogawa Fault is attributed to an unaltered young lava layer (Fig. 6.28).

Comparing the 3D resistivity model and the geothermal reservoir model shown in Fig. 6.26, the shape of the lower boundary of the clay cap and the contour of 200 °C exhibit similarity with the resistivity distribution in the vertical resistivity sections of the 3D model (Fig. 6.29).

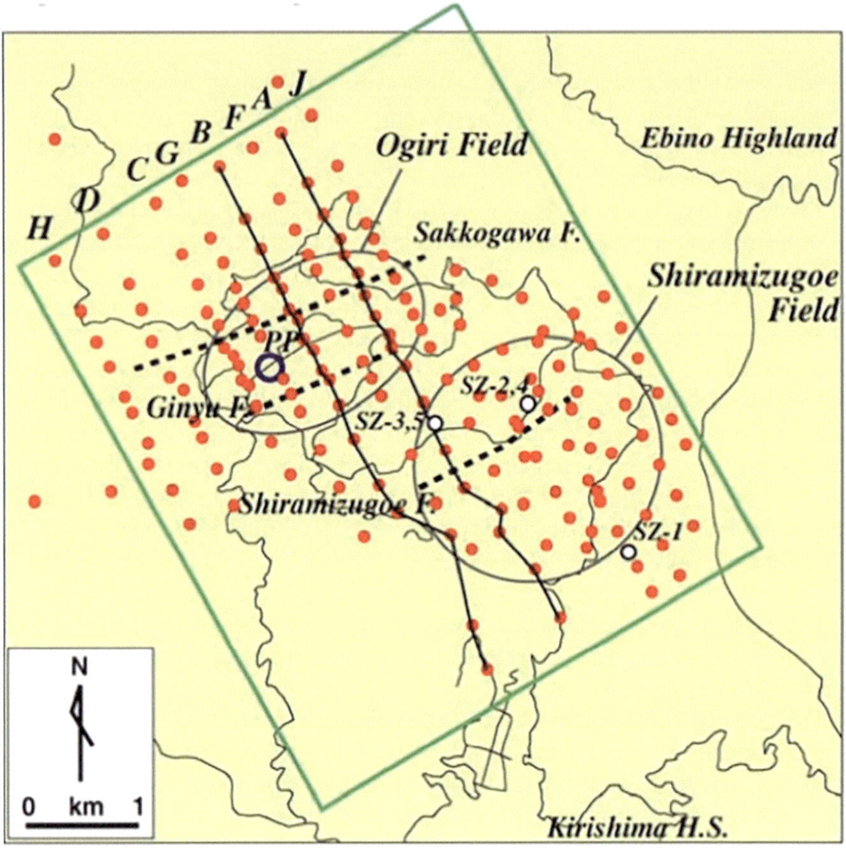

Fig. 6.27 MT stations conducted in and around the Ogiri geothermal field and the Shiramizugoe geothermal field with locations of cross-section of Line-A and Line-B (Uchida 2005)

6.1.5 Yamagawa Geothermal Field at the Southern Edge of the Kyushu Island

6.1.5.1 Introduction

The Yamagawa geothermal power plant is situated in the Fushime area at the southern edge of the Satsuma peninsula in Kyushu Island, making it Japan's southernmost geothermal power plant. A flat vegetation field surrounds the plant and is close to the seashore (Fig. 6.30). The field is situated inside the Ata caldera, which is one of volcanic-tectonic depression zones of the Kagoshima graben and encircles volcanic features such as Kaimon-dake, lake Ikeda, lake Unagi, and Yamagawa-bay (Okada et al. 2000). Geothermal exploration activities in the Fushime area, including geological, geophysical (gravity), and exploratory well drilling, began in

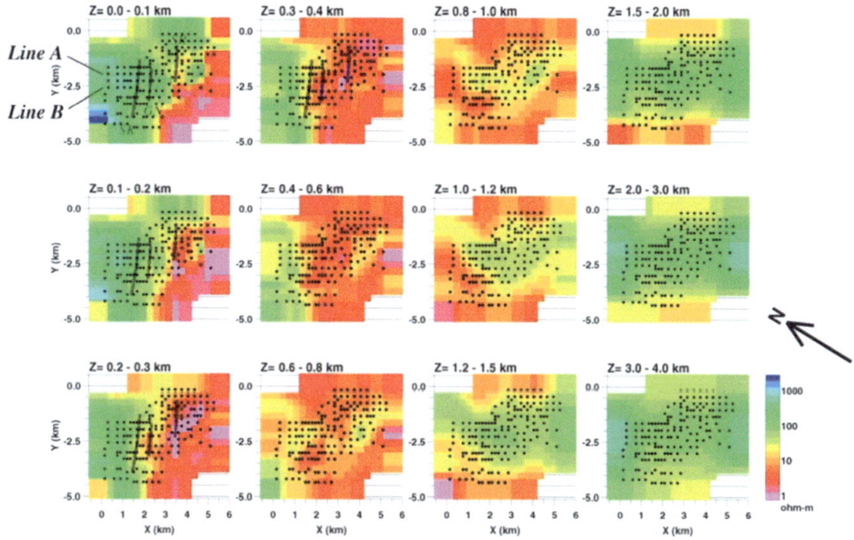

Fig. 6.28 Depth-slice resistivity sections of the 3D inversion process result made on the MT data obtained in the Oogiri and Shiramizugoe field (x direction is 150 degrees clockwise from North) (Uchida 2005)

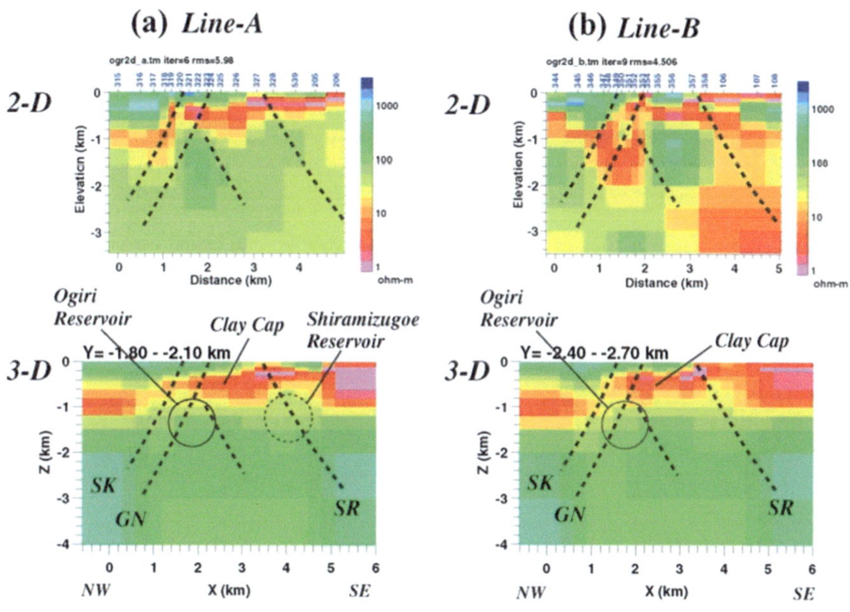

Fig. 6.29 Vertical resistivity sections computed by 2D and 3D inversion models along **a** Line-A and **b** Line-B. Thick dashed lines are estimated faults. Large circles are estimated location of the production zones (Uchida 2005)

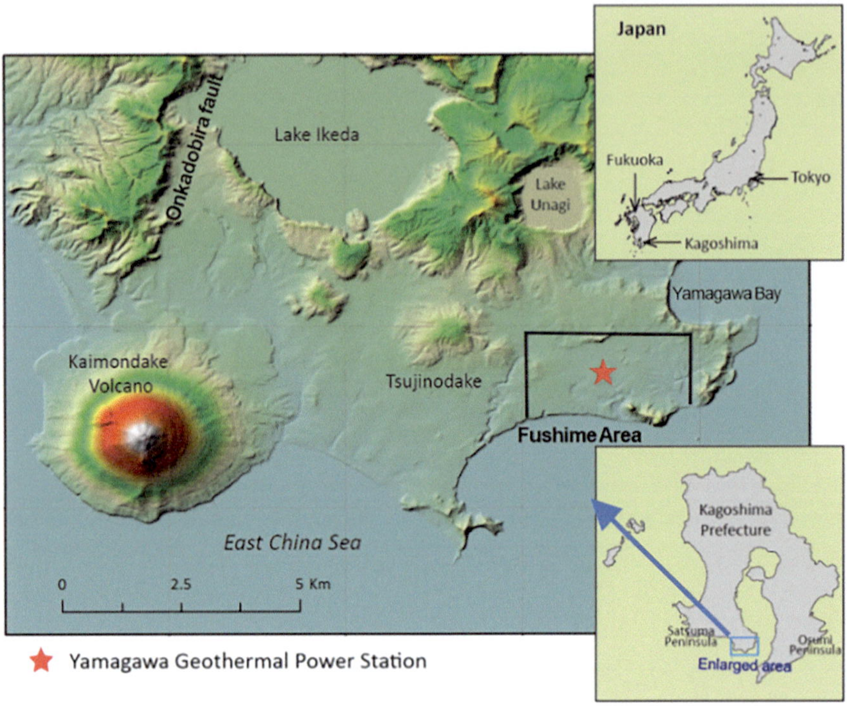

Fig. 6.30 Location of the Fushime geothermal field and the Yamagawa geothermal power plant (Wulaningsih et al. 2017)

1962 and continued until around 1988. These activities were primarily conducted by governmental organizations such as the Geological Survey of Japan, universities, and NEDO.

In parallel with these governmental exploration activities, Japan Petroleum Exploration Co., Ltd (JAPEX) embarked on a geothermal exploration venture in 1977. By 1988, after confirming business feasibility, JAPEX established a wholly owned subsidiary, JAPEX Geothermal Kyushu Ltd (JGK), to which the business rights were transferred. In 1995, JGK inaugurated commercial operation of the Yamagawa geothermal power plant with an installed capacity of 30MWe. Similar to the business structure of the Takigami and Oogiri plants, JGK managed the geothermal resource development, while the plant operation was a joint effort with Kyushu EPC. JGK was responsible for the geothermal field management and steam supply, while Kyushu EPC handled the energy conversion. In 2005, JAPEX decided to sell the assets and business rights for the Yamagawa geothermal power plant to Kyushu EPC. Since then, Kyushu EPC has managed all plant assets and operations, including steam supply and energy conversion (Fig. 6.31).

At the Yamagawa geothermal power plant, two-phase fluid is separated under high pressure and temperature conditions due to constraints related to scale precipitation.

Fig. 6.31 Picture of the Yamagawa geothermal power plant and plant facilities (left and foreground) and the Takeyama volcanic neck (pyroxene dacite) (right background) (Photograph by K. Tagomori on 17 Nov 2008)

Fig. 6.32 Overview of the Yamagawa Binary Power Station (ORC type) manufactured by Fuji Electric (Kondo et al. 2019)

With other mitigation methods available, Kyuden Mirai Energy Co., Inc. (a subsidiary of Kyushu EPC) installed a binary power generator with a rated output of 4.99 MWe in 2018 to utilize the extractable energy in the brine. The binary plant has been in commercial operation since then (Kondo et al. 2019) (Fig. 6.32).

6.1.5.2 Geothermal Systems of the Yamagawa Geothermal Field

According to statistics published by TENPES (2021), 35 wells have been drilled for production and reinjection purposes, and the power plant's utilization factor is about 50%. The lower utilization factor is due to unexpected issues such as Pb/Zn enriched sulfide scale precipitation in production wells and silica scale precipitation in reinjection wells (Sakuma 1999). As of March 2018, there were 8 production wells and 9 reinjection wells in operation, with depths ranging from 1,800 to 2,100 m for production wells and from 1,400 to 2,500 m for reinjection wells.

In 2015, JOGMEC entrusted JAPEX to conduct a seismic survey in cooperation with Kyushu EPC to enhance the assessment of the Yamagawa geothermal field's reservoir structure, combining this new information with data from past studies.

The main features of the Yamagawa geothermal systems include:

- Reservoir temperatures: Approximately 300 °C at depths around 1,700 m, with some production wells in the northern part of the production recording temperatures over 350 °C.
- Temperature inverted zone: Recognized at depths from 800 to 1,200 m where formation temperature is around 150 °C. Brine from these depths sometimes descends behind the casing walls to the casing shoe, adversely affecting well productivity.
- Reservoir type: Water-dominated, with geothermal fluids originating from seawater diluted by groundwater during the ascent (Okada et al. 2000),
- Predominant geothermal reservoir structure: Fractures associated with intrusive dacite located near the center of the field at around 1,500 m depth and high permeability zone related to faults trending NW–SE along the Takeyama-Tsujinodake tectonic line (Okada et al. 2000),
- Geological features: The field is covered by thick volcanic products, and some dyke intrusions aligned NW–SE from Tsujinodake to Takeyama dome, acting as local heat sources connected to geothermal manifestations such as hot springs and fumaroles in the Fushime geothermal field (Fig. 6.33),
- Mineralization: In deeper reservoirs with high temperatures over 250 °C, quartz, albite and chlorite are found. Nearer to the surface and in zones below 160 °C, smectite is widely distributed. Transition zones contain smectite/chlorite, mixed layer with sericite and smectite/sericite in the relatively shallower part, and in the lower part of the transition zones at 180–200 °C to 250 °C, quartz, albite, and wairakite are present (Fig. 6.34).

6.1.5.3 Applied Geophysical Exploration in the Yamagawa Geothermal Field

Since initial exploration and development of the Yamagawa field was carried out by JAPEX whose main business is oil and natural gas development, geophysical methods used in the Yamagawa field followed those for the oil/gas resource. The field's location on flat land close to the seashore might have influenced the geophysical process. This book introduces findings from advanced well-logging technologies based on a paper by Okada et al. (2002). To study the distribution, types, and characteristics of fractures in the Yamagawa field, and their relationship with other rock physical properties and tectonic settings, electric image logging surveys (FMI) and focusing resistivity logging surveys (Dual Lateral Log) were deployed in three selected wells. Although drilling core samples were not acquired in these wells, cores from an adjacent well provided useful information for analyzing the log data.

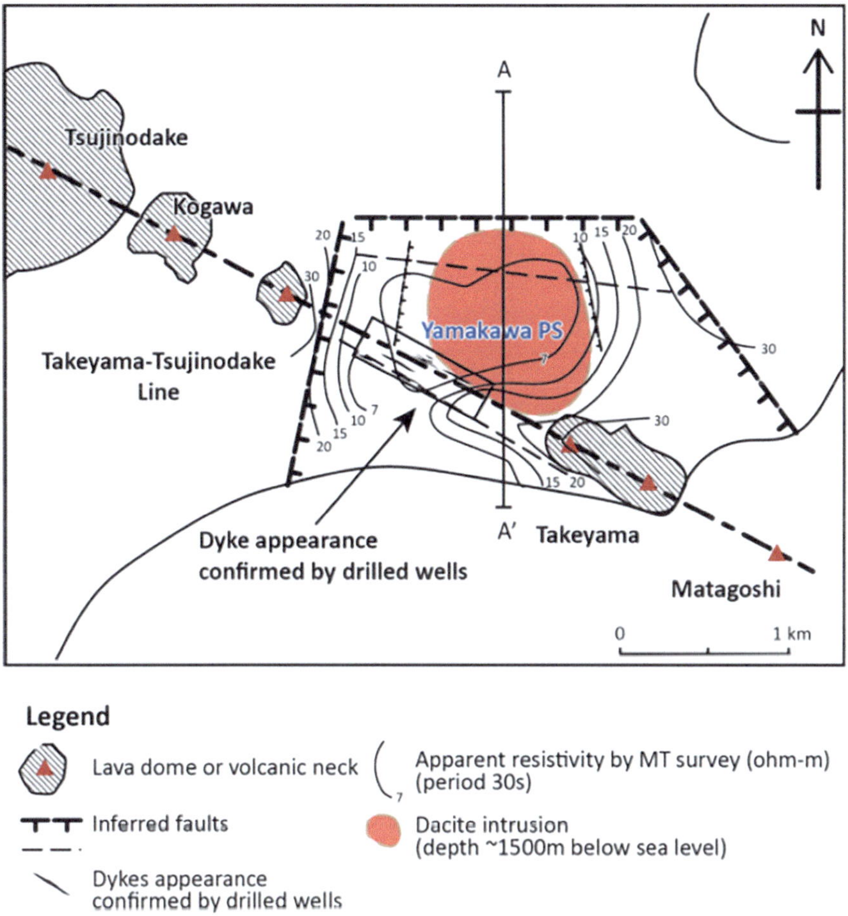

Fig. 6.33 Geologic features such as lava domes, volcanic neck, faults, and dikes along with resistivity counters in the Yamagawa power plant and surrounding area (Wulaningsih et al. 2017)

The lithology over the logging interval mainly consists of pyroclastic rocks and dacite intrusions, with productive fracture zones found within the dacite, which appears as a resistive feature in the images. The image analysis process involved identifying fracture features in the images and categorizing them into five fracture types and one structural feature, referring to Figs. 6.35 and 6.36,

1. Drilling-induced fracture: Appears almost parallel to the borehole axis.
2. Conductive fracture: Seen as black sine curves.
3. Minor fracture: Small-scale fractures.
4. Mega fracture: Large-scale conductive fractures with large aperture and extended reach.

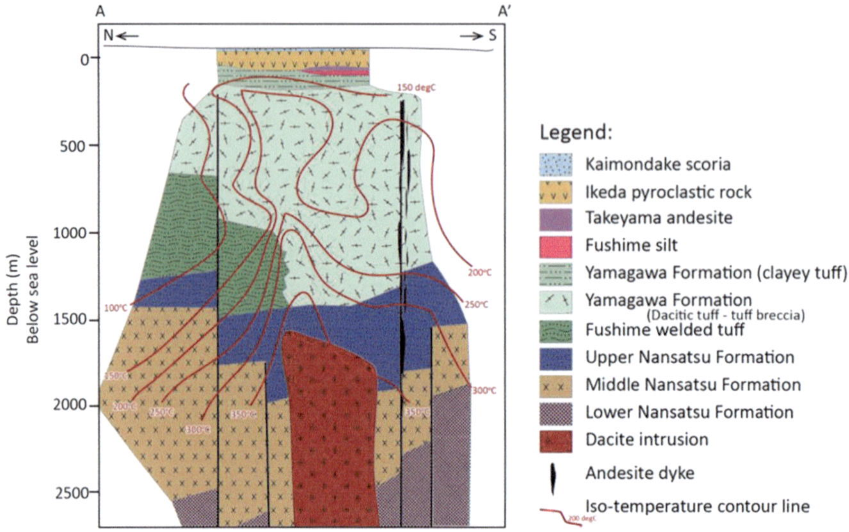

Fig. 6.34 Geological cross-section along with subsurface temperature distribution of the Yamagawa geothermal reservoir (Wulaningsih et al. 2017)

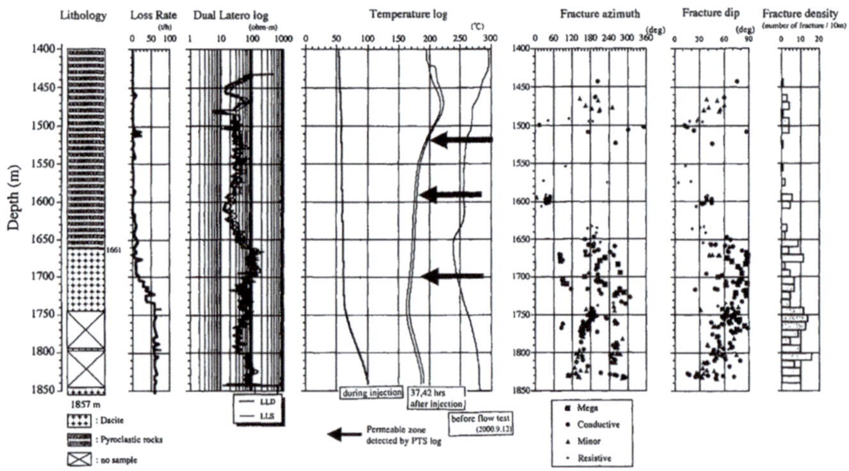

Fig. 6.35 Composite well logging chart including lithology, resistivity, temperature and FMI analyzed data on a well bearing mega fracture in the Yamagawa geothermal field (Okada et al. 2002)

5. Resistive fracture: Sealed with resistive minerals and seen as light-colored sine curve features.

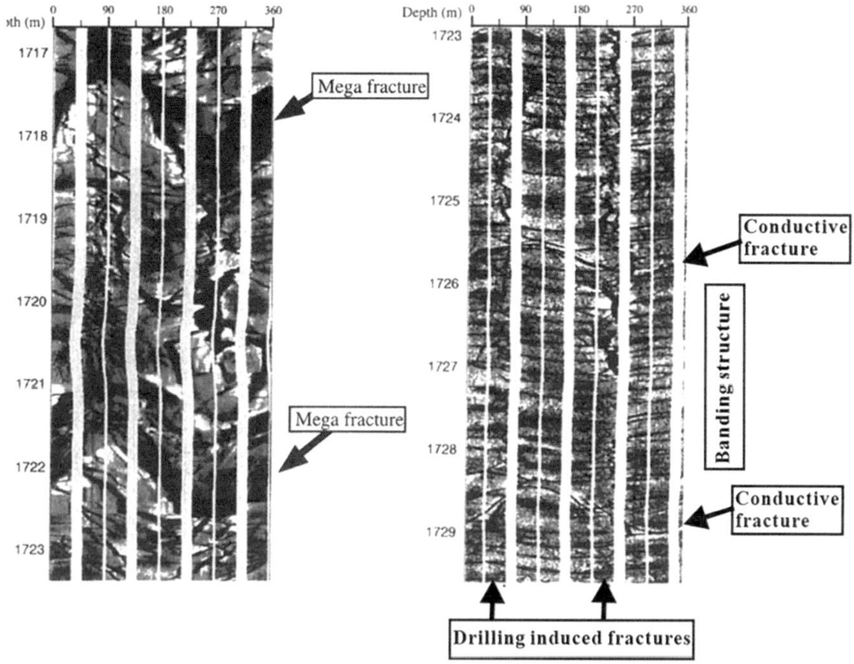

Fig. 6.36 FMI image log in the section bearing mega fracture (left) and that in the section bearing conductive fractures, banding structure and drilling induced fractures (right) done in the Yamagawa field. Lithology of both section is dacite (Okada et al. 2002)

6. Banding structure: Appears only in the intrusion and resembles thin conductive fractures with much higher density, likely related to flow structure in the dacite intrusion

Fractures formed by cooling contraction exhibit high permeability. In the Yamagawa wells, these north–south trending fractures with a vertical dip near the northern edge of the intrusion are well developed. The regional stress regime, characterized by WNW-ESE trending fractures, aligns with the maximum horizontal stress direction in the area. Mega fractures have a more significant impact on the well permeability and productivity than fracture density. By analyzing the characteristics of mega fractures through image logs and integrating resistivity data, PTS logging data, and other subsurface information, we can significantly enhance future well-siting and refine reservoir models.

6.2 Challenges in Ultra-High-Temperature Geothermal Fields in the World

In this section, we summarize four ultra-high-temperature geothermal fields and/or projects worldwide: Kakkoda in Japan, Kilauea Lava Lake in the USA, DESCRAMBLE project in Italy, and Iceland Deep Drilling Project (IDDP) in Iceland.

6.2.1 Kakkonda, Iwate, Northeastern Japan

6.2.1.1 Early History Towards the Kakkonda Geothermal Power Developments

The Kakkonda geothermal area was called the Takinoue geothermal area (However, the power station's name is Kakkonda) before 1984 when the Unit 2 project started. The Japan Metals & Chemicals, Co., Ltd. (JMC) commenced the full-scale geothermal surveys in 1969 at the Takinoue area, highlighting the attention to the continued geothermal development area after the success of the Matsukawa power plant. Before the rise of the curtain of full-scale surveys by JMC, the Geological Survey of Japan, AIST (GSJ) had already started basic surveys such as geological, electrical, and temperature surveys in 1957 as part of nationwide geothermal and hot spring surveys. When we first entered the field, a test borehole of GSJ named GSR-2 whose depth is 404 m emitted steam and hot water intermittently.

JMC selected the Takinoue area for the following three reasons. The first reason is that geothermal manifestations such as hot springs and fumaroles are more violent in comparison with those of the Matsukawa area. The second reason is watching the result of steam and hot water of GSR-2 in addition to GSJ's surveys. The third reason is that JMC predicted the shallower geothermal reservoir about 700 m deep compared with the Matsukawa area of about 1,000 m drilling, because the Miocene strata crop out directly along the Kakkonda River. This means to reduce the drilling expense.

Before the drilling of exploration wells started, the conclusion of a fundamental contract of the co-development was exchanged between Tohoku Electric Power Co., Inc. (TEP) and JMC which was in the change of steam supply to TEP including surveys, drillings, separators, and steam transmission in 1970. However, the most difficult problem took place for the development of this area although obtaining approval for development from the National Parks Council in 1971. Geothermal hot water containing 0.5 ppm and over of As in effluent standard must be reinjected into underground without fail which was enacted by the Prime Minister's Office Ordinance on June 21, 1971. JMC had to prepare for reinjection wells and sites instead of hot water utilization hereby. However, JMC adopted the worst reinjection measure the shallow reinjection setting near the production zone without the consideration of breakthrough, resulting in the cooling of the production reservoir fluid (see Chap. 12). The countermeasure causes a large-scale steam attenuation afterward.

Apart from anything else, the first exploration well 201 of 700 m deep was successfully drilled in 1972. Continuing 5 exploration, 11 production, and 15 reinjection wells were drilled from 1972 to 1977. A 50 MW Unit 1 power plant was put into operation in 1978 and fell into disuse in 2022 after 42 years of power generation.

JMC proposed the geothermal co-development plan for an additional 50 MWe Unit 2 to TEP in 1977 before starting of power generation of a 50 MWe Unit 1 and closed a contract of encouraging conclusion between the above two companies in 1982. The first production well of Unit 2 successfully resulted in 1984 on which Tohoku Geothermal Energy Co., Ltd. (TGE) transferred the development from the above two companies.

However, the development plunged into darkness because of consecutive four production wells targeting the shallow reservoir ranging from 1,000 m to 2,000 m ended in failure. The countermeasure for steam production challenged the drilling of a deep-seated reservoir around 3,000 m deep in 1988. The first deep drilling started to aim for 3,000 m deep based on the following three prospects.

(1) A test slim hole of 2,126 m in depth recorded a higher temperature than 260–°C, and several small-scale lost circulations during drilling, resulting in steam production of 11 t/h.
(2) The study of the existing wells of hydrothermal and metamorphic minerals such as wairakite, prehnite, biotite, and cordierite predicted the existence of a high-temperature heat source beneath the shallow reservoir.
(3) Re-evaluation and re-analyses of the seismic reflection, micro-seismic, and gravity surveys predicted high permeable strata below 2,000–3,000 m.

On 12th July 1989, the first deep well encountered full-scale lost circulation during drilling at 2,799 m around the "Kakkonda Granite" which is the heat source and discrete magma chamber, leading to steam production of 63 t/h (6 MWe/a well; wellhead pressure is 0.6 MPaG). The continuing four deep wells ensured a voluminous steam of high temperature and high pressure. However, the generation scale of Unit 2 diminished from 50 to 30 MWe because of a large-scale steam attenuation of Unit 1 by reinjecting cool water breakthrough, a 90 MWe total steam production by the evaluation of reservoir simulation, and the above-mentioned unsuccess results of shallow wells. In 1996, the Kakkonda Unit 2 power plant was put into operation after the conclusion of a fundamental contract between TEP and TGE, and the approval of power generation by the Electricity Business Act.

The deep well has the following advantages and disadvantages. The first advantage is that we are capable of obtaining steam of high temperature and high pressure being of benefit to power generation. The second advantage is that a small amount of hot water accompanied by steam due to high specific enthalpy contributes greatly to a small amount of reinjection, resulting in a lowering of temperature reduction of the reservoir. Besides, the small amount of hot water also contributes to a small lowering of reservoir pressure decreasing.

On the other hand, the deep well has the following three disadvantages.

(1) Large-scale drilling systems including a mud cooling system, a top drive system, and the frequently repeated performance of three-cone bits are required to tolerate high temperatures (Saito et al. 1998). This means an expensive drilling cost.
(2) To perform the deep drilling, it must drill through the shallow reservoir composed of many fractures causing lost circulation during drilling. The countermeasure for the lost circulation is especially difficult in the case of the existence of a fluid production zone owing to decreasing reservoir pressure.
(3) The acidic hot water ejected from the deep-seated reservoir influences the corrosion damage of the casing and transmission line. Moreover, the hot water contains an amount of chemical content. As a result, chemical scales such as SiO_2, NaCl, $FeCl_2$, and so on adhere to the generation turbine and transmission line in the case of imperfect separation between steam and hot water.

6.2.1.2 Kakkonda Deep Geothermal Resources Survey Project from 1992 to 2000

In Japan, the New Energy and Industrial Technology Development Organization (NEDO) conducted the Deep Geothermal Resources Survey Project in the Kakkonda geothermal field from the fiscal year 1992 to 2000 (A Japanese fiscal year starts in April and ends in March). The planning of this project was based on the successful experiences of the development of the Kakkonda Unit 2 power plant. Therefore, NEDO contracted with the Japan Metals and Chemicals Co. Ltd. (JMC) and collaborative companies as an entrustment contractor. NEDO also cooperated with the Geological Survey of Japan, AIST, from a scientific research point of view.

Geothermal power developments in Japan were not steadily expanding because of conflicts with national parks settled in almost all the active Quaternary volcanoes and numerous pre-existing hot spring resort conglomerates. One practical solution to this problem is to find and utilize deeper reservoirs beneath the geothermal fields concerned. The drilling depths of Japanese geothermal wells were increasing year by year, indicating a considerable demand for drilling into deep reservoirs (Fig. 6.37; Saito 1993).

A primary subject of this project was drilling the deepest geothermal well in Japan to a depth as great as 4000 m in the Kakkonda geothermal field (Fig. 6.38) to obtain fundamental knowledge about deep geothermal systems and to establish strategies regarding exploitation methods (Muraoka et al. 1998a).

The NEDO exploration well WD-1a was spudded in the uppermost reach of the drilling pad of the Kakkonda geothermal power plant on January 5, 1994. The well WD-1a was drilled to a depth of 1,505 m by a 3,000 m class rig between January and June 1994 and 3,729 m by a 5,000 m class rig between January and July 1995 (Fig. 6.39; Saito et al. 1998).

A measurement-while-drilling (MWD) tool, a mud-cooling system to cool the return mud, a top drive system (TDS), and many other techniques available in those days were used to overcome the predicted high-temperature drilling. CO_2 gas was

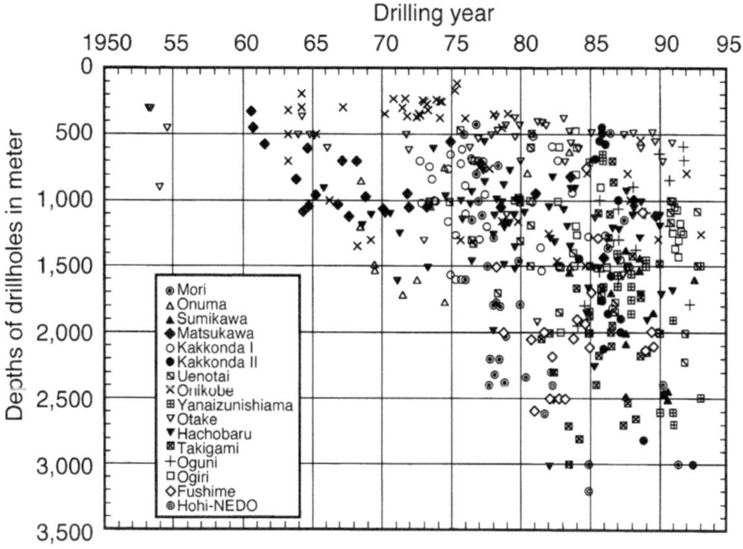

Fig. 6.37 Drilling depths of geothermal wells in Japan between 1950 and 1993 (Saito 1993). Drilling year indicates the start of drilling operations

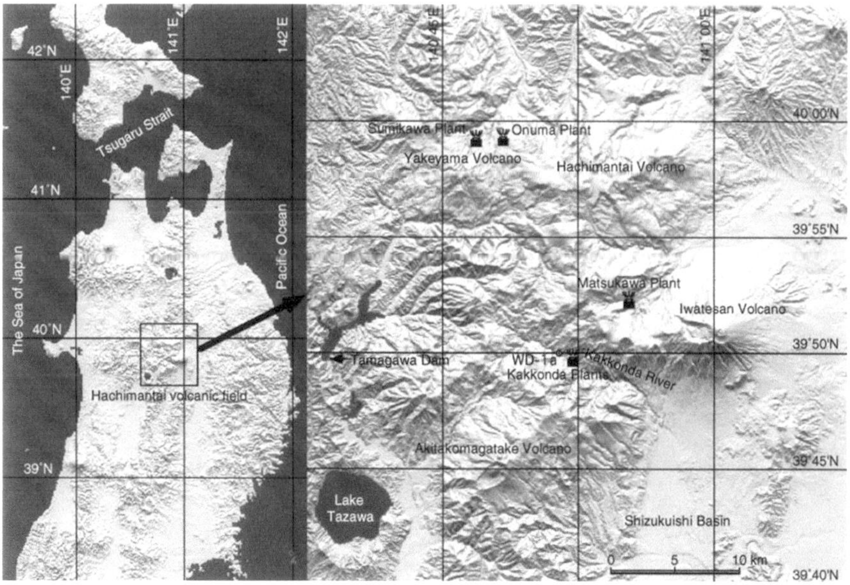

Fig. 6.38 Shaded relief map of the Hachimantai volcanic field. The map was produced by the 50 m mesh DEM data published by the Geographical Survey Institute, Japan (Muraoka et al. 1998a)

Fig. 6.39 Drilling operation of the well WD-1a commenced again by the 5,000 m-class rig "110 UE" in January 1995

ejected with the returned mud. At a depth of 3,642 m, mud with high H_2S gas returned to the surface. The borehole temperature was thought to be too high. Then, although the master plan depth was 4,000 m, the drilling operation was terminated at a depth of 3,729 m for safety concerns (Saito et al. 1998).

The well WD-1a penetrated a neo-plutonic intrusion called Kakkonda Granite from a depth of 2,860 m (Fig. 6.40). K-Ar ages of separated minerals from the Kakkonda Granite shows the cooling ages of 0.24–0.11 Ma for hornblende, 0.21–0.02 Ma for biotite, and 0.14–0.01 Ma for potassium feldspar (Doi et al. 1998). These are the youngest ages for granite in the world (Doi et al. 1998).

Recovered borehole temperatures were too high to be measured by the conventional Kuster tools. Then, the Kuster tools measured the borehole temperatures within the short standing time, and the Horner method extrapolated the equilibrium temperatures. Melting checks by the chemical compound tablets with known melting point temperatures and minimum homogenization temperatures of fluid inclusions were also used (Fig. 6.41; Ikeuchi et al. 1998). All the data, the Horner method, chemical tablets, and fluid inclusions suggest that the bottom hole temperature was at least higher than 500 °C, a record of geothermal wells in the world as of 1998.

The temperature curve shallower than the depth of 3,100 m is controlled by the boiling point temperature curve of water (Fig. 6.41). This part is further divided into the shallower low-pressure reservoir for Unit 1 and the deeper high-pressure reservoir

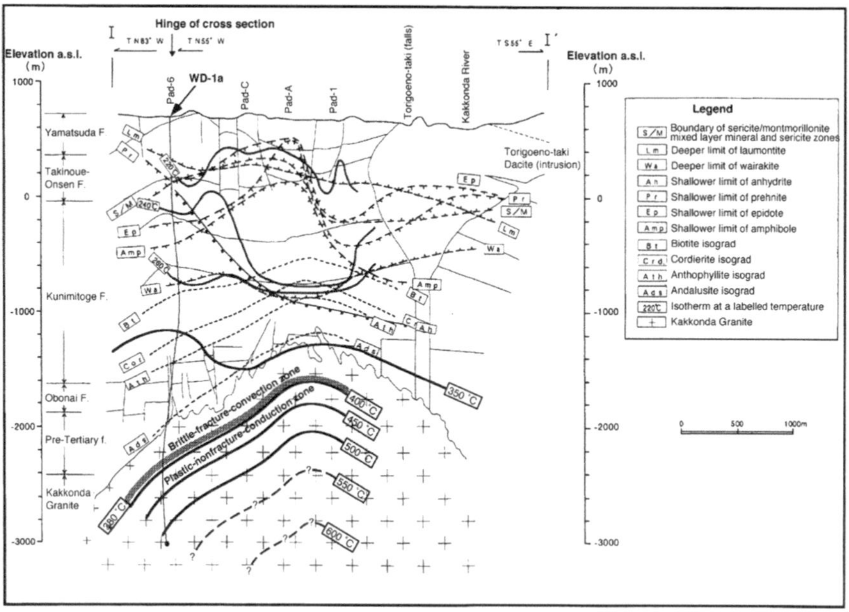

Fig. 6.40 Trajectory of the well WD-1a in the NW-SE geological cross-section, modified after Kato and Sato (1995) by Muraoka et al (1998a)

for Unit 2. Going to the deeper side of the temperature inflection point 380 °C at a depth of 3,100 m, the borehole temperature abruptly increases with depth. This part exceeds the boiling point temperature curve of water and is ascribed to the thermal conduction zone of the cooling temperature profiles of magma (Fig. 3.21). It is known that the minimum melting point temperature in the quartz-albite-orthoclase-H_2O granitic system (the petrogeny's residua system) ranges from 630 to 780 °C in the pressure range of the crust (Luth et al. 1964). Therefore, the bottom hole temperature 500 to 550 °C means that the Kakkonda Granite was only 80 to 280 °C cooled down from the minimum melting point temperature of the granitic magma. This is consistent with very young K–Ar ages of the Kakkonda Granite (Doi et al. 1998).

Fractures of the well WD-1a were analyzed using FMI data and spotted cores taken from the 13 depths (Kato et al. 1998). FMI is the abbreviation for Schlumberger's Fullbore Formation MicroImager, and can image the micro resistivity variation on the borehole walls. We here remark on the high-frequency FMI fractures below a depth of 1,800 m, concentrated in the contact metamorphic aureoles of the Kakkonda Granite (Fig. 6.42; Muraoka et al. 1998a). Most of them are low-angle fractures (Kato et al. 1998).

Contact metamorphism is a dehydration reaction on the side of the host rocks because the magma intrusions thermal-conductively heat the host rocks. Particularly, isograds of anhydrous metamorphic minerals such as andalusite, clinopyroxene,

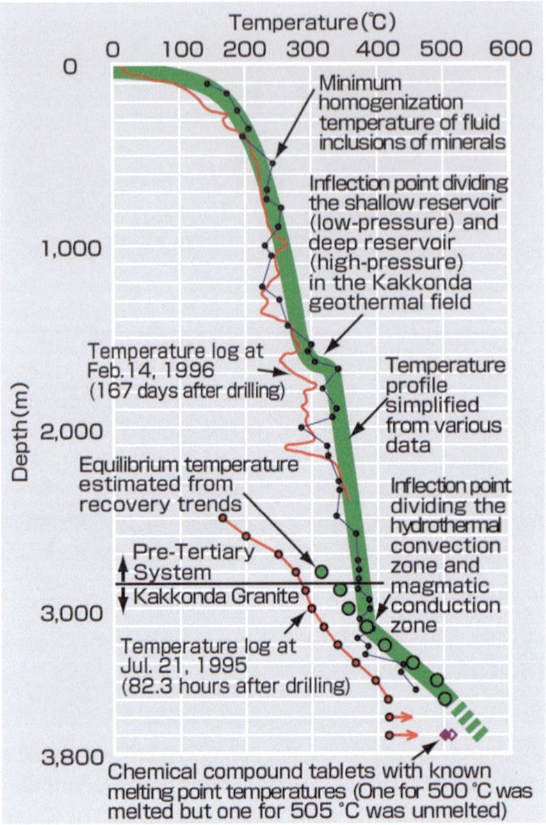

Fig. 6.41 Temperature profiles of the well WD-1a in the Kakkonda geothermal fields, simplified from Ikeuchi et al. (1998) by Muraoka et al. (2013). The estimated temperature was higher than 500 °C at a bottom-hole depth of 3,729 m

garnet, and cordierite near the intrusion could form strong dehydration zones. The frequency of the FMI fractures seems to be increasing towards these higher dehydration zones and the contact with the intrusion (Fig. 6.42). Therefore, these fractures might have been generated in the water weakening by the high pore water pressure induced by the contact metamorphic dehydration reactions. FMI logs were only achieved above a depth of 2,653 m because of the temperature limitation. However, the dense fractures were observed down to the intrusion contact in the normal resistivity logs (Fig. 6.42) and spotted core samples.

To clarify the water contents of the contact metamorphic aureole of the Kakkonda Granite, Muraoka and Ohtani (2000) analyzed 186 samples of cuttings taken from every 20 m interval along the well WD-1a as shown in Fig. 6.43. H_2O (−) is called surface moisture water and is not necessarily important. H_2O (−) only increases towards the high porosity shallow depth. H_2O (+) is called crystalline water and could reflect the dehydration results of the contact metamorphic aureoles. H_2O (+) evidently decreases from the biotite isograd at a depth of 1,610 m to the contact of the Kakkonda Granite at a depth of 2,860 m. The H_2O (+) content descends a stairway downward in the cordierite isograd and clinopyroxene + garnet isograd. Thus, the

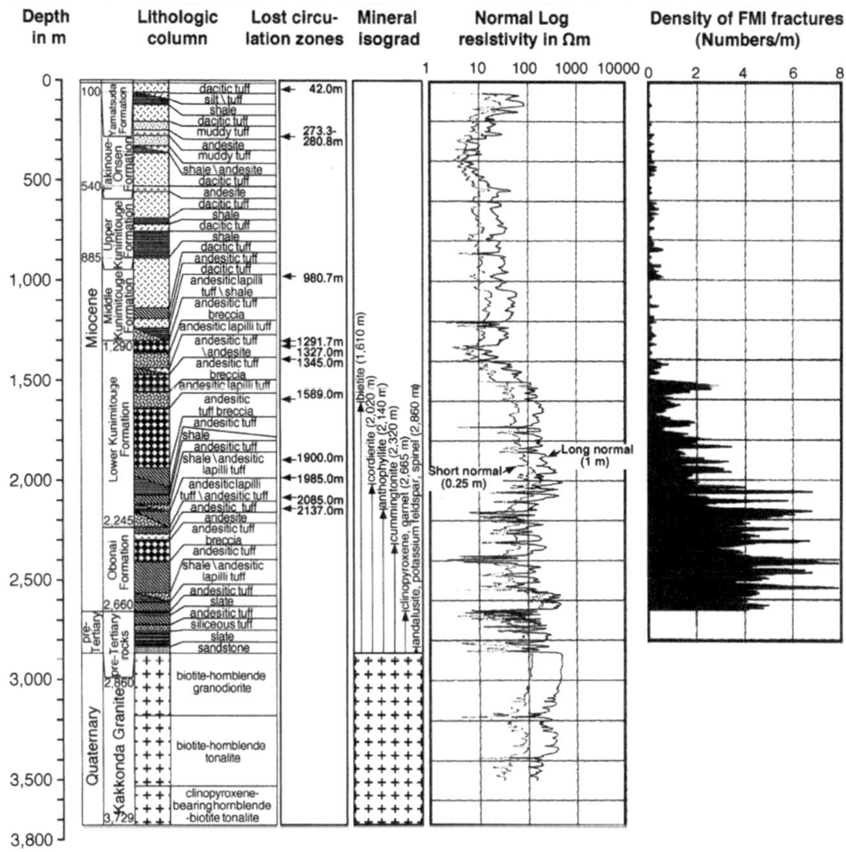

Fig. 6.42 Frequency of fractures in the contact metamorphic aureoles in the Kakkonda Granite in the well WD-1a (Muraoka et al. 1998a). See Kato et al. (1998) for the fracture orientation analyses

high-density FMI fractures at a depth between 1,610 and 2,860 m could result from the water weakening by the dehydration reactions of the contact metamorphic aureole, which was induced by the magma intrusion of the Kakkonda Granite.

Since the 1980s, the rheological significance of the brittle-plastic transition has been often discussed (Brace and Kohlstedt 1980; Kirby 1983; Masuda and Omori in Chap. 4 of this book). It was applied to the possibility of a bottom of the hydrothermal circulations (Fournier 1987). The brittle-plastic transition was proposed to be a temperature range from 370–400 °C in quartz- dominant rocks in tectonically active regions (Fournier 1991). The brittele-plastic transition is systematically discussed in Chap. 4 of this book.

The temperature inflection point at a temperature of 380 °C and a depth of 3,100 m immediately reminds us of the possibility of the brittle-plastic transition (Fig. 6.41). Fractures are abundant above the inflection point but very rare below the inflection point, as seen in the spotted cores. Permeability is high above the inflection point,

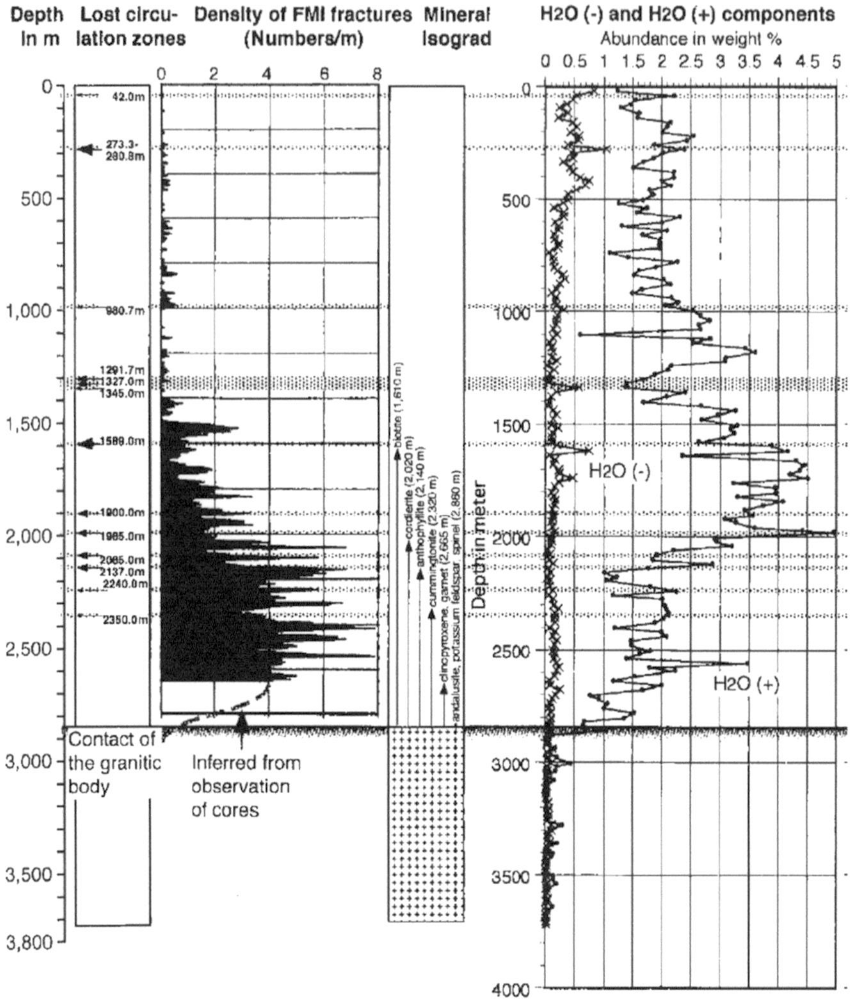

Fig. 6.43 Profiles of water contents in cutting samples in the contact metamorphic aureole of the Kakkonda Granite along the well WD-1a (Muraoka and Ohtani 2000)

as known in the Unit 1 and Unit 2 reservoirs, but very low below the point without any lost circulation zones (Fig. 6.42). The permeability contrast could instead be a cause of the temperature inflection itself. All these phenomena are ascribed to the inflection point that divides the lower isotropic-stress plastic zone and the upper anisotropic-stress brittle zone. Likewise, the temperature inflection point divides the lower thermal conduction zone and the upper hydrothermal convection zone. Thus, Muraoka et al. (1998a) emphasized that the well WD-1a was probably the first to penetrate the brittle-plastic transition completely. This conclusion is illustrated in

Fig. 6.40 where the 380 °C isotherm separates the brittle-plastic transition, fracture-dominant versus non-fracture zones, and hydrothermal convection versus thermal conduction zones.

During drilling, the WD-1a experienced discharge of non-condensible gas such as CO_2 and H_2S below 3,350 m. The WD-1a did not encounter productive fractures within the Kakkonda Granite because most of the zone is plastic. Then, Kasai et al. (1998) used a reverse circulation method to collect fluid samples from 3,708 m near the well bottom using the drill pipes as straw after a standing time of 196 h. Obtained fluids were 1.5 m^3 of hypersaline metal-rich brine, which was 38.5 wt% NaCl equivalent (Table 6.3; Kasai et al. 1998). Halite and other chloride salts immediately crystallized at the surface, where the volume ratio of liquid versus solid was about 13:2. The mixing model with tritium estimated the dilution ratio by the river water, and the original salinity was estimated to be 55 wt% NaCl equivalent (Kasai et al. 1998). It is summarized that the two phases of hypersaline brine with probably 55 wt% salinity and non-condensible CO_2 and H_2S gases are stored in the interstitial pores of the Kakkonda Granite.

The theory of lithosphere strength can be applied to the well WD-1a (Muraoka et al. 1998b; Muraoka 2012; Muraoka et al. 2013). The theory consists of Byerlee's law in a brittle region and a power law creep equation in a plastic region. The Byerlee's

Table 6.3 Chemical analysis of soluble solids precipitated at the surface collected from 3,708 m of the well WD-1a by reverse circulation (Kasai et al. 1998)

Components	Concentration in mg\kg
Cl	484000
Br	300
I	60
SO_4	<100
Na	256000
K	14000
Ca	4400
Mg	10
Fe	24000
Al	<1
Mn	4900
Zn	960
Pb	210
Cd	<1
Cu	<1
Au	<1
Ag	<1
SiO_2	2100
B	910
Residue	178000

law in a brittle region is expressed by the following equation (Brace and Kohlstedt 1980):

$$\overline{\sigma}_1 \approx 5\overline{\sigma}_3, \overline{\sigma}_3 < 110\text{MPa}$$

$$\overline{\sigma}_1 \approx 3.1\overline{\sigma}_3 + 210, \overline{\sigma}_3 > 110\text{MPa} \tag{6.1}$$

where

$\overline{\sigma}_1 =$ the maximum principal stress

$\overline{\sigma}_3 =$ the minimum principal stress.

The power law creep is expressed by the following equation (Brace and Kohlstedt 1980):

$$\dot{\varepsilon} = A(\sigma_1 - \sigma_3)^n e^{\left(\frac{-H}{RT}\right)} \tag{6.2}$$

where

$\dot{\varepsilon} =$ strain rate, s^{-1}

$A =$ material constant specified to the rock or mineral species

n = stress exponent

$H =$ activation enthalpy, J mol^{-1}

$R =$ gas constant, J K^{-1} mol^{-1}

$T =$ temperature, K.

Most of these parameters are reasonably determined on the well WD-1a. The strain rate $\dot{\varepsilon}$ is given to be 10^{-12} s^{-1} because of the active compressive tectonic region (Fournier 1991). Taking the data from quartz diorite close to the Kakkonda Granite (Kirby 1983), the stress exponent n is given to be 2.4, and activation enthalpy H is given 219 kJ mol^{-1}. The temperature T is given by the temperature profile of the well WD-1a in Fig. 6.41. The material constant A varies from 10^{-19} to 10^{49} according to the given material, and only this parameter cannot be determined a priori. Then, we considered it through trial and error as shown in Fig. 6.44 (Muraoka et al. 1998b; Muraoka 2012; Muraoka et al. 2013). When we choose $10^{1.85}$ and $10^{1.35}$ for A, a plastic zone appears at a depth of the tight fracture zone from 2.0 to 2.8 km. It is contradictory. When we choose $10^{0.35}$ and $10^{-0.15}$ for A, the maximum strength profile appears at a depth of the magmatic intrusion, the Kakkonda Granite, which is a cause of the plastic zone. It is contradictory, too. In such squeezing, we can determine the material constant A to be about $10^{0.85}$. Then, the strength curve on the plastic field is drawn, as shown in Fig. 6.44.

Previously, a pressure-depth conversion factor of 26 MPa/km was used (Muraoka et al. 1998b; Muraoka 2012; Muraoka et al. 2013), which was revised to 25 MPa/km in Fig. 6.44 according to the result of Sect. 3.2. The four points of stress ratio measurements of the differential strain curve analysis (DSCA) on the core samples (New Energy and Industrial Technology Development Organization 1996) are well explained by the model of the strength profile drawn by Byerlee's law and the power law creep equation as shown in Fig. 6.44. Particularly, stress ratios among σ_1, σ_2, and

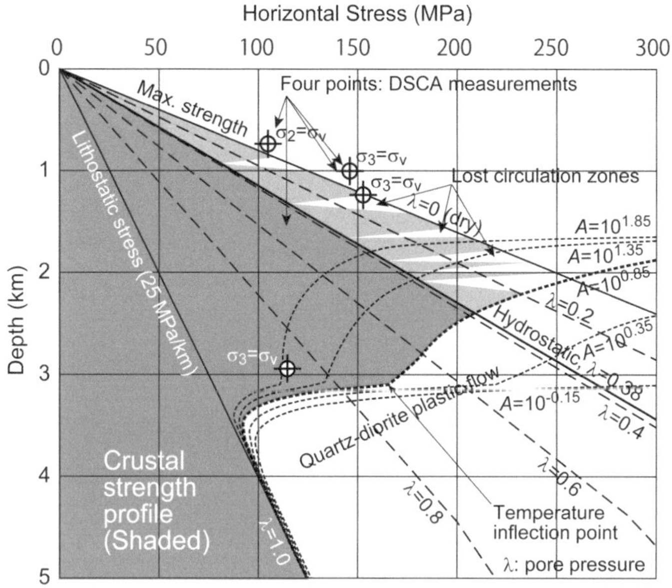

Fig. 6.44 Rock strength profile of the well WD-1a in the Kakkonda geothermal field, northeastern Japan (Revised from Muraoka et al. 1998b; Muraoka 2012; Muraoka et al. 2013). The strength profile is drawn on the shallower brittle zone by Byerlee's law and on the deeper plastic zone by the power law creep. Four marks with circles and crosses show differential strain curve analysis (DSCA) measurement results on the core samples from the WD-1a well (New Energy and Industrial Technology Development Organization 1996)

σ_3 of the DSCA data at the greatest depth 2,938 m show 1.00, 0.93 and 0.82, respectively (New Energy and Industrial Technology Development Organization 1996), which suggest a nearly isotropic condition. The dramatic strength weakening and isotropic stress field at the greatest depth of 2,938 m demonstrate the existence of the plastic region in Fig. 6.44. Three points of DSCA stress ratio measurements in the brittle field are close to the dry line ($\lambda = 0$). This is also reasonable because the core sampling could have been performed only from impermeable zones with no lost circulation. These observations strongly support that the well WD-1a penetrated the brittle-plastic transition (Muraoka et al. 1998a).

The practical brittle-plastic transition is ascribed to the temperature inflection point at a depth of 3,100 m and a temperature of 380 °C (Muraoka et al. 1998b). Two equations, the Byerlee's law ($\lambda = 0.38$) in a brittle region and a power law creep in a plastic region, graphically intersect at a depth of 2,400 m and a temperature of 360 °C (Fig. 6.44). This point may also be called the brittle-plastic transition in terms of equations. However, this point is situated in a zone of very high concentration of low-angle fractures in the depth interval from 1,770 to 2,860 m (Fig. 6.42). This fracture zone likely reflects not only dehydration weakening but also the stress concentration to the maximum strength zone of the bottom of the brittle crust layer under the

WNW-ESE regional contraction tectonics (Muraoka and Ohtani 2000). Therefore, this intersection point is rather recognized as a maximum strength zone.

The Deep Geothermal Resources Survey Project in the Kakkonda geothermal field from 1992 to 2000 brought us numerous scientific topics on high-temperature geothermal systems. Therefore, we obtained fundamental knowledge about deep geothermal systems, our half goal. However, we did not necessarily succeed in establishing strategies regarding exploitation methods in the deep geothermal systems, which were our remaining half goal because we encountered a plastic region. Thus, we needed another breakthrough in exploiting deep geothermal systems.

6.2.1.3 Current Efforts After the Great East Japan Earthquake in 2011

Since the 1990s, the Japanese government gradually leaned towards large-scale atomic power generation rather than renewable geothermal energy. In 1997, the Japanese government decided to exclude geothermal steam flash power generations from the definition of "new energy" in "Act on Special Measures Concerning the Promotion of New Energy Usage". This implies the government's termination of geothermal R&D within a few years, and actually, the geothermal R&D by the government ceased in Japan in March 2002.

However, the Great East Japan Earthquake in March 2011 (Mw = 9.0) dramatically changed this situation at a tremendous cost by producing 146,520 evacuees from the crisis at the Fukushima no.1 nuclear power plant. Facing this national energy crisis, geothermal energy was reevaluated. Then, the solid governmental support for geothermal developments and R&D was revived in Japan in 2012.

Looking at the world, geothermal R&D was widely focused on enhanced or engineered geothermal systems (EGS; Tester et al. 2006). The concept of EGS is as follows: If the temperature is not enough, it can be engineered by deep drilling. If the permeability is not enough, it can be engineered by hydraulic fracturing. If the water is deficient, it can be engineered by water injection. However, the EGS has two bottlenecks in practical use: one is that the recoverability of injected water is about 50% or less than that in fracture-dominant regions such as Japan, which inevitably requires replenishing large volumes of injected water throughout the power generation operation, and the other is that the injected water raises pore fluid pressures in crustal rocks, causing induced earthquakes.

Then, Muraoka et al. (2013, 2014) proposed the plastic zone EGS power generation based on the experience of the Deep Geothermal Resources Survey Project in Kakkonda (Fig. 6.45). Suppose we develop an artificial brittle fracture reservoir by hydraulic fracturing in the plastic zone, as expected in the Kakkonda geothermal field. In that case, the plastic enclosure will avoid injected water loss and reduce the propagation of the seismic intensity of induced earthquakes. Therefore, the plastic zone EGS power generation resolves two bottlenecks of EGS simultaneously. However, this is still only a proposal.

Recently, the Supercritical Geothermal Project has been actively conducted by NEDO and FREA, AIST (Asanuma et al. 2021). From 2021 to 2023, a special issue,

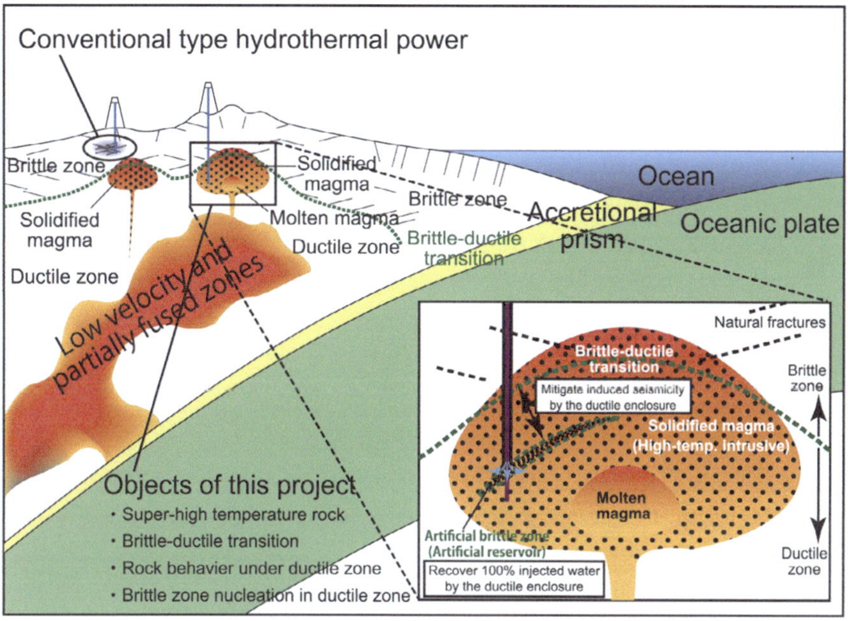

Fig. 6.45 Concept of the plastic zone EGS power generation (Slightly modified from Muraoka et al. 2013)

"Supercritical Geothermal", composed of 19 articles, was also published on *Geothermics.* This project aims to use natural supercritical fluids for geothermal power generation, but the demonstration of the presence of supercritical fluid resources is still underway.

6.2.1.4 Strategy of the Greater-Depth Drilling

Strategy of Ultra-High Temperature Drilling

Temperature limitations range in conventional drilling tools and materials from 150 °C to 220 °C excluding mud dispersant. In contrast, a geothermal well WD-1a drilled by NEDO (New Energy and Industrial Technology Development Organization), founded by the Japanese Government, reached a depth of 3,729 m in the Kakkonda geothermal field, where the bottom hole formation temperature is more than 500 °C (Figs. 6.41 and 10.7). The following three strategies were essential for ultra-high temperature greater-depth drilling:

(a) To overcome the temperature limitations and cool the borehole and the down-hole drilling tools, a large-scale drilling mud cooling system was adopted. Two open cooling towers, one closed mud cooler, and a 500 kl pit were set in the WD-1a (Saito et al. 1998).

(b) Every 27 m-long drill pipe stand using top drive systems was adopted to speed up the round trip for bottom hole assembly cooling.
(c) Very thin mud (ex. 3% bentonite, 0.1% high-temperature dispersant, 1% lubricant, and caustic soda) was used to avoid mud gelation by high temperature.

The result of the measured temperature data at 2,700 m in the depth of the Kakkonda K6-3 well recorded as follows (Saito 1993):

- Drilling fluid temperature measured at the flow line—70 °C
- Bottom hole during drilling temperature within one hour after circulation ceased by the thermometer—152 °C
- Formation temperature after 9 days standing time—330 °C.

Countermeasures against lost circulation during drilling in shallower fractures reservoirs

It must pass through shallower fractures and shallower reservoirs to reach greater depth. The complete lost circulation and gush out of a fluid flow during drilling frequently happen in Japan.

(a) The drilling and borehole damages occur in the shallower fractures (less than about 500 m in depth) as stated in Sect. 11.2.6 (geneses of six "shallow" fractures). Civil engineering's air-hammer and pre-drilling methods are sometimes adopted as countermeasures for these damages. However, the most dangerous phenomenon is the thermal and gas blowout which occurs in the case of high pressure with higher temperature than mud fluid pressure (refer to Sect. 13.4; in situ boiling). Saito (1993) describes the countermeasures for the blowout at 277 m in depth (17″1/2 borehole) of the well K6-2 in Kakkonda and points out the importance of BOP (blowout preventer), water injection through sleeve valve, filling by sand and gravel, and cementing. He gives an example of the two-stage BOP (29″1/2-500 BOP plus 20″-2000 BOP), and emphasizes a prompt shut up of BOP.
(b) Complete lost circulation during drilling occurs frequently in the shallow reservoir composed of many fractures ranging in depth from about 500 m to 2,000 m in Japan, resulting in impossible deeper drilling. The reservoir consists mainly of volcanic products (ex., tuff breccia, and lapilli tuff), as shown in Figs. 9.1 and 11.9, Table 11.2. These rocks tend to collapse and slough due to poor sorting and weak adherence of mutual rock particles. Countermeasures against the complete lost circulation during drilling are given below:

    - Adoption of aerated mud drilling.
    - Repetition of cementing.
    - Slow drilling using thinner mud water.
    - Complete blockage for fracture zone by casing alteration to shallow depth.
    - Setting the drilling pad on the outer surface of the shallow reservoir and adopting high-angle directional drilling.

It is better to prohibit the enforcement of drilling by massive pumping, which would result in the stagnation of drilling tools because the reservoir rocks tend to collapse due to the weak adherence of rock particles in Japan.

6.2.2 Drilling into the Kilauea Lava Lake in the USA

In volcanological research rather than geothermal exploration, drillings into the Kilauea Iki lava lake in the USA were worth describing the natural analog to drilling into the magma chamber and its extraordinary high-temperature borehole record.

The 1959 summit eruption was one of the largest eruptions in the history of the Kilauea volcano. The eruption intermittently occurred from November 14 to December 20, 1959 (Richter and Moore 1966). Most of the voluminous lava flowed out along the southern flanks of the Kilauea volcano. Some lava flows filled pre-existing Kilauea Iki, Alae, and Makaopuhi Craters, forming molten lava lakes (Fig. 6.46). Among them, Kilauea Iki lava lake was the most accessible, and this lake has been drilled repeatedly as the lake has cooled and solidified; in support of a long-term study of the processes controlling basalt crystallization and fractionation (Helz and Thornber 1987).

Shortly after the end of the 1959 summit eruption the Hawaiian Volcano Observatory of the U.S. Geological Survey initiated a program of drilling and thermal

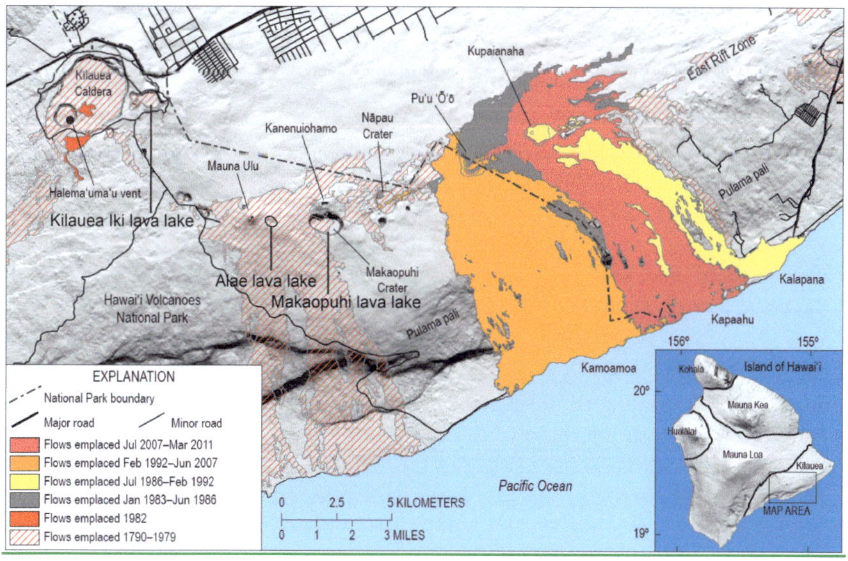

Fig. 6.46 Map showing the extent of lava flows of the ongoing Kilauea East Rift Zone eruption, quoted from Helz et al. (2014). Only the names of the three lava lakes were added. The 1959 lava flows were almost concealed by lava flows from 1790 to 1979

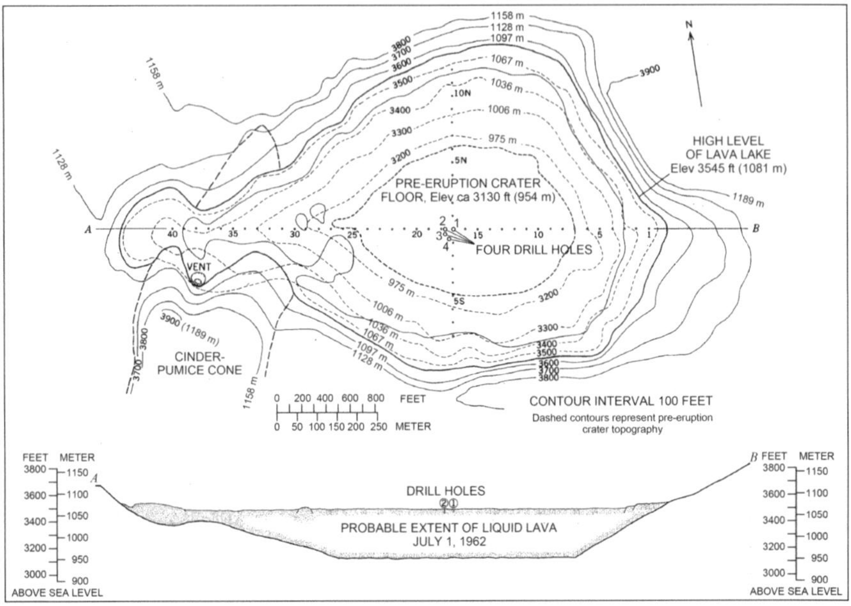

Fig. 6.47 Map and section of Kilauea Iki crater before and after the 1959 summit eruption showing the location of U.S. Geological Survey drill holes. Small open circles show the location of permanent reference stations on the surface of the lava lake (Richter and Moore 1966). Partly redrawn, such as the addition of metric representation

measurement. The U.S. Geological Survey drilled four holes between April 1960 and June 1962 (Fig. 6.47) and detailed the physical, petrographic, and chemical data obtained from studies of core and melt; collected at known temperatures (Richter and Moore 1966). The University of Hawaii and the Lawrence Radiation Laboratory of the University of California cooperatively joined the project.

Drill Hole 1 (Fig. 6.47) penetrated the solidified crust and reached the liquid lava beneath it on 8 September 1960, nine months after the end of the eruption. The base of the solidified crust was observed at a depth of about 6.83 m, and the temperature at the base was about 1,065 °C, measured by the optical pyrometer (Richter and Moore 1966). Drill Hole 2 (Fig. 6.47) penetrated the solidified crust and reached the liquid lava beneath it on 5 October 1961, twenty-two months after the end of the eruption. The base of the solidified crust was observed at a depth of about 11.95 m, and the temperature at the base of the crust was about 1,106 °C measured by the chromel–alumel thermocouples (Fig. 6.48; Richter and Moore 1966).

Most of the temperatures obtained in the drill holes were measured with sheathed chromel–alumel thermocouples and a slide-wire potentiometer with a reading accuracy of 0.005 millivolts. The sheathed thermocouples consisted of chromel–alumel pairs, swaged with powdered MgO filler, in a thin-wall 310-stainless-steel case. Unsheathed thermocouples were used successfully for the first measurements in Drill Hole 1. However, after melt penetration, a few seconds of exposure to the extremely

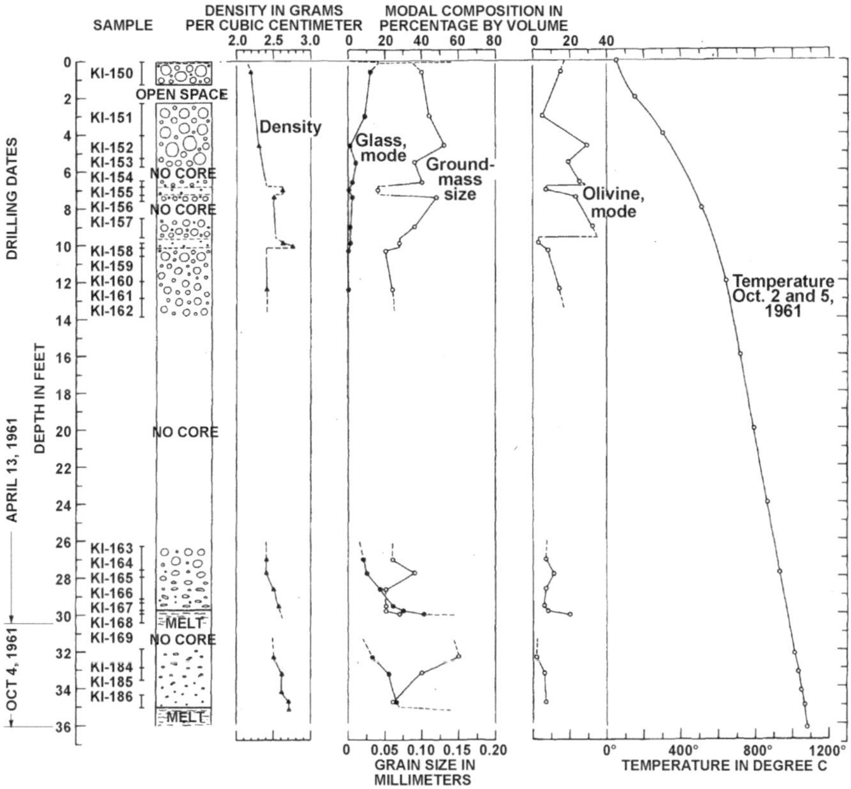

Fig. 6.48 Composite section through Drill Hole 2 with pertinent mineralogical and physical data. The relative size, shape, and abundance of the vesicles in the drill core are illustrated diagrammatically by pattern. Dashed lines denote apparent textural and chemical discontinuities (Richter and Moore 1966). A discrepancy in the temperature at the base of the crust in this figure from the text description is probably derived from the slight depth difference taken at 11.03 m (36.2 feet), where 1,081 °C

hot SO_2-rich bottom-hole vapors would disintegrate the alumel element. In Drill Hole 3, a compound 5-pair chromel–alumel thermocouple was used with junctions spaced 3 feet apart. Laboratory calibrations, using the melting-point technique and, latterly, a comparison with a National Bureau of Standards calibrated platinum-platinum-90, rhodium-10 thermocouple, indicate an accuracy of about one-half of 1 percent at 500 °C and about 1 percent at 1,000 °C for the thermocouple temperatures (Richter and Moore 1966).

Bottom-hole temperatures were also measured, whenever conditions permitted, with incandescent filament-type optical pyrometers. In the shallower holes (Drill Hole 1 and first penetration of Drill Hole 2), optical-pyrometer temperatures were within 5 °C of those obtained by chromel–alumel thermocouples. However, a more significant divergence was observed in the deeper holes, with the optical-pyrometer values consistently higher by as much as 10 °C. The effectiveness of the optical

Fig. 6.49 Photographs of drilling operations at Kīlauea Iki lava lake in 1975 (Tilling et al. 2014). **A**. General view of the drill rig. It and other heavy equipment were lowered into the crater by helicopter. B. The top of drill hole 75–2. **B** (Enlarged). View looking down the drill hole (~6 cm in diameter) from the surface to the bottom hole (~40 m), where still-molten 1959 lava glows at ~1,110 °C (USGS photographs by Robin T. Holcomb)

pyrometer is easily recognized when we see the drilling operation in Fig. 6.49, mainly looking down the hole bottom (C) (Tilling et al. 2014). There is no liquid condensate because of the extraordinary high-temperature drilling into the crust of the basalt lava lake. Surface and near-surface (less than 1 ft) drill-hole temperatures were measured with mercury thermometers (Richter and Moore 1966). The Kilauea Iki lava lake was drilled repeatedly in 1967, 1975 (Fig. 6.49), 1976, 1979, 1981, and 1988 (Tilling et al. 2014). Their activities followed the cooling trajectory after the lava lake of the 1959 summit eruption and are omitted here in detail.

6.2.3 The DESCRAMBLE Project in Italy

The DESCRAMBLE project was conducted by ENEL Green Power and its consortium from May 2015 to April 2018 in the Larderello-Travale geothermal field in Italy. The project aimed to drill geothermal resources at extremely high temperatures in continental crust conditions to demonstrate novel drilling techniques and investigate supercritical geothermal resources (Bertani et al. 2018). The project coordinator was Ruggero Bertani, who passed away in June 2018, immediately after the drilling process. The results of the DESCRAMBLE project, which are open to the public, are, therefore, so limited, except for those of Bertani et al. (2018).

As described in Chap. 1, deep reservoirs in the Larderello-Travale geothermal field consist of the early-Paleozoic metamorphic basements at depths between 2,500 m and 3,500 m. These deep reservoirs are characterized by a temperature greater than 300–350 °C and pressure around 60–70 MPa (Arias et al. 2010). The heat source is the widely extending Quaternary granitic batholith beneath these deep reservoirs. Some deep wells are known to exceed 400 °C. Ferrara et al. (1985) reported that some deep wells in San Vito exceeded 3,000 m, and the bottom temperatures exceeded 420 °C. This estimation was confirmed because the bottom temperatures were sufficiently high to melt zinc (melting point = 419.58 °C) (Fournier 1991).

The presence of a deep, strong amplitude and almost continuous seismic marker, K-horizon, has long been identified by both 2D and 3D seismic surveys below the vapor-dominated reservoirs in the Larderello-Travale geothermal field since the late 1970s (de Franco et al. 2019; Fig. 6.50). The test site of the DESCRABLE project was chosen in the Lago area, the Larderello-Travale geothermal field, which was the shallowest K-horizon area delineated from the 2D and 3D seismic surveys (Bertani et al. 2018). In the DESCRABLE project, an existing well, Venelle-2, was planned to be deepened from 2.2 km to 3–3.5 km. The high seismic impedance of a seismic marker, K-horizon, resembling a bright spot, was interpreted as a physically perturbed layer, possibly comprising super-critical fluids (de Franco et al. 2019). The exploratory well San Pompeo-2, drilled in 1979 to cross the K-horizon, indicated high-temperature and high-pressure evidence, with temperature >400 °C and pressure >24 MPa extrapolated to a depth of 2,930 m (de Franco et al. 2019).

Drilling activities started on 28th April 2017. After plugging the existing open hole, deviation started with drilling in sliding and rotary mode (kick-off point at 1,054 m). The aim was to keep an inclination lower than 10° (pseudo vertical well). To reduce the risk of drilling water losses, various squeezes of clogging materials and barite were performed to clog the absorbent zone. However, after each job, the pressurization tests gave negative results. Drilling started again with water and controlled parameters, reaching the final depth of 2,900 m. A coring (9 m) was performed at a depth of 2,830 m. Then, a series of static temperature logs were made using ENEL Green Power's Kusters and synthetic fluid inclusion measurements by the National Research Council in Italy (CNR). At a depth of 2,900 m, further coring (9 m) and a static temperature log were achieved. To realize a new leak-off test, a 6″ swellable packer was run in a hole and set at 2,898 m, but the attempt to put the tool failed, and the packer was pulled out. A last static temperature log was performed using the tool designed for the project. It was decided to abandon the well drilling temporarily at a 2,900 m depth (Bertani et al. 2018).

The results of the temperature and pressure logs in the well Venelle-2 deep are summarized as follows: As electrical wireline cables rated to 450 °C are unavailable, the SINTEF PT tool was developed in the DESCRAMBLE project to log based on

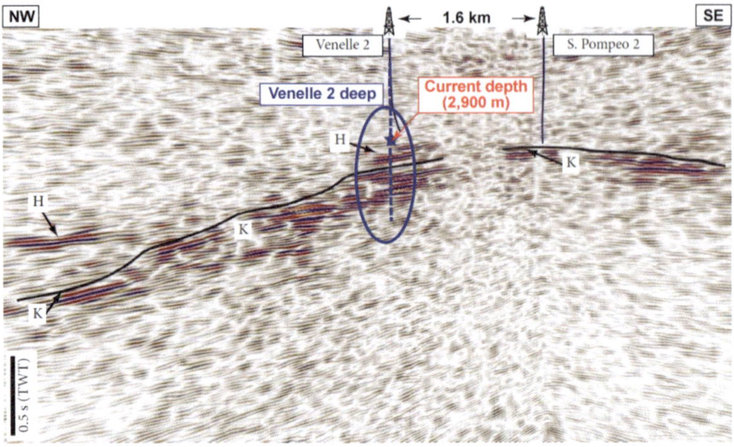

Fig. 6.50 Seismic reflection profiles, the K-horizon, and the well Venelle-2 trajectory along the NW-SE section in the Lago area in the Larderello-Travale geothermal field (de Franco et al. 2019). The vertical axis is two-way travel time (TWT). The well Venelle 2 deep was depicted from Bertni (2017)

internal memory and powered by high-temperature batteries. The maximum temperature measured by the SINTEF PT tool is 372.9 °C at a depth of 2,610 m. In addition, a mechanical Kuster tool obtained the highest temperature record, 504 °C (Bertani et al. 2018).

However, before reaching the K-horizon, high-pressure fluids unexpectedly encountered and induced a well blow-out. At the top of granite intrusions, amounts of tourmaline-quartz breccia and vein fragments, typical of high-temperature magmatic-hydrothermal systems, erupted (Bertani et al. 2018).

The drilling has not proved the existence of a reservoir or supercritical fluid. However, it has reached the supercritical conditions of pure water and investigated the thick seismic reflectors.

6.2.4 Iceland Deep Drilling Project (IDDP) in Iceland

The well NJ-11, drilled in the Nesjavellir geothermal field in Iceland in 1988, was the deepest well in the Nesjavellir geothermal field at that time. It was drilled to a depth of 2,265 m in basalt on the upper northeastern flank of the Hengill volcano, where a young fissure swarm cuts the volcano. Temperature measurements were made after the well was completed and during cold-water injection into the wellhead, showing that the injected water did not penetrate below about 1,200 m. They also show that the high-temperature fluid >381 °C entered near the bottom of the well and flowed upward and out into the formation (Fournier 1991).

After that, the Iceland Deep Drilling Project (IDDP) was launched as a long-term project to explore high-temperature geothermal resources in Iceland (https://www.iddp.is). When the IDDP consortium was formed in the year 2000, three geothermal fields in Iceland were chosen as suitable locations to search for supercritical resources: Krafla in the northeast of Iceland, Hengill in the west, and Reykjanes in the southwest (Fig. 6.51; Friðleifsson et al. 2003, 2020).

The first attempt to drill into a supercritical reservoir was made in 2008–2009 in the Krafla caldera as the well IDDP-1. However, the well did not attain supercritical fluid pressures because drilling was suspended at too shallow a depth (Elders et al. 2011). This was because 900 °C rhyolite magma intercepted drilling at a depth of only 2,072 m. The IDDP-1 was completed with a liner set above the rhyolite intrusion. When the well was tested, it produced superheated 452 °C steam at 14 MPa pressure and had a flow rate and pressure sufficient to generate about 35 MWe. After two years of flow testing, unfortunately, the well casing collapsed, and the well was abandoned (Friðleifsson et al. 2013).

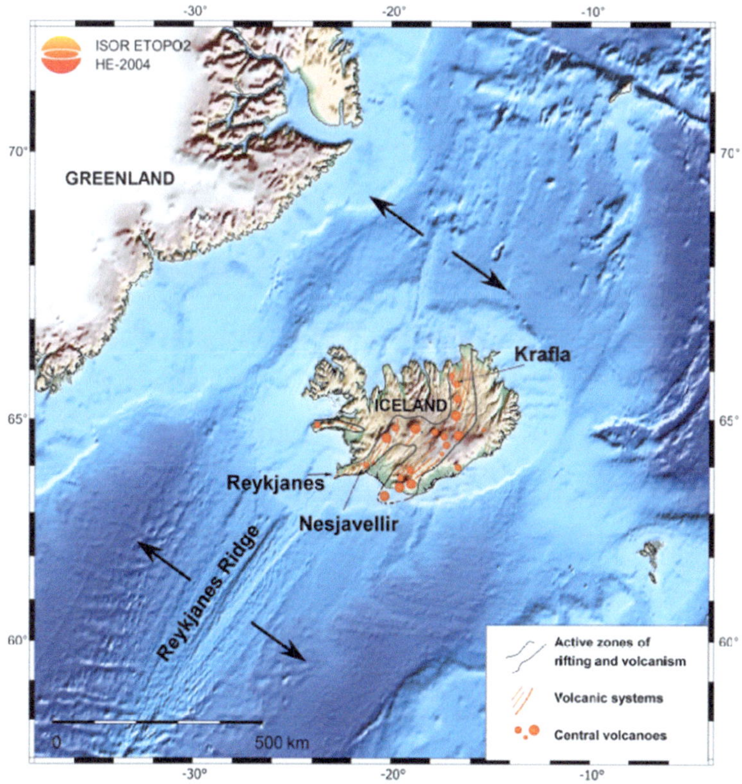

Fig. 6.51 Active rift zones in Iceland and three candidate sites for the Iceland Deep Drilling Project (IDDP) (Friðleifsson et al. 2020)

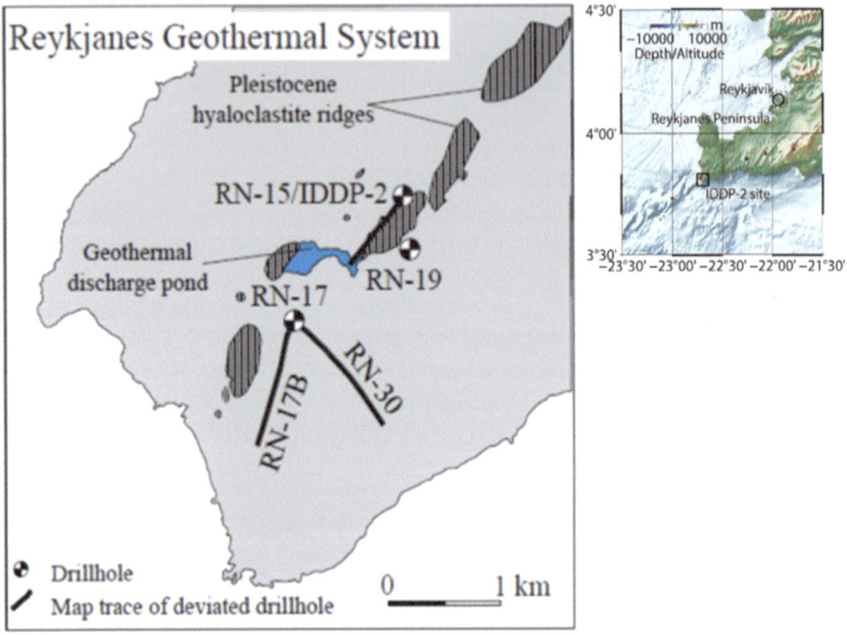

Fig. 6.52 Map showing the location of drill holes in the Reykjanes geothermal field where spot cores have been obtained by IDDP, i.e. RN-17B, RN-19, RN-30, and RN-15/IDDP-2 (Friðleifsson et al. 2013)

In 2013, the IDDP began planning to drill a deep exploratory/research well, IDDP-2, in the Reykjanes geothermal field in southwestern Iceland, where HS Orka (an electric power company based in the Reykjanes peninsula) generates up to 100 MWe of electric power (Friðleifsson et al. 2013). From a scientific perspective, this location is of great interest because the Reykjanes peninsula is the landward extension of the Mid-Atlantic Ridge (Fig. 6.52; Friðleifsson et al. 2013).

The well IDDP-2 was drilled into the Reykjanes geothermal field, Iceland, to a depth of 4.659 m by deepening the 3 km deep vertical well RN-15 (Friðleifsson et al. 2017, 2020). The Iceland Drilling Company (IDC) began deepening the well RN-15/IDDP-2 on 11 August 2016, using the 350-ton drilling rig (Fig. 6.53) with an electric top drive system (MH PTD-500-AC). The well RN-15/IDDP-2 reached a depth of 4,659 m on January 25, 2017. At only six days after the stop of the lost circulation, the measured bottom hole temperature was 426 °C, and the fluid pressure was 34 MPa (Fig. 6.54; Friðleifsson et al. 2017). The critical point for seawater is known to be about 406 °C and 29.8 MPa (Bischoff and Rosenbauer 1988), and the salinity of the fluids of the Reykjanes geothermal reservoirs is usually less than that of seawater (Friðleifsson et al. 2020). This is not strange. As described in Sect. 1.8.2, the mafic crust's chloride content is minimal, resulting in chloride-poor reservoir fluids commonly observed in this region. Thus, there is a high probability that the well IDDP-2 encounters supercritical geothermal conditions.

Fig. 6.53 Drilling rig of the well RN-15/IDDP-2, Reykjanes, Iceland (Photograph by J. Kasahara on October 1 2017)

However, no reservoir fluids have been sampled by this time, and information on the temperature and pressure at the feed zones has not been well determined (Óskarsson 2021). A significant problem encountered during the drilling of the IDDP-2 well in 2016–2017 was a total loss of circulation below a 2.5 km depth that continued to the final depth (Friðleifsson et al. 2021). The total circulation loss continued throughout the drilling and the subsequent 1.5 years of reinjection tests. The total circulation loss could not be cured with lost circulation materials, nor by 12 successive attempts to seal the loss zone with cement. Therefore, drilling had to continue with only intermittent cutting returns between 3,000 and 3,180 m depth and no return of drill cuttings from deeper than 3,180 m. An amount of rock equal to the total volume of the drilled well (~60 m^3) disappeared as drill cuttings into this loss zone (Friðleifsson et al. 2020). The extension tectonics in a sheeted dike complex along the Mid-Atlantic Ridge and the thick, brittle layer associated with the mafic rock composition give an extraordinarily continuous high-permeability zone even in such a deep geological environment at a depth interval of 3.0–4.6 km (Weisenberger et al. 2019).

The IDDP-2 has not proved the existence of a reservoir or supercritical fluid. The following steps are to conduct flow testing and fluid sampling to determine the chemical and thermodynamic properties of the formation fluids. The practical use for power generation is not necessarily optimistic in some simulations (Driesner 2021; Lamy-Chappuis et al. 2023). However, the IDDP's efforts provided enormous new insights into developing the next-generation geothermal power globally.

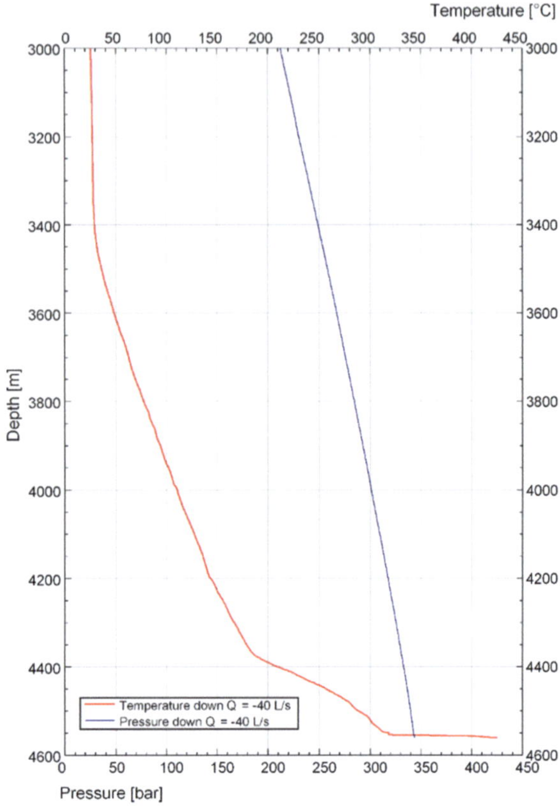

Fig. 6.54 Temperature and pressure log to a depth of 4,560 m in IDDP-2 after six days to stop the drilling water. As can be seen from the temperature profile, the well is far from thermal equilibrium because of the cooling by the cold drilling water. When thermal recovery is complete, temperatures will likely exceed the estimated 426 °C measured at the bottom of the hole. The PT logging was done with a K-10 logging tool, which was calibrated to only 380 °C (Friðleifsson et al. 2017)

6.2.5 *Implications for Future Geothermal Development*

Silicic magma is the ultimate replenishment material in the silicic crust region, like subduction zones and active continental margins. As silicic magma and silicic crust comprise abundant chloride, as discussed in Sect. 1.8.2, it inevitably accelerates hypersaline brine evolution in the high-pressure and high-temperature environment at greater depths by the steam loss, as seen in the well WD-1a. These brines cause the pressure and temperature of the critical point of fluids to rise dramatically (Fournier 1987; Driesner and Heinrich 2007). Supercritical fluids might not be easy to obtain in the consuming plate boundaries.

Mafic magma is the ultimate replenishment material in the mafic crust region, such as ocean ridges and continental rifts. It sometimes allows a low-chloride water environment but forms an unusual recharge environment with high permeability due to the oceanic plate-producing extensional tectonics, as seen in the well IDDP-2. This intensive water recharge might be probably explained by the hydration with hydrothermal alteration of ocean crust in the model of axial valleys of ocean ridges

(Fig. 1.3), and the hydrated ocean crust macroscopically serves for dehydration-induced magmatism in the subduction zones (Fig. 2.7). Supercritical fluids might also not be easy to obtain in the divergent plate boundaries.

Therefore, it is considered that some techniques of engineered geothermal systems, including artificial water injection, are, more or less, required for future geothermal developments in the high-pressure and high-temperature environment at greater depths. It is quite certain that the heat source at greater depth could be efficiently utilized.

References

Arias A, Dini I, Casini M, Fiordelisi A, Perticone I, Aiuto P (2010) Geoscientific feature update of the Larderello-Travale geothermal system (Italy) for a regional numerical modeling. In: Proceedings world geothermal congress 2010, Bali, Indonesia, pp 25–29

Asanuma H, Mogi T, Tsuchiya N, Watanabe N, Naganawa S, Ogawa Y, Fujimitsu Y, Kajiwara T, Osato K, Shimada K, Horimoto S, Sato T, Yamada S, Watanabe K (2021) Japanese supercritical geothermal project for drastic increase of geothermal power generation in 2050. In: Proceedings world geothermal congress 2020+1, Reykjavik, Iceland, p 5

Bertani R (2017) Keynote speech: drilling toward supercritical fluid: the descramble experience at Ladrellero. In: Proceedings of IMAGE final conference, novel approach for geothermal exploration. Akureyri, Iceland

Bertani R, Büsing H, Buske S, Dini A, Hjelstuen M, Luchini M, Manzella A, Nybo R, Rabbel W, Serniotti L, The DESCRAMBLE Science and Technology Team (2018) The first results of the DESCRAMBLE project. In: Proceedings, 43rd workshop on geothermal reservoir engineering Stanford University, Stanford, California, SGP-TR-213

Bischoff JL, Rosenbauer RJ (1988) Liquid-vapor relations in the critical region of the system NaCl-H_2O from 380 °C to 414 °C: a refined determination of the critical point of seawater. Geochim Cosmochim Acta 52:2121–2126

Brace WF, Kohlstedt DL (1980) Limits on lithospheric stress imposed by laboratory experiments. J Geophys Res 85:6248 6252

de Franco R, Petracchini L, Scrocca D, Caielli G, Montegrossi G, Santilano A, Manzella A (2019) Synthetic seismic reflection modelling in a supercritical geothermal system: an image of the K-Horizon in the Larderello Field (Italy). Geofluids 8492453:21. https://doi.org/10.1155/2019/8492453

Doi N, Kato O, Ikeuchi K, Komatsu R, Miyazaki S, Akaku K, Uchida T (1998) Genesis of the plutonic-hydrothermal system around quaternary granite in the Kakkonda geothermal system, Japan. Geothermics 16:552–589

Driesner T (2021) Supercritical and superhot geothermal resources—some fundamental insights. In: Proceedings world geothermal congress 2020+1, Reykjavik, Iceland, p 6

Driesner T, Heinrich CA (2007) The system H_2O-NaCl. I. correlation formulae for phase relations in temperature-pressure-composition space from 0 to 1000 °C, 0 to 5000 bar, and 0 to 1 X_{NaCl}. Geochim Cosmochim Acta 71:488–4901

Elders WA, Friðleifsson GÓ, Zierenberg RA, Pope EC, Mortensen AK, Guðmundsson Á, Lowenstern JB, Marks NE, Owens L, Bird DK, Reed M, Olsen NJ, Schiffman P (2011) Origin of a rhyolite that intruded a geothermal well while drilling in a basaltic volcano, at Krafla, Iceland. Geology 39:231–234

Ferrara GC, Luccioli F, Palmerini GC, Scappini U (1985) Update report on geothermal development in Italy. In: Stone C (ed) International volume of 1985 international symposium on geothermal energy, geothermal resources council transaction, pp 95–105

Fournier RO (1987) Conceptual models of brine evolution in magmatic-hydrothermal systems. U.S. Geol Surv Prof Pap 1350:1487–1506

Fournier RO (1991) The transition from hydrostatic to greater than hydrostatic fluid pressure in presently active continental hydrothermal systems in crystalline rock. Geophys Res Lett 18:955–958

Friðleifsson GÓ, Ármannsson H, Árnason K, Bjarnason IÞ, Gíslason G (2003) IDDP feasibility report. Iceland deep drilling project, part I geosciences—site selection. Orkustofnun, Report, OS-2003-007, p 103. http://www.os.is/gogn/Skyrslur/OS-2003/OS-2003-007.pdf.

Friðleifsson GÓ, Elders WA, Bignall G (2013) A plan for a 5 km-deep borehole at Reykjanes, Iceland, into the root zone of a black smoker on land. Sci Drill 16:73–79. https://doi.org/10.5194/sd-16-73-2013

Friðleifsson1 GÓ, Elders WA, Zierenberg RA, Stefánsson A, Fowler APG, Weisenberger TB, Harðarson BS, Mesfin KG (2017) The Iceland deep drilling project 4.5km deep well, IDDP-2, in the seawater-recharged Reykjanes geothermal field in SW Iceland has successfully reached its supercritical target. Sci Drill 23:1–12. https://doi.org/10.5194/sd-23-1-2017

Friðleifsson GÓ, Elders WA, Zierenberg RA, Fowler APG, Weisenberger TB, Mesfin KG, Sigurðsson Ó, Níelsson S, Einarsson G, Óskarsson F, Guðnason, EA, Tulinius H, Hokstad K, Benoit G, Nono F, Loggia D, Parat F, Cichy SB, Escobedo D, Mainprice D (2020) The Iceland deep drilling project at Reykjanes: drilling into the root zone of a black smoker analog. J Volcanol Geotherm Res 391:106435. https://doi.org/10.1016/j.jvolgeores.2018.08.013

Friðleifsson GÓ, Albertsson A, Stefánsson A, Þórólfsson, G, Mesfin KG, Matthíasdóttir KV, Sigurðsson K, Sigurðsson Ó, Gíslason Þ, Elders WA, Zierenberg RA, Bali E, Guðnason EÁ, Óskarsson F, Weisenberger TB (2021) The IDDP-2 DEEPEGS demonstrator at Reykjanes—overview. In: Proceedings of world geothermal congress 2020+1, Reykjavik, Iceland, p 11

GSI (Geographical Survey Institute) (1997) Digital Map 50 m Grid (Elevation) Nippon-III (CD-ROM)

Goko K (1995) Geological analysis and evaluation of the Ogiri geothermal structure in the Kirishima geothermal area. Resour Geol 45(1):41–52

Goko K (2000) Structure and hydrology of the Ogiri Field, West Kirishima geothermal area, Kyushu, Japan. Geothermics 29:127–149

Helz RT, Thornber CR (1987) Geothermometry of Kilauea Iki lava lake, Hawaii. Bullet Volcanol 49:651–668. https://doi.org/10.1007/BF01080357

Helz RT, Clague DA, Sisson TW, Thornber CR (2014) Petrologic insights into basaltic volcanism at historically active Hawaiian volcanoes. In: Chapter. 6 Poland MP, Takahashi TJ, Landowski CM (eds) Characteristics of Hawaiian volcanoes: U.S. geological survey professional paper vol 1801, pp 237–292

Horikoshi T, Takayama T, Takeshita K, Goko K, Yoshizawa H (2005) An analysis of the geothermal structure of the Ogiri geothermal field based on the surveys and operational data of the Ogiri Power Station after its commencement. Resour Geol 55(1):25–38 (in Japanese with English abstarct)

Idemitsu Kosan (2024) https://www.idemitsu.com/jp/business/geothermal/takigamipowerplant.html

Ikeuchi K, Doi N, Sakagawa Y, Kamenosono H, Uchida T (1998) High-temperature measurements in well WD-1a and the thermal strucuture of the Kakkonda geothermal system, Japan. Geothermics 27:591–607

Inagaki H, Tsutsumi S, Soeda Y, Sofyan Y, Saito H (2022) Three-dimensional MT inversion in the Hatchobaru geothermal field, central Kyushu, southwestern Japan, In: Proceedings of the 8th Indonesian international geothermal convention & exhibition (IIGCE)

Itoi R, Kumamoto Y, Tanaka T, Takayama J (2010) History matching simulation of the Ogiri geothermal field, Japan. In: Proceedings of world geothermal congress 2010, Bali, Indonesia, pp 25–29

JMA (Japan Meteorological Agency) (2013) National catalogue of the active volcanoes in Japan (4th edn). https://www.data.jma.go.jp/vois/data/tokyo/STOCK/souran_eng/menu.htm

Kamata H (1989) Volcanic and structural history of the Hohi volcanic zone, central Kyushu, Japan. Bullet Volcanol 51:315–332

Kasahara J, Hasada Y, Mikada H, Fujise Y (2024) Seismic exploration at Hatchobaru-Otake geothermal field, Kyushu. Ext Abstr Soc Explor Geophys Jpn 115–117 (in Japanese)

Kasahara J, Hasada Y, Furuya S, Shimazaki M, Mikada H, Ohnuma H, Fujise Y (2025) DAS and DTS geothermal study in the Takegami geothermal field in northeastern Kyushu, Japan. In: Mikada, Zdanov, Kasahara (eds) Active geophysical monitoring, 3rd edn. Elsevier Pub.

Kasai K, Sakagawa Y, Komatsu R, Sasaki M, Akaku K, Uchida T (1998) The origin of hypersaline liquid in the quaternary Kakkonda granite, sampled from well WD-1a, Kakkonda geothermal system, Japan. Geothermics 27:631–645

Kato O, Doi N, Sakagawa Y, Uchida T (1998) Fracture systems in and around well WD-1, Kakkonda geothermal field, Japan. Geothermics 27:609–629

Kato O, Sato K (1995) Development of deep-seated geothermal reservoir bringing the quaternary granite into focus in the Kakkonda geothermal field, Northeast Japan. Resour Geol 45:131–144 (in Japanese with English abstract)

Kirby SH (1983) Rheology of the lithosphere. Rev Geophys Space Phys 21:1458–1487

Kondo S, Oyama H, Yamada S (2019) Yamagawa binary power station geothermal ORC plant in Japan. Geothermal Resources Councils Transactions 43:127–132

Kyuden Mirai Energy Co., Inc., (2024) Brochure: introduction to geothermal power station of Kyuden Mirai Energy Co., Inc

Lamy-Chappuis B, Yapparova A, Driesner T (2023) Simulations of the IDDP-2 well, Reykjanes, Iceland, and its behavior in different operation scenarios. Geothermics 114:102790. https://doi.org/10.1016/j.geothermics.2023.102790

Luth WC, Jahns RH, Tuttle OF (1964) The granite system at pressure of 4 to 10 kilobars. J Geophys Res 69:759–773

Momita M, Tokita H, Matsuda K, Takagi H, Soeda Y, Tosha T, Koide K (2000) Deep geothermal structure and the hydrothermal system in the Otake-Hatchobaru geothermal field, Japan. In: Proceedings of the 22nd New Zealand geothermal workshop, pp 257–262

Muraoka H (2012) Geothermal energy (Chap. 35). In: Lackner M, Sajjadi B, Chen WY (eds) Handbook of climate change mitigation and adaptation. Springer, Cham. https://doi.org/10.1007/978-3-030-72579-2_35

Muraoka H, Tateno M, Okubo Y (1998b) Some aspects of the brittle-plastic transition as a depth limit of hydrothermal circulation. In: Abstract of 1998 annual meeting of the geothermal research society of Japan, p 3

Muraoka H, Asanuma H, Ito H (2013) Understanding geothermal systems in ductile zones and their perspective for power generation with engineered geothermal system technologies. J Geography (Chigaku Zasshi) 122:343–362 (in Japanese with English abstract)

Muraoka1 H, Asanuma H, Tsuchiya N, Ito T, Mogi T, Ito H, The Participants of the ICDP/JBBP Workshop (2014) The Japan beyond-brittle project. Sci Drill 17:51–59. https://doi.org/10.5194/sd-17-51-2014

Muraoka H, Ohtani T (2000) Profiling of the Kakkonda geothermal system by bulk rock chemistry of the WD-1a. Rep Geol Surv Jpn 284:35–55 (in Japanese with English abstract)

Muraoka H, Uchida T, Sasada M, Yagi M, Akaku K, Sasaki M, Yasukawa K, Miyazaki S, Doi N, Saito S, Sato K, Tanaka S (1998) Deep geothermal resources survey program: igneous, metamorphic and hydrothermal processes in a well encountering 500 °C at 3729 m depth, Kakkonda, Japan. Geothermics 27:507–534

NEDO (New Energy and Industrial Technology Development Organization) (1996) FY 1995 report of the deep geothermal resources survey project, p 887p (in Japanese)

Okada H, Yasuda Y, Yagi M, Kai K (2000) Geology and fluid chemistry of the Fushime geothermal field, Kyushu, Japan. Geothermics 29:279–311

Okada H, Yamada Y (2002) Fracture distribution in and around intrusive rocks in the Fushime geothermal field, Japan: evidence from the FMI logging. In: SPWLA 43rd annual logging symposium. Oiso, Japan

Óskarsson F (2021) Composition of reservoir fluids in well IDDP-2. In: Proceedings world geothermal congress 2020+1, Reykjavik, Iceland, p 7

Richter DH, Moore JG (1966) Petrology of the Kilauea Iki Lava Lake, Hawaii. U.S. Geological Survey Professional Paper 537-B, p 26

Saito S (1993) A trend of geothermal drillings in Japan. J Geotherm Res Soc Jpn 15:104–110 (in Japanese)

Saito S, Sakuma S, Uchida T (1998) Drilling procedures, techniques and test results for a 3.7 km deep, 500°C exploration well, Kakkonda, Japan. Geothermics 27:573–590

Sakuma K (1999) Development and present status of the Yamagawa geothermal field. J Geotherm Res Soc Jpn 21:291–297 (in Japanese with English abstract)

Tagomori K, Honda M, Nagano H, Akiyoshi M, Haruguchi K, Ushijima K (1997) Geothermal structure in the Hatchobaru geothermal field derived from detailed gravity and resistivity structure. Chinetsu 34(4):1–17 (in Japanese with English abstract)

Takayama J, Itoi R, Morita S (2014) Eighteen-year history of reservoir development in the Ogiri geothermal field, Japan. Geotherm Resour Counc Trans 38:107–111

Takenaka T, Furuya S (1991) Geochemical model of the Takigami geothermal system, northeast Kyushu, Japan. Geochem J 25:267–281

Takenaka T, Goto H, Yamamoto Y, Furuya S (1995) Exploration and development of the Takigami geothermal system, Kyushu, Japan. Resour Geol 45(6):361–376 (in Japanese with English abstract)

Tester JW, Anderson BJ, Batchelor AS, Blackwell DD, DiPippo R, Drake EM, Garnish J, Livesay B, Moore MC, Nichols K, Petty S, Toksöks MN, Veatch Jr RW (2006) The future of geothermal energy: impact of enhanced geothermal systems on the United States in the 21st Century. In: Prepared by the Massachusetts Institute of Technology, under Idaho National Laboratory Subcontract No. 63 00019 for the U.S. Department of Energy, Washington, DC, USA, p 358. http://geothermal.inel.gov/publications/future_of_geothermal_energy.pdf.

TENPSE (Thermal and Nuclear Power Engineering Society) (2021) Current status and trends of geothermal power generation in 2020, p 141 (in Japanese)

Tilling RI, Kauahikaua JP, Brantley SR, Neal CA (2014) The Hawaiian volcano observatory—a natural laboratory for studying basaltic volcanism. In: Poland MP, Takahashi TJ, Landowski CM (eds) Characteristics of Hawaiian volcanoes: U.S. Geol Surv Prof Pap 1801:1–64

Tsutsumi S, Wulaningsih T, Soeda Y, Kiyota Y, Matsuda K, Oshima S, Taguchi S, Tsukada Y (2021) Geology, mineralogy and fluid geochemistry of wells discharging acid fluids in Hatchobaru geothermal field, Kyushu Island, southwestern Japan. In: Proceedings, world geothermal congress 2020+1, Reykjavik, Iceland, p 8

Uchida T (2005) Three-dimensional magnetotelluric investigation in geothermal fields in Japan and Indonesia. In: Proceedings world geothermal congress 2005 Antalya, Turkey

Ushijima K, Mustopa EJ, Jotaki H, Mizunaga H (2005) Magnetotelluric soundings in the Takigami geothermal area, Japan. In: Proceedings world geothermal congress 2005 Antalya, Turkey

Weisenberger TB, Harðarson BS, Mesfin KG, Einarsson GM, Níelsson S, Zierenberg RA, Friðleifsson GÓ (2019) The Iceland deep drilling project at Reykjanes—4.5 km deep drilling into supercritical conditions. In: Proceedings 44th Workshop on Geothermal Reservoir Engineering Stanford University, Stanford, California, SGP-TR-214

Wulaningsih T, Tsutsumi S, Inagaki H, Umee Y, O'Brien J (2017) A 3D model of the Yamagawa geothermal system: Insights into reservoir structure and future field management. In: Proceedings 39th New Zealand geothermal workshop

Chapter 7
Geothermal Exploration Using the Optical Fiber Sensing: Distributed Temperature Sensing (DTS) Technology

Naotsugu Ikeda

Abstract This chapter explores the application of distributed temperature sensing (DTS) technology in geothermal exploration, focusing on its ability to monitor transient and transitional temperature phenomena during events such as water injection or cessation within geothermal wells. Using Raman scattering in optical fibers, DTS enables precise measurement of temperature variations over time, making it especially effective to identify location and property of highly permeable fractures. It enhances the characterization of reservoir properties, fluid flow, and fracture behavior, improving geothermal well management. Some of the technical challenges in QC issues are also addressed.

Keywords Distributed temperature sensing (DTS) · Raman scattering · Transient and transitional temperature phenomena · Crossflow identification · Fracture type geothermal reservoir characterization · Hydrogen-induced signal attenuation

7.1 Introduction

In conventional production logging, the wellbore fluid is typically assumed to remain in a steady or pseudo-static state throughout the survey, with certain exceptions such as pressure transient tests. Another exception is the Radio Active Tracer Log, where multiple gamma-ray logging passes are conducted to track the gamma-ray peak of an ejected radioactive isotope tracer (Gore and Terry 1956). This method provides high-resolution measurements of flow velocities and low-speed flow profiles.

However, in reality, the state of wellbore fluid often undergoes transient or transitional phenomena. Temperature recovery tests capture transient phenomena over a long timescale, while transitional phenomena can be observed during the initiation or cessation of water injection, where temperature profiles exhibit drastic changes. To accurately capture these transient or transitional phenomena, it is necessary to

N. Ikeda (✉)
Welma Co.Ltd., Fukuoka, Japan
e-mail: n-ikeda@welma.co.jp

J. Kasahara et al. (eds.), *Challenges in Geothermal Development*, Advances in Geological Science, https://doi.org/10.1007/978-981-95-0553-1_7

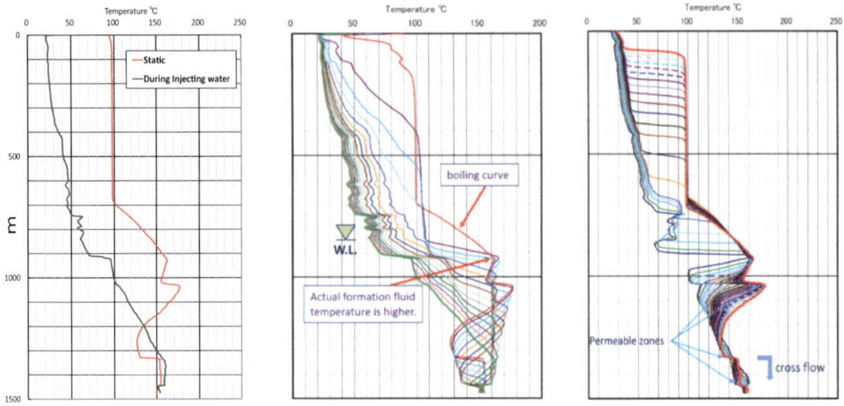

Fig. 7.1 Left: temperature log under static conditions and injecting water in a production well, Center: DTS recorded every 64 s when starting water injection, Right: DTS recorded after stopping the injection

acquire temperature profiles for the entire section simultaneously at short time intervals. Distributed temperature sensing (DTS) can achieve this by recording successive and instantaneous temperature profiles over the entire wellbore, enabling the capture of transient or transitional temperature phenomena.

During the development of the DTS logging application, it was discovered that data acquired during transient or transitional phenomena is highly valuable for delineating reservoir properties. The strength of DTS becomes evident even through simple visual comparison. Figure 7.1 presents a set of DTS logs alongside a conventional temperature log. For the DTS survey, temperature profiles were recorded every 1.5 min during a short injection test, followed by a fall-off period. The dramatic changes in temperature profiles, which appear transiently after each event, provide critical insight into the physical properties of permeable fractures and formation fluids.

The analysis of transient temperature phenomena offers the advantage of requiring minimal intervention time and/or water volume, such as the amount injected during the test. This chapter will explain the process in detail.

7.2 DTS Measurement Principle

The principle of DTS measurement is based on the temperature-dependent nature of Raman scattering of laser pulse, combined with depth information derived from optical time domain reflectometry (OTDR). As the laser pulse propagates through the optical fiber, light scattering, including Raman scattering, occurs due to interaction with the molecule lattice of the fiber. The backscattered light returns is captured and processed by the surface instrument using OTDR techniques. The returned light is

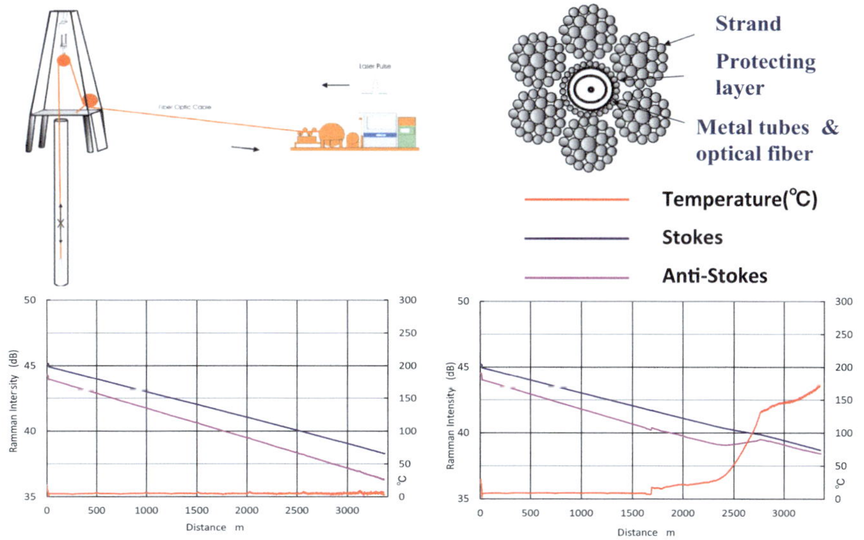

Fig. 7.2 Upper left: DTS setup image, Upper right: DTS cable design, Bottom left and right; The relationship between the received Raman signals intensity and the processed temperature distribution (NEDO 2001)

separated into discrete time windows, allowing the measurement of Raman signals for each time segment, which corresponds can be converted to a depth segment under assumption of a constant speed of light in the fiber (Fig. 7.2 upper left).

In geothermal well environments, DTS measurements face significant challenges, primarily due to hydrogen adsorption at elevated temperatures, which deteriorates the optical fiber's transmission characteristics. Studies, such as Iino et al. (1990) and Stone (1987), have established a strong correlation between the rate of hydroxyl adsorption and the partial pressure of hydrogen. As a mitigation method, nitrogen flushing to reduce partial pressure of hydrogen was found effective (Ikeda 2003). Figure 7.2 (upper right) shows a cross-sectional view of a DTS cable designed for geothermal well applications.

Single-mode fibers are predominantly used in telecommunication due to their low optical attenuation and minimal signal distortion (phase dispersion). For DTS, while single-mode fibers are used for long-distance measurements, less expensive graded-index multimode fibers are more cost effective and generally used over shorter distances (below 3~4 km).

During laser propagation in the optical fiber, various optical losses occur due to interactions between light and the molecular lattice. These include Rayleigh scattering, Brilloin scattering, and Raman scattering as illustrated in Fig. 8.1.1 (referenced from another chapter).

Rayleigh scattering is an elastic process (~10^{-15} s) where the photon energy and frequency remain unchanged. It is caused by the interaction, particularly at locations with structural inhomogeneities (e.g., polarizability or density fluctuations).

Its intensity is significantly higher than inelastic scatterings due to its simpler elastic nature and higher probability of occurrence.

Brillouin scattering is an inelastic scattering, whose intensity is typically 10^{-5}~10^{-6} of Rayleigh scattering. A small frequency shift (GHz range) occurs due to the doppler effect caused by interaction with acoustic waves in the medium. As it sensitive to strain and temperature at the scattering sites, it can be used to measure these parameters.

Raman scattering is an inelastic scattering process (~10^{-12} s) due to interactions between photons and optical phonons. Phonons, while not actual particles, are treated as quasi-particles in quantum mechanics to describe vibrational energy states of molecules. This interaction can result in either the creation of a phono, or annihilation of a phonon, leading to distinct energy shifts in the scattered light. Raman scattering intensity is typically, 10^{-6} to 10^{-8} of Rayleigh scattering and the frequency shift is order of THz.:

- Stokes Scattering: The incident photon loses energy by creating a phonon, resulting in scattered light with a lower frequency ($\nu_{\text{scattered}} = \nu_{\text{incident}} - \nu_{\text{phonon}}$).
- Anti-Stokes Scattering: The incident photon gains energy by annihilating a phonon, resulting in scattered light with a higher frequency ($\nu_{\text{scattered}} = \nu_{\text{incident}} + \nu_{\text{phonon}}$).

Phonons follow the Bose–Einstein distribution, which describes the average number of phonons n(ω) at a given energy level and temperature:

$$n(\omega) = \frac{1}{\left(e^{\frac{hbar\omega}{K_B T}} - 1\right)} \tag{7.1}$$

where: $\hbar$ is the reduced Planck constant,ω is the angular frequency of the phonon mode, K_B is the Boltzmann constant, T is the absolute temperature.

As it is inferred from Eq. 7.1, the intensities of Raman scatterings (both Stokes and Anti-Stokes) are temperature dependent due to the population of phonons at the scattering sites. The relationships are as follows:

In Stokes scattering, the incident photon loses energy by creating a phonon, resulting in scattered light with a lower frequency. The intensity is proportional to the sum of the pre-existing thermal phonon population and a quantum contribution:

$$I_S \propto n(\omega) + 1 \tag{7.2}$$

where the additional $+1$ accounts for the creation of a new phonon, adding a quantum state to the system.

In Anti-Stokes scattering, the incident photon gains energy by annihilating a thermally excited phonon, leading to scattered light with a higher frequency. Its intensity is directly proportional to the existing phonon population. As temperature increases, the number of thermally excited phonons increases, leading to a significant rise in anti-Stokes scattering intensity.

$$I_{AS} \propto n(\omega) \tag{7.3}$$

The ratio of the intensities of Stokes (I_S) to anti-Stokes (I_{AS}) scattering is given by:

$$\frac{I_{AS}}{I_S} = e^{-\frac{hbar\omega}{K_B T}} \tag{7.4}$$

The relationship demonstrates that the intensity ratio depends on the temperature at the scattering site. As temperature decreases, the phonons population ($n(\omega)$) decreases, reducing the intensity of anti-Stokes scattering relative to Stokes scattering.

DTS Measurement

Light attenuation during the round trip in fiber optic is compensated for before processing the intensity ratio of Stokes and anti-Stokes signals to derive temperature profiles accurately. Figure 7.2 (bottom) illustrates a comparison between the intensities of Stokes and anti-Stokes signals, as well as the resulting temperature profiles under both constant and varying temperature conditions. By addressing technical challenges and leveraging advanced technologies, DTS provides a robust solution for geothermal exploration, enabling real-time, high-resolution temperature profiling essential for effectively understanding and managing geothermal reservoirs.

7.3 DTS Application for Analyzing Geothermal Reservoir Properties

7.3.1 Temperature Recovery Test

The geostatic temperature profile of a geothermal well is estimated through a temperature recovery test. This test typically involves multiple temperature log runs at specific time intervals while the well remains in a static condition. Assuming conductive heat transfer dominates, the heat recovery can be modeled using the diffusivity equation, with a solution provided by Dowdle and Cobb (1975). Regardless of the theoretical model used to derive the geostatic temperature, having more data improves the accuracy of predictions and model fitting. DTS is particularly advantageous in this context, as it can acquire continuous temperature profiles of the entire borehole at sampling intervals as short as one minute.

To evaluate the performance of DTS for temperature recovery tests, an experiment was carried out where water at 13°C was injected into a long-dormant geothermal well at a rate of 800 L per minute for three hours. After the water injection stopped, temperature recovery was monitored. DTS was placed near the bottom of the well, and the temperature profile was monitored every 1.5 min (Fig. 7.3) throughout the test, except for the period from ST = 24 h to ST = 48 h when the tool was retrieved for serving another test program. With the massive amount of data available, the advantage of DTS is clear, as it allows for confident regression line drawing for the Horner analysis (Fig. 7.4).

In this particular test, the temperature profile returned to the initial state by the end of the test at ST = 72 h (Fig. 7.5). This convenient situation prompted further exploration of the spatially and temporally abundant temperature data. It became evident that a similar diffusivity equation is used for modeling pressure propagation in porous media, where enhanced interpretation techniques have been developed. The most notable is the Horner plot derivative. For this particular dataset, the geostatic temperature for a given time interval is computed by linear regression of the Horner plot using DTS temperature data for the time window of 4 h (n_{0min}~n_{4hrs}). This computation is repeated until the time window reaches the end of the recovery test (n_{68hrs}~n_{72hrs}). The processed result for five representative depths (one depth in a 13 3/8″ casing, three depths in a 9 5/8″ liner, and one depth in a 7″ liner) is shown in Fig. 7.6. An anomaly highlighted by a dark blue circle (A) is caused by measurement disruption from ST = 24 to 48 h when the tool was pulled out of the hole for measurement in another well. It is clear that the computed temperature in any depth asymptotically reaches the respective geostatic temperature. The rate of the asymptotic trend depends on the hole size, with larger the hole size taking longer to reach temperature equilibrium (B). The computed temperature at 1,600 m does not

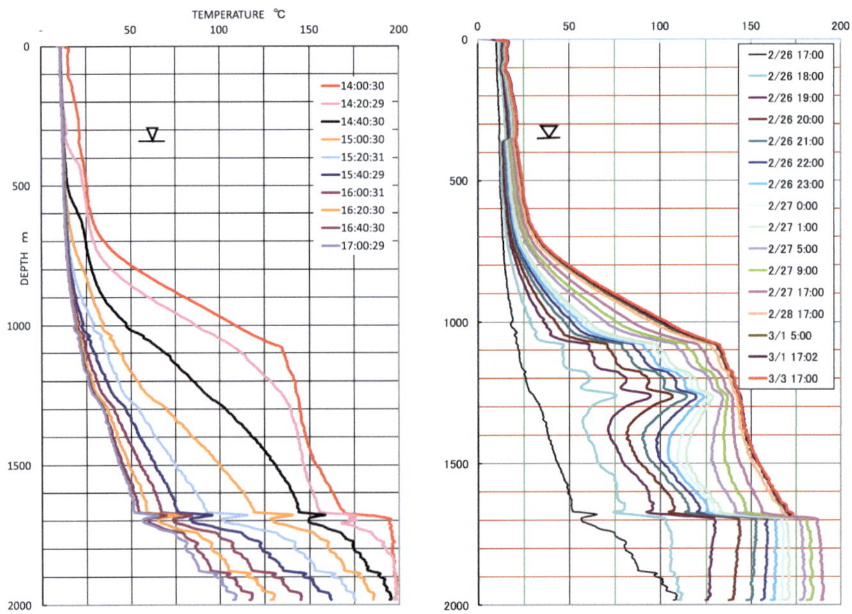

Fig. 7.3 Left: Transition of DTS temperature profiles during water injection, Right: Transition of DTS temperature profiles after stopping water injection (NEDO 2001)

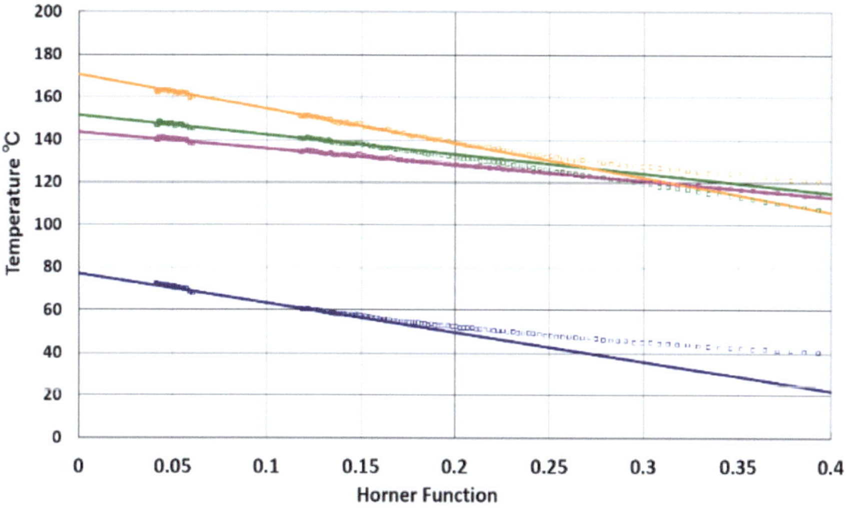

Fig. 7.4 Horner plots for some selected depths (NEDO 2001)

follow an asymptotic trend after 9 h or so (C). This anomaly is possibly caused by convective flow, as it is located close to 7″ liner where the temperature is uniformly high, implying the presence of strong convective flow within the liner.

Among other applications to be explored, the temperature recovery derivative curve can be used for the following:

Fig. 7.5 Geostatic temperature profile derived from the Horner plot model and DTS data (NEDO 2001) *7″ liner section not computed because once retrieved tool could not go there

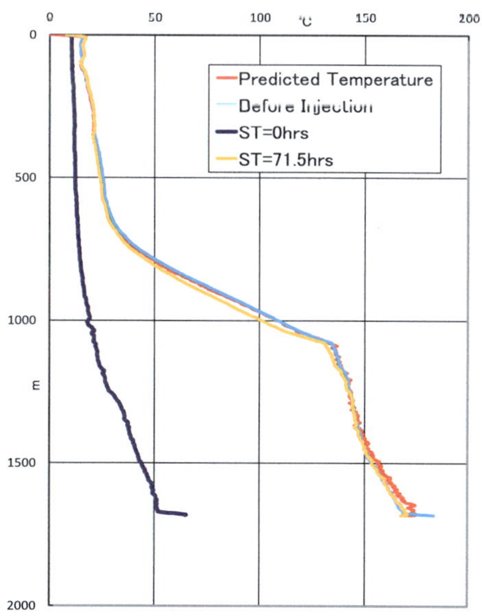

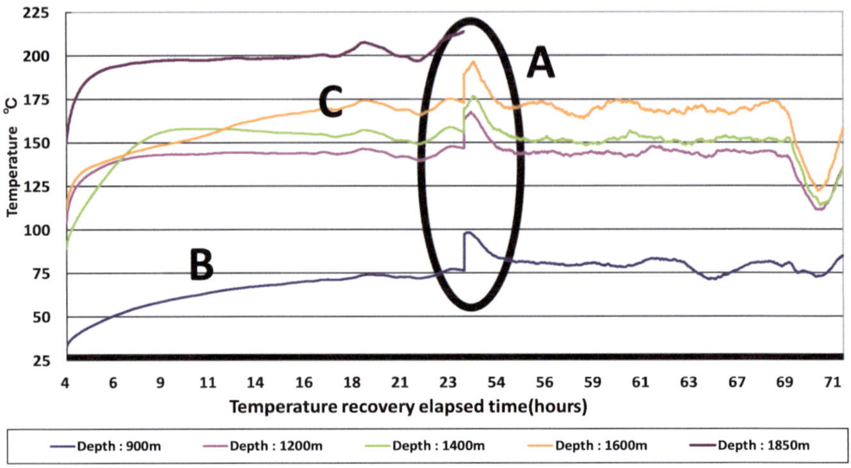

Fig. 7.6 Derivative of Horner plots for some selected depths (NEDO 2001)

- Optimizing the time required for the temperature recovery test.
- Assessing whether the temperature recovery follows the conductive heat transfer model
- If a section of the borehole is under strong influence of convective flow, it may allow preliminary estimates of the geostatic temperature using early-time data.

7.3.2 Flow Speed Profile by DTS

In a well under static conditions, injecting cold water, starting steam production or other interventions that cause a change in fluid flow will result in transitional changes in the temperature profile. Initially, for a certain period, the temperature profile changes reflect the fluid movement because the fluid in the well tends to retain its original temperature for some time. In other words, these transitional changes indicate fluid movement. This interpretation technique is similar to radioactive tracer logs (R.A. tracer logs), where DTS traces temperature profiles instead of gamma-ray profiles to sense fluid movement. Like R.A. tracer logs, the DTS method excels in detecting small flow speeds.

The interpretation principle is depicted in Fig. 7.7, where DTS temperature profiles were recorded every 1.5 min during water injection. The temperature boundary between the hot brine and the cold injected water is seen descending at regular intervals. These intervals narrow below 300 m and show no movement below 400 m due to fluid loss at these depths. Curve correlation techniques can be applied to derive the fluid speed profile (Fig. 7.7 V). Any local temperature features can be used to trace the flow profile. As the flow proceeds, the unique temperature features eventually disappear due to surrounding heat and non-uniform flow speed across

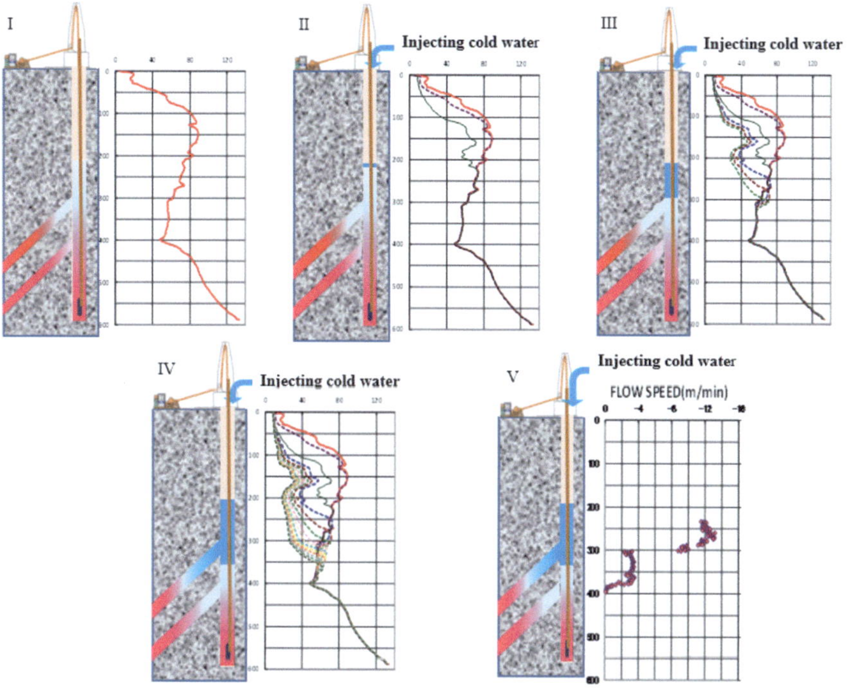

Fig. 7.7 (I) Original temperature profile, (II) starting water injection, (III) injected water reached shallow feed point, (IV) injected water passed shallow feed point, (V) flow speed profile computed by curve resemblance processing (NEDO, 2001 with modification)

the wellbore cross-section. The water level can be identified by deflection of the temperature curve caused by the boundary between the cold water and less cooled air section, observable only during the early stages of the event.

7.3.3 Fracture Characterization and Identification of Crossflow by the DTS Data

(1) *Hydrostatic Pressure Computationby Temperature Profile in a Water Filled Well*

Hydrostatic pressure in a well filled with fresh water can be computed using the following equation (Eq. 7.5).

$$P(z) = P_{water\_level} + \int_{water\_level}^{z_o} \rho(z)dz \quad (7.5)$$

where *P(z)*: hydrostatic pressure, P_{water_level}: pressure at water level, *z*: vertical depth, $\rho(z)$: water density under the condition of the depth z.. z_o: a given vertical depth

The borehole pressure profile can be computed by utilizing temperature profile data and directional data of the well, along with knowledge of the compressed water table. A test was conducted to assess the accuracy of this method by comparing DTS data with quartz gauge data under static conditions and during water injection in a directional well. The results showed a reasonably good match as illustrated in Table 7.1 and Fig. 7.8.

The DTS's temperature and derived pressure data showed some deviation from the conventional log data, especially where the temperature profile changes rapidly. This discrepancy is attributed to the response time of conventional temperature sensors in the air due to their small heat capacity and the slower temperature response of the quartz gauge. Using Eq. 7.5, DTS enables the computation of pressure profiles during transient and/or transitional phenomena, provided the water level can be identified and assuming negligible flow friction loss and mineral content in the borehole fluid.

Table 7.1 Computed hydrostatic pressure and quartz gauge data (NEDO, 2001 with revised computation)

Depth (meters)	Pressure (kg/cm^2A)		Remarks
	DTS	Quartz gauge	
1202	89.15	87.4	Water revel: 210 m
1379	101.19	99.59	

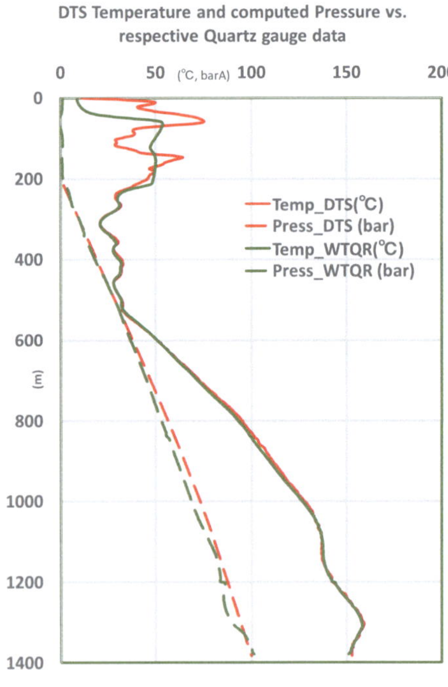

Fig. 7.8 DTS Temperature and computed pressure vs. Quartz gauge data (NEDO 2001)

(2) *DTS Application where Natural Crossflow is present between Permeable Fractures*

Fluid flow to and out of fractures easily occurs in the geothermal wells and massive crossflows between permeable layers are often present. The temperature profile is relatively flat where fluid flow is present. If external disturbances, such as water injection, are introduced to alter the flow profile, the temperature profile exhibits transitional changes, and the flow profile can be deduced as explained above. Proper manipulation of DTS data can identify flow profiles and borehole pressure profiles. Once the flow rate under a given fluid pressure is known, it is possible to calculate pore fluid pressure and injectivity index for individual permeable beds and deduce pore fluid temperature.

This allows for the characterization of each fracture, enabling well-to-well correlation of the permeable fractures. An example of this is shown in the sequence of a temperature recovery test (Fig. 7.9 left) followed by a short water injection (Fig. 7.9 center) and temperature recovery (Fig. 7.9 right), with temperature profiles recorded by DTS.

In the static temperature profiles (Fig. 7.9 left), three distinctive unconformities at 1,115, 1,495, and 1,540 m indicate highly permeable beds at these depths, with crossflows between these zones causing the temperature unconformities. Conventional temperature logs provide limited analysis of flow or fracture properties, often resulting in mere assumptions.

To analyze the fluid flow profile during injection, fresh water was injected at the end of the temperature recovery test while monitoring the temperature profiles with DTS (Fig. 7.9 center). Successive temperature profiles were recorded every 65 s. After water injection began, the fluid level stabilized after rising a few meters (Fig. 7.12).

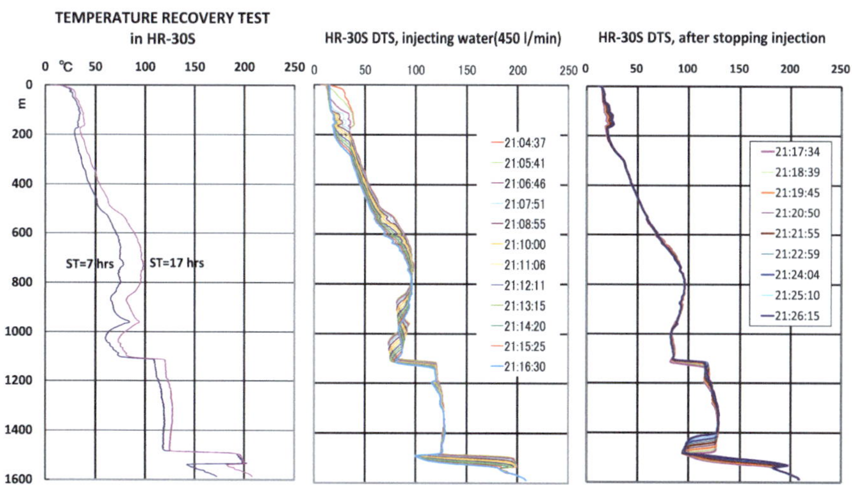

Fig. 7.9 Left: static temperature profile (st = 7 h and 17 h), Center: DTS recorded during injecting water, Right: DTS recorded after stopping water injection in HR-30S (Ikeda 2000)

Localized temperature anomalies, such as peaks and boundaries, shifted downward due to the fluid flow. For example, the temperature peak initially observed at 960 m moved downward at a speed corresponding to the injection flow rate inside the casing, indicating minimal fluid loss above 1,115 m. The flow rate slowed significantly at 1,115 m, suggesting that most of the fluid entered the fracture at that depth.

An intriguing phenomenon was the sudden appearance of a cold-water mass at 1,495 m, which gradually spread downward to 1,540 m, with no fluid movement observed below 1,540 m. The cause of the cold-water mass was initially unclear. Flow rates above and below these three fractures are computed as previously described. When the temperature boundary at 1,495 m reached 1,540 m, water injection was stopped. Post-injection DTS profiles (Fig. 7.9 right) showed the water level returning to its original static condition, suggesting that the fluid flow profile, including cross-flows, quickly reverted to the pre-injection state. Successive temperature profiles indicated that hot fluid at 204 °C was being produced from 1,540 m, while fluid at 94°C was being produced from 1,495 m, mixing to 124 °C and exiting at 1,115 m. These transitional temperature profiles were expected to dissipate and eventually converge to the static condition shown in Fig. 7.9 left.

The process is illustrated schematically in Fig. 7.10. During injection (Fig. 7.10 left), most descending flow exited through the fracture at 1,115 m, while some fluid continued flowing further downward. The pore pressure at 1,495 m exceeded the hydrostatic pressure during injection, resulting in fluid production at a slower rate. Conversely, the pore pressure at 1,540 m was lower than the hydrostatic pressure, causing fluid from above and from 1,495 m to exit at 1,540 m.

(3) *Characterizing Individual Permeable Fractures and Estimating Fracture Planes*

Flow rates into or out of each fracture, along with borehole pressures at each fracture location (showing differential pressures between the borehole and the formation's

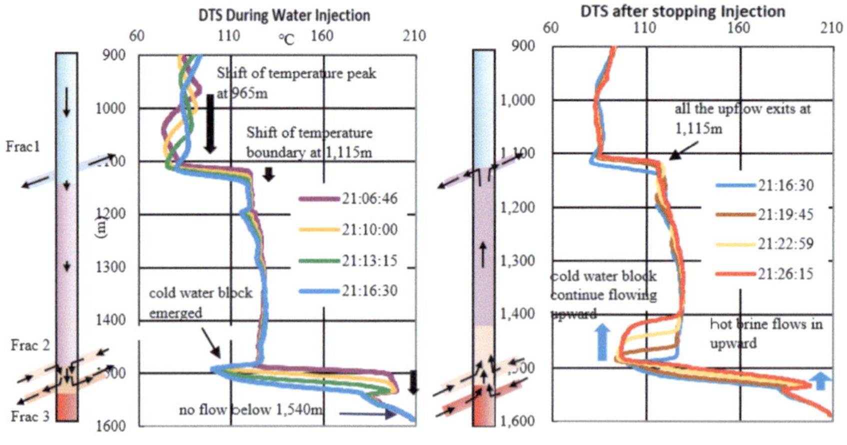

Fig. 7.10 Characterizing permeable fractures and elucidating reservoir structure in HRS-30S (Ikeda 2000)

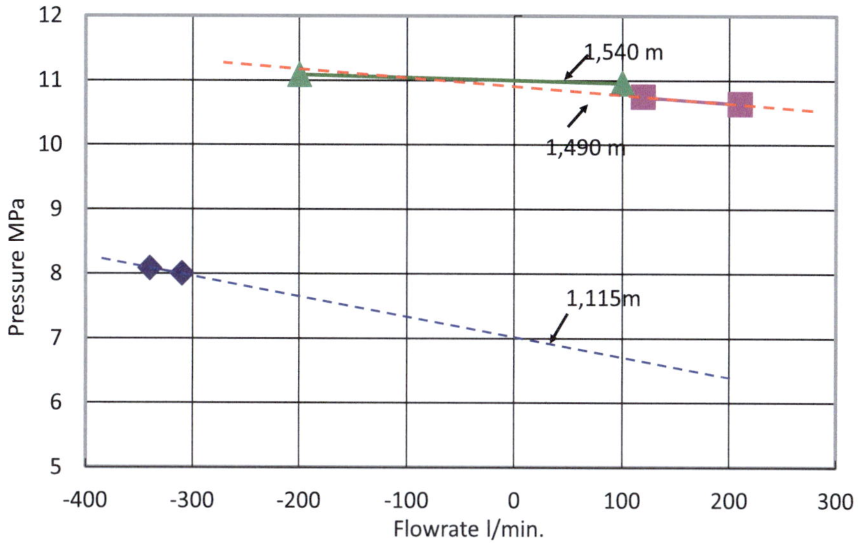

Fig. 7.11 Hydrostatic pressure vs. Flowrate for each fracture in HR-30S (Ikeda 2000)

intrinsic pressure), are plotted in Fig. 7.11. By interpolating and extrapolating these data points, the pore pressure and injectivity index of each fracture can be determined. Additionally, the formation fluid temperature of each fracture can also be estimated from these measurements.

The computed results are presented in Table 7.2. The pore fluid temperature for the fracture at 1,115 m could not be determined due to it having the lowest pore pressure among the major fractures, which resulted in no fluid production. DTS logs reveal that fluid flow, whether inward or outward, occurred in very thin layers, suggesting that fractures are the primary production mechanism. This is much clearer compared to the data obtained from spinner measurements.

The accuracy of this interpretation technique has some limitations, as discussed in Sect. 7.1 above. Since it relies on state changes caused by events such as water injection and secession, the influence of transient phenomena on this interpretation technique must also be considered. In practice, the high permeability of productive fractures in geothermal wells typically causes the transient period to end quickly, while successive transitional temperature profiles remain clearly defined for some time (Fig. 7.12). However, if transient phenomena persist for an extended period, flow analysis using successive DTS data during the transient period can be applied to a different interpretation technique, as described in Sect. 7.4 below.

Characterizing individual permeable fractures allows for the correlation of fractures across multiple wells. A coherent correlation of fracture characteristics in more than three wells can provide a basis for defining the fracture plane. While comprehensive fracture characterization catalogs in a geothermal field are limited, unique

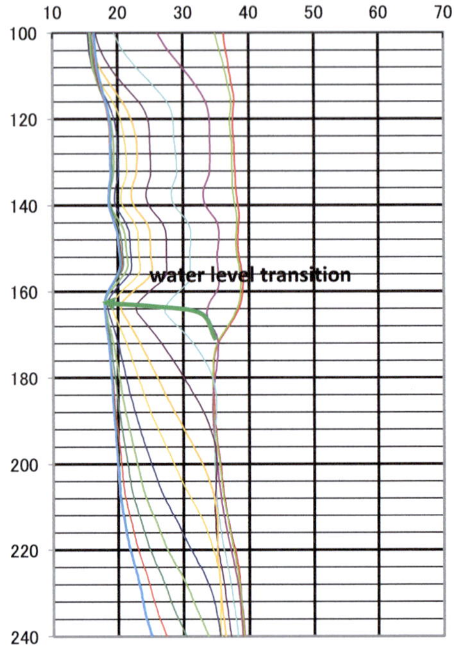

Fig. 7.12 DTS temperature profiles near water level during water injection in HR-30S (Ikeda 2000)

Table 7.2 Interpreted physical properties of permeable fractures in HR-30S (Ikeda 2000)

Flow condition	at static condition		during injection		Pore fluid pressure	Pore fluid Temperature	Injectivity
Water level	168 m		160 m				
Depth (TVD)	Pressure (Mpa)	Flowrate (l/min).	Pressure (Mpa)	Flowrate (l/min).	(MPa)	(°C)	(l/sec/MPa)
1115 m (989.1 m)	8	-310	8.08	-340	7.17	N/A	~6
1490 m (1276.3 m)	10.63	210	10.74	120	10.82	~94	~14
1540 m (1314.6 m)	10.96	100	11.09	-200	10.99	~204	~38

static temperature profiles in well HR-30S were also found in the adjacent well HR-28S. The fractures, as indicated by deflections of the static temperature profiles in both wells, show a strong correlation, suggesting that the respective fracture pairs may belong to the same fracture planes (Fig. 7.13).

However, the two fractures at 1,490 and 1,540 m in HR-30S, though separated by only 50 m, have pore fluid temperature that differs by about 110°C. This strongly suggests that the permeable structures corresponding to these two fractures are unlikely to intersect within a short distance from the well. Given that the shallowest fracture in HR-28S is also located near the others, it is reasonable to assume that the fracture planes associated with the fractures found in both HR-28S and HR-30S are likely running more or less in parallel.

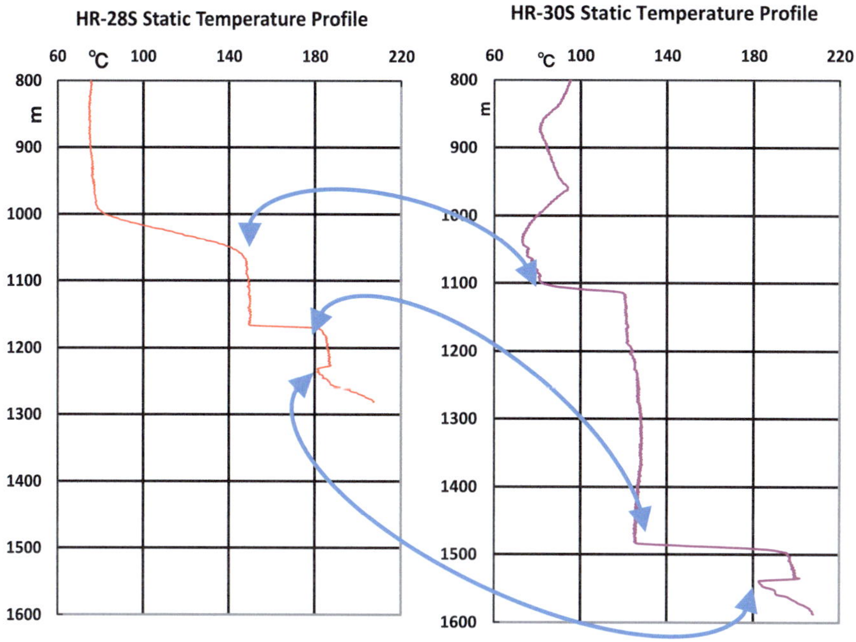

Fig. 7.13 Static temperature profiles and probable correlation of fractures in HR-28S and HR-30S in the Hatchobaru geothermal field. (Ikeda 2000)

Chemical tracer tests offer valuable insights for evaluating fracture connectivity and estimating the location of permeable fracture planes. Figure 7.14 provides a 3D view of the wellbore traces for production wells, four re-injection wells, and a remote production well in a geothermal field. The circles in the figures represent permeable fractures identified through DTS or PTS (Pressure Temperature Spinner logs).

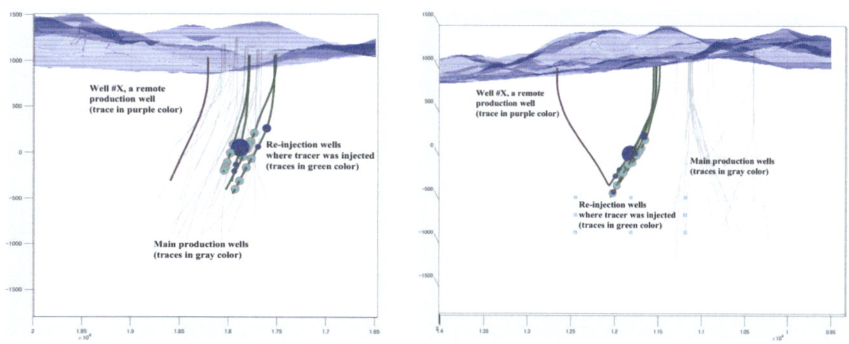

Fig. 7.14 3D view of the well bore traces in a geothermal field with circles representing major feed points (Left: view from the south, Right: view from the east), (wells are color-coded: gray for production wells, green for re-injection wells, purple for a remote well)

Among the re-injection wells where chemical tracers were injected, these four wells showed a strong correlation with the remote production well, but little correlation with the main production wells, as illustrated in Fig. 7.14. The tracer test results are presented in two ways.

- The faster the tracer reaches a particular well, the redder the color of the cylinder
- The higher the peak tracer concentration, the bigger the size of the cylinder

In summary, the tracer test revealed a high-permeability connection between the four re-injection wells and the remote production well, while showing minimal connection with the main production wells (Fig. 7.15).

These fractures exhibit high permeability, suggesting they have wide apertures and substantial extensions. The overall structure of the fracture planes was estimated using the least squares method, incorporating feed points from the four re-injection wells that demonstrated a strong permeability connection with the remote production well.

The computed results are shown in Fig. 7.16. The predicted strike and dip angle of the permeable plane are 59 degrees and −5.6 degrees, respectively, indicating an East–West orientation. Notably, the extension of this plane intersects with the remote production well near its bottom, known as the main feed point. Although the existence of this West–East permeable structure has not been universally confirmed, the methodology for predicting the structure of permeable fracture planes is robust and promising for identifying fracture-dominated geothermal reservoirs.

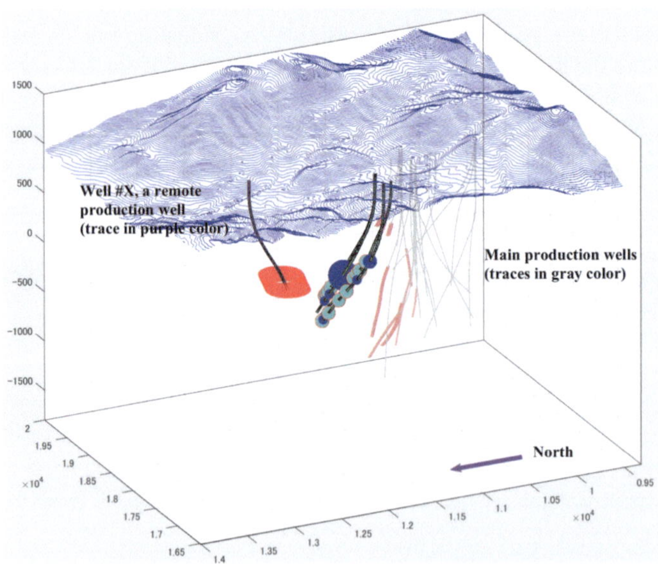

Fig. 7.15 3D view of the well bore traces in a geothermal field with a tracer test response

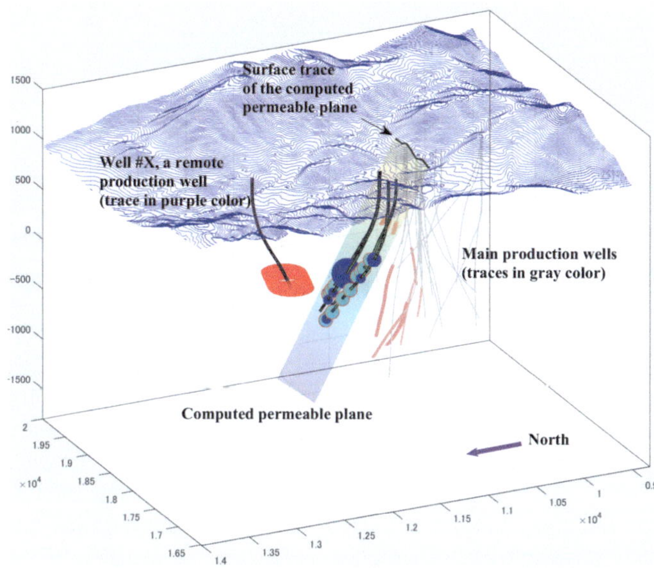

Fig. 7.16 3D view of the well bore traces in a geothermal field with a predicted permeable plane derived by correlating feed points

(4) *Permeability Computation by Rate-Variable Convolution Technique Based on the DTS Data*

The DTS interpretation technique introduced in Sect. 7.2 is effective when the transient state subsides quickly. In this section, I present an analysis of a transient phenomenon captured by DTS. The use of the rate-variable convolution technique for deriving well permeability is well-established in oil wells (e.g., Stewart et al. 1983; Simmon 1986). To more accurately account for the after-flow effect in transient test interpretation and enhance precision, the rate-variable convolution technique was introduced. This method utilizes data from simultaneous downhole pressure and flow measurements. The pressure response of a variable rate system can be expressed by the following equation, as shown in NEDO (1993);

$$P_w = P_i + m\left[\sum_{i=1}^{n}(q_i - q_{i-1})\log(t - t_{i-1}) + q_n\left\{\log\left(\frac{4k}{\varphi\mu c r_w^2}\right) - \gamma\log(e) + 2S\log(e)\right\}\right] \tag{7.6}$$

where P_w is the wellbore pressure, P_i is the initial pressure, q_i is the current flowrate, $q_{i\text{-}1}$ is the flowrate during the previous step, all under downhole condition. t is current time and $t_{i\text{-}1}$ is the time at the last change of the flowrate. The subsequent terms after q_n account for wellbore storage and skin effect, which are not discussed in this study.

The coefficient m in case of re-injection is given by;

$$\frac{kh}{\mu} = \frac{1}{\log e} \frac{1}{4\pi m} \tag{7.7}$$

where k is permeability, h is the thickness of the permeable bed and μ is fluid viscosity.

According to Eq. 7.6, downhole pressure has a linear relationship with the term $\Sigma(q_i - q_{i-1})\log(t - t_{i-1})$, and the slope of this relationship yields the coefficient m, allowing for calculation of kh using Eq. 7.7.

In some geothermal wells, pressure transient phenomena are observed by DTS measurements during the start of injection and fall-off periods. An example of this is shown in Fig. 7.17, where DTS temperature charts are captured at the start and after cessation of water injection. These charts highlight the transitional movement of the water level, identified by the temperature boundary between the injected cold water and the uncooled air section. In both cases, the water level is clearly defined, making it possible to compute the pressure profile in the well at any given time. Simultaneously, the downhole flow rate can be derived from the temperature profiles, enabling the application of the variable rate convolution technique.

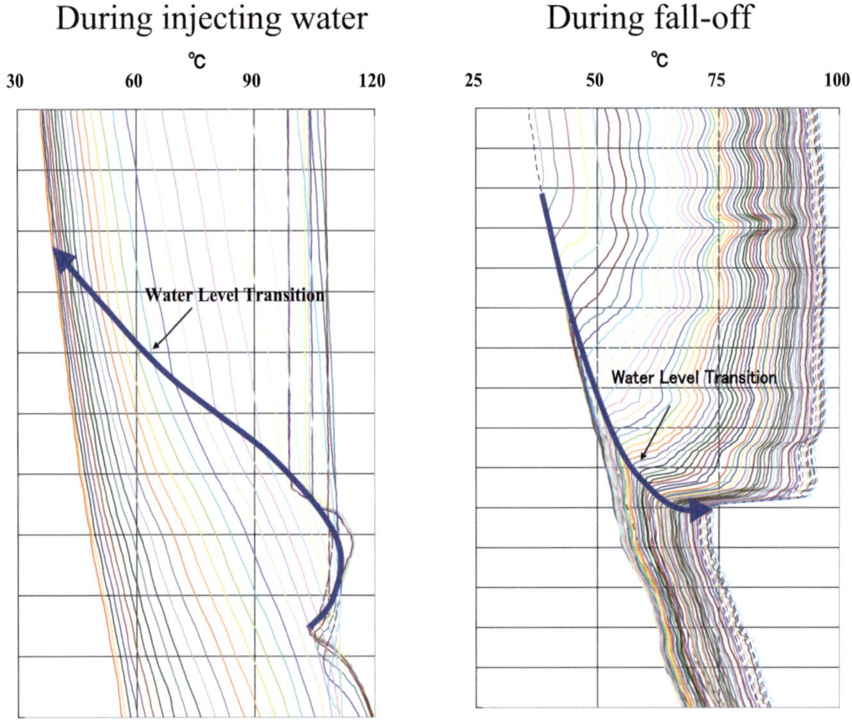

Fig. 7.17 Temperature profiles and water level transition during and after water injection (Ikeda 2003)

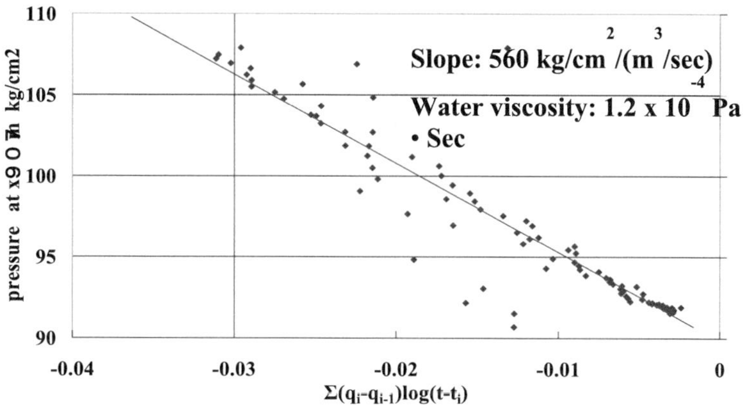

Fig. 7.18 Variable-rate convolution function vs. downhole pressure computed from DTS data, incorporating both the initiation of water injection and the subsequent fall-off data (Ikeda 2003)

The results, shown in Fig. 7.18, exhibit a coherent linear relationship between the two parameters, validating the use of the above equation in this example. From the slope, *kh* was calculated to be approximately 0.4 Darcy·m, assuming the water viscosity of 1.2×10^{-4} Pa·sec under the given conditions. This value aligned remarkably well with the value obtained from the curve-matching method used in a full-scale pressure transient test.

(5) *Fracture Property Analysis in a Well with Natural Crossflow*

The DTS data shown in Fig. 7.19 was recorded at the start of water injection. The transitional temperature profiles revealed the presence of a natural downward crossflow within the well. The relationship between the well-fluid properties and the transitional temperature profiles is straightforward and clearly illustrated in the schematics provided in the figure.

In terms of interpretation, this example is similar to the case introduced in Sect. 7.1. However, the transitional temperature profile, particularly the one running in parallel to the original profile below 470 m, is unique and aligns perfectly with the predicted model. This phenomenon can only be captured using DTS. The overall temperature profiles, including this distinctive feature, provide strong support for the validity of this interpretation technique in analyzing the flow profile and fracture properties.

(6) *Identifying Flow Profiles in Wells with Massive Crossflow Between Fractures with Slightly Different Temperatures*

The interpretation of DTS flow profiles requires careful attention to subtle but critical details in the data. The example presented includes DTS temperature data recorded at the start of water injection (Fig. 7.20) and after the injection ceased (Fig. 7.21).

At first glance, the temperature profile changes between 1,185 and 1,360 m appear insignificant on a compressed scale. However, when examined on a magnified scale, these changes clearly indicate a strong upward flow between these depths. Under

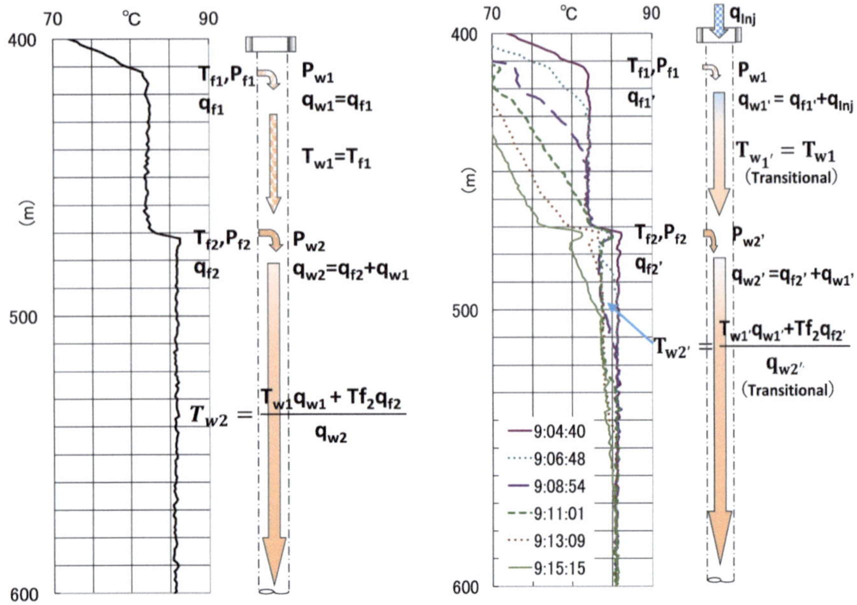

Fig. 7.19 DTS temperature profiles (Left: under static conditions, Right: at the start of water injection)

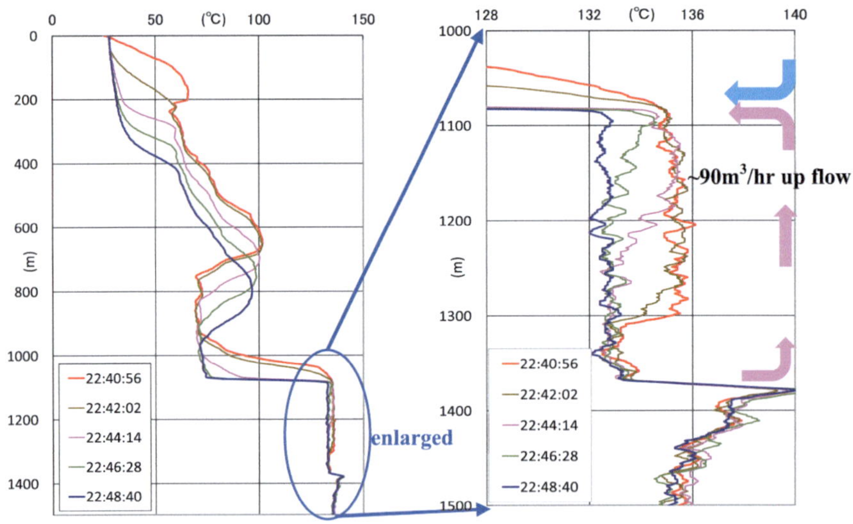

Fig. 7.20 DTS temperature profiles at the beginning of water injection

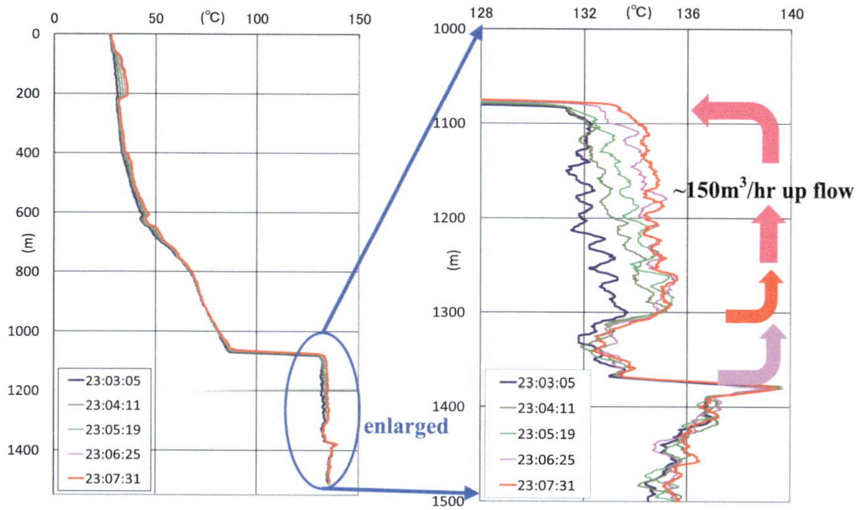

Fig. 7.21 DTS temperature profiles after cessation of water injection

static conditions, inflow was observed from the two lower zones (1,302 and 1,380 m). During water injection, inflow from the 1,302-m zone ceased, while inflow from the 1,380-m zone continued.

Despite the small temperature difference between the pore fluids in the 1,380-m and 1,302-m zones, the flow profile between these zones can still be clearly distinguished through careful analysis.

(7) *Downhole Geyser-Like Phenomena Captured by DTS*

This section highlights another intriguing phenomenon observed through DTS. In this example, DTS was employed to monitor an aging geothermal production well, while injecting water at a steady rate of 150 L per minute. To accurately capture the rapid fluctuations in downhole temperature profiles, a fast reading interval of 24 to 25 s was chosen, as shown in Fig. 7.22.

Despite the continuous water injection during the survey, a vapor zone was detected above the fluid zone, with its interval fluctuating between 80 and 100 m. However, the top of the vapor zone remained stable at around 85 m. Above the vapor zone, temperatures dropped sharply due to the cooling effect of the injected water.

Successive DTS temperature recordings revealed a cyclic, pulsation-like pattern in the localized section of the temperature profiles. Throughout the observation period, the fluid level (assumed to be the liquid head) oscillated, rising and falling by approximately 40 m in less than 50 s, with an overall oscillating cycle of just under 3 min. During each cycle, the upper side temperature state persisted for about 30 s, while the lower temperature state lasted for around 50 s. This cyclical behavior was recorded consistently at regular intervals indicating a recurring process.

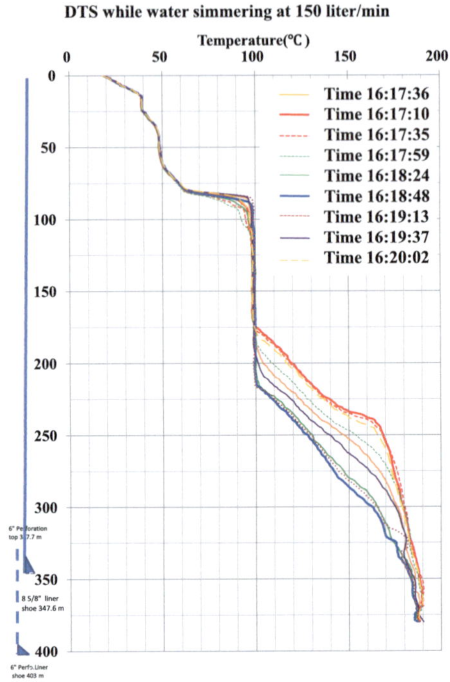

Fig. 7.22 Pulsed fluctuations of temperature profiles in an old production well captured by DTS

To further understand the mechanism behind this phenomenon, the changes in temperature profiles observed during water injection at a rate of 300 L per minute in another production well (Fig. 7.1 center) offer valuable insights.

Figure 7.23 (left) presents two successive temperature profiles, extracted from Fig. 7.1, recorded during the initial phase of water injection at 22:58:02 and at 23:03:25. Computed pressure profiles, derived using Eq. 7.5, were plotted for each temperature profile under the assumption that the well below the respective water levels was filled with liquid water.

At first glance, the transition between these two temperature profiles suggests that the fluid level dropped from 718 to 843 m. However, this interpretation seems improbable, given that the DTS survey was conducted while water was being injected at a rate of 300 L per minute. To investigate further, water saturation pressures corresponding to the temperature profiles were calculated as shown in Fig. 7.23 (center). Notably, the two pressure profiles recorded at 23:03:25—one assuming the well was filled with water below 843 m and the other assuming water saturation pressure—align well between 843 and 925 m.

Based on these observations, the anomalous section of the temperature profile recorded at 22:58:02, which deviates between 718 and 925 m, likely represents the presence of a two-phase fluid. Therefore, the actual liquid water level is expected to be slightly above 925 m. Additionally, the volumetric water–vapor ratio was calculated from the pressure gradient of the saturation water pressure profile, corresponding to

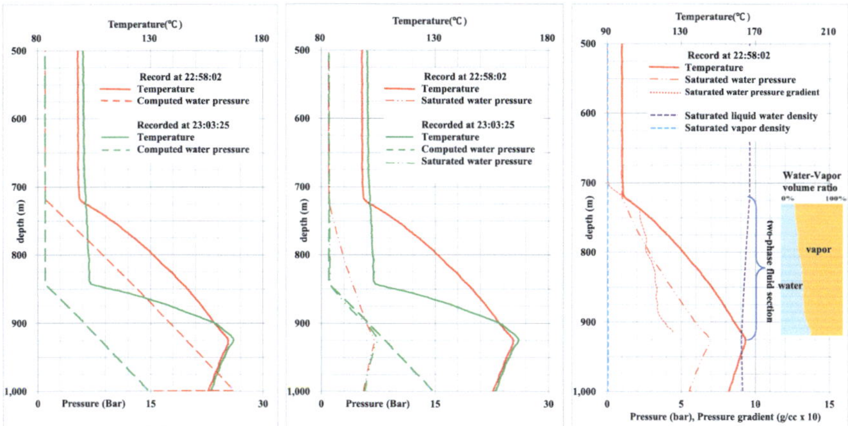

Fig. 7.23 DTS temperature and various computed data profiles derived from Fig. 7.1(left): temperature and pressure (assuming filled with liquid water) for both recordings, (center): plus; water saturation for both recordings, (right): plus; pressure gradient, densities of saturated water and vapor, and volumetric ratio of water and vapor in two-phase section, less; water-filled pressure only for the later recording

the bulk fluid density under static conditions. By using the bulk fluid density along with the temperature-dependent densities of saturation water and vapor, the volumetric ratio across the two-phase section was computed A schematic representation of the calculation is provided in Fig. 7.23 (right).

As analyzed, the significant and rapid temperature fluctuations along with corresponding fluid movements under near-boiling conditions, suggest phase transitions between saturated water and a two-phase mixture. The cyclic temperature variations depicted in Fig. 7.22 appear to correlate with these fluid phase changes. A similar approach is taken to analyze this phenomenon. Two temperature profiles are selected: one representing the higher fluid level state at 16:17:35 and the other representing the lower fluid level state at 16:18:48. For each profile, two corresponding pressure profiles are computed—one assuming the borehole is filled with water below the fluid level, and the other assuming the saturated water pressure (Fig. 7.24(left)).

The pressure profiles for the lower fluid level state at 16:18:48 exhibit near-parallel behavior between 280 and 350 m, indicating that this section is likely filled with saturated liquid water. Between 215 and 280 m, the saturated water pressure profile becomes significantly steeper than the computed water-filled pressure, suggesting the presence of either a two-phase mixture or subcooled liquid water. To maintain pressure consistency with the lower zone, this section is likely in a two-phase state, with the actual liquid level estimated to be around 245 m. Figure 7.24(center) schematically illustrates this analysis.

In contrast, for the higher fluid level state at 16:17:48, the saturated water pressure profile starts to deviate from the water-filled pressure profile below the expected fluid level at 172 m. However, the two curves converge near 245 m (marked with a circle

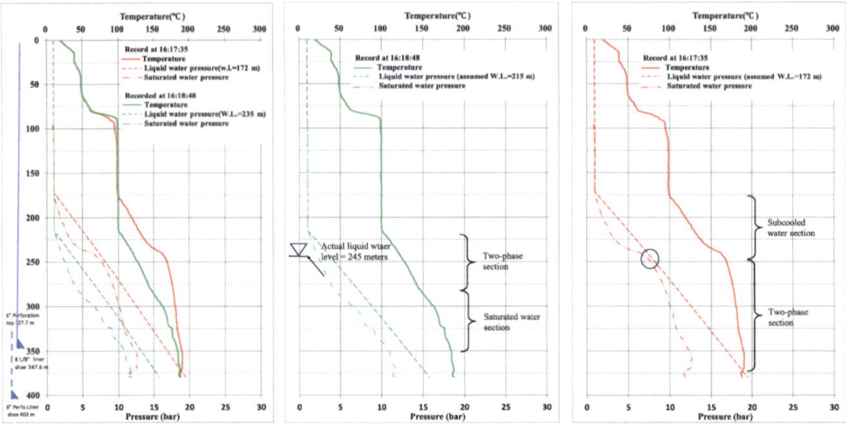

Fig. 7.24 DTS data recorded in an aged production well, exhibiting oscillating temperature profiles. Left: profiles of 1, Temperature, 2. Pressure (assuming filled with liquid water), and 3. Saturation pressure for the two conditions. Center: same set of data profiles for the lower fluid level, Right: same set of data profiles for the higher water level

in Fig. 7.24center), indicating that the borehole between 172 and 245 m is filled with subcooled water, as both subcooled water and saturated liquid water have similar densities. Below 245 m, the saturated water pressure steepens again, suggesting that the section is in a two-phase state.

Based on the analysis of these two oscillating states, the rapid and large temperature fluctuations can be attributed to phase transitions between two-phase and saturated water within this zone. If this interpretation holds, the oscillation causes pressure fluctuation of approximately 6 bars within about 50 s at 245 m in depth. In other words, alternating flashing and condensing occur regularly in this section of the borehole. This phenomenon is likely driven by the balance between heat input from the deeper reservoir and cooling from water injection, similar to the mechanism observed in natural geysers. While this analysis is inferred from the DTS temperature data, it should be verified using pressure log data.

Regardless of the accuracy of the analysis presented here, this type of rapid temperature fluctuation can only be captured through the use of DTS technology.

7.4 Future Prospects of DTS Technology

1. Enhanced Measurement with Logging Sonde: Connecting a logging sonde equipped with sensors—such as temperature, pressure, spinner, and caliper—to the end of a DTS cable can significantly improve measurement quality and enhance the interpretation capacity of DTS data. Ideally, the DTS fiber optic

cable should be used to transmit data to the surface, a goal that could be achieved by developing suitable optical drivers and high-temperature batteries.
2. Transient Temperature Phenomena: The use of DTS in capturing transient phenomena during and after water injection has been explored. However, few DTS surveys have been conducted under producing conditions. Future surveys in such situations may provide new insights and generate additional value.
3. Challenges and Anomalies: Besides the known hydrogen-induced issues, other characteristics, such as spatial resolution and response time for specific DTS cable designs, have not been thoroughly cataloged. Temperature anomalies, such as sudden jumps or drops of a few degrees, have been observed when surveys in environments with rapidly changing temperatures, like those encountered during water injection. These anomalies may result from sections of the DTS cable resting against the borehole wall while other sections float in the air or water.

7.5 Revisiting DTS Measurement Principle and Causes of Problems

7.5.1 Scattering and Attenuation Mechanism of Light in Fiber Optics

As previously noted, multiple factors contribute to the attenuation of light as it propagates through an optical fiber cable. Since DTS relies on measuring Raman backscattering intensity using Optical Time Domain Reflectometry (OTDR) technology, maintaining uniform attenuation characteristics along the fiber length is crucial for accurate temperature measurement.

The principal types of attenuation (scattering) in optical fibers are summarized below and illustrated in Fig. 7.25:

- Rayleigh Scattering: The type of loss is caused by microscopic density fluctuations in the molecular lattice of the fiber. It is inversely proportional to the fourth power of the wavelength.
- Silica Infrared Absorption (Infrared Region): Intrinsic loss caused by vibrational absorption of Si–O bonds in the silica lattice, whose fundamental and overtone transitions absorb light in the infrared spectrum.
- Chemical Bonds: This loss occurs due to absorption by chemical bonds, such as hydrogen, OH- ions, and chloride ion bonds, which absorb light and degrade transmission quality.
- Impurity Absorption: Light absorption is caused by impurities in the fiber, such as transition metals like iron, copper, chromium, and nickel, which absorb specific wavelengths.

Minor and secondary loss mechanisms including the following:

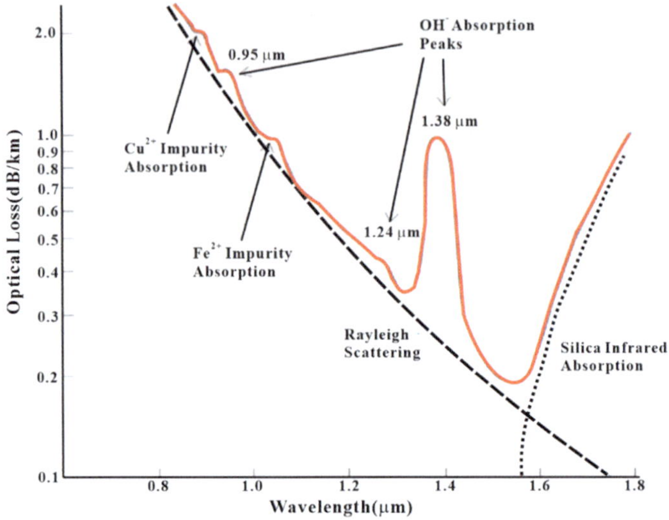

Fig. 7.25 Causes of transmission loss on the fiber optics (adapted from Hecht 2006)

- Waveguide Imperfections: Losses arise from imperfections in the fiber core and cladding or from micro-bending of the fiber, which disrupts the light path and causes scattering.
- Ultraviolet absorption: At short wavelengths (< 0.4 μm), absorption occurs due to **electronic transitions** in the silica band structure, forming the ultraviolet absorption edge.

7.5.2 Hydrogen Issues in DTS Measurement

The most significant challenge for optical fibers in high-temperature environments, such as geothermal well logging, is the degradation of light propagation caused by the adsorption of hydrogen (reversible) and hydroxyl groups (irreversible) (Sumitomo Electric Industries (SEI) 1995). This issue is exemplified in Fig. 7.26, where water intrusion into a ruptured capillary tube led to poor data quality. The compromised section was confirmed by repeating the measurement after pulling the cable up by 100 m. Over a stretch of approximately 300 m from the bottom of the cable, discrepancies between the two measurements were observed, indicating damage caused by hydrogen infiltration.

Hydrogen poses a major challenge for DTS performance in elevated temperature environments due to its high diffusivity. Two processes contribute to light absorption caused by hydrogen.

1. Dissolution Hydrogen: The process is considered reversible. When hydrogen partial pressure decreases, dissolved hydrogen diffuses out. Normally, diatomic

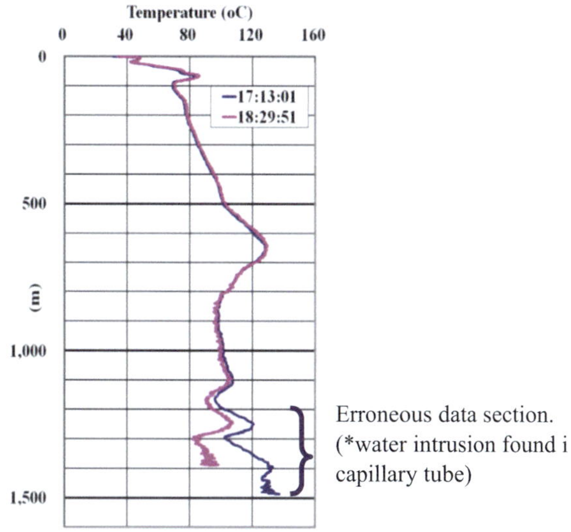

Fig. 7.26 Temperature profiles recorded at two different depths revealed DTS problem

hydrogen is optically transparent at low pressure and does not absorb infrared radiation. However, at high partial pressure or in solid form, hydrogen absorbs infrared rays due to broken molecular symmetry.

2. Formation of OH Groups by Hydrogen: This is an irreversible process. The transmission loss caused by OH groups depends on the number of defects in the glass molecular bonds. Dopants used to control the refractive index can introduce irregularities in the molecular structure of the glass, leading to the formation of defects that are prone to bonding with hydrogen, resulting in increased light absorption.

7.5.3 Source of Hydrogen Causing Problems

Several sources of hydrogen are present:

- Fiber Coating: Polyimide, silicon resin, etc., decompose at high temperatures.
- Lubricant: Lubricant used in the manufacture of metal tubes.
- Ambient Hot Water: Geothermal wells are filled with high-temperature hot water and steam, and hydrogen from the hot water diffuses through the metal and dissolves in the optical fiber.
- Cable Material: If the fiber optic cable is made of different metals, hydrogen may be generated by galvanic corrosion reaction with moisture (Fig. 7.27).

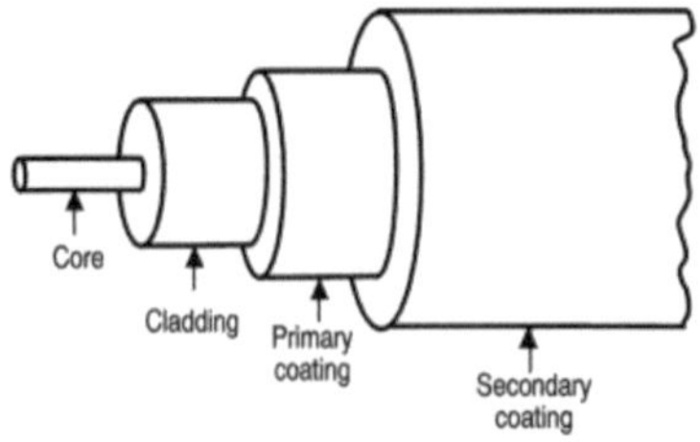

For the high temperature application,
Polyimide is popularly used as the coating material.

Fig. 7.27 Basic structure of the fiber optic used for DTS sensor (NEDO 2003)

7.5.4 Mitigation Methods for the Hydrogen Induced Deterioration

Certain effective measures have been proposed, some of which have been successfully implemented to suppress this hydrogen adverse effect (SEI, 1995; Ikeda 2003).

Mitigation Method 1: Nitrogen Circulation

Nitrogen circulation is used to reduce the partial pressure of hydrogen/hydroxyl molecules. Successful performance was confirmed by the continuous survey under 275ºC for 43 h, where the optic-fiber characteristics did not change as shown in Fig. 7.28. This method may be valid as far as the temperature is below 300°C and the exposure to the environment is limited to a few days.

Mitigation Method 2: Looped Fiber-Optic Cable

This method involves taking measurements alternately at both ends of the looped fiber-optic cable as depicted in Fig. 7.29.

Fig. 7.28 Comparison of Raman intensity distribution before and after going through a continuous survey under 275 degrees Celsius for 43 h. (Ikeda 2003)

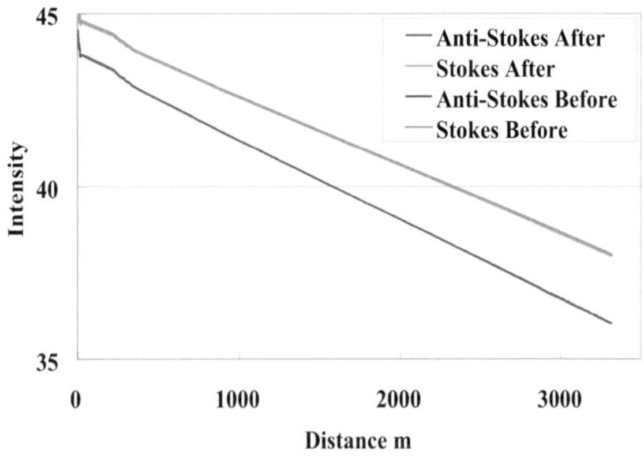

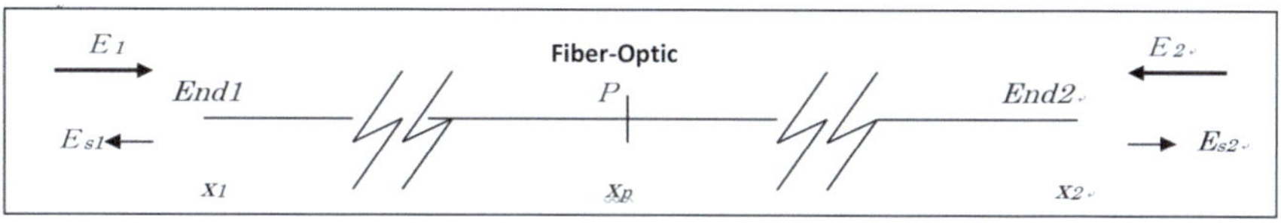

Fig. 7.29 Schematic diagram depicting relevant parameters for the looped fiber optic DTS

E_1 andE_2 are the intensities of incident laser pulses from *End1* (x_1) and *End2* (x_2). Backscattered light generated at $P\ (x_p)$ is detected at *End1* and *End2*. Now, $\alpha\ (x)$ is defined as the attenuation factor of the fiber-optic. Then,

Light attenuation is expressed as:

$$-dE = \alpha(x)Edx \tag{7.8}$$

Based on *Eq. 7.8*, E_{s1} is derived as follows:

$$E_{s1} = E_1 \cdot \eta(x_p) \cdot \exp\left\{-2\int_{x_1}^{x_p} \alpha(x)dx\right\} \tag{7.9}$$

where $\eta(x_p)$ is the Raman backscattering coefficient,

In the same way,E_{s2} is derived as follows:

$$E_{s2} = E_2 \cdot \eta(x_p) \cdot \exp\left\{-2\int_{x_p}^{x_2} \alpha(x)dx\right\} \tag{7.10}$$

Multiplying *Eq. 7.9* by *Eq. 7.10*:

$$E_{s1} \cdot E_{s2} = E_1 \cdot E_2 \cdot \{\eta(x_p)\}^2 \cdot \exp\left[\left\{-2\int_{x_1}^{x_p} \alpha(x)dx\right\} + \left\{-2\int_{x_p}^{x_2} \alpha(x)dx\right\}\right]$$

$$= E_1 \cdot E_2 \cdot \{\eta(x_p)\}^2 \cdot \exp\left[\left\{-2\int_{x_1}^{x_2} \alpha(x)dx\right\}\right]$$

Attenuation factor over the entire cable length, which is measurable.

Appropriate Raman scattering coefficient ($\eta(x_p)$) can be obtained regardless of the irregular loss distribution, given the signal strength is above a certain level).

The Other Mitigation Candidates

While alternative solutions may be available, the following options could enhance high-temperature performance in optical fiber systems.

1. Pretreatment: Hydrogen purging during production to eliminate residual hydrogen in the system
2. Single-Mode (S.M) Fiber with Pure Quartz Core and Fluoridated Cladding.
3. Graded-Index (GI) Optical Fiber with GeO_2-Doped Core and Fluoridated Cladding.

Single-mode (S.M) fiber operates at a longer wavelength (e.g., 1.55um), where inherent signal loss is lower. When meticulously manufactured, S.M. fiber can minimize impurities and defects in the molecular lattice, which, although costly, offers the potential for significantly lower signal losses and improved resistance to light absorption caused by hydrogen and OH^-ions. Several studies have evaluated fluorine's effectiveness as a dopant, and the following were identified as the likely causes (Ikeda 2003);

1. Fluorine replaces OH or Cl groups without inducing light loss in the transparent regions of quartz glass.
2. Si–F bonds are stronger than Si–O bonds, resulting in a larger optical energy transition gap.
3. The formation of Si–F bonds enhances the stability of the glass molecular lattice, reducing structural disorder and reducing Anderson local electron sites in the band, which shifts the Urbach edge to the shorter wavelengths.
4. Fluorine promotes structural relaxation, allowing glass to form at lower temperatures. This reduced structural irregularities, further shifting the Urbach edge to shorter wavelengths.

However, doping with fluorine also presents technical challenges. For example, adding large amounts of fluorine is difficult due to the production of volatile Si-F_4. Additionally, fluorine doping produces a relatively low refractive index contrast, making it challenging to manufacture stable optical waveguides. Its high diffusivity in the fiber also complicates the creation of a graded index. As a result, fluorine is often used in conjunction with other dopants, such as germanium and phosphorus to achieve better performance.

7.5.5 DTS Survey QC Process

Adequate quality control is essential for any type of measurement system. In the case of DTS, one of its key limitations is the lack of a robust calibration process or a reliable method to ensure measurement accuracy. Since the entire length of the cable functions as the sensor, the physical size of the DTS sensor is inherently large, making the precision more challenging. To address these limitations and improve measurement accuracy, the following quality control (QC) processes should be implemented, based on the author's experience:

1. Repeatability Check: Perform repeated measurements using different sections of the cable to verify consistency,

2. Temperature Check: Conduct checks at the known temperature points, such as pure saturated steam zones, to confirm accuracy.
3. Independent Verification: Use a memory-type or optical driver-equipped logging tool connected to the bottom of the DTS cable to provide independent, corroborative data.

Acknowledgements I sincerely thank the companies listed below for generously providing the data used in the following figures and tables, enabling their inclusion in this publication.

Figure 7.1, 7.3 , 7.4, 7.5 and 7.7. 7.6, 7.8 and Table 7.1: Courtesy of Idemitsu Kosan Co., Ltd.

References

Dowdle WL, Cobb WM (1975) Static formation temperature from well logs—an empirical method. J Petrol Technol 27(11):1326–1330

Gore GL, Terry LL (1956) Radio active tracer techniques. J Pet Technol

Hecht J (2006) Understanding fiber optics. Laser Light Press

Iino A, Kuwabara M, Kokura K (1990) Mechanisms of hydrogen-induced losses in silica-based optical fibers. J Lightwave Technol 8(11):1675–1679

Ikeda N (2000) Fractured reservoir management by fiber optic distributed temperature measurement. In: Proceedings SPWLA of Japan

Ikeda N (2003) Fracture reservoir characterization by fiber-optic distributed temperature log. In: Proceedings, 28th workshop on geothermal reservoir engineering. Stanford University

NEDO (1993) Well Test Manual, (in JPN)

NEDO (2001) NEDO project report (010019185), (in JPN)

NEDO (2003) NEDO project report (100000371), (in JPN)

Simmon JF (1986) Interpretation of underbalanced surge pressure data by rate-time convolution. In: SPE 61st annual technical conference and exhibition in New Orleans, SPE15477

Stewart G, Wittmann MJ, Meunier D (1983) After-flow measurement and deconvolution in well test analysis. In: SPE 58th annual technical conference and exhibition in San Francisco, SPE12174

Stone J (1987) Interactions of hydrogen and deuterium with silica optical fibers: a review. J Lightwave Technol 5(5):712–733

Sumitomo Electric Industries (SEI) (1995) Joint study report on distributed temperature measurement system for geothermal wells. SEI Internal Report

Chapter 8
Latest Geothermal Exploration and Development

Junzo Kasahara and Yoko Hasada

Abstract This chapter discusses the latest technologies in geothermal exploration, focusing particularly on the application of optical fiber sensing in boreholes, the full-waveform inversion (FWI) method, the P-S conversion, the use of natural earthquakes, and the gravity tensor.

Keywords DTS · DAS · Waveform inversion · P-to-S conversion · S-to-P conversion · Gravitational tensor

8.1 Optical Fiber Sensing

8.1.1 Backscattering in Optical Fiber and DTS and DAS

Distributed Fiber Optic Sensing (**DFOS**) is an advanced technology used for measuring temperature, strain, and dynamic strain (vibration) at multiple distributed points along an optical fiber. When light is introduced into the fiber, a portion of it scatters due to microscopic fluctuations of the molecular lattice properties, with scattered light returning to the point of insertion. There are three primary types of scattering: Raman, Brillouin, and Rayleigh scatterings (Hartog 2017) (see Fig. 8.1). Temperature measurements use Raman scattering, which is an inelastic scattering process where incident light interacts with a material, exchanging energy with the material's vibrational, rotational, or other low-frequency modes, including phonons in a solid. This process causes the scattered light to have a wavelength different from that of the incident light. This inelastic interaction either creates a phonon (Stokes scattering having longer wavelength and losing energy) or annihilates a phonon

---

J. Kasahara (✉)
Shizuoka University, Shizuoka, Japan
e-mail: kasahara.junzo@shizuoka.ac.jp

Y. Hasada
The Engineering Advancement Association of Japan, Tokyo, Japan
e-mail: hasada.yoko@daiwatansa.co.jp

J. Kasahara et al. (eds.), *Challenges in Geothermal Development*, Advances in Geological Science, https://doi.org/10.1007/978-981-95-0553-1_8

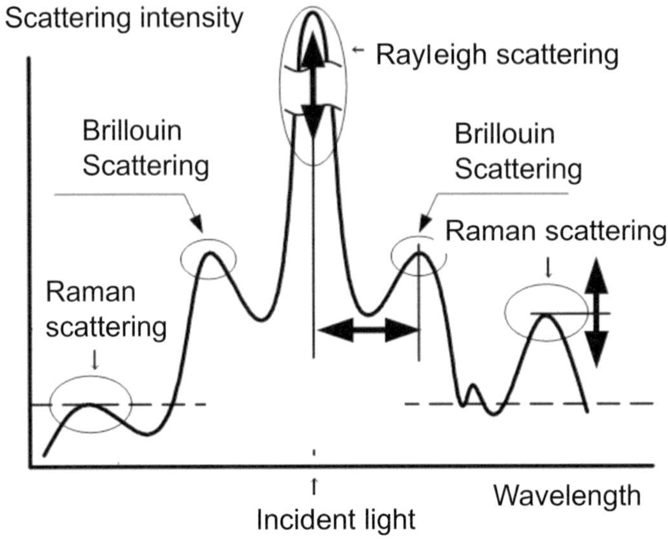

Fig. 8.1 Scattering modes of light in the optical fiber. The temperature and seismic measurements are used by the Raman and Rayleigh backscattering modes of inserted laser light, respectively

(Anti-Stokes scattering having shorter wavelength and adding energy). The intensity ratio between Stokes and Anti-Stokes scattering is dependent on the temperature at the scattering site, as the thermal population of the phonons follows Bose–Einstein (B–E) statistics.

Distributed Temperature Sensing (**DTS**) utilizes this Raman scattering phenomenon to measure temperature at the distributed points along the fiber. The typical wavelength used for DTS is 1,380 nm in Raman mode, with a resolution of 0.1 °C at each meter. The maximum temperature recorded in our geothermal measurements was 294.03 °C at 2, 058 m along the borehole (Kasahara et al. 2025).

Seismic (acoustic) measurements leverage interferometry of Rayleigh scattering, which arises from fluctuations in the fiber's molecular or lattice properties. The technique, known as Distributed acoustic-sensing **(DAS)**, enables the detection of seismic vibrations by measuring the strain they induce in the fiber. As these seismic vibrations cause local strain in the fiber, the inherent scattering sites within the fiber material lead to fluctuations in the backscattered light. DAS then employs interferometric techniques to detect phase shifts and intensity variations resulting from these fluctuations, typically using an **interrogator unit**. Several key terms are integral to DAS measurements. "**Ping rate**" is the frequency at which light pulses are injected into the fiber. This rate is controlled by the fiber's length and the desired "**noise floor** level". "**Gauge length**" is the spatial averaging of the DAS data, typically set to 16.4 m. "**Data sampling rate**" determines the frequency at which seismic waves are sampled. "**Ping rate**" is the repetition rate of input laser pulses. Like the conventional exploration seismic equipment, 1 kHz can be used.

If the "**ping rate**" is 20 kHz, 20 kHz is decimated to 1 kHz to enhance the S/N. According to the OptaSense ODH-4 interrogator unit, "**Noise floor**" is the level of background noise, often characterized as 3.3 $n\varepsilon/\sqrt{Hz}$ for this unit with a "**Dynamic range** of approximately 90 dB (OptaSense datasheet 2022). For measurements using a 3 kmoptical fiber, a DAS interrogator unit with a wavelength of 1,550 nm used, with settings of 25 kHzping rate, 1 kHz sampling rate, and 16.4 mgauge length. Further details on these settings are described in 8.1.2.

8.1.2 Application of DTS and DAS Methods to the Geothermal Exploration

The concept of the new geothermal exploration using DFOS technology is illustrated in Fig. 8.2. In this approach, optical fiber is installed in geothermal boreholes, and seismic sources at the surface generate seismic waves. Ground surface tests have confirmed the effectiveness of optical fiber for seismic measurements (Hasada et al., 2018; Kasahara et al. 2022a, 2024b). The seismic waves can be reflected or diffracted at geothermal reservoirs, which are typically filled with fluids, and the reflected/diffracted seismic waves can be captured by the optical fiber. Geothermal reservoirs can then be estimated using techniques such as reverse-time migration or back propagation of the observed seismic waveforms. Temperature measurements along the borehole are also recorded.

DAS detects both direct and reflected seismic arrivals from the sources and geothermal reservoirs. When the seismic arrivals are recorded sufficiently distributed observation points, time-reversal calculations can backpropagate these arrivals to focus the reflected waves at the reflection points. In geothermal fields, volcanic tuffs and soft sediments typically cover the near-surface layers, resulting in significant seismic wave attenuation. By positioning seismic sensors in the borehole, the seismic attenuation from surface layers can be reduced, improving the quality of seismic data needed to characterize deeper subsurface structures.

Kasahara et al. (2019b) used optical fiber with a polyimide coating within dual SUS tubes (Fig. 8.3). The fiber was wound onto a winch for deployment. Continuous measurements were taken over approximately two weeks at each site. The seismic source emitted a frequency sweep from 10 to 75 Hz over 30 s. Observed waveforms from 500 sweeps were stacked with optimal weights, and spike noises were removed during the stacking process. Figure 8.4 shows an example of DAS and DTS temperature profiles obtained from the **Medipolisgeothermal field**. In the IK-4 borehole (NEDO, 2008, 2009, 2010), the temperature measured at a depth of 914 m was 272.8 °C.

Between 2018 and 2023, Kasahara et al. (2024b) conducted geothermal explorations using optical fibers at seven locations across Japan (Fig. 8.5). Figure 8.6 shows the horizontal deviation of boreholes and the temperate profiles measured by the DTS at these locations, with the maximum temperature recorded at 288 °C

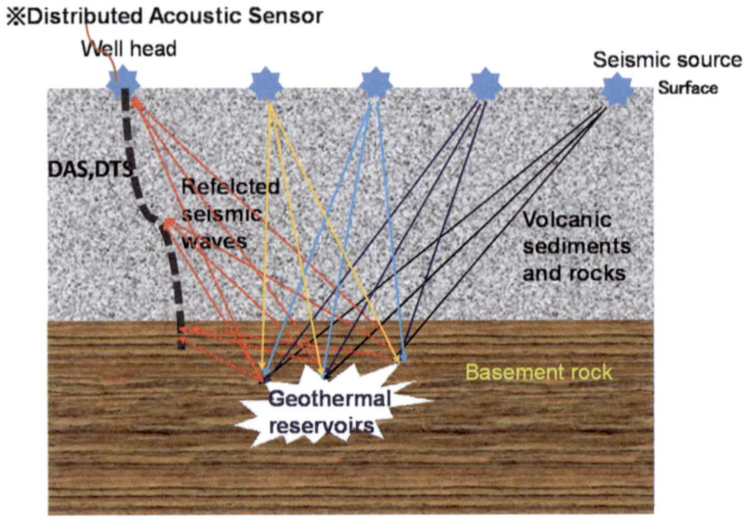

Fig. 8.2 Conceptual diagram showing the use of Distributed Temperature Sensing (**DTS**) and Distributed Acoustic Sensing (**DAS**) methods with optical fiber in a geothermal borehole. Seismic waves generated by seismic sources can be reflected or diffracted by geothermal reservoirs, and the DAS records these reflected seismic waves. By employing reverse-time migration or backpropagation, the reflector locations can be imaged, providing *insights into* physical reservoir parameters through waveform analysis

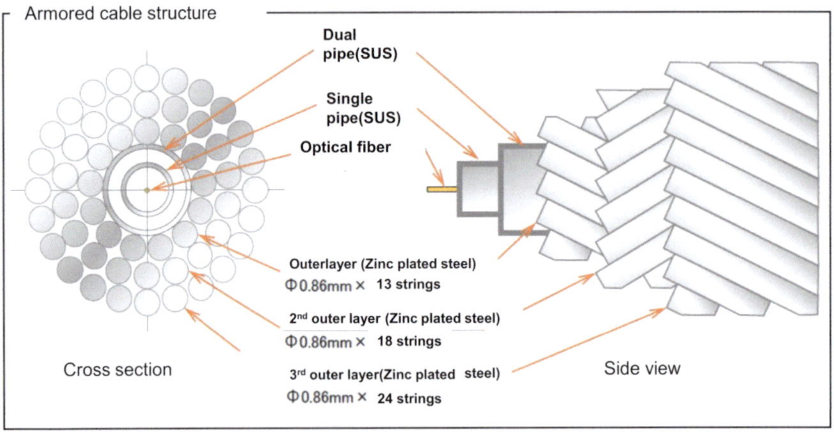

Fig. 8.3 Optical fiber system used for the DAS and DTS measurements (provided by WELMA Co. Ltd.)

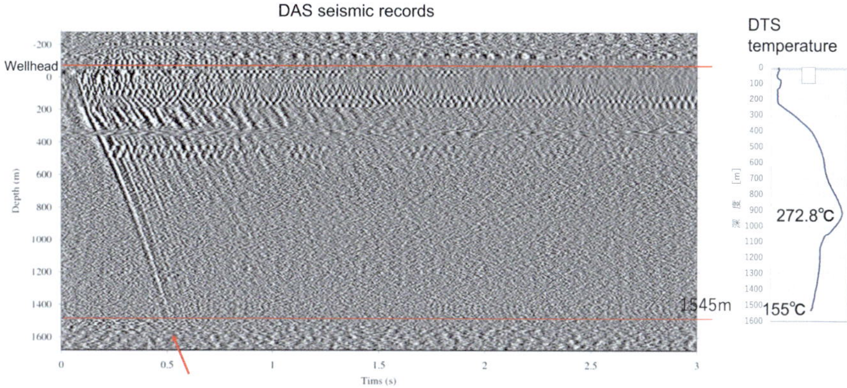

Fig. 8.4 DAS records at zero-offset (left) and temperature profile (right) in the IK-4 geothermal borehole in Medipolis (Kasahara et al. 2019b). The vertical axis is depth in meters. The horizontal axes of left and right are the elapsed time in seconds and temperature in °C, respectively. The bottom of the borehole was 1,750 m. The red arrow shows P first arrivals

(Kasahara et al. 2024b). The case of Otake-Hachobaru is presented in Kasahara et al. (2024a).

Using the compressional direct arrivals called P first arrivals in the DAS records, subsurface P-wave velocity (*Vp*) distributions were calculated, as shown in Fig. 8.7.

Fig. 8.5 The location map of DAS evaluation tests and geothermal measurements at seven sites in Japan using optical fiber systems in boreholes. Locations include (1): Res. Inst. of Electric Power, (2): Medipolis, (3): Ohnuma, (4): Mori, (5): Sumikawa, (6): Takigami, (7): Kijiyama and Hatchobaru—Otake (not shown in the map)

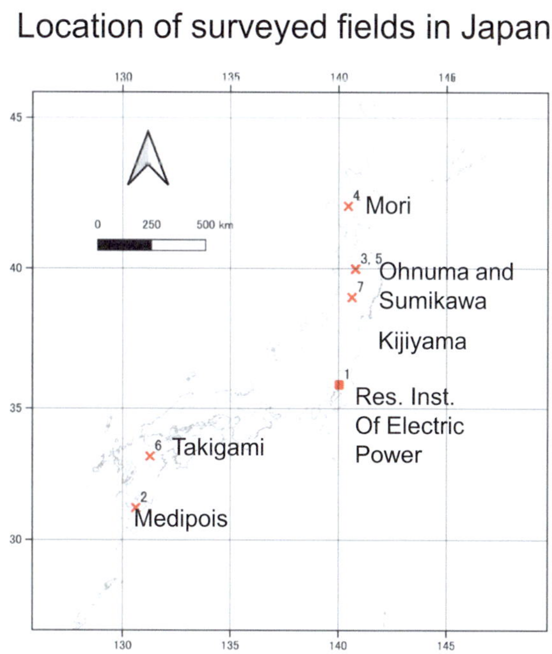

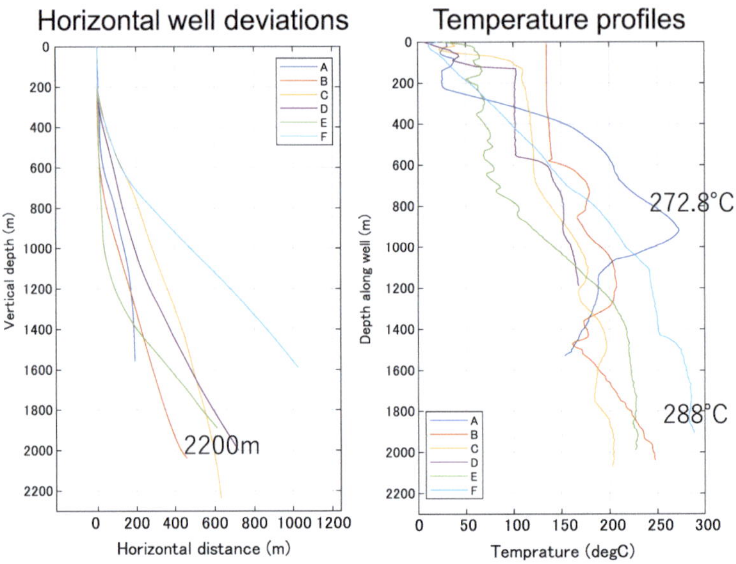

Fig. 8.6 The horizontal deviation (left) and the temperature profiles of the seven borehole measurements by Kasahara et al. (2024b). The vertical axes are borehole depths

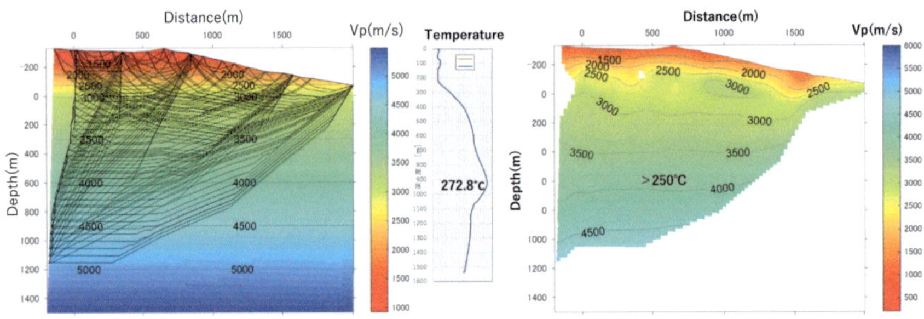

Fig. 8.7 Ray paths (left), temperature profile (middle), and Vp distribution in the Medipolis geothermal field (Kasahara et al. 2019b)

Kasahara et al. (2022b) applied a frequency-wavelength (f-k) filter to separate downgoing and upgoing waves. The upgoing waves, primarily consisting of reflected waves, were used to generate migration images and reflection profiles. Figure 8.8 illustrates the f-k filtering technique applied to the DAS records from the Ohnuma geothermal field (Kasahara et al. 2020b, 2022b).

Kasahara et al. (2019b, 2022b, 2022c) interpreted the geothermal structures in the survey region using the temperature profile from the Ohnuma geothermal borehole (Fig. 8.9). The DTS temperature profile, shown in Fig. 8.10 was compared with previous data obtained in 2013 (Mitsubishi informal data), showing similar temperature distributions. DAS data, collected from the seismic sources, were used

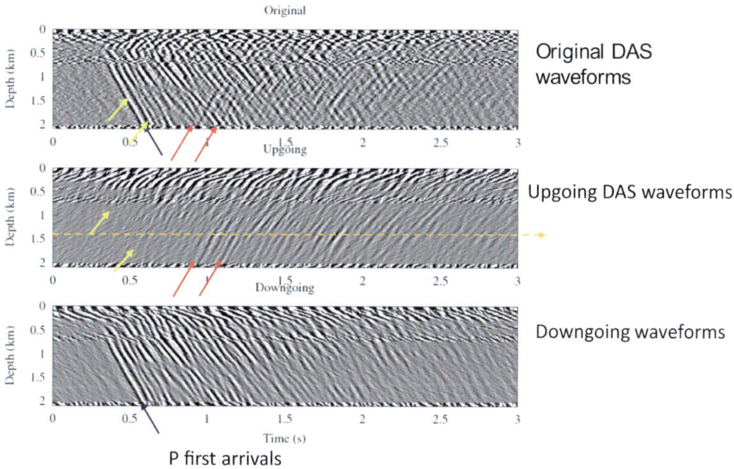

Fig. 8.8 Separation of upgoing (reflected) (middle) and downgoing (direct) (bottom) arrivals from the original DAS records (top) using f-k filtering in the Ohnuma geothermal field (Kasahara et al. 2022b). The vertical axis represents depth in km, while the horizontal axis represents elapsed time in seconds. Blue and red arrows indicate direct P-wave and reflected arrivals, respectively, with Yellow arrows highlighting P-wave reflections at shallow depths

to generate 3D reflection images shown in Fig. 8.11. The intense seismic reflectors identified at depths of 2.8–3.0 km depth are interpreted as potential geothermal reservoirs, as these are likely associated with fluid-filled fracture systems.

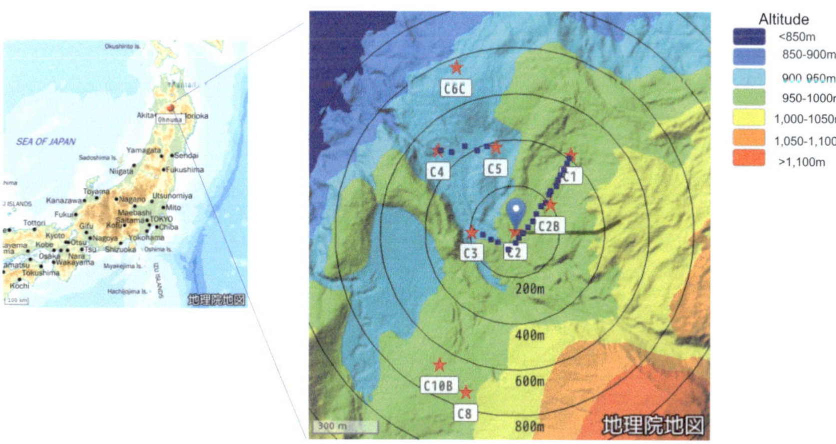

Fig. 8.9 (Left) Map of Japan and the location of the Ohnuma geothermal field. (Right) Detailed map of the Ohnuma survey area, showing the borehole trajectory (thick gray line), and the locations of seismic sources (red stars) and surface seismometers (blue squares)

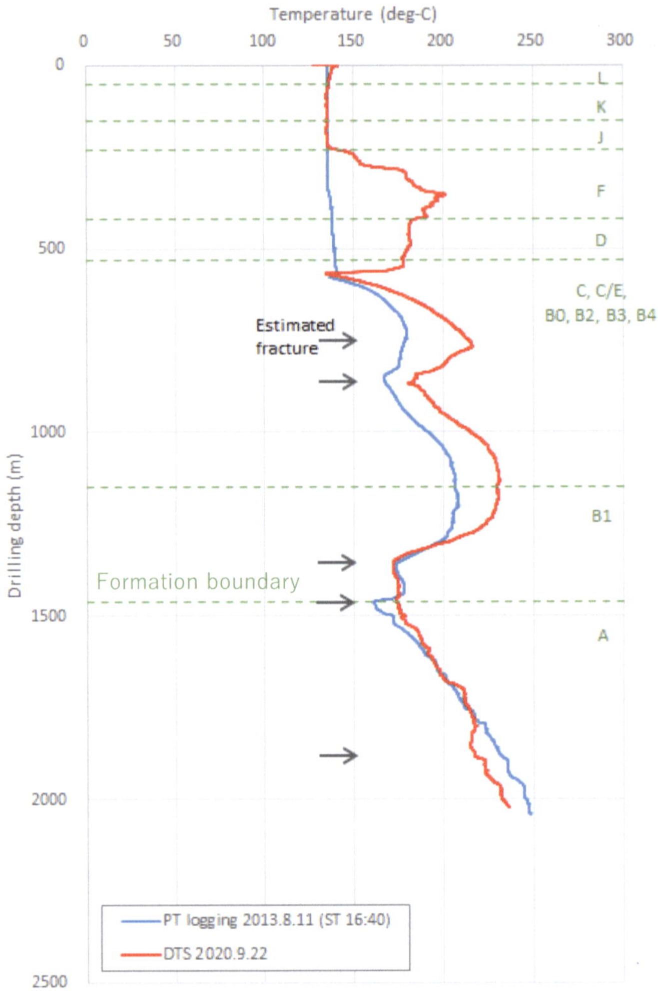

Fig. 8.10 Temperature profile of the O-13R borehole in the Ohnuma geothermal field, obtained in 2013 (blue line) and 2020 (red line) (Kasahara et al. 2022b)

Chapter 14.2 describes results from the Mori geothermal field in Hokkaido, Japan, and Chap. 6.2 discusses the findings from the Takigami geothermal field.

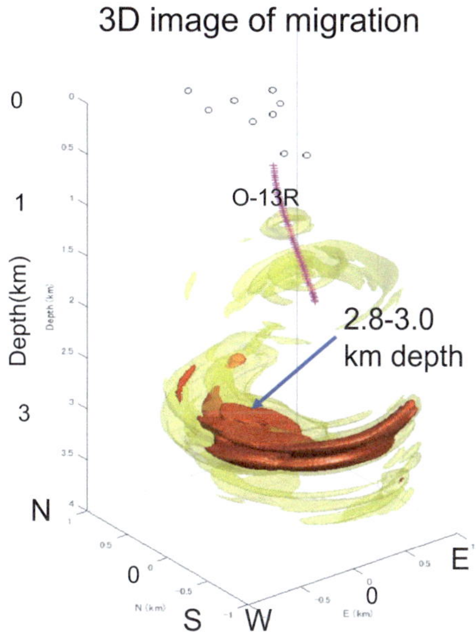

Fig. 8.11 Three-D views of seismic reflection intensity in the Ohnuma geothermal field (Kasahara et al. 2022b)

8.2 Full-Waveform Inversion (FWI) Approach to the Geothermal Exploration

This section explores the application of the full-waveform inversion (FWI) method for imaging geothermal reservoirs, focusing on its potential to detect supercritical water zones.

8.2.1 Method and Simulation Models

Seismic reflection surveys commonly rely on the seismic migration techniques for subsurface imaging. Recently, Full-Waveform Inversion (**FWI**) method has been employed as an advanced technique capable of high-resolution imaging. FWI operates similarly to **reverse-time** migration (RTM) or **backpropagation**, utilizing the **reciprocal principle of Green's function**. Although this method has been applied to 3D seismic data, it is distinct from the time lapse method. Kasahara and Hasada (2016) used the backpropagation technique of residual waveforms to image temporally changing zones. While this method provides excellent images of time-variant zone, it does not estimate physical properties at the target zone. Several key studies have contributed to FWI development, including Tarantola (1984, 1996), Virieux and Operto (2009), Tromp et al. (2005), Alkhalifah (2016), and Schuster (2017). Kasahara et al. (2019a) applied FWI to investigate a supercritical water reservoir, building upon Tromp et al.'*s* adjoint method. In this approach, sensitivity kernels for compressibility, rigidity, and density are computed through backpropagation.

The **misfit kernels** for physical parameters are defined as follows:

Three-component waveform data:

Let $\mathbf{d}(\mathbf{x_r}, t)$ be the observed data at r = 1..N receiver stations, and $\mathbf{s}(\mathbf{x_r}, t, \mathbf{m})$ be the synthetics waveform, where $\mathbf{m}$ is the model vector and $\mathbf{N}$ is number of elements.

The perturbation in the displacement field $f_i^\dagger$ is given as:

$$f_i^\dagger(\boldsymbol{x}, t) = \sum_{r=1}^{N} [s_i(\boldsymbol{x}_r, T - t) - d_i(\boldsymbol{x}_r, T - t)]\delta(\boldsymbol{x} - \boldsymbol{x}_r) \tag{8.1}$$

The adjoint waveform fields $s_k^\dagger$ is then defined as:

$$s_k^\dagger(\boldsymbol{x}', t') = \int_0^{t'} \int_V G_{ki}(\boldsymbol{x}', \boldsymbol{x}; t' - t) f_i^\dagger(\boldsymbol{x}, t) d^3\boldsymbol{x} dt \tag{8.2}$$

where G_{ki}: the Green function.

The misfit kernels for density (ρ,) rigidity (μ) and bulk modulus (κ) are expressed as:

$$K_\rho(\boldsymbol{x}) = -\int_0^T \rho(\boldsymbol{x}) \boldsymbol{s}^\dagger(\boldsymbol{x}, T - t) \cdot \partial_t^2 \boldsymbol{s}(\boldsymbol{x}, t) dt \tag{8.3}$$

$$K_\mu(\boldsymbol{x}) = -\int_0^T 2\mu(\boldsymbol{x}) D^\dagger(\boldsymbol{x}, T - t) : D(\boldsymbol{x}, t) dt \tag{8.4}$$

$$K_\kappa(\boldsymbol{x}) = -\int_0^T \kappa(\boldsymbol{x}) \left[\nabla \cdot \boldsymbol{s}^\dagger(\boldsymbol{x}, T - t)\right] [\nabla \cdot \boldsymbol{s}(\boldsymbol{x}, t)] dt \tag{8.5}$$

Here, D is the traceless strain deviator and $D^\dagger$ is its adjoint waveform and ":" denotes the inner product between matrices.

Misfit kernels of $K_{\alpha,}$ and K_β are defined as:

$$K_\rho' = K_\rho + K_\kappa + K_\mu, \quad K_\beta = 2\left(K_\mu - \frac{4}{3}\frac{\mu}{\kappa} K_\kappa\right), \quad K_\alpha = 2\left(\frac{\kappa + \frac{4}{3}\mu}{\kappa}\right) K_\kappa \tag{8.6}$$

Three simulation cases were examined by Kasahara et al. (2019a, 2020a).

Simulation 1 involves an active seismic source placed downhole at 2 km depth with the goal of imaging a supercritical water reservoir located at 4 km-deep, with dimensions 1 km-long and 200 m-thick. Changes in physical properties were assumed to be $\Delta Vp = -5\%$, $\Delta Vs = -5\%$ and $\Delta\rho = -2\%$. The optical fiber DAS system as was considered to be the seismic receiver.

Simulation 2 uses natural earthquakes as seismic sources surround an igneous intrusion, an idealized setup, to assess the feasibility of passive seismic sources for imaging (Fig. 8.13).

Simulation 3 (Fig. 8.14) is a more realistic scenario reflecting the geographical conditions of the Medipolis geothermal field. Kasahara et al. (2020a) applied a vertical single force applied at a point 6 km from the left side of the model. The DAS in the well and the surface geophone array are assumed to be installed at the Medipolis field. The horizontal components of the DAS systems were specifically considered, anticipating the introduction of a helical fiber.

8.2.2 Results of FWI simulations

8.2.2.1 Results of Simulation 1 for the Case of Buried Active Sources

The study model is illustrated in Fig. 8.12. Figure 8.15 (top) shows waveforms recorded at all receiving locations: surface seismometers and DAS systems in two vertical boreholes and one horizontal borehole. Residual waveforms before and after physical property changes are depicted in Fig. 8.15 (bottom). Minimum waveform changes in the surface seismometer array suggest limited contribution of surface arrays to imaging. Horizontal borehole DAS contribution was examined and similarly found not significant.

Reconstructed images for three sources located at 1, 3, and 5 km from the borehole demonstrates satisfactory retrieval of the modeled zone (Figs. 8.16, 8.17, and 8.18). The best results were achieved for the source at 3 km (Fig. 8.17). The source location affects the reconstruction image of each physical property. Iterative inversion yielded *Vp, Vs,* and density values of 4%, 4%, and 2%, respectively, matching the assumed location and thickness. Results for the other source locations (Figs. 8.16 and 8.18) show varying degrees of success, with *Vs* is poorly resolved for the source at 2 km depth and 1 km distance from the borehole, although *Vp* and density values were reasonably retrieved (Fig. 8.16). When the source is located at 2 km depth and 5 km distance from the borehole, while *Vp* is not well retrieved, *Vs* and density values are retrieved to some extent (Fig. 8.18).

8.2.2.2 Results of Simulation 2 for Natural Earthquakes as Seismic Sources

The model 2 examines the feasibility of natural earthquakes for reservoir imaging, assuming an igneous intrusion is assumed (Fig. 8.13) and seven natural earthquakes as passive seismic sources. While prior natural seismic activities beneath the Medipolis

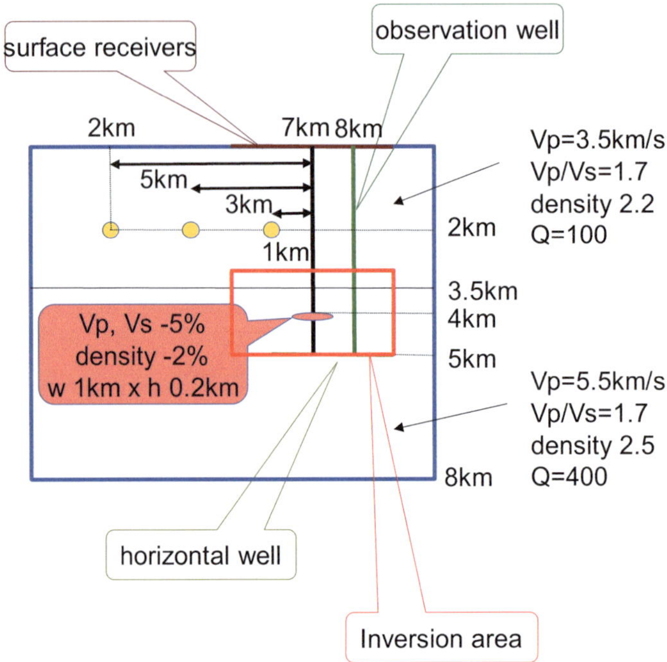

Fig. 8.12 Model of simulation 1. A rectangular (1 km in width and 0.2 km-thick) supercritical water reservoir is assumed at 4 km depth. Three seismic sources at 1 km, 3 km, and 5 km from the drilling borehole are examined. Three DAS systems are considered in the drilling borehole, the observation borehole, and the horizontal borehole beneath the assumed reservoir. A ground surface seismic array is also used. *Vp, Vs*, and density changes are tested in the simulation. (After Kasahara et al., 2019a)

geothermal field was identified (Kasahara et al. 2019b), appropriate activity was absent during the 2018 feasibility study in southern Kyushu geothermal field.

Simulation results (Fig. 8.19) show relatively low *Vp* imaging within the intrusion and strong smearing artifacts. *Vs* changes were imaged only near the intrusion top, while density imaging results aligned with *Vp*. The study suggests that while earthquake energy exceeds that of active sources, imaging with natural earthquakes is challenging due to uncontrollable source locations and characteristics.

8.2.2.3 Results of Simulation 3

The synthetic waveforms generated for the 4 km-deep reservoir model of simulation 3 (Fig. 8.14, Kasahara et al. 2020a) are shown in Fig. 8.20. FWI simulation used synthetic residual waveforms (Fig. 8.21) assuming time lapse changes. PP, PS and SS reflections were observed at depths exceeding 3 km in DAS records, with surface geophones capturing reflections at approximately 3–5 km.

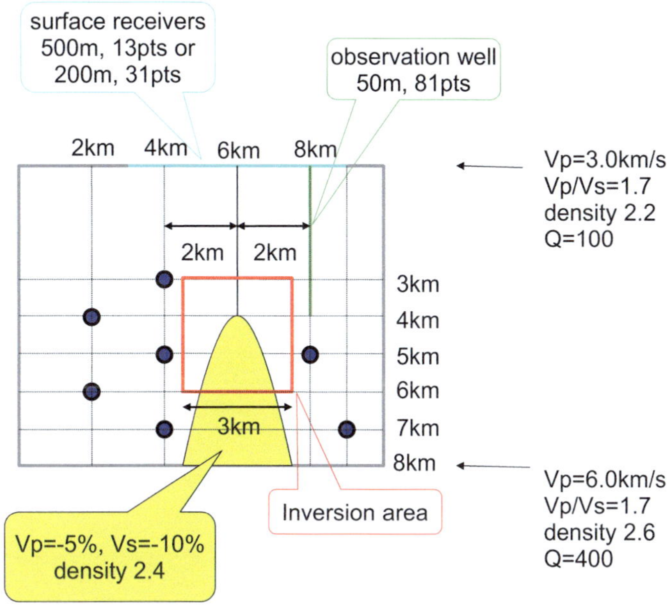

Fig. 8.13 Model of simulation 2. Seven natural earthquakes surrounding an igneous intrusion. Ground surface seismometer array and DAS in the borehole down to 4 km depth are used for seismic receivers

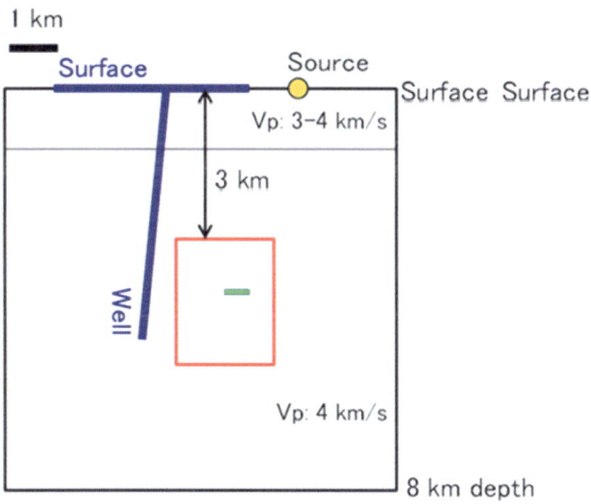

Fig. 8.14 Model of simulation 3 to evaluate the feasibility of detecting a 4 km-deep reservoir through P-to-S conversion. The green rectangle represents a 500 m-long and 100 m-thick rectangular reservoir located at a depth of 4 km. No density change is assumed. A vertical single force (yellow circle) is applied at 6 km from the left edge of the model. The *Vp/Vs* ratio in the reservoir is 3. The red rectangle outlines the area used in the FWI calculation. (After Kasahara et al. 2020a)

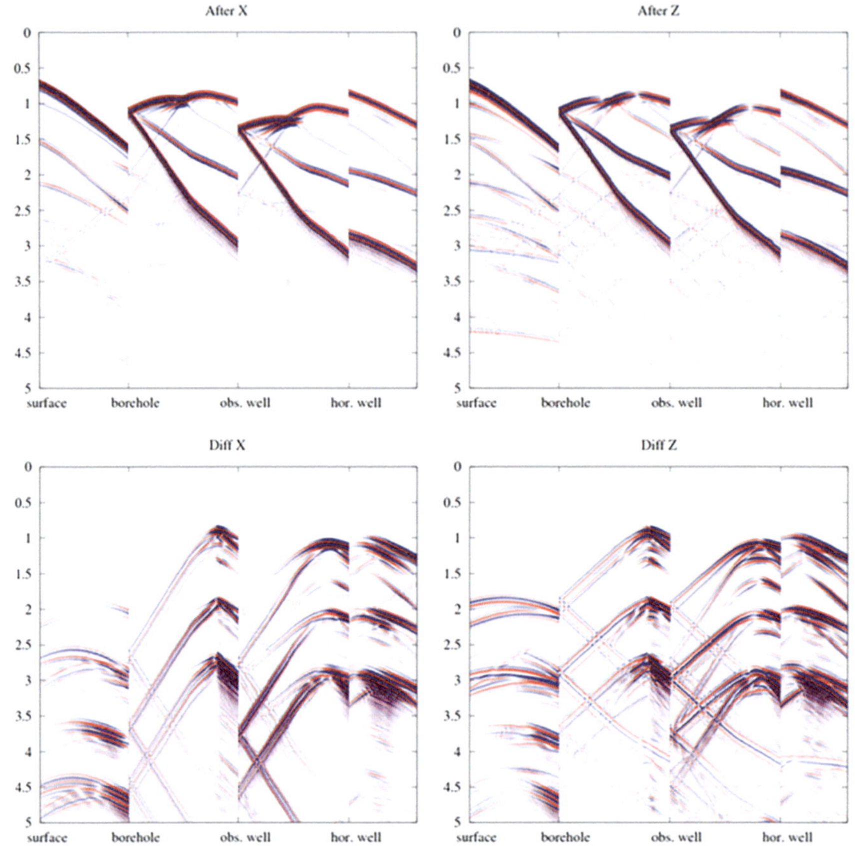

Fig. 8.15 (Top) horizontals (left) and vertical (right) component waveforms recorded by surface geophones, DAS in the drilling borehole, DAS in the observed borehole, and DAS in the horizontal borehole for each diagram. (Bottom) the residual waveforms of horizontal and vertical components before and after the change. The vertical axis of each diagram: travel time in seconds. The source is located 3 km from the drilling well. (After Kasahara et al. 2019a)

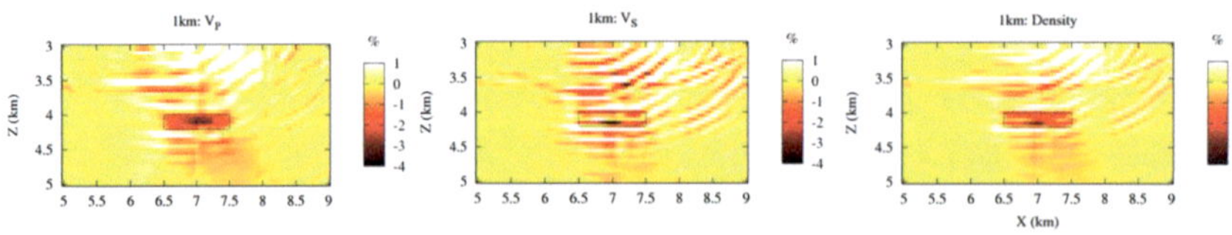

Fig. 8.16 Simulation results for P-wave velocity, S-wave velocity, and density (left to right). A single source located 1 km from the drilling borehole was used with receivers at the surface, in two vertical boreholes, and a horizontal borehole at a depth of 5 km depth are assumed. The input temporal changes (−5% in velocities and −2% in density) are reasonably recovered. (After Kasahara et al. 2019a)

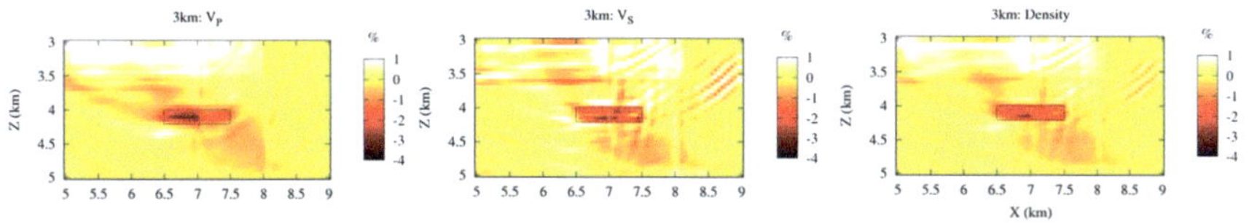

Fig. 8.17 Simulation result for the source location at 3 km distance from the borehole. Left to right are *Vp, Vs,* and density

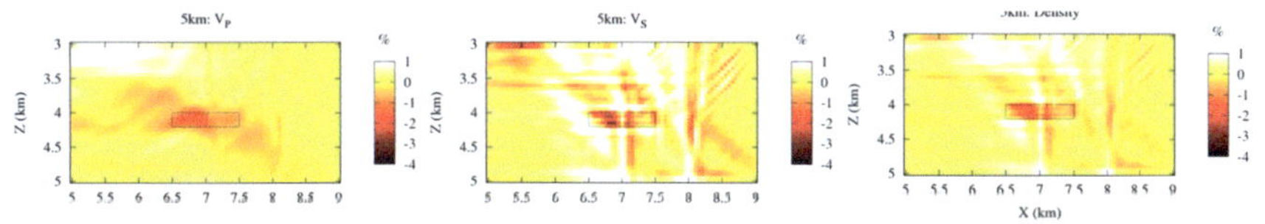

Fig. 8.18 Simulation result for the source location at 5 km distance from the borehole. Left to right are *Vp, Vs,* and density

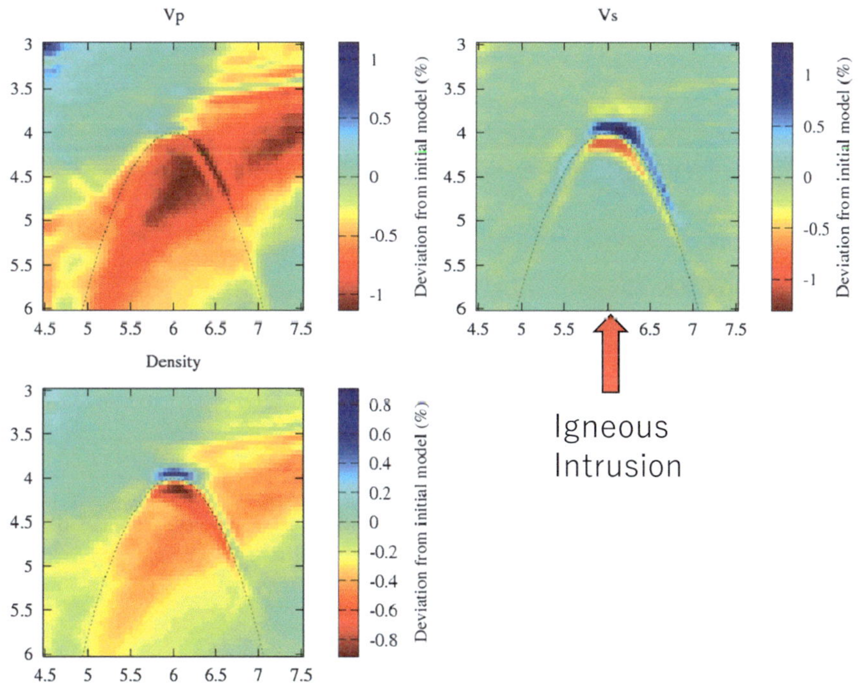

Fig. 8.19 Simulation result for natural earthquake sources shown in Fig. 8.13. Top left: *Vp*, Top right: *Vs,* and Bottom: density

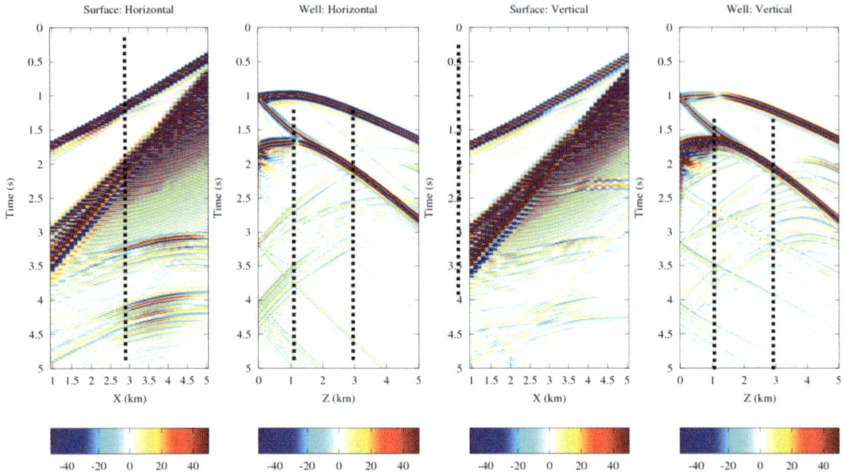

Fig. 8.20 Synthetic waveforms for the model shown in Fig. 8.14. A vertical single force is applied. From left to right: Horizontal component at the surface, a horizontal component in the well, a vertical component at the surface, and a vertical component in the well. (After Kasahara et al. 2020a)

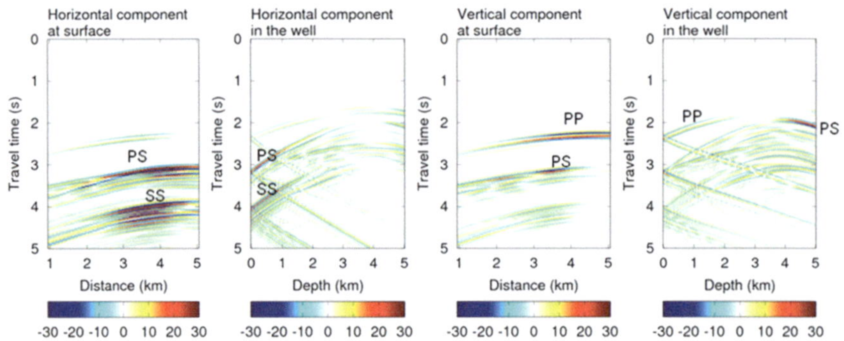

Fig. 8.21 Residual waveforms in the wells and ground surface geophones between the model without and with reservoir. Relatively small P and large PS phases are observed in the horizontal components. $Vp/Vs = 3$ in the reservoir, obtained by forward modeling to generate the P-to-S conversion. PP, PS, and SS reflected phases are observed in the reservoir. PP, PS, and SS are reflected phases at the reservoir. (After Kasahara et al. 2020a)

Inversion was carried out in two stages. The first stage mainly focused on the location and shape. Figure 8.22 shows the result of the first stage. The location of the assumed reservoir appears to be well retrieved. Using the results of the sensitivity kernels in the first stage shown in Fig. 8.23, the second stage models were generated. Inversion results (Fig. 8.24) demonstrated reasonable reconstruction of the reservoir's location and shape, although physical properties were not well-recovered.

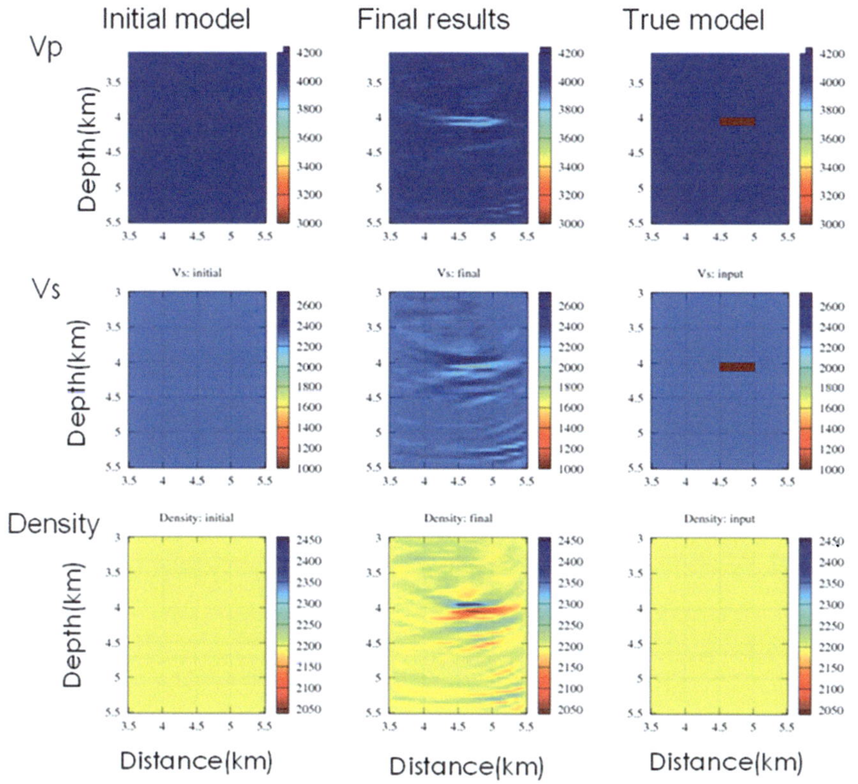

Fig. 8.22 The first stage's result is shown in the central column. The true model is shown in the right column, and the initial model is shown on the left. No density change is assumed. (After Kasahara et al. 2020a)

8.2.3 Discussion and Conclusions

Kasahara et al. (2019a) carried out two simulations to evaluate imaging of supercritical water reservoirs using FWI. Simulation 1 demonstrated accurate retrieval of location, shape, and physical properties (*Vp, Vs,* and density) of the reservoir using borehole seismic sources and dense DAS arrays. However, the individual contributions of each receiver type have not been fully investigated.

Simulation 2 highlighted challenges with natural earthquakes due to uncontrollable source conditions, despite their high energy. Although a noise test was not carried for the simulation, background noise can be separated from the source signal using the spectral comb method described in Kasahara and Hasada (2016). In addition, stacking data for a long duration drastically enhances the S/N. We could use several weeks for imaging. Practically, more quantitative evaluation will be needed in the future.

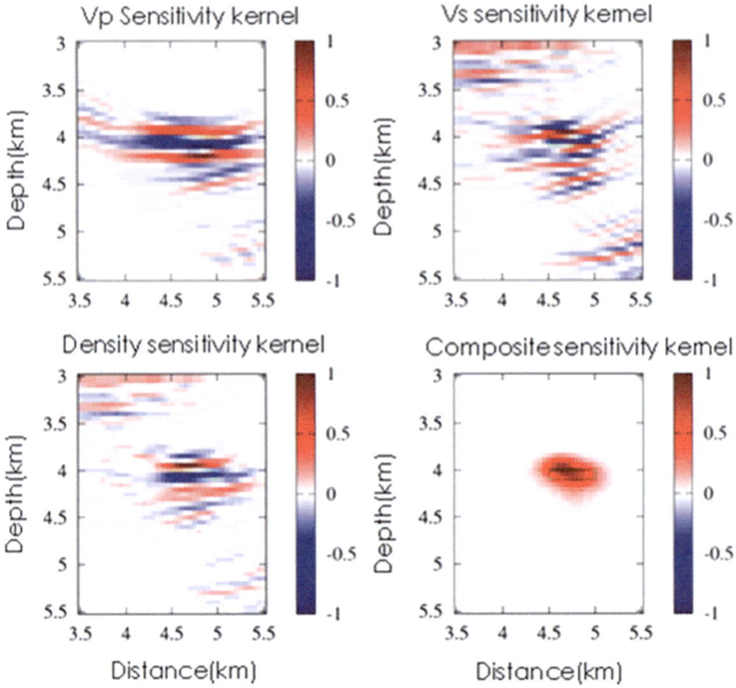

Fig. 8.23 Sensitivity kernels at the first stage of the FWI model. By combining the three sensitivity kernels, the location and shape of the second stage were developed, as shown at the bottom right. (After Kasahara et al. 2020a)

In the real situation, many factors affect the imaging quality, such as the distribution of supercritical zones. A low-velocity zone might scatter seismic waves. The geology and fractures will also strongly affect the imaging.

Simulation 3 underscored the feasibility of detecting deep reservoirs using FWI with active sources. However, geological and structural complexities, such as fractures and low-velocity zones, may introduce additional challenges. Future studies should emphasize quantitative noise evaluations and incorporate real-world complexities.

In conclusion, full-waveform inversion combined with DAS arrays shows promise for imaging fluid zones in geothermal reservoirs using both active and passive seismic sources.

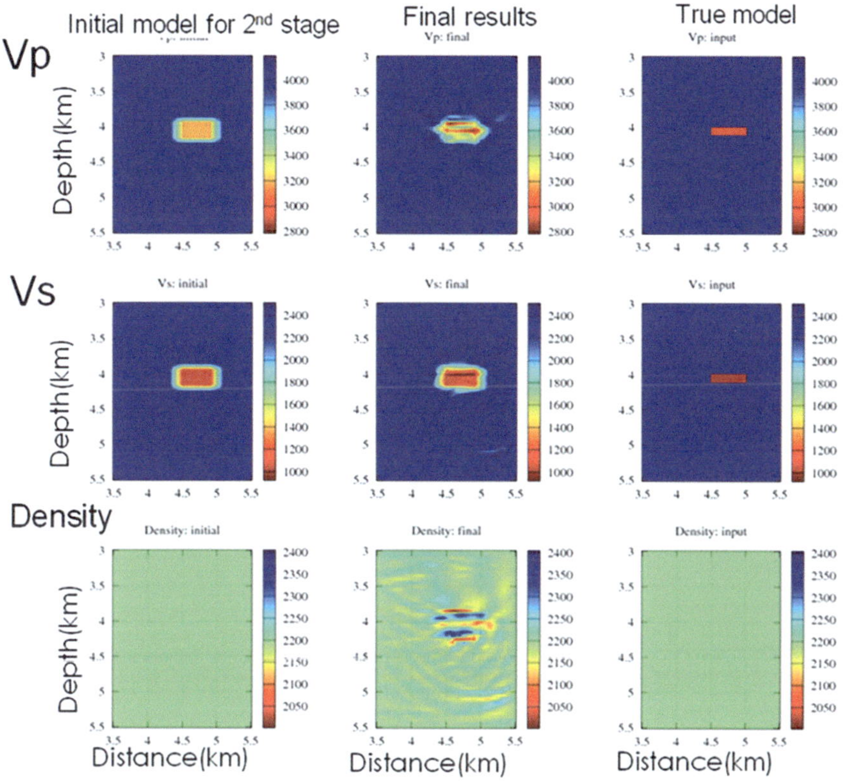

Fig. 8.24 Final FWI results for simulation 3 are shown in the central column. The true model is shown in the right column. The initial model of the second stage is shown in the left column (After Kasahara et al. 2020a)

8.3 Use of P-To-S Conversion

8.3.1 Case of Medipolis

Kasahara et al. (2019b) installed a fiber optic cable to conduct DASs and DTSs measurements at a depth of 977 m. Additionally, 20 sets of three-component (3C) geophones were deployed along a 2 km-long east–west (EW) line. Over four and a half days, they observed seven natural earthquakes via DASs and ground surface geophones (Table 8.1 and Fig. 8.25). The largest earthquake with a magnitude of 5.2, occurred west of Tanegashima at a focal depth of 123 km (denoted as "A" in Fig. 8.25). Figure 8.26 shows the records for this earthquake (EQ5 in Table 8.1) as captured by DASs and ground surface geophones.

Analysis of the seismic records from the 3C ground surface geophones revealed large amplitude arrivals on the north–south (NS) and east–west (EW) components

Table 8.1 List of earthquakes observed by DASs for four and a half days

	Origin time	Latitude	Longitude	Depth (km)	Magnitude	Area
EQ1	2018/11/20/03:16 28.7	31° 23.6′N	130° 37.1′E	10	1.4	C
EQ2	2018/11/20/19:34 52.2	29° 21.5′N	129° 48.4′E	65	4	D
EQ3	2018/11/20/22:02 42.3	31° 24.2′N	130° 34.3′E	14	0.9	C
EQ4	2018/11/20/23:09 01.5	31° 30.0′N	130° 42.4′E	119	2.3	E
EQ5	2018/11/21/04:09 49.9	30° 24.0′N	130° 9.0′E	123	5.2	A
EQ6	2018/11/21/04:59 29.1	30° 26.9′N	130° 9.1′E	123	3.1	B
EQ7	2018/11/21/17:49 18.0	31° 19.3′N	130° 42.7′E	22	2.7	F

approximately 0.8 s after the P-wave first arrivals. The up-down (UD) component exhibited only weak P-wave arrivals (Fig. 8.26). Conversely, DAS records did not show these phases or S-wave arrivals because the incident angles of the observed earthquakes were near vertical (Table 8.1). DASs predominantly detected vibrations aligned with the fiber's elongation direction. Another similar phase for EQ2 is shown in Fig. 8.27.

8.3.2 *Forward Modeling to Evaluate P-To-S Conversion*

The characteristics observed in Figs. 8.26 and 8.27 are frequently encountered in marine seismic studies. These P-to-S converted phases occur when P-waves transition from a hard rock base into soft marine sediment (Kasahara et al. 1982). In such case, P-to-S conversion delays the signal by approximately 2 s and generates large amplitudes due to the thick marine sediment layers. Typical seismic velocities for the hard rock base and marine sediments are for the hard rock base, *Vp* is approximately 3.5–6 km/s, and for the marine sediments, *Vp* is ~2 km/s and *Vs* is 100–300 m/s, respectively. At the Medipolis geothermal field, the seismic velocity profile differs, with a *Vp* value of approximately 3–4 km/s between the depth of 200 and 1,000 m and a typical *Vp/Vs* ratio of 1.75 (Kasahara et al. 2019b).

To examine the cause of the P-to-S conversion at the Medipolis, the delay time (~0.8 s) from the P-wave first arrival to the P-to-S converted phase was estimated. A high *Vp/Vs* zone ($Vp/Vs = 3$) was assumed at a depth of 4 km with a thickness of 500 m (Fig. 8.28). Theoretical waveforms generated by deep seismic sources corresponding to the plane wave incident at a depth of 5 km were computed using the seismic velocity profile derived from DAS observations (Kasahara et al. 2019b).

The calculated waveform for the model (Fig. 8.28) is shown in Fig. 8.29, which revealed that the P-to-S converted phase is prominent on the horizontal components within the four km-deep reservoir. The vertical component shows intense P first arrivals and weak scattered signals.

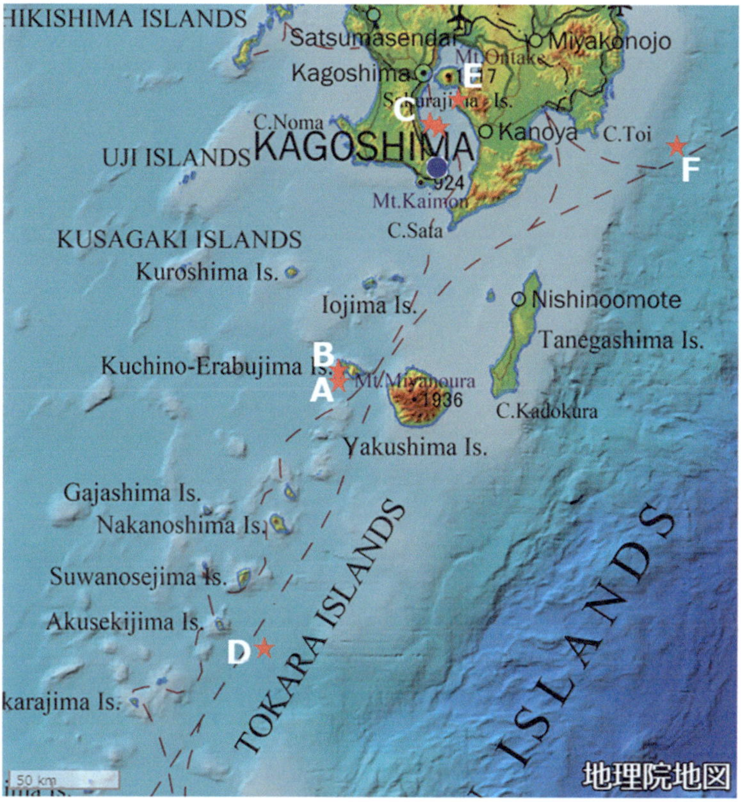

Fig. 8.25 Earthquakes observed by DASs for four and a half days (stars). Blue circles denote the location of DASs at the Medipolis field

The waveforms at 8.8 km (shown in Fig. 8.30) were extracted to examine the P-to-S conversion amplitude. The weak P first arrival and successive intense P-to-S conversion arrival were observed on the horizontal component. In contrast, the vertical component shows only P arrival. The calculated waveforms can explain the observed characteristics, such as weak P first arrivals and intense P-to-S converted phases on the horizontal components when the incident seismic wave is slightly oblique to the high *Vp/Vs*. zone. The waveforms in the well and ground surface were also estimated.

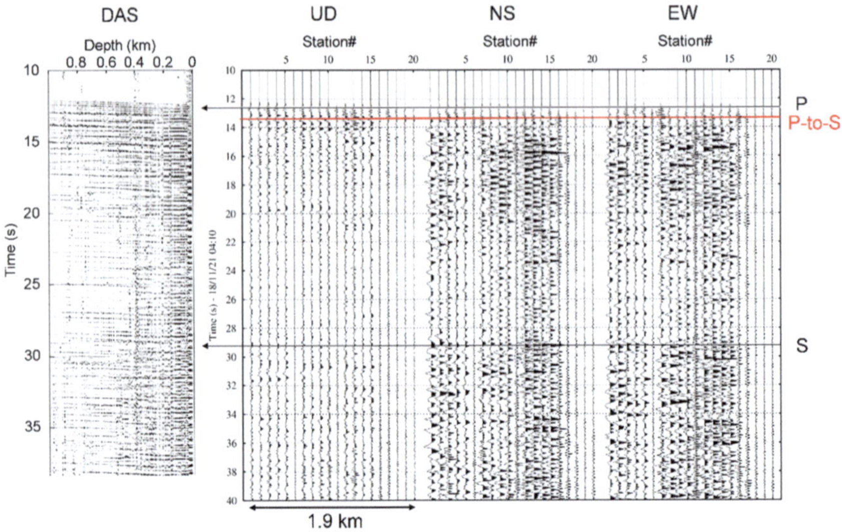

Fig. 8.26 Possible P-to-S conversion arrivals seen in EQ5 (M = 4.0) ("A" in Fig. 8.25) observed by the ground surface geophones (Kasahara et al. 2020a). (Left) DASs records. The vertical axis denotes the arrival time in seconds and the horizontal axis denotes the distance. (Right) Ground surface geophone records of UD, NS, and EW components. Each ground surface seismic station is located at 100 m spacing. Large amplitude arrivals are recognized in the NS and EW components approximately 0.8 s after the first P arrivals observed in the UD component records

8.3.3 Discussion and Conclusions

The seismic study at the Medipolis geothermal field identified a distinct phase on the horizontal component of the ground surface geophones, attributed to P-to-S conversion. Theoretical seismograms, based on a *Vp/Vs* = 3 zone at a depth of 4 km with a 20° dip angle, closely matched the observed waveforms by the ground surface geophones.

As discussed in 8.2, the possibility of detecting any P-to-S converted origin using the FWI method was also evaluated. A high *Vp/Vs* zone at a depth of 4 km with a thickness of 500 m was assumed. FWI suggests retrieving the location and shape of high *Vp/Vs* zones if we use the combination of active seismic source(s), DASs in the well, and the ground surface geophones. However, the retrieval of physical properties was limited. Because the current FWI model reflects more realistic field conditions like those in Medipolis, the limited aperture of the observed system might cause poor retrieval of the physical properties. Idealized FWI modeling, where it assumed a more extensive DAS system, yielded good physical properties (Kasahara et al. 2019a).

A previous study at Medipolis in 2019 also identified high *Vp/Vs* zones at approximately 4 km depth. DAS Vertical Seismic Profile (DAS-VSP) exhibited several distinct upgoing phases. Based on the forward modeling and migration of upward

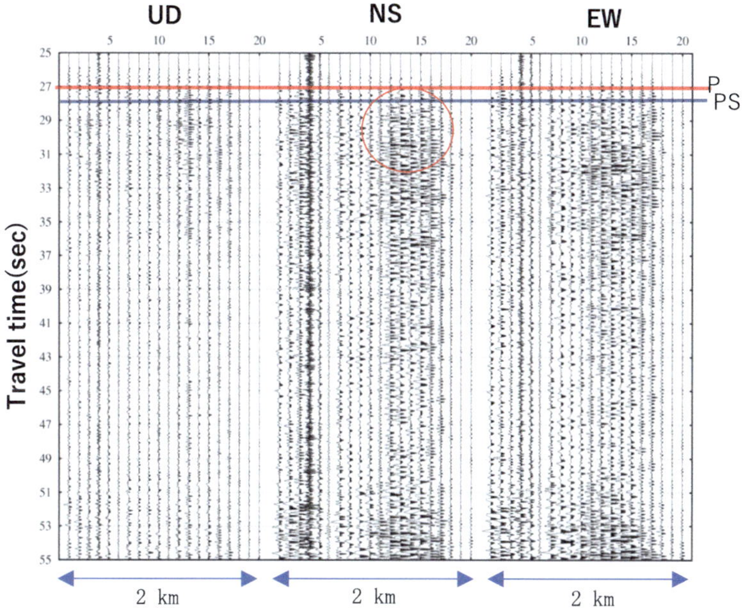

Fig. 8.27 P-to-S conversion for the earthquake, *i.e.*, EQ2 (M = 4.0) ("D" in Fig. 8.25) (Kasahara et al. 2020a). The time delay in P-to-S converted phase from P first arrival is approximately 0.8 s. The vertical axis denotes the travel time

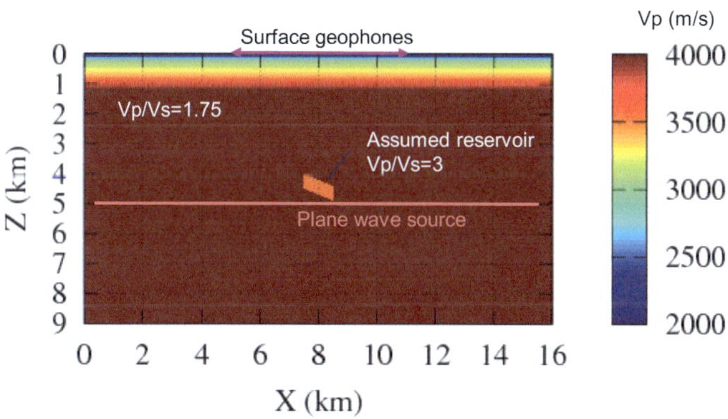

Fig. 8.28 Model to examine the P-to-S conversion. *Vp/Vs* ratio in the reservoir is assumed to be 3, with a reservoir dip of 20° and a plane wave source at a depth of 5 km. (After Kasahara et al. 2020a)

phases, it was estimated that the *Vp/Vs* = 3 zone should be present at a depth of 3.6 km (Kasahara et al. 2020a). Although the depth of high *Vp/Vs* zones obtained by the present P-to-S conversion analysis is slightly greater than 3.6 km, the discrepancy between the two studies can be explained by the uncertainty near-surface *Vs* distribution.

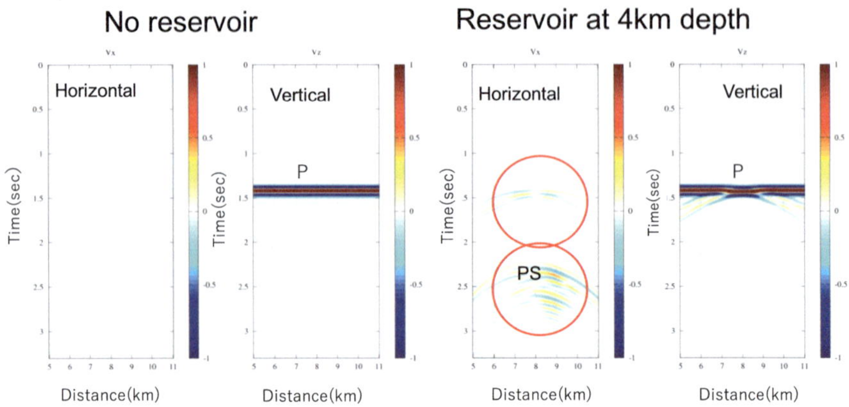

Fig. 8.29 Calculated waveforms of ground surface geophones using the model shown in Fig. 8.28 (Kasahara et al. 2020a). Incident wave is assumed as plane wave at a depth of 5 km. Horizontal axis: distance (km) and vertical axis: travel time (s). (Left two) no reservoir case. (Right two) presence of a 4 km-deep reservoir (Fig. 8.28)

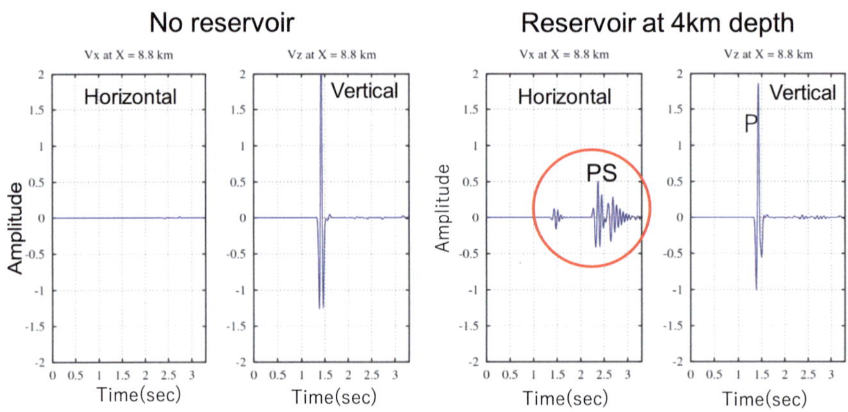

Fig. 8.30 Waveforms at 8.8 km from the left edge of the model shown in Fig. 8.28 (Kasahara et al. 2020a). It is noted that the P-to-S conversion arrival is observed on the horizontal component. Still, only intense P first arrival is seen on the vertical component for the 4 km-deep reservoir model

They considered the meaning of $Vp/Vs = 3$. Using the numerous *Vp* and *Vs* measurements under high pressure and temperature by Kern (1978), Kern and Richter (1981), and Kern et al. (1997), most quartz-free rocks showed minimal temperature dependence of *Vp* and *Vs*. Rocks, including quartz minerals, demonstrated large temperature dependence on *Vp* due to that phase change from α to β quartz (Kern and Richter 1981). *Vp/Vs* remained constant at approximately 1.75 for quartz-free rocks, and $Vp/Vs = 3$ was not explained. The serpentine rock exhibited significant decreases in *Vs* and *Vp* at temperatures higher than 500 °C (Kern and Richter 1981). The amount of velocity decrease at high temperatures was explained by the dehydration of antigorite, a high-temperature serpentine mineral. The velocity decrease in *Vs* was larger than that in *Vp*. Water saturation in rocks might be a cause of high *Vp/Vs*.

In addition, the *Vp/Vs* is related to the Poisson's ratio "v" by the following equation.

$$\nu = \frac{1}{2}\frac{(V_P/V_S)^2 - 2}{(V_P/V_S)^2 - 1} \tag{8.7}$$

Using the above equation, *Vp/Vs*= 3 corresponds to 0.467 for the Poisson's ratio. Because the Poisson's ratio of water is 0.5, the suggested zone at a depth of approximately 3.6–4 km could be filled by water. Winkler and Nur (1982) obtained *Vp/Vs* > 2 for the Massillon sandstone if the sandstone was 100% saturated by water.

Considering the above discussion, high *Vp/Vs* suggests the presence of high-water saturation zones at a depth of approximately 4 km. Because the temperature at the depth of 914 m is 264 °C in the IK-4 well at the Medipolis geothermal field, the zone around 4 km-deep might be close to the supercritical state if it is filled with water.

8.4 Use of Microearthquakes

Microearthquake activities are often associated with volcanic processes or geothermal production and reinjection. Monitoring these activities is essential for safe geothermal power plant operation, as it helps prevent earthquake-induced damage. Beyond safety considerations, the spatial distribution of microearthquake hypocenters provides valuable insights into the flow paths of geothermal fluid.

A surface seismic network covering the geothermal field is used for hypocenter determination. In addition to the observation using the surface seismometer network, we can use borehole DAS data. Figure 8.31 shows DAS records of a microearthquake in a geothermal field.

Readings of P and S arrival times are used to determine the hypocenter distribution. The *Vp* structure must be known. If we assume *Vp/Vs*=1.73, we can get the *Vs* structure. However, the *Vp/Vs* is higher than 1.73 at shallow depth. The origin time can be estimated using P arrivals and S-P times (the Wadachi diagram (Utsu, 1984)). Figure 8.32 shows the S-P times of the DAS record and surface seismometer readings. If we use permanent surface seismometer readings to determine the origin time, it gives an earlier time. The reason is that the near-surface layer has a slower *Vs*, suggesting a higher *Vp/Vs* ratio.

The hypocenter determination has often been done using the least squares method using the P and S arrival times. However, the hypocenter determination using the least square approach is unstable due to non-linearity. It is suggested that the grid search method using the P arrival time readings and the estimated origin times can provide the optimum hypsometers. Hypocenters minimizing the travel time residuals between calculated and observed travel times (O–C) are determined (Figs. 8.33 and 8.34).

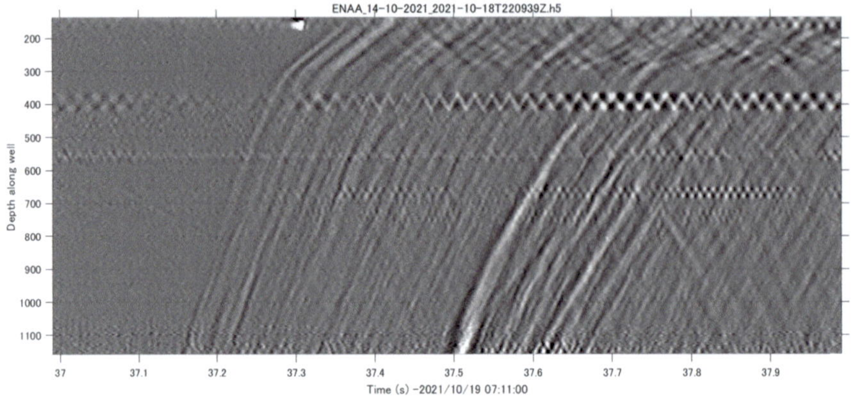

Fig. 8.31 A DAS record of a microearthquake observed in a geothermal borehole. The vertical axis shows the depth along the borehole, and the horizontal axis shows the travel time in seconds

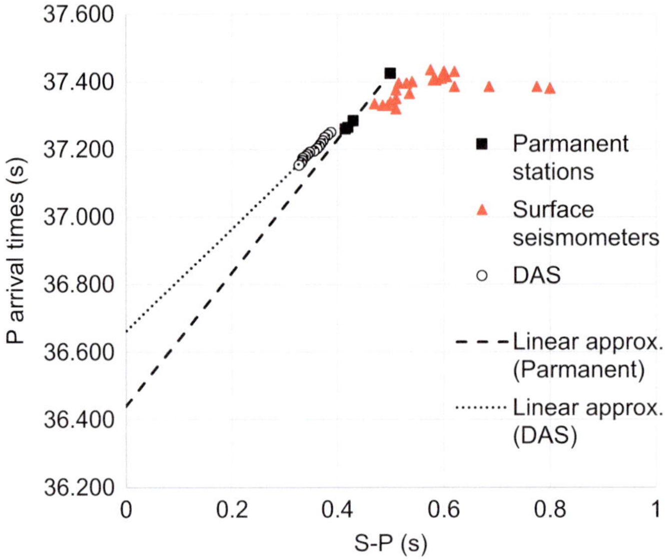

Fig. 8.32 The vertical and the horizontal axes are the P arrival and S-P times for the earthquake shown in Fig. 8.31. The origin time estimation using the DAS and the permanent surface time readings provides different ones due to higher *Vp/Vs* ratios near the surface layer

Other examples of a different geothermal field are shown in Figs. 8.35, 8.36, 8.37 and 8.38. The calculated travel times of P and S are nicely fitted to the observer DAS and surface seismometer records. The travel time residuals are a few ten milliseconds. The horizontal and vertical errors of the hypocenter could be less than 50 m.

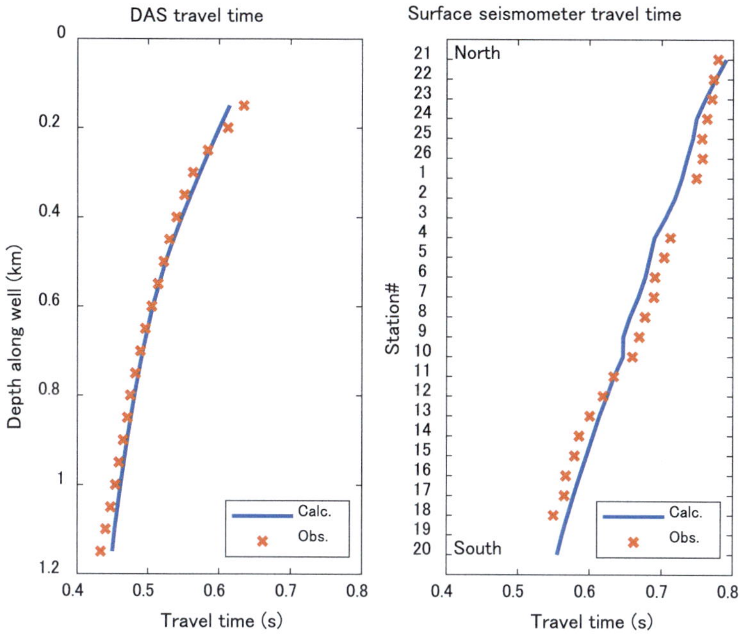

Fig. 8.33 Fitting of the optimized hypocenter for DAS and surface seismometers for the earthquake shown in Fig. 8.31. Left: DAS P readings. Right: surface seismometers

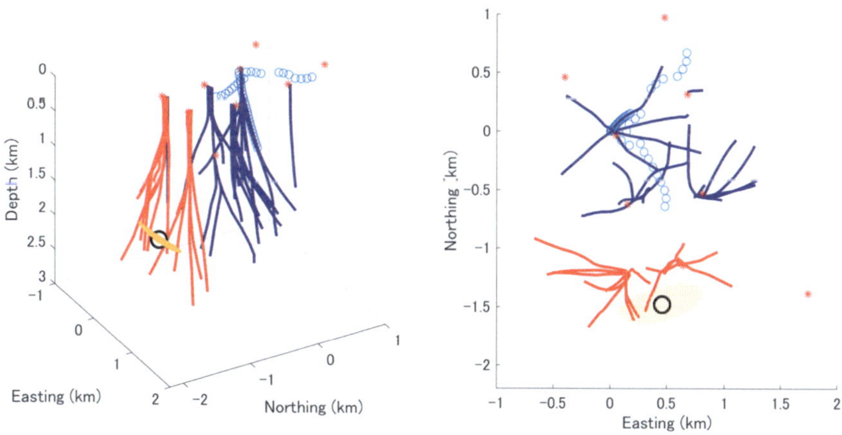

Fig. 8.34 The estimated hypocenter (black circle) for the earthquake is shown in Fig. 8.31. The red and blue lines are production well and reinjection trajectory. Light blue circles are temporary surface seismometers, and red asterisks are permanent stations

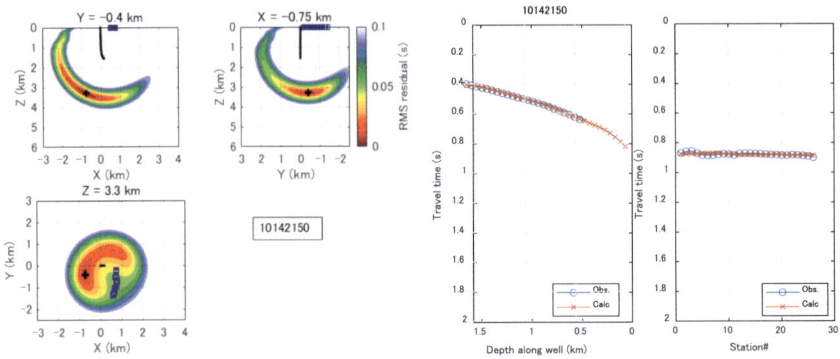

Fig. 8.35 The earthquake of October 14 at 21:50 in 2023. Right two diagrams; fittings of DAS and surface seismometer P first arrival readings, Left: hypocenter with O-C residuals

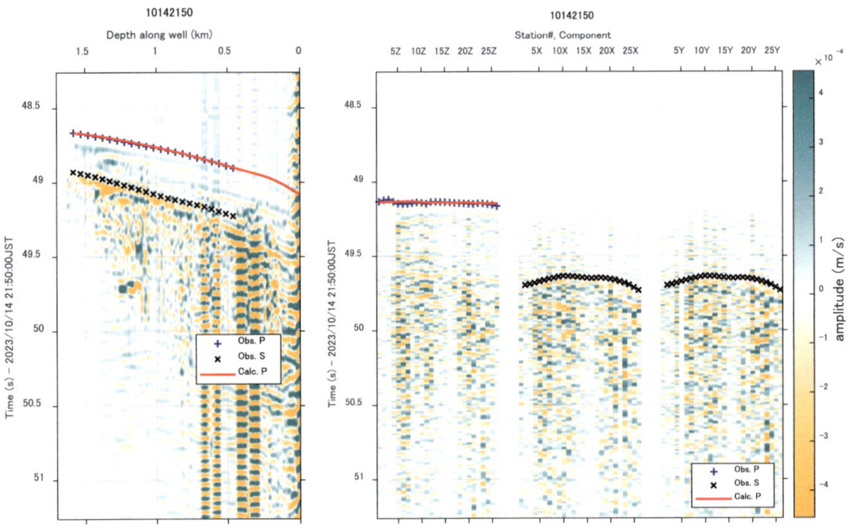

Fig. 8.36 DAS and surface seismometer records for the earthquake of October 14 at 21:50 in 2023. Fittings of calculated P and S arrival times to DAS records (left) and surface seismometer records of vertical, E-W, and N-S components. Twenty-six surface seismometers were located along the line with 50 m spacing

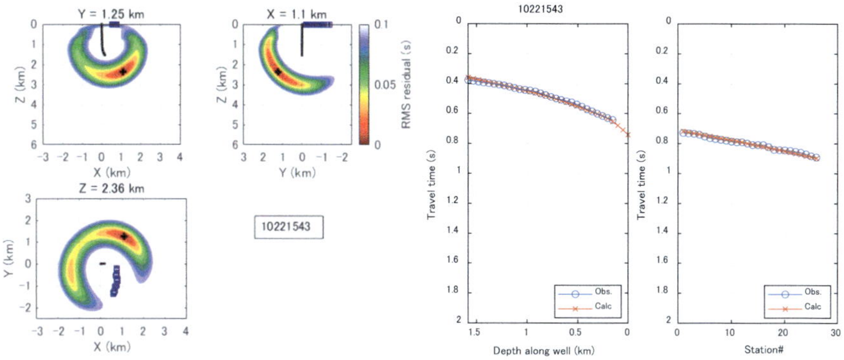

Fig. 8.37 The earthquake of October 22 at 15:43 in 2023. Right two diagrams; fittings of DAS and surface seismometer P first arrival reading, Left: hypocenter with O-C residuals

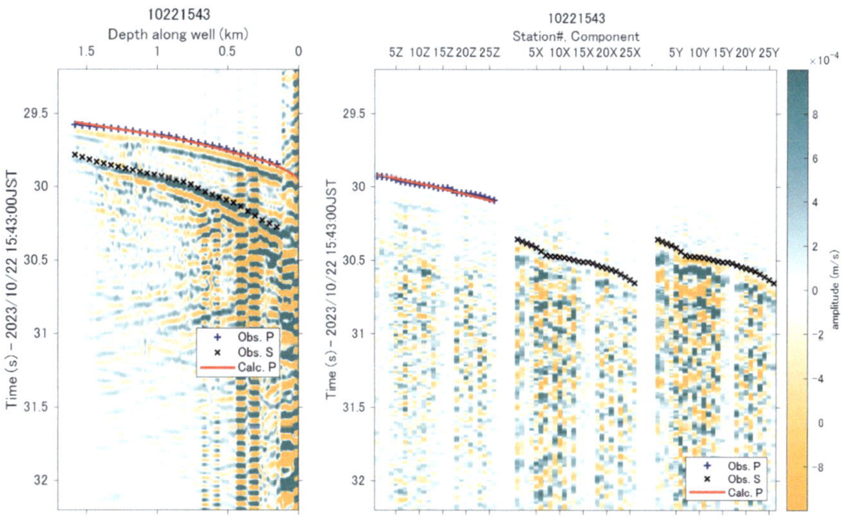

Fig. 8.38 The DAS and surface seismometer records for the earthquake of October 22 at 15:43 in 2023. Fittings of calculated P and S arrival times to DAS records (left) and surface seismometer records of vertical, E-W, and N-S components. Twenty-six surface seismometers were located along the line with 50 m spacing

8.5 Gravity and Gravitational Tensor (FMG) Approach to the Geothermal Exploration

8.5.1 Basic Equations

Gravity has been used to map density distribution in the ground. Recently, gravity derivatives, the second derivatives of gravitational potential, have been used. This section shows the method using the gravity and gravity tensor following Kasahara et al. (2010, 2011).

The instrument to measure the gravity tensors is known as the "gradiometer," developed by Bell Aerospace (now Lockheed Martin) during the 1970s and'80 s for US military purposes (Jekeli 1993). The usefulness of the gravity tensor for the exploration of oil, gas, and mineral resources has been recognized because the gravity tensor is proportional to the cube of the distance to the anomalous mass (Hammond and Murphy 2003). On the other hand, the gravity is proportional to the square of the distance to the anomalous mass.

This section describes the analysis method using the inversion method. Wan and Zhdanov (2008) proposed the gravitational inversion, but we propose a new method using the Laplacian eigenfunction proposed by Saito (2008).

Gravitational potential and gravity tensors are defined by (8.8) and defined by (8.9)–(8.11), respectively (Wan and Zhdanov 2008).

$$U(\mathbf{r}) = \gamma \iiint_D \frac{\rho(\mathbf{r}')}{|\mathbf{r}' - \mathbf{r}|} \mathbf{dv}' \tag{8.8}$$

$$\mathbf{g} = \begin{bmatrix} g_{xx} & g_{xy} & g_{xz} \\ g_{yx} & g_{yy} & g_{yz} \\ g_{zx} & g_{zy} & g_{zz} \end{bmatrix} \tag{8.9}$$

$$g_{\alpha\beta}(\mathbf{r}) = \frac{\partial^2}{\partial\alpha\partial\beta} U(\mathbf{r}), \quad \alpha, \beta = x, y, z \tag{8.10}$$

$$g_{\alpha\beta}(\mathbf{r}) = \begin{cases} 3\gamma \iiint_D \frac{\rho(\mathbf{r}')(\alpha-\alpha')(\beta-\beta')}{|\mathbf{r}'-\mathbf{r}|^5} dv', & (\alpha \neq \beta) \\ \gamma \iiint_D \frac{\rho(\mathbf{r}')}{|\mathbf{r}'-\mathbf{r}|^3} \left\{ 3\frac{(\alpha-\alpha')^2}{|\mathbf{r}'-\mathbf{r}|^2} - 1 \right\} dv', & (\alpha = \beta) \end{cases} \tag{8.11}$$

8.5.2 Methods for Analysis

Firstly, we introduce forward modeling and then explain model fitting.

(1) **Forward Modeling**

The model density anomaly is assumed in three-dimensional space. The principle is shown in Figure 8.39 Gravitational tensors are calculated by equations (8.9)–(8.11) (Figs. 8.40 and 8.41).

(2) Inversion Using Hyperparameters

Hyperparameters perform one of the inversions. The basic equations are (12)–(16). "λ","α" and "β" are hyperparameters that constrain damping and horizontal and vertical smoothness, respectively.

$$\Delta \mathbf{g} = \mathbf{K} \Delta \rho \tag{8.12}$$

$$\mathbf{K} = \left[\mathbf{k}_{xx}^{\mathrm{T}} \; \mathbf{k}_{xy}^{\mathrm{T}} \; \mathbf{k}_{xz}^{\mathrm{T}} \; \mathbf{k}_{yy}^{\mathrm{T}} \; \mathbf{k}_{yz}^{\mathrm{T}} \; \mathbf{k}_{zz}^{\mathrm{T}} \right]^{\mathrm{T}}, \tag{8.13}$$

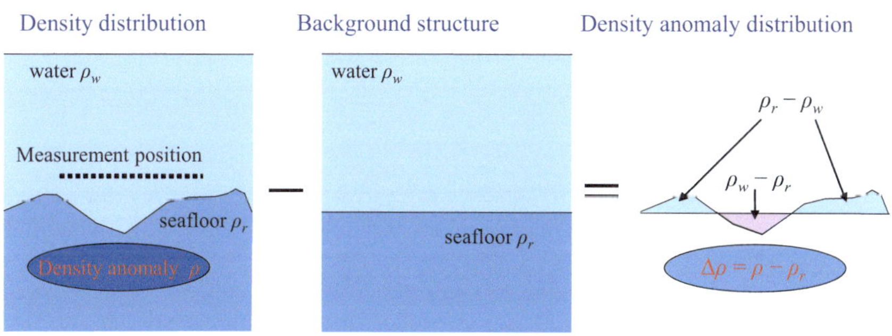

Fig. 8.39 Density anomaly calculation. (After Kasahara et al., 2010)

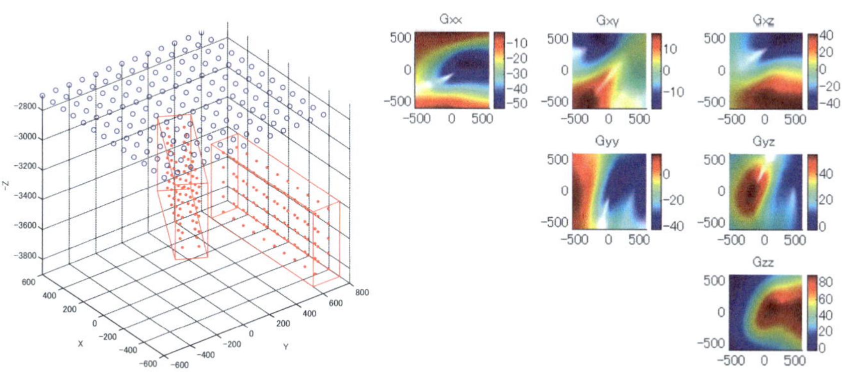

Fig. 8.40 **a** An example of a density anomaly model. Two cuboid density anomalies are approximated by mass points (red dots). Blue circles denote the points where the tensor gravity is calculated. **b** Calculated tensor gravity components Gxx, Gxy, Gxz, Gyy, Gyz, and Gzz. The unit of a gravitational tensor is Eötvös (1E = 10^{-9}Gal/cm = 10^{-9} s^{-2}). Unit of vertical and horizontal axes are meters. (After Kasahara et al. 2010)

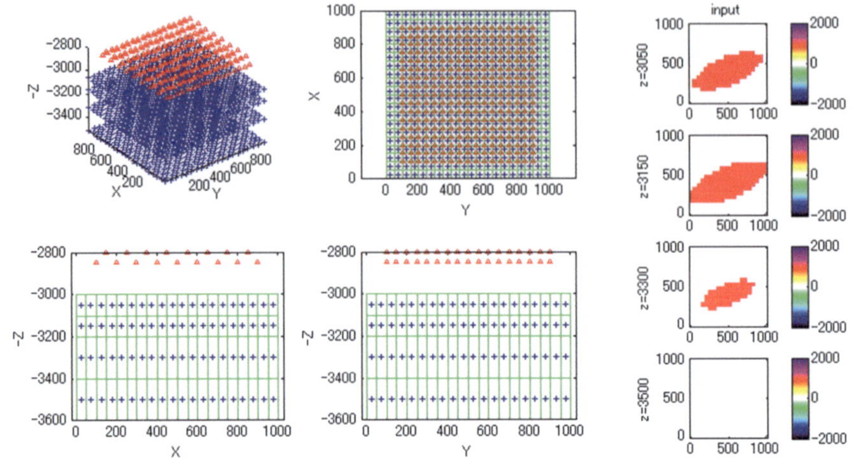

Fig. 8.41 **a** An initial model for 3D inversion test. Unit is meter. **b** Four layers contain density anomalies (top view). The unit of density is kg/m^3. (After Kasahara et al. 2010)

$$\left(k_{\alpha\beta}\right)_{nm} = \begin{cases} 3\gamma \dfrac{(\alpha_n-\alpha'_m)(\beta_n-\beta'_m)dv_m}{|\mathbf{r}'_m-\mathbf{r}_n|^5}, & (\alpha \neq \beta) \\ \gamma \dfrac{dv_m}{|\mathbf{r}'_m-\mathbf{r}_n|^3}\left\{3\dfrac{(\alpha_n-\alpha'_m)}{|\mathbf{r}'_m-\mathbf{r}_n|^2} - 1\right\}, & (\alpha = \beta) \end{cases} \tag{8.14}$$

$$|\Delta\mathbf{g} - \mathbf{K}\Delta\rho|^2 \rightarrow \min \tag{8.15}$$

$$|\Delta\mathbf{g}-\mathbf{K}\Delta\rho|^2 + \lambda|\Delta\rho|^2 + \alpha\left|\left(\partial^2/\partial x^2 + \partial^2/\partial y^2\right)\Delta\rho\right|^2 + \beta\left|\partial^2/\partial z^2\Delta\rho\right|^2 \rightarrow \min \tag{8.16}$$

where λ is the damping parameter, α is the horizontal smoothness parameter, and β is the vertical smoothness parameter.

The quality of retrieved images can be equally obtained by damping or spatial smoothness factors, as shown in Fig. 8.42. The hyperparameters can be selected to minimize ABIC (Akaike's Bayesian information criterion). Due to the nature of the gravitational tensor, the deeper part is less resolved than the shallow part. Constraints on the bottom of density anomalies could be obtained from seismic data.

(3) **Eigenfunction Inversion**

The inversion method is introduced for the density anomaly in the ground by the eigenfunction of the Laplacian operator proposed by Saito (2008), which is the eigenfunction of the Laplace equation under a boundary condition. A set of basic functions expresses the model with fewer parameters. The Laplacian eigenfunctions $\boldsymbol{\phi}_m(\mathbf{r})$ (Saito 2008) are obtained by solving the eigenvalue problem of the following $L(\mathbf{r}, \mathbf{r}')$, which is the fundamental solution of 3D Laplacian.

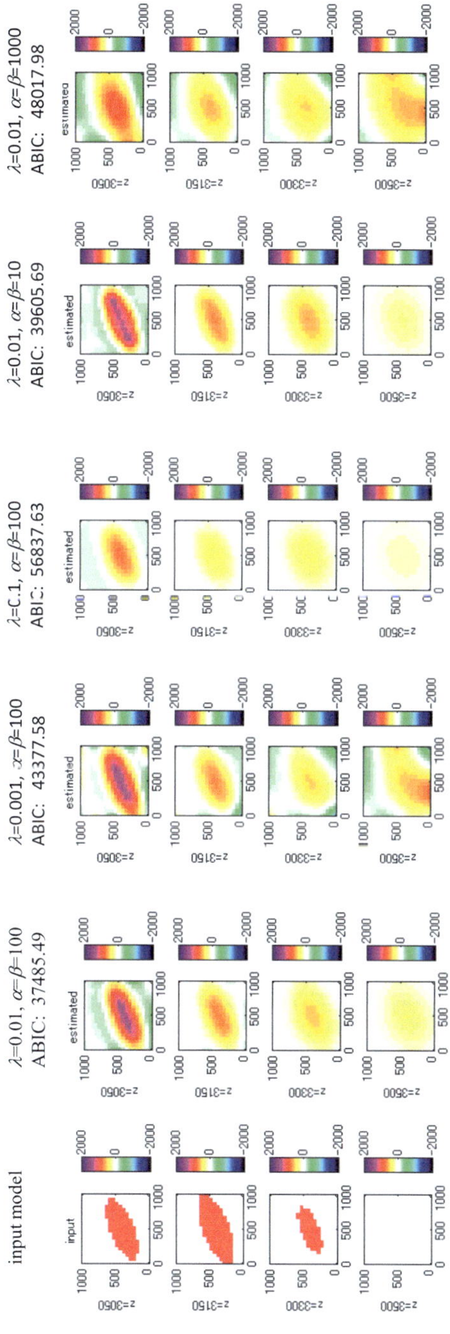

Fig. 8.42 Inversion tests by three kinds of hyperparameters. λ: damping parameter, α: horizontal smoothness parameter, and β: vertical smoothness parameter. (After Kasahara et al. 2010)

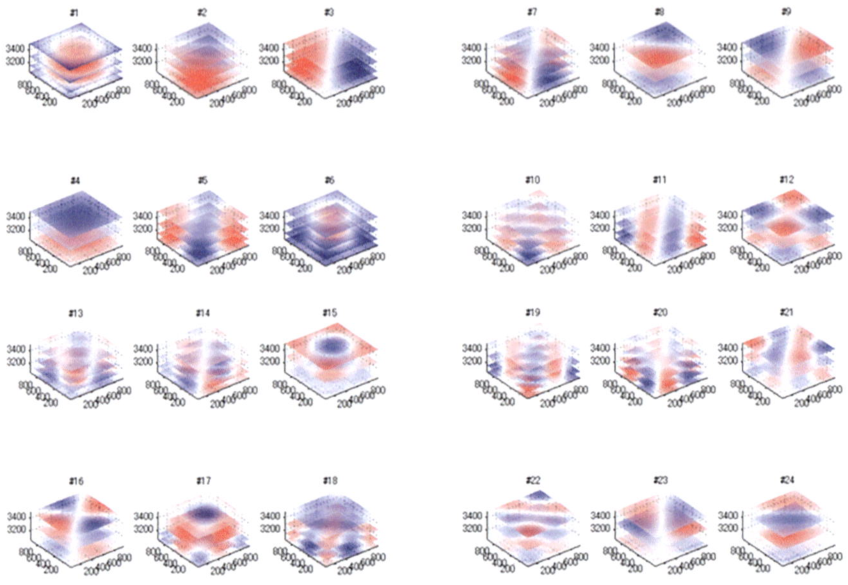

Fig. 8.43 Modes of Laplacian eigenfunction in 3D space. (After Kasahara et al., 2010)

$$L(\mathbf{r}, \mathbf{r}') = \frac{1}{4\pi |\mathbf{r} - \mathbf{r}'|} \tag{8.17}$$

The modes of Laplacian eigenfunction in 3D space are illustrated in Fig. 8.43. The eigenfunctions are sorted in the descending order of the eigenvalues.

The density anomaly distribution $\Delta\mathbf{g}(\mathbf{r}_m)$ in the model is expanded using Laplacian eigenfunction $\boldsymbol{\phi}_m(\mathbf{r}_m)$.

$$\Delta\mathbf{g}(\mathbf{r}_m) = \sum_{j=1}^{J} a_j \boldsymbol{\phi}_m(\mathbf{r}_m) \tag{8.18}$$

Lower-order modes correspond to global structure, and higher-order modes correspond to the fine structure of underground structures. The model order J should be taken to give the required spatial resolution.

Inversion using the new model parameter $\boldsymbol{a} = \{a_j\}$ is estimated to minimize the following function,

$$\left|\Delta\mathbf{g}(\mathbf{r}_m) - \mathbf{K}\boldsymbol{\phi}_m\boldsymbol{a}\right|^2 + \mu|\boldsymbol{a}|^2 \rightarrow \min \tag{8.19}$$

where μ is the hyperparameter for damping.

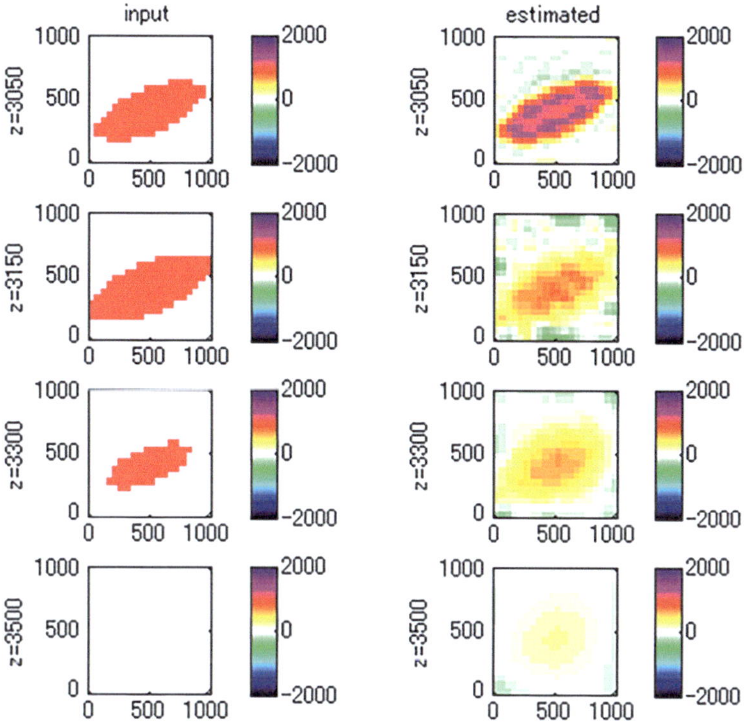

Fig. 8.44 Shows the Model and result of inversion using a Laplacian eigenfunction. The structure used 400 unknowns and a 1,600-block model. (After Kasahara et al. 2010).

Figure 8.44 shows the result of inversion using the Laplacian eigenfunction method. As shown in Fig. 8.42, we can obtain a similar resolution to the hyperparameter method described before, and the computation is much faster than the hyperparameter inversion.

8.5.3 Example of Survey Data

JOGMEC conducted the surveys of gravity and gravity tensors in Kyushu (JOGMEC 2014). Two results of gravitational tensors in Kuju (Figs. 8.45 and 8.46) and Ogiri (Fig. 8.47 and 8.48) are shown with gravity data. g_{DD} and g_D are vertical components. g_{DD} and g_D maps show similar high and low areas. gEW and gNS are thought to be sensitive to fault distribution and geological discontinuity. O'Breien et al (2005) used the gravity tensor measurements in Gulf of Mexico.

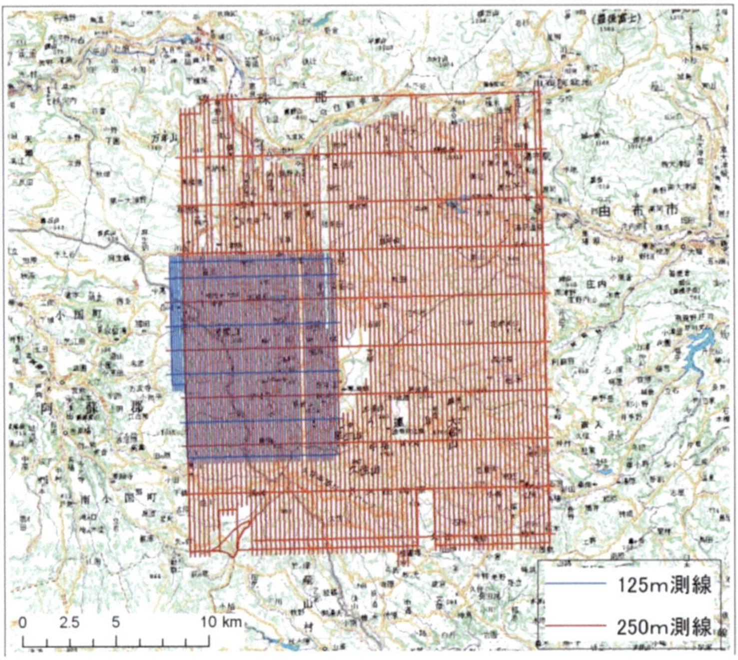

Fig. 8.45 Map of the gravitational survey in Kuju (JOGMEC 2014)

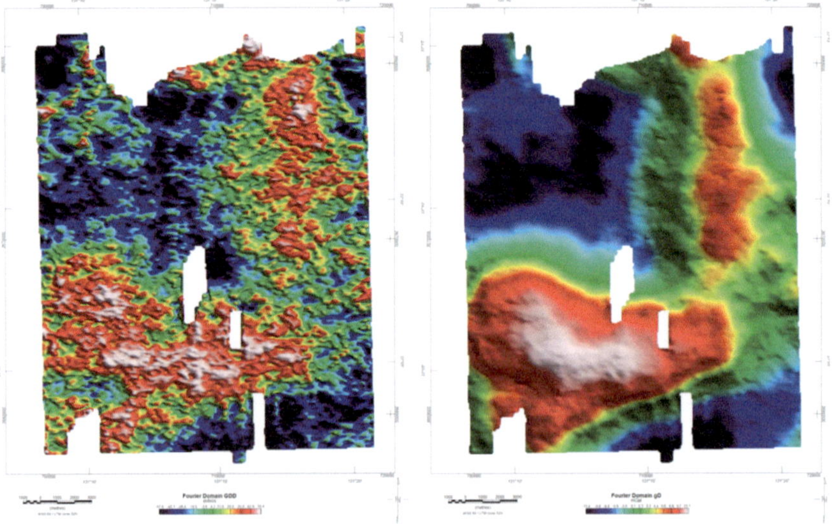

Fig. 8.46 Gravity and gravity (g_D) (right) derivatives (g_{DD}) (left) map in Kuju area, in Kyushu, Japan (JOGMEC 2014)

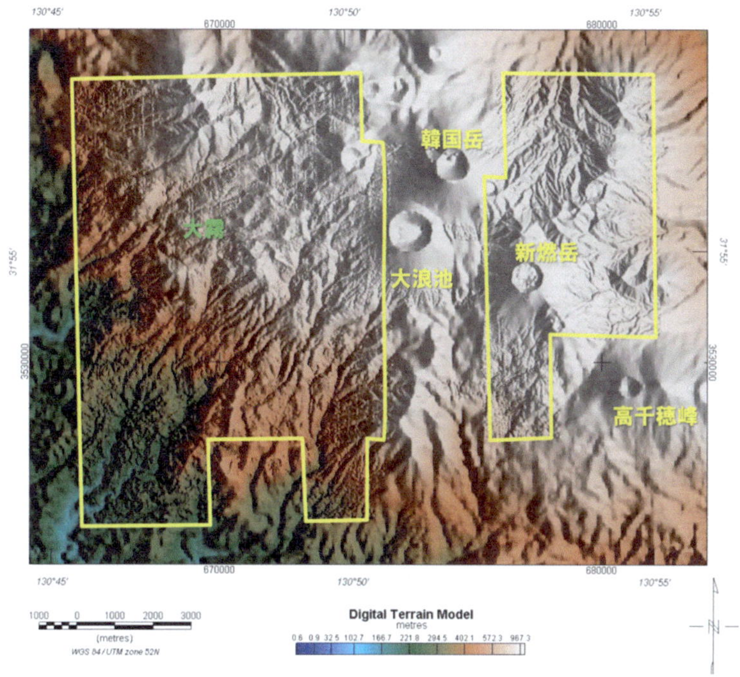

Fig. 8.47 Map of the gravitational survey in Ogiri, Kirishima area (JOGMEC 2014)

8.5.4 Discussion and Conclusions

Kasahara et al. (2010, 2011) presented two inversion methods: hyperparameter inversion and Laplacian eigenfunction inversion. The Laplacian eigenfunction inversion is based on mode decomposition, offering the advantage of avoiding local minima. This is because lower-order modes represent the best solutions on a global scale, and higher-order modes mean better solutions on a fine scale. They obtained similar results using several examples, such as the hyperparameter inversion, using faster computational time.

Gravitational tensor data can provide better resolution in top-of-density anomaly, but it needs to constrain the deeper limit. The seismic method could lower the limit. By this nature of the two methods, the joint inversion of seismic data and gravitational tensor data could give the best constraint on density anomaly mapping for deeper limit and precise upper resolution, respectively.

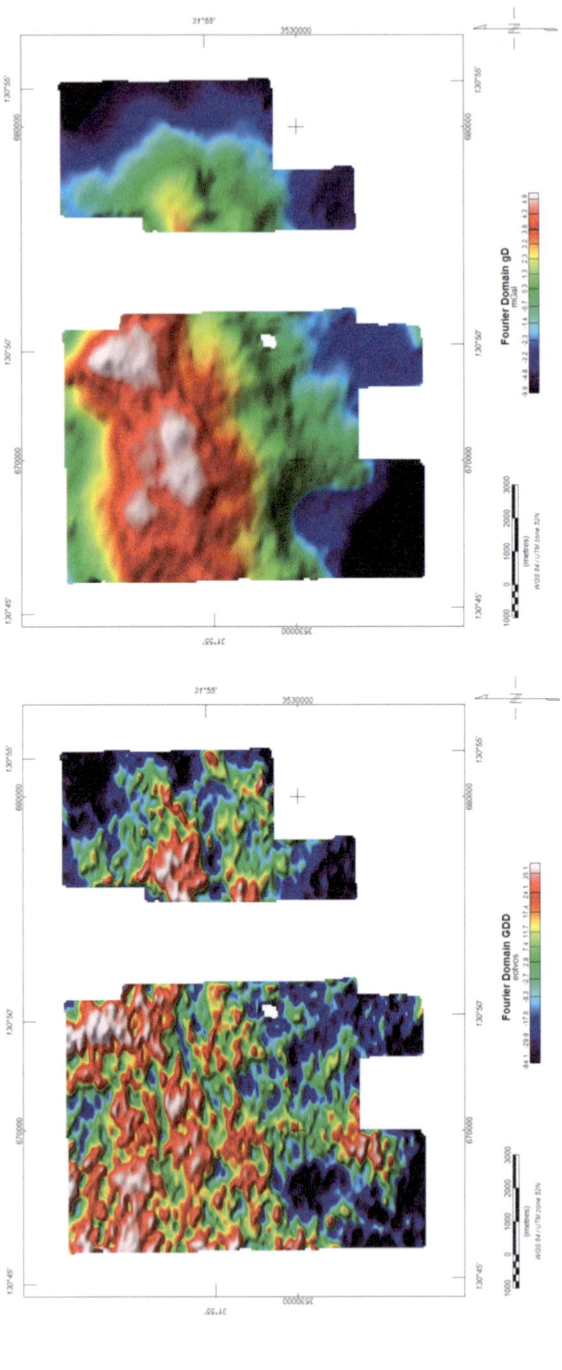

Fig. 8.48 Gravity and gravity (g_D) (right) derivatives (g_{DD}) (left) map in Ogiri, Kirishima area, in Kyushu, Japan. (JOGMEC 2014)

References

Alkhalifah TA (2016) Full waveform inversion in an anisotropic world, Revised Edn. EAGE, p 197

Hammond S, Murphy CA (2003) Air-FTG™: bell geospace's gravity gradiometer—a description and case study. ASEG Prev 105:24–26

Hartog A (2017) An introduction to distributed optical fibre sensors. CRC Press, p 442

Hasada Y, Kasahara J, Kawashima H, Yamauchi YY, Sugimoto Y, Yamaguchi T, Kubota K (2018) Comparison of the records by optical fiber DAS (Distributed Acoustic Sensor) and geophone using natural earthquakes. JIn: PGU 2018 annual meeting abstract

Jekeli C (1993) A review of gravity gradiometer survey system data analyses. Geophysics 58:508–514

JOGMEC (2014) Measurement method and data processing of AGG. In: JOGMEC geothermal division report meeting in 2014. https://geothermal.jogmec.go.jp/initiatives/achievement/file/7.pdf

Kasahara J, Nagumo S, Koresawa S, Nishi Y, Sugimoto H (1982) A linear trend of hypocenter distribution in the outer slope region of the Japan Trench revealed by the OBS array-Preliminary report. Bull Earthq Res Inst Univ Tokyo 57:83–104

Kasahara J, Hasada Y, Kondo H, Hayato, Tsuruga K (2010) Underground imaging by the 3D gravity tensor inversion method. In: 123th society of exploration geophysicists Japan Fall meeting extended abstract. pp 123–125

Kasahara J, Hasada Y, Kondo H, Tsuruga K (2011) A new 3D gravitational tensor inversion for imaging of density distribution in the ground. In: EAGE extended abstract, Viena, Austria

Kasahara J, Hasada Y (2016) Time lapse approach to monitoring oil, gas, and CO_2 storage by seismic method. Elsevier Pub., p 201

Kasahara J, Yamaguchi T, Sugimoto Y, Kawashima H, Yamauchi Y, Hasada Y, Kubota K (2018) Evaluation of fiber optic DAS as a dense seismic array for continuous monitoring of civil engineering structures. In: JPGU 2018 annual meeting abstract

Kasahara J, Hasada Y, Yamaguchi T (2019a) Seismic imaging of supercritical geothermal reservoir using full-waveform inversion method. In: PROCEEDINGS, 43rd workshop on geothermal reservoir engineering, February 11–13, SGP-TR-214. Stanford University, Stanford, California

Kasahara J, Hadasa Y, Kuzume H, Yaaguchi T, Mikada H (2019b) Time-lapse approach to monitor temporal changes in the supercritical water reservoir. In: PROCEEDINGS, 43rd workshop on geothermal reservoir engineering, February 11–13, SGP-TR-214. Stanford University, Stanford, California

Kasahara J, Hasada Y, Kuzume H (2020a) Possibility of high Vp/Vs zone in the geothermal field suggested by the P-to-S conversion. In: PROCEEDINGS, 45th workshop on geothermal reservoir engineering, February 10–12, SGP-TR-216. Stanford University, Stanford, California

Kasahara J, Hasada Y, Kuzume H, Mikada H, Fujise Y (2020b) The second seismic study at the geothermal field in southern Kyushu, Japan using an optical fiber system and surface geophones, PROCEEDINGS, 45th Workshop on Geothermal Reservoir Engineering. Stanford University, Stanford, CA

Kasahara J, Hasada Y, Kuzume H, Ohnuma H, Mikada H, Fujise Y (2022a) A field experiment of temperature-tolerant distributed acoustic sensor in deep geothermal reservoir prospecting. Lead Edge 41:331–337

Kasahara J, Hasada Y, Ohnuma H, Mikada H, Fujise Y (2022b) Seismic study at the Sumikawa geothermal field in Akita Pref. In: Proceedings of the Society of Exploration Geophysicists, Japan, Spring Meeting Japan using a geothermal well

Kasahara J, Hasada Y, Kuzume H, Mikada H, Fujise Y (2022c) DAS-Seismic study in the Ohnuma geothermal borehole in Honshu, Japan. J Geothermal Res Soc 44:73–84

Kasahara J, Hasada Y, Mikada H, Ohnuma H, Fujise Y (2024a) Geothermal exploration using distributed acoustic sensing in a borehole in Otake-Hatchobaru geothermal field in Japan. In: Japan geoscience union 2024 annual meeting, abstract

Kasahara J, Hasada Y, Mikada H (2024b) Geothermal reservoir characterization using distributed acoustic sensing from vertical seismic profiling in six geothermal fields in Japan. In: Distributed acoustic sensing in borehole geophysics, AGU monograph

Kasahara J, Hasada Y, Suzuki M, Takahashi T, Tatsunokuchi H, Tatsunokughi H, Fujise Y (2025) The second geothermal exploration using the DAS in the Kijiyama geothermal borehole, Yuzawa, Akita. In: Preliminary report, 150th Society of Exploration Geophysicists Japan Fall Meeting Extended Abstract.

Kern H (1978) The effect of high temperature and high confining pressure on compressional wave velocities I quartz-bearing and quartz-free igneous and metamorphic rocks. Tectonophysics 44:185–203

Kern H, Richter A (1981) Temperature derivatives of compressional and shear wave velocities in crust and mantle rocks a 6 kbar confining pressure. J Geophys 49:47–56

Kern H, Liu B, Popp T (1997) Relationship between anisotropy of P and S wave velocities and anisotropy of attenuation in serpentine an amphibolite. J Geophys Res 102(B2):3051–3065

NEDO (2008, 2009, 2010) Report of geothermal development promotion, East of Ikeda Lake No. C-2–10, first, second and third phases

O'Brien J, Rodriguez A, Sixta D, Davies MA, Houghton P (2005) Resolving the K-2 salt structure in the Gulf of Mexico: an integrated approach using prestack depth imaging and full tensor gravity gradiometry. Lead Edge 24:404–409

OptaSense (2022) Datasheet of ODH-4

Saito N (2008) Data analysis and representation on a general domain using eigenfunctions of Laplacian. Appl Comput Harmon Anal 25:68–97

Schuster GT (2017) Seismic inversion. Investigations in geophysics, no 20, SEG

Tarantola A (1984) Inversion of seismic reflection data in the acoustic approximation, Geophysics, 49:1259–1266

Tarantola A (1996) A strategy for nonlinear inversion of seismic reflection data, Geophysics, 51(10):893–1903

Tromp J, Tape C, Liu Q, (2005) Seismic tomography, adjoint methods, time reversal and banana-doughnut kernels. Geophys J Int 160(1):195–216

Utsu T (1984) Seismolgy. Kyoritu Pub

Virieus J, Opeto S (2009) An overview of full-waveform inversion in exploration geophysics. Geophysics, 74:WCC1–WCC26

Wan L, Zhdanov MS (2008) Focusing inversion of marine full-tensor gradiometry data in offshore geophysical exploration. SEG Las Vegas 2008 Annual Meeting

Winkler KW, Nur A (1982) Seismic attenuation: effects of pore fluids and frictional sliding. Geophysics 4:1–15

Part II
Geological Investigation and Well Targeting for Geothermal Development, Japan

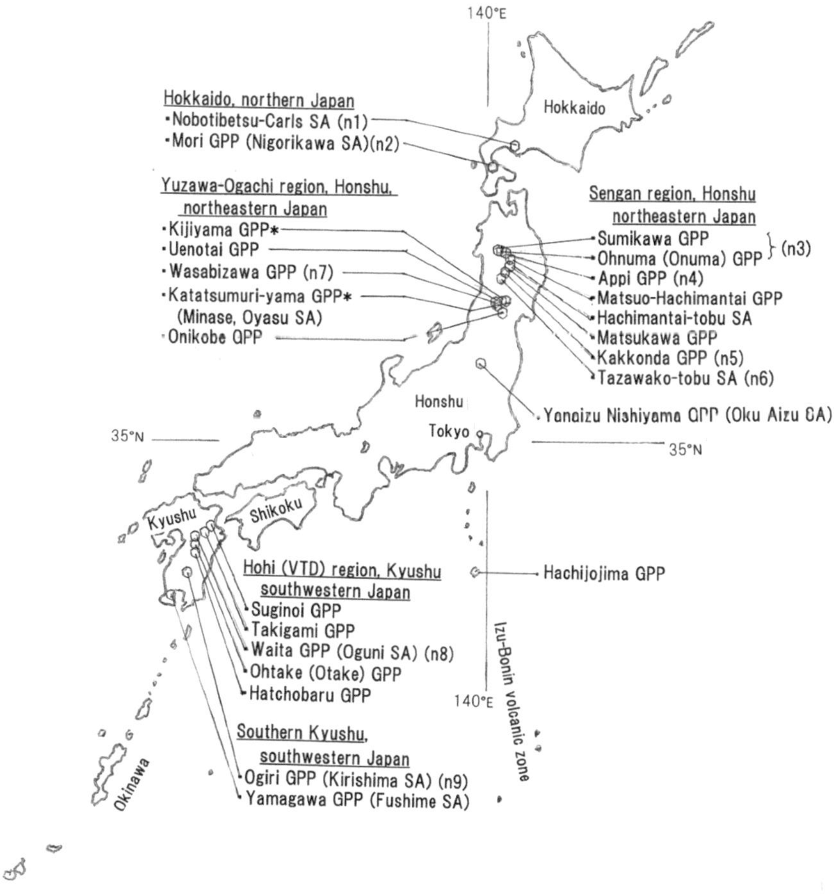

List of Geothermal Power Plants (GPPs) and Survey Areas (SAs) Referenced in Part II (* indicates a project under construction). Notes: (n1) Includes part of the Iburi Survey Area. (n2) Includes part of the Kaminoyu–Santai Survey Area. (n3) Refers to the northern Hachimantai area. (n4) Includes the Ashiro Survey Area. (n5) Formerly known as Takinoue. (n6) The Nyuto-onsen Survey Area is within this region. (n7) Includes the Akinomiya Survey Area. (n8) Includes the Waita Survey Area. (n9) Refers to the Shiramizugoe and Kurino–Tearai Survey Areas. These correspond to the promising locations of 23 geothermal areas in Japan analyzed for permeability–thickness (kh), geological features, production performance, and drilling records, as shown in Fig. 11.6.

Chapter 9
Geothermal Systems and Geothermal Reservoirs

Ko Sato

Abstract Cold meteoric water infiltrates deep underground, where it is heated through interaction with magmatic intrusions. Driven by thermal buoyancy, the heated fluid ascends through permeable zones. A three-dimensional geothermal convection system comprises a high-temperature zone and a recharge area. Within such systems, geothermal reservoirs are hot and permeable layers that can be economically exploited to produce fluid or heat, typically found at depths of up to 3 km or more.

Keywords Geothermal system · Upflow zone · Anion index · Stress field · Horst-Graben structure

9.1 Structural Setting of Geothermal Systems at Active Volcanoes

Figure 9.1 presents a conceptual model of a geothermal system developed beneath an active volcano in a region of pronounced topographic relief, using the Akita-Komagatake volcano and its surrounding area as a representative example. Geothermal systems in silicic volcanic terrains are relatively uncommon in Japan. In such settings, the heat source is typically attributed to a magmatic intrusion—an isolated magma chamber—positioned directly beneath the active volcanic edifice. The geothermal system is primarily structured around fracture networks that extend downward into the deep-seated basement rocks. These include features labeled SS, C1, f3, f4, Fc, and Fd, which correspond to those illustrated in Figs. 14.33–14.36, with the exception of SS, which is unique to Fig. 9.1.

Geothermal fluids ascend vertically from deeper zones to the surface through permeable fractures, forming a concentrated vertical flow path known as the upflow zone. High-temperature, immature acidic fluids are generated either by conductive

---

K. Sato (✉)
Socio Engineering Co., Ltd., Tokyo, Japan
e-mail: satoukou6450@gmail.com

J. Kasahara et al. (eds.), *Challenges in Geothermal Development*, Advances in Geological Science, https://doi.org/10.1007/978-981-95-0553-1_9

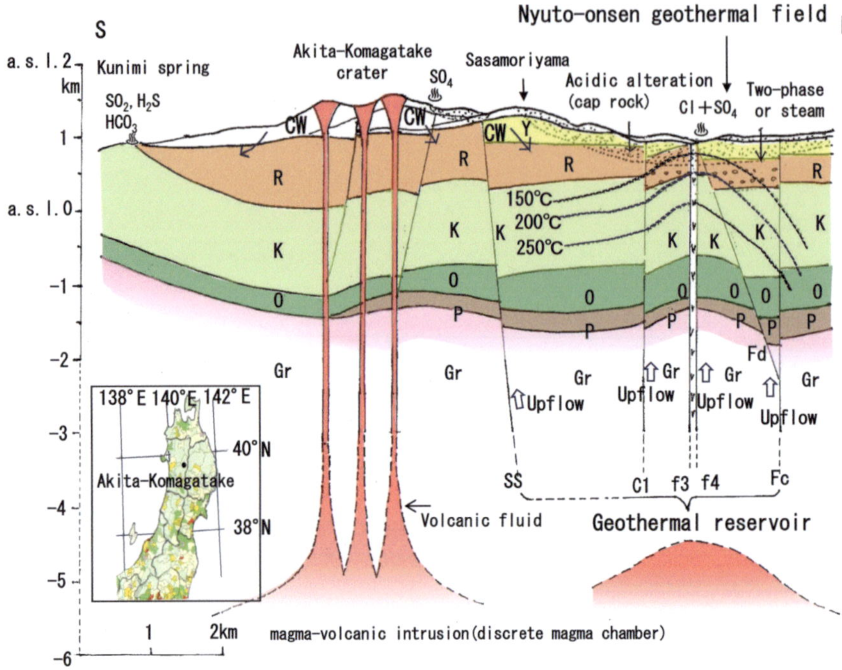

Fig. 9.1 Schematic cross-sectional profile of a geothermal system associated with an island-arc andesitic volcano, exemplified by the Akita-Komagatake active volcano and the Nyuto-Onsen geothermal field. The north–south (N–S) section corresponds to the plane illustrated in Fig. 14.30. The vertical scale below −2 km a.s.l. is arbitrary. The geological units are denoted as follows: "Y"—Yamatsuda Formation (sandstone, mudstone, tuff, coal seam); "R"—Koshitomaesawa and Ryukawa Formations (sandstone, mudstone, tuff); "K"—Kunimitoge Formation (green tuff, mudstone, andesite, basalt); "O"—Obonai Formation (altered volcanics); "P"—pre-Tertiary formation (silicified rocks); "Gr"—Cretaceous granitic rocks; "v"—intrusive rocks; and "CW"—meteoric water recharge zones

heating of percolating meteoric water or through the input of high-temperature steam and magmatic gases.

As these fluids rise, they undergo water–rock interactions, becoming neutralized and transforming into near-neutral pH alkali chloride waters. These waters subsequently mix with acidic SO_4-type hot waters, heated by steam near the volcanic summit, and with HCO_3-type waters found near the base of the volcano. A schematic representation of this typical geothermal structure—classified by the major anion composition (Anion Index; AI) of hot spring water—is presented by Noda (1987) in Fig. 9.2.

The Anion Index (AI) is calculated using the following equation:

$$AI = 0.5 \times \left(\frac{SO_4}{Cl + SO_4} + \frac{Cl + SO_4}{Cl + SO_4 + HCO_3} \right)$$

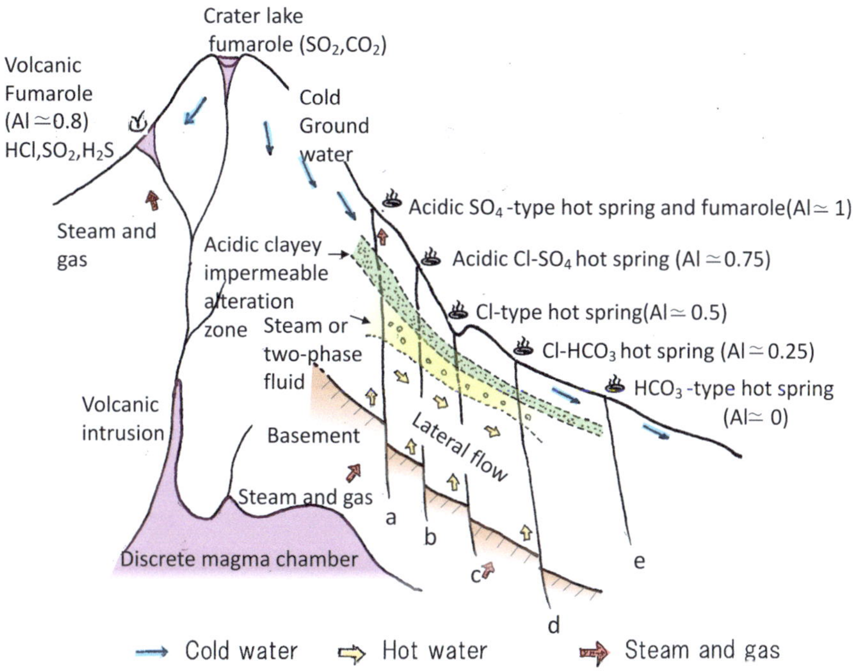

Fig. 9.2 Schematic model of a geothermal system associated with an island-arc andesitic volcano in high-relief topography, illustrating the Anion Index (AI). The figure is not to scale. Fractures in the geothermal reservoir are labeled "a" through "e." (modified from Noda, 1987)

where all concentrations are expressed in equivalents. The AI values—typically 0, 0.25, 0.5, 0.75, 0.8, and 1—characterize the relative dominance of anions in geothermal fluids. The geothermal reservoir in this system, as depicted in Fig. 9.1, contains a network of fractures labeled "a" through "e." These subsurface structures significantly influence the emergence and type of hot spring manifestations observed at the surface. A comparison between Figs. 9.1 and 9.2 reveals that only partial surface manifestations—specifically SO_4-type and Cl + SO_4-type hot waters—are represented in Fig. 9.1.

Impermeable clay-rich acidic alteration zones overlie geothermal systems and serve as effective cap rocks. These zones play a vital role in sealing the system by inhibiting heat dissipation from upwelling geothermal fluids and preventing the intrusion of cold meteoric water. Cap rocks typically develop at the base of volcanic edifices, where groundwater stagnation promotes prolonged hydrothermal alteration. In contrast, they are rarely formed in recharge areas near steep volcanic slopes, except where mudstone is present. In the Sengan geothermal area, for example, Takashima et al. (2001) reported that cap rocks are absent in zones affected by the latest volcanic activity (0.13 Ma), suggesting that younger volcanic systems tend to lack cap rock development due to persistent groundwater recharge.

When planning the location of production wells, drilling near the summit of an active volcano should be avoided for the following reasons:

(1) The summit of an active volcano typically serves as a groundwater recharge zone, where cold meteoric water infiltrates downward.
(2) The formation of hydrothermal alteration zones and the dissolution and deposition of vein minerals (e.g., quartz, calcite) require prolonged hydrothermal activity (see Sect. 11.1.5).

As a result, cap rock and drusy quartz formations are generally absent near volcanic summits. However, geothermal reservoirs may still exist beneath composite volcanoes that overlie older volcanic sequences. A representative example is the Hachijojima volcano, which is part of the Izu–Bonin (Ogasawara) volcanic arc.

9.2 Relationship Between Magma Activity and Stress Field

Figure 9.3 schematically illustrates the representative relationships between magmatic activity and stress fields.

CASE A represents a scenario without magmatic intrusion. In this case, the local stress field (σ_1') is consistent with the regional stress field (σ_1). If σ_1 ($\approx \sigma_1$') corresponds to a horizontally oriented east–west (EW) stress field, then strike-slip faults trending northeast (NE) and northwest (NW), as well as extensional fractures trending EW, are typically generated.

CASE B involves a scenario with magmatic intrusion, where the local stress field is modified: σ_1' is oriented vertically, and σ_2' horizontally, while σ_1 remains horizontal in the regional stress field. As a result, normal faults and extensional fractures are likely to form in the direction of σ_2'.

CASE C corresponds to the case of vigorous magma intrusion. Here, the local stress field is further altered such that σ_1' becomes vertical and σ_3' becomes horizontal, despite σ_1 still being horizontal in the regional field. This change results from pressure associated with magmatic inflation along the σ_3' direction. Consequently, normal faults and extensional fractures preferentially develop in the direction of σ_3.

9.3 Upflow Zones Accompanied by Horst-Graben Structures

Figure 9.4 presents the horst–graben structure in the Oguni geothermal field, situated within the Hohi Volcano-Tectonic Depression (VTD) region of Kyushu, Japan. This cross-section is constructed based on lithostratigraphic correlations of similar rock sequences (Fujita and Abe, 1988). The boundaries between horsts and grabens are characterized by crushed zones, which act as permeable conduits facilitating the upward migration of geothermal fluids. In this figure, the "Na" fault is confirmed

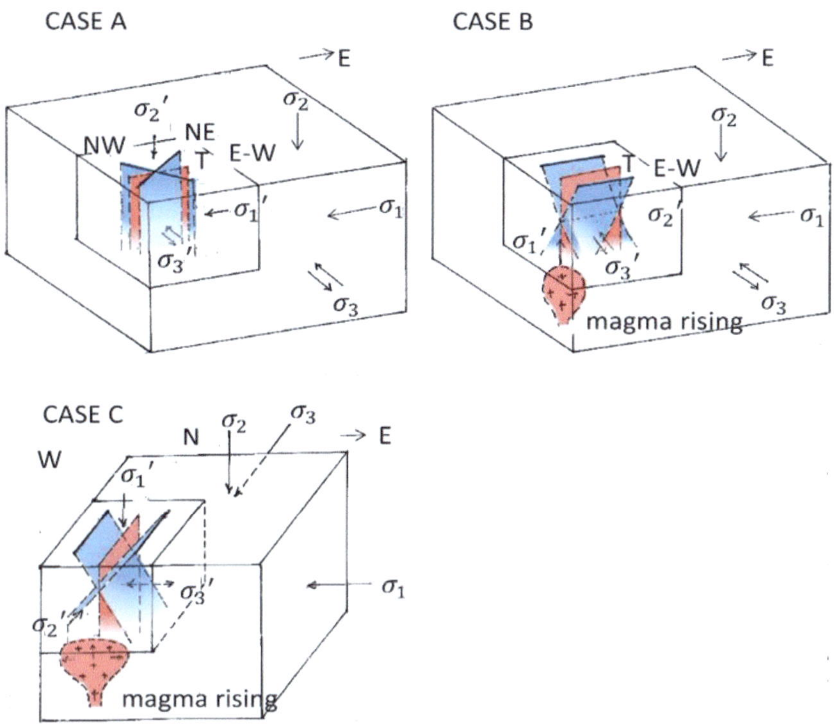

Fig. 9.3 Schematic diagram illustrating the relationship between magma activity and the stress field. Blue planes represent normal and strike-slip faults and associated fractures. Red planes (T) indicate extensional faults and fractures

by lost circulation events encountered during drilling at wells DY-1 and KW-1. The Takenoyu Fault ("Ta") has been validated through extensive subsurface surveys, whereas the "Wa" fault is inferred from stratigraphic correlations and temperature distribution patterns.

This structural configuration, although formed prior to the deposition of the Kuju volcanic rocks, continues to influence present-day upflow zones, as evidenced by the formation temperature anomalies in the figure. According to Tamanyu (1985, 1993), horst–graben structures in this region are attributed to differential tectonic movements, including fault-block uplift and subsidence, occurring between approximately 6 Ma and 0.5 Ma.

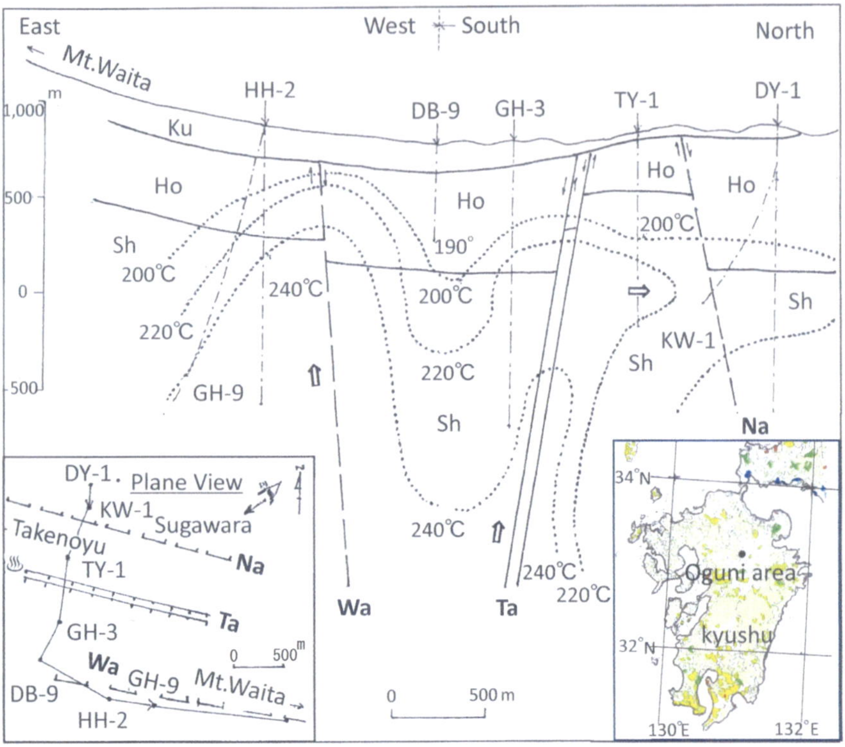

Fig. 9.4 An example of a horst–graben structure and the associated upflow zone in the Oguni geothermal area, located in the Hohi Volcano-Tectonic Depression (VTD) region, Kyushu, Japan. The faults labeled "Na," "Ta," and "Wa" are normal faults formed under a stress regime dominated by vertical σ_1, induced by magmatic activity (see Fig. 9.3, CASE B; Table 12.2). Geological units are indicated as follows: "Ku" refers to the Kuju volcanic rocks (Late Pleistocene to Holocene), "Ho" to the Hohi volcanic rocks (Early Pleistocene), and "Sh" to the Shishimuta Group (Pliocene, pre-Kusu volcanic products). Fault boundaries serve as permeable pathways for ascending geothermal fluids. HH-2, DB-9, GH-3, TY-1, DY-1, and KW-1 indicate well locations (Sato et al., 2019)

References

Fujita T, Abe M (1988) Survey of Oguni geothermal field, Kyushu dist. Japan. J Jap Geotherm Ener Assoc (Chinetsu) 25(4):1–28 (in Japanese with English abstract)

Noda T (1987) Anion index as an indicator of geothermal activity. J Geotherm Res Soc Jap 9(2):133–141 (in Japanese with English abstract)

Sato K, Ito N, Tsukuda K (2019) Geothermal geological survey and production well target (Original title in Japanese). E-pix, Ofunato Printing Inc., Iwate, p 155. ISBN 978-4-901602-67-9 (in Japanese)

Takashima I, Ohkawa K, Miura K, Saiki K, Murakami H (2001) Alteration and thermoluminescence ages of altered and volcanic rocks in the Hachimantai Appi geothermal field, NE Japan—Proposal for selection of promising geothermal areas by alteration survey, dating, and sensing. J Geotherm Res Soc Jap 23(4):251–261 (in Japanese with English abstract)

Tamanyu S (1985) Stratigraphy and geologic structures of the Hohi geothermal area, based mainly on the borehole data. Rept Geol Surv Jap 264:115–142 (in Japanese with English abstract)
Tamanyu S (1993) Geotectonic developing history in the Beppu-Kuju graben, central Kyushu, Japan. Mem Geol Soc Jap 41:93–106 (in Japanese with English abstract)

Chapter 10
Understanding of High-Temperature Zones in Geothermal Reservoirs

Narimitsu Ito and Ko Sato

Abstract In the early stages of geothermal exploration, geochemical surveys are typically conducted prior to drilling, using silica and cation geothermometers as summarized in Table 10.1. During the drilling phase, borehole temperature profiles can be categorized into six types, as illustrated in Fig. 10.5, with particular emphasis on fluid convection patterns inferred from geophysical anomalies associated with subsurface fractures. In addition, fluid inclusion analysis is recommended to improve the estimation of deep reservoir temperatures at this stage.

Keywords Geothermometers · Deep subsurface temperature · Subsurface temperature profiles · Lateral flow · Fluid inclusion

10.1 Assumption of the High-Temperature Zone Before Drilling

10.1.1 Geothermometers and Their Inaccuracy

During the initial phase of geothermal exploration—prior to the drilling of test wells—underground fluids are typically sampled from hot springs. Geochemical analysis of these fluids provides insight into the origin, flow paths, mixing behavior, and temperature of subsurface geothermal fluids. Developing a preliminary geochemical model based on hot spring data is essential to inform subsequent exploration decisions.

Numerous geothermometers have been developed to estimate subsurface temperatures by analyzing the chemical composition of hot spring waters, and they are

N. Ito (✉)
NEWJEC Inc., Osaka, Japan
e-mail: itonr@newjec.co.jp

K. Sato
Socio Engineering Co., Ltd., Tokyo, Japan
e-mail: satoukou6450@gmail.com

J. Kasahara et al. (eds.), *Challenges in Geothermal Development*, Advances in Geological Science, https://doi.org/10.1007/978-981-95-0553-1_10

widely used in geothermal exploration. Table 10.1 summarizes representative examples. Solute geothermometers are among the most commonly applied geochemical methods for estimating reservoir temperatures and can be broadly categorized into two types: those based on solubility equilibria and those based on cation exchange equilibria. The silica geothermometer belongs to the former category, while cation geothermometers fall under the latter.

These methods assume that chemical equilibrium is maintained within the reservoir and during the upward migration of geothermal fluids through hydrothermal convection. Among them, silica geothermometers—based on the solubility of silica minerals—are often favored for their simplicity and practicality. However, it is important to recognize that silica geothermometers offer only limited accuracy. The temperatures they yield should be interpreted as rough estimates due to the multiple assumptions involved. Common silica-based geothermometers include the quartz-no steam loss, quartz-maximum steam loss, and chalcedony models. Each assumes conditions such as sustained chemical equilibrium during fluid ascent and negligible effects from mixing or dilution.

Yoshida (2000) illustrates the inaccuracy of temperature estimates derived from geothermometers, as shown in Fig. 10.1. In this example, hot water A (100 °C) and hot water B (250 °C) are mixed in equal proportions by weight to produce hot water C. The resulting silica concentration and enthalpy of water C are 259 mg/

Table 10.1 Typical silica and cation geothermometers (modified from D'amore and Panichi 1987; Arnórsson et al. 1983). T: absolute temperature (K), t: Celsius temperature (°C), where t = T-273.15; [SiO_2]: total silica concentration (mg/kg); Na, K, Ca: ion concentration (mg/kg)

Geothermometer	Equations	Restrictions
Quartz—no steam loss	$T = \frac{1309}{5.19-\log[SiO_2]}$	t = 0–250 °C
	$T = \frac{1164}{4.90-\log[SiO_2]}$	t = 180–300 °C
Quartz—maximum steam loss	$T = \frac{1522}{5.75-\log[SiO_2]}$	t = 0–250 °C
	$T = \frac{1498}{5.70-\log[SiO_2]}$	t = 180–300 °C
Chalcedony	$T = \frac{1032}{4.69-\log[SiO_2]}$	t = 0–250 °C
	$T = \frac{1112}{4.91-\log[SiO_2]}$	t = 25–180 °C
α—Cristobalite	$T = \frac{1000}{4.78-\log[SiO_2]}$	t = 0–250 °C
β—Cristobalite	$T = \frac{781}{4.51-\log[SiO_2]}$	t = 0–250 °C
Amorphous silica	$T = \frac{731}{4.52-\log[SiO_2]}$	t = 0–250 °C
Na/K	$T = \frac{1217}{\log(Na/K)+1.483}$ (Fournier)	t > 150 °C
	$T = \frac{855.6}{\log(Na/K)+0.8573}$ (Truesdell)	t > 150 °C
	$T = \frac{933}{\log(Na/K)+0.993}$ (Amorsson)	t = 25–250 °C
Na–K-Ca	$T = \frac{1647}{\log(Na/K)+\beta\left\{\log\left(\sqrt{Ca}/Na\right)+2.06\right\}+2.47}$	t = < 100 °C, β = 4/3 t> = 100 °C, β = 1/3

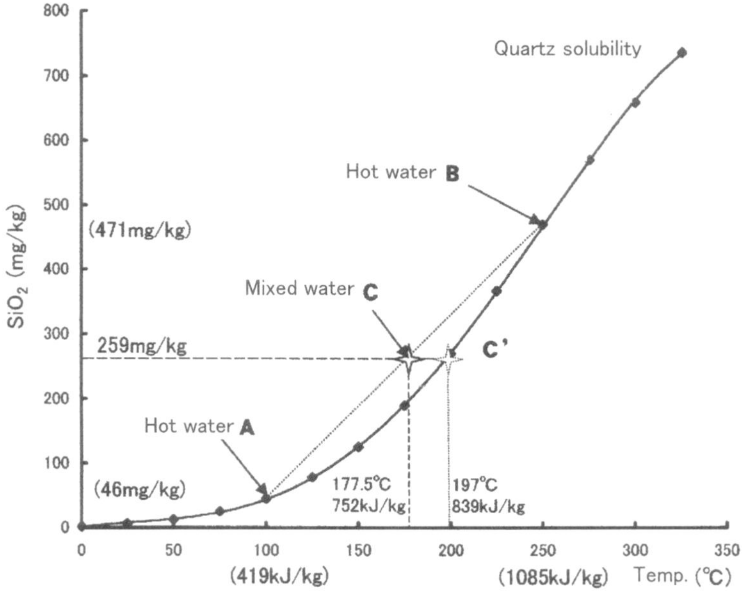

Fig. 10.1 Relationship between SiO_2 concentration and silica geothermometer of mixed hot water (Yoshida 2000)

kg and 752 kJ/kg, respectively, corresponding to a true temperature of 177.5 °C. However, when the temperature is estimated using the silica concentration alone, the result is 197 °C—approximately 20 °C higher than the actual value. This discrepancy demonstrates that mixing of fluids in geothermal systems can introduce significant errors in geothermometric temperature estimations.

Furthermore, Arnórsson (1975) reported a similar mixing effect on silica-based temperature estimations in Icelandic warm spring waters. When high-temperature water with high-chloride concentrations mixed with low-temperature water containing lower chloride levels, the chalcedony geothermometer overestimated the temperature by approximately 40 °C compared to the actual temperature calculated from the mixing ratio. Arnórsson also observed that, although silica-based temperature estimates generally agree with measured temperatures in high-temperature well waters, they tend to overestimate temperatures by 10–40 °C in thermal spring waters ranging from 15 to 100 °C.

Additionally, widely used alkaline geothermometers—such as the Na–K and Na–K–Ca types—are based on empirical statistical relationships, typically derived from least-squares regression between chemical composition and measured well temperatures. Consequently, these geothermometers should not be expected to yield precise temperature values, but rather serve as approximate indicators of subsurface thermal conditions.

10.1.2 Geochemical Model Based on Concentration and Temperature

Despite the known limitations of geothermometers, constructing a reliable geochemical model is essential during the early stages of geothermal exploration, when other types of data are often scarce. To enhance accuracy, geothermometers should not be used in isolation but rather in conjunction with complementary information from multiple sources.

In geothermal fields, deep, high-temperature, high-salinity fluids often mix with shallow, low-temperature, low-salinity waters. This mixing process is the primary factor controlling the chemical composition and temperature of geothermal fluids. Therefore, the first step in constructing a geochemical model is to provide a rational explanation of the mixing mechanism. Key dissolved constituents commonly analyzed in geothermal hot springs include sodium, potassium, calcium, magnesium, chloride, sulfate, and silica. Geochemical investigations are based on the concentrations of these components.

Additionally, the hydrogen and oxygen isotopic compositions of water are essential for elucidating the origin and mixing behavior of geothermal fluids, and they are widely used in hot spring research (Matsubaya et al. 1973). Incorporating isotopic data contributes to a more accurate and comprehensive geochemical model. The temperature of hot spring water is influenced not only by mixing of fluids but also by conductive heat transfer and steam heating. These effects can be estimated through analysis of dissolved component concentrations. Since water density varies with temperature, it is necessary to convert geothermometer-derived temperatures into specific enthalpy values and adjust the concentrations of chemical constituents accordingly to represent reservoir conditions. Therefore, geochemical analyses should be expressed in terms of specific enthalpy (kJ/kg) and solute concentrations (mg/kg).

The relationship between chloride concentration and specific enthalpy in hot spring waters is useful for evaluating the mixing of deep high-temperature fluids with cold groundwater, as well as steam–liquid separation (boiling) processes occurring underground. This method enables estimation of the temperature and chloride concentration of deep geothermal fluids that are otherwise inaccessible during the early stages of exploration (Fig. 10.2). For more accurate interpretation, comprehensive chemical data on both chloride and silica concentrations in hot spring waters are essential. Therefore, regions with a high density of hot springs offer significant advantages for geothermal resource assessment.

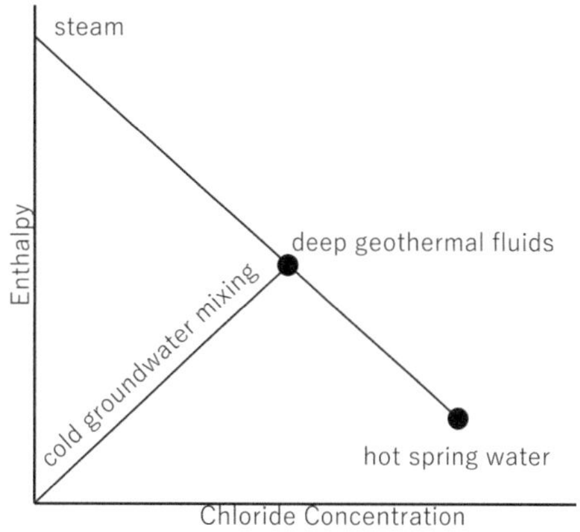

Fig. 10.2 Enthalpy-chloride mixing diagram (modified from Henley 1984)

10.1.3 An Example Model Involving Boiling in the Transfer Process of Geothermal Liquid—the Nigorikawa Geothermal System, Japan

By analyzing the chloride concentration distribution across multiple hot springs, it is possible to estimate not only the subsurface temperature profile using geochemical thermometry, but also the flow paths of geothermal fluids and steam–liquid separation processes occurring within the reservoir, as illustrated in Fig. 10.2. In the Nigorikawa Basin, a distinctive spatial variation in chloride concentration has been observed among the hot springs (Yoshida 1982). Further details on the geological setting and geothermal characteristics of the Nigorikawa system are provided in Chap. 14.2.

Interestingly, both low- and high-chloride hot springs in the Nigorikawa region exhibited similar temperatures based on both measured and silica geothermometer values, although the low-chloride springs showed slightly elevated concentrations of CO_2 and HCO_3^-. To explain these observations, a conceptual model of geothermal fluid movement involving underground steam separation was developed (Fig. 10.3).

Following the commencement of geothermal power plant operations, it was observed that chloride-free thermal water now exists within the system, while deep parent geothermal water continues to discharge at the Nigorikawa hot spring. This phenomenon was interpreted as the result of mixing between native geothermal fluid and reinjected wastewater, influencing both the chloride concentration and enthalpy of the circulating hydrothermal fluid (Fig. 10.4; Yoshida 1991).

Although hot spring surveys initially yielded limited insight during the early stages of exploration, subsequent analysis—incorporating not only geochemical temperature estimates but also the concentrations of conservative elements such as chloride—enabled the inference of subsurface fluid flow. This approach has proven useful for estimating the behavior of high-temperature fluids within geothermal reservoirs.

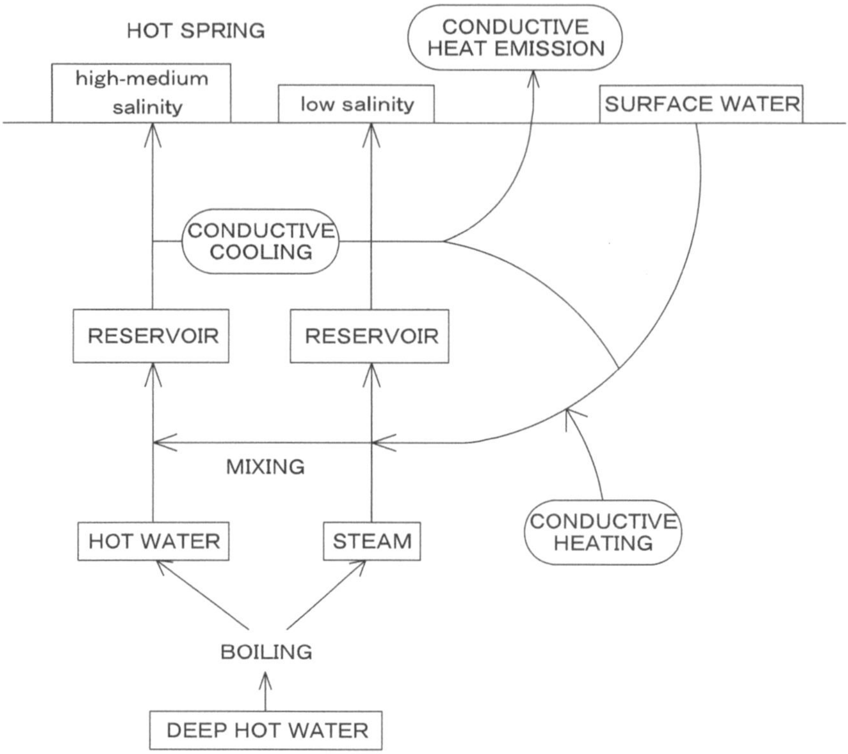

Fig. 10.3 Conceptual hydrothermal system comprising two distinct reservoir types in the Nigorikawa Basin (Yoshida 1982)

Fig. 10.4 Relationship between chloride concentration and specific enthalpy of reservoir fluids in the Nigorikawa geothermal system (Yoshida 1991)

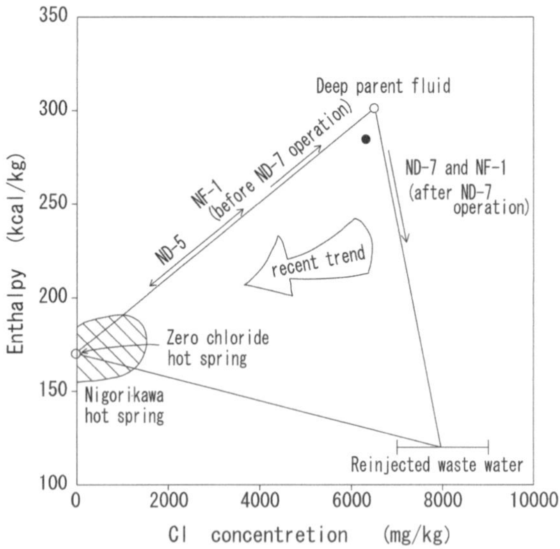

10.2 Estimating Deep Subsurface Temperature Using Shallow Well Data

Estimating the temperature of deep geothermal reservoirs based solely on shallow well data is inherently challenging. In addition to the approaches introduced in the preceding section, several other methods have been proposed to address these limitations:

(1) Temperature Gradient Extrapolation: Extending the temperature gradient below the depth of shallow wells is a straightforward method. However, it often becomes unreliable at greater depths due to discrepancies between measured and extrapolated temperatures, especially when ignoring the effects of convective, conductive, and lateral heat flow patterns.
(2) Geostatistical Analysis: Geostatistical techniques have been applied to actual geothermal fields in Japan. For instance, Tamanyu et al. (1995) used temperature logging via the relaxation method to differentiate between conductive and convective regimes in the Hohi VTD region of Kyushu. Shoji (2004) applied regression models—including power, exponential, and logarithmic functions—to estimate deep temperatures in the Yanaizu-Nishiyama field, finding the logarithmic model to be most accurate based on data from 41 wells. While Kriging is also widely employed, it is inherently limited to interpolation and not suitable for extrapolating beyond measured depths. Despite the utility of these methods, Japan's complex geological structures remain a fundamental constraint on geostatistical applications.
(3) Neighboring Temperature Profile Method: In geothermal fields where lateral fluid flow is dominant, temperature profiles from adjacent wells are often referenced to estimate deep subsurface temperatures. However, for such estimations to be reliable, supporting evidence of comparable geological structures, fracture networks, and fluid flow conditions between the wells is essential.
(4) Activity Index Method: Hayashi (1982, 1999) proposed the activity index as a means of evaluating subsurface thermal conditions based on geothermal well data and surface manifestations. The index is calculated as the ratio of the observed thermal gradient to the theoretical boiling point curve of water. When the maximum temperature closely follows the boiling point curve, the deep subsurface temperature is inferred to be high. In contrast, if the temperature profile follows the conductive gradient, the reservoir temperature is considered low.
(5) Geophysical Anomalies: Anomalies identified through geophysical surveys—such as steep gravity gradients, abrupt changes in electrical resistivity, or seismic reflectivity contrasts—often correspond to convective heat transport and are indicative of fracture networks. In contrast, conduction-dominated systems typically exhibit smooth, continuous temperature increases, reflecting the influence of low-permeability or impermeable formations.

10.3 Various Profiles of Subsurface Temperature

Figure 10.5 presents six schematic subsurface temperature profiles (temperature–depth curves), categorized as A1–A3 and B1–B3, based on thermal behavior and hydrogeological conditions:

A1: The temperature profile closely follows the fluid's boiling point curve, indicating convective heat transfer facilitated by permeable fractures—a characteristic feature of active geothermal systems.
A2: A modified version of A1 is observed when the groundwater level drops. A steam phase and a two-phase fluid mixture (steam and liquid) appear above the lowered water table.
A3: In static conditions with multiple fluid inflow zones, downward flow continues inside the borehole. If upflow occurs between these feed points, the temperature curve becomes convex, suggesting elevated subsurface temperatures.
B1: Groundwater infiltration occurs through shallow fractures or porous strata. Below an impermeable aquiclude, the temperature rises steeply, then transitions into a conductive gradient until reaching a deep, permeable reservoir.
B2: Characterized by lateral movement of high-temperature fluid in the shallow zone, with cooler water percolating downward in the intermediate zone.
B3: Represents an isothermal profile resulting from a pressurized upflow system or a vapor-dominated geothermal reservoir.

Various types of subsurface temperature surveys are commonly conducted in geothermal fields, ranging from shallow measurements using one-meter drill augers to deep investigations with 500 m exploration drilling rigs. To accurately estimate deep subsurface temperatures, a comprehensive analysis of temperature profiles—such as those discussed in Sects. 10.1 and 10.2—is essential.

These studies support the prediction of geothermal reservoir temperature conditions, particularly in areas where the groundwater level is relatively high and the temperature gradient follows the boiling point curve of water (profiles A1, A3, B2, and B3 in Fig. 10.5).

In contrast, when the groundwater level is deeper, estimating deep subsurface temperatures becomes more difficult, as shown in profile A2. Nevertheless, geothermal reservoirs can still develop even when shallow subsurface temperatures are relatively low. This underscores the limitation of relying solely on conductive heat gradients for accurate prediction of temperature variations in the subsurface (profile B1 in Fig. 10.5).

The cross-section shown in Fig. 10.6, which also appears as Fig. 9.4, illustrates shallow lateral fluid flow in the Waita geothermal field, located in the Hohi VTD region. This flow pattern corresponds to type B2 in Fig. 10.5. The diagram highlights fluid migration along the basal conglomerate of unconformable strata. The stratigraphic and structural symbols in Fig. 10.6 are as follows: PA, PB, and PC represent altered volcanic rocks of the Pliocene age (pre-Kusu stage); b through e indicate formations of the Kusu Group (early Pleistocene); a, f, and g correspond to Hohi volcanic rocks (early Pleistocene). Boreholes are labeled DY-1, DW-2, DB-4, and DW-6. Identified faults include the Shibayakata-toge Fault (f1), the Takenoyu Fault (f3), and the Nakadake Fault (f4).

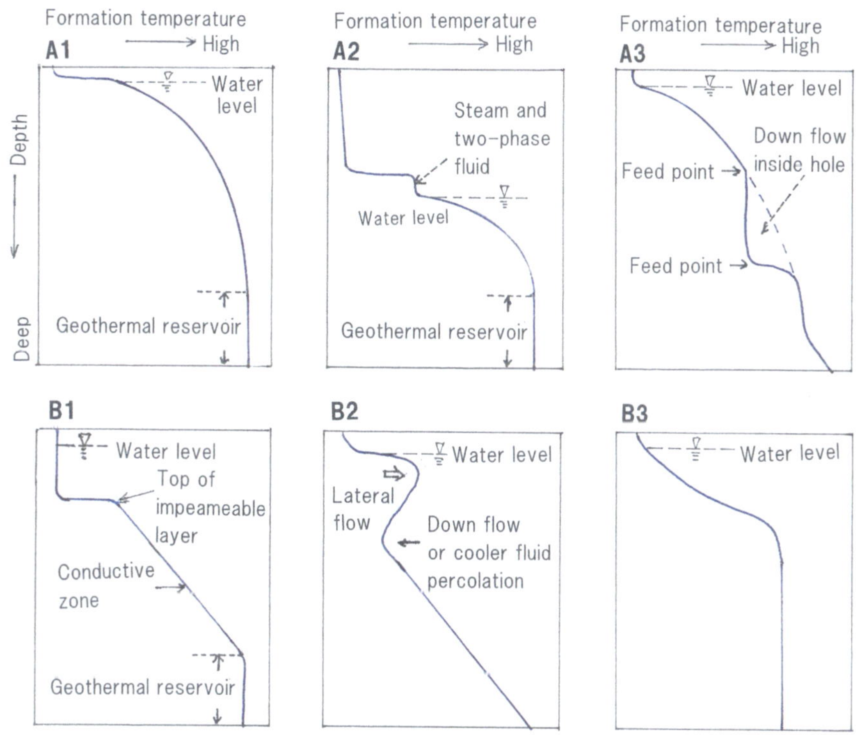

Fig. 10.5 Six schematic profiles illustrating typical subsurface temperature–depth distributions observed in geothermal fields (Sato et al. 2019)

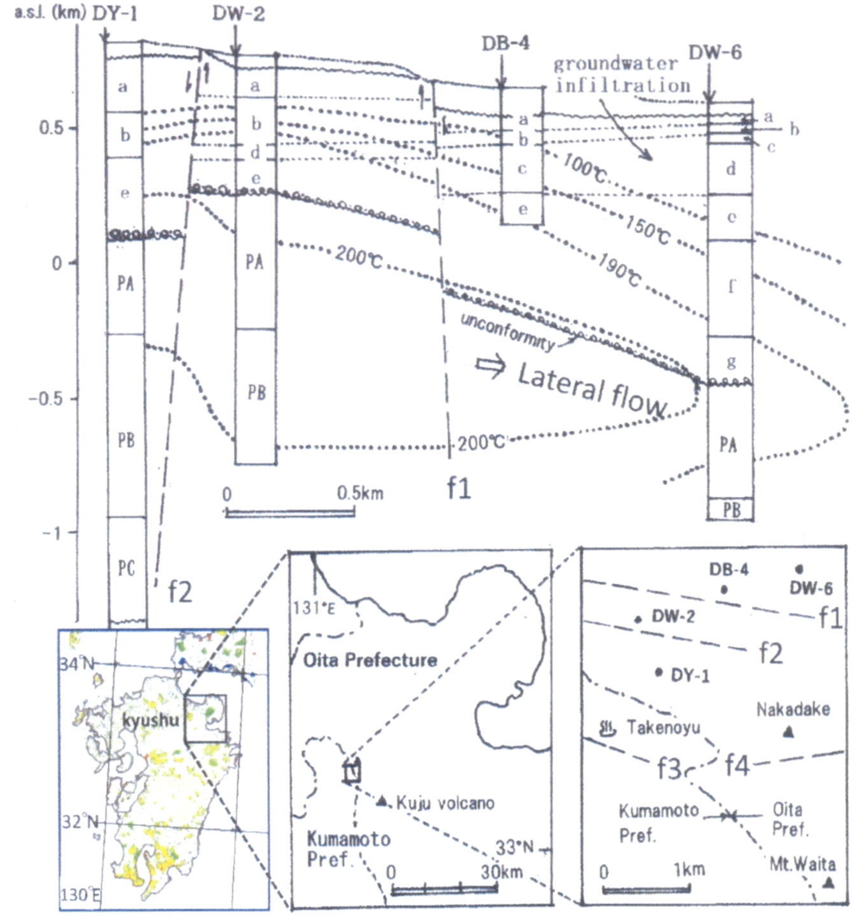

Fig. 10.6 Example of a lateral flow system in the Waita geothermal field, located within the Hohi VTD region, Kyushu. Symbols are explained in the main text (Sato et al. 2022)

10.4 Temperature Survey Inside a Borehole

Temperature measurements conducted within boreholes provide essential data for evaluating geothermal reservoirs. Formation temperature is typically assessed using static downhole measurements after a sufficient standing time—the period during which the well is left undisturbed following drilling operations.

A widely used method for estimating formation temperature is Horner's method, which accounts for transient thermal recovery. When interpreting borehole temperature data, two critical factors must be taken into account:

(1) Severe loss of circulation of drilling mud can result in delayed thermal recovery, as the invading mud disturbs the natural temperature profile.

(2) Incomplete or missing records of mud circulation during drilling may lead to erroneous interpretations of temperature data.

Therefore, it is essential to evaluate two key time intervals recorded during drilling operations: (1) the time elapsed from the end of drilling to the cessation of mud circulation (t), and (2) the time from the cessation of mud circulation to the temperature measurement (θ, or standing time).

The formation temperature at each depth can be extrapolated using Horner's method, applying the ratio $\theta/(t + \theta)$. It should be noted that temperatures observed during steam or hot water production tests reflect flowing fluid conditions, which may differ significantly from static formation temperatures. Zones of anomalous permeability—often indicating feed points such as fractures—can be identified through downhole temperature logging, including during injection tests and PTS logging (Pressure, Temperature, and Spinner), either during or after drilling.

10.5 Fluid Inclusion Survey

Crustal fluids can become trapped within minerals during crystal growth or the healing of microcracks, forming what are known as fluid inclusions. These inclusions are classified into two types: primary and secondary.

Primary inclusions are captured during the initial crystallization of the host mineral. In contrast, secondary inclusions form within healed micro-fractures that develop after the crystallization process, typically due to stress-induced cracking followed by fluid infiltration. Consequently, secondary inclusions are geologically younger than primary ones.

The minimum homogenization temperature of secondary inclusions generally correlates well with the present-day formation temperature (Fig. 10.7; see Roedder 1984). By examining secondary inclusions in minerals retrieved from drill cuttings, it is possible to estimate in-situ formation temperatures without relying on downhole temperature logging. From this perspective, fluid inclusion analysis provides a valuable, real-time method for evaluating reservoir conditions during drilling operations.

Additionally, fluid inclusion analysis provides valuable insights into the thermal evolution of geothermal reservoirs. By comparing the characteristics of primary and secondary inclusions, it is possible to reconstruct past temperature changes and evaluate fluid mixing processes, including contamination by meteoric water or magmatic fluids—reflected in salinity (wt.%) data.

The symbols in Fig. 10.7 are defined as follows:

- Open squares: liquid-rich inclusions (homogenization temperature)
- Black squares: polyphase inclusions (homogenization temperature)
- Open circles (blue): liquid inclusions (salinity)
- Open triangles: polyphase inclusions (salinity)
- Red squares: vapor-rich inclusions (salinity)

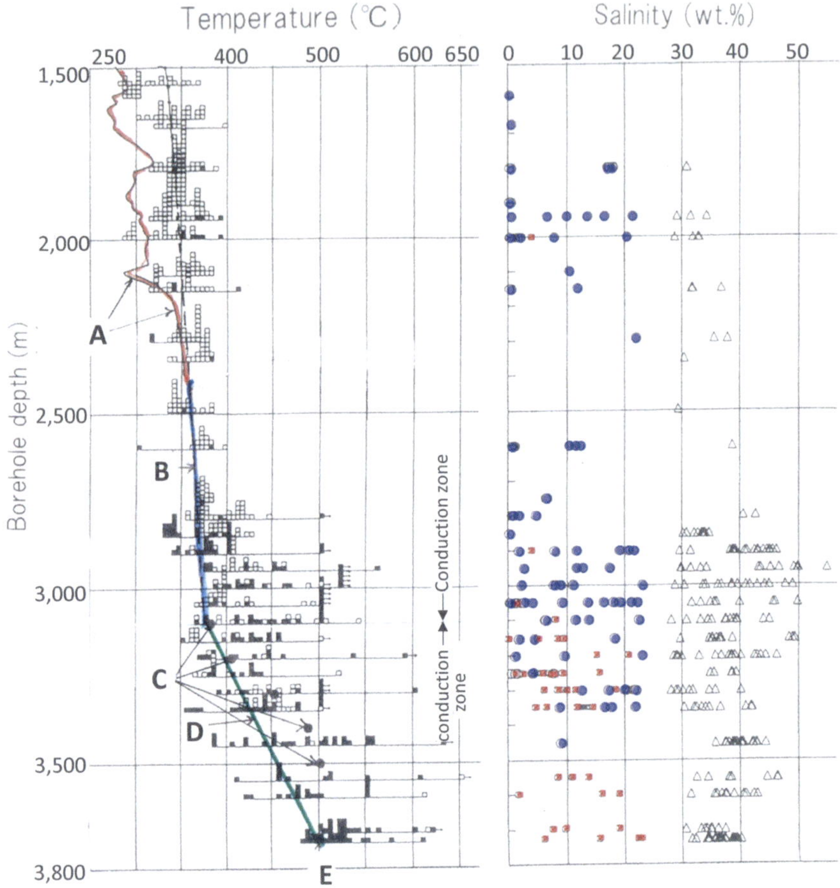

Fig. 10.7 Results of microthermometric analysis of fluid inclusions from the WD-1a well in the Kakkonda geothermal area (adapted from Komatsu et al. 1998). Symbols are interpreted as follows: A (red wavy curve): Formation temperature of the WD-1a well after one year of standing time, B (blue line): Extrapolated temperature profile from Well-18, C (black dot): Formation temperature estimated using Horner's method, D (green line): Estimated temperature at a depth of 3,700 m based on temperature melting tablet data (Ikeuchi et al. 1998), E: The melting tablet melted at 500 °C but remained solid at 510 °C

References

Arnórsson S (1975) Application of the silica geothermometer in low-temperature hydrothermal areas in Iceland. Am J Sci 275:763–784

Arnórsson S, Gunnlaugsson E, Svavarsson H (1983) The chemistry of geothermal waters in Iceland. III. Chemical geothermometry in geothermal investigations. Geochim Cosmochim Act 47:567–577

D'amore F, Panichi C (1987) Geochemistry in geothermal exploration. In: M. Ungemach P (eds) Applied geothermics, economides. John Wiley & Sons, Ltd., Chichester, 69–89, 238

Hayashi M (1982) The target for exploration of deep high-temperature geothermal reservoirs. J Geotherm Res Soc Japan 4(2):81–90 (in Japanese with English abstract)

Hayashi M (1999) Proposed Geothermal Index and its application to geothermal exploration. J Geotherm Res Soc Japan 21(1):29–35 (in Japanese with English abstract)

Henley R (1984) Chemical structure of geothermal systems. In: Robertson (JM) Fluid-mineral equilibria in hydrothermal systems, 9–28

Ikeuchi K, Doi N, Sakagawa Y, Kamenosono H, Uchida T (1988) High temperature measurements in well WD-1A and the thermal structure of the Kakkonda geothermal system Japan. Geothermics 27(5/6):591–608

Komatsu R, Ikeuchi K, Doi N, Sasaki M, Uchida T, Sasada M (1998) Characteristics of the Quaternary Kakkonda Granite and geothermal system clarified by fluid inclusion study of deep investigation well, Kakkonda Japan. J Geotherm Res Soc Japan 20(3):209–224 (in Japanese with English abstract)

Matsubaya O, Sakai H, Kusachi I, Satake H (1973) Hydrogen and oxygen isotope ratios and major element chemistry of Japanese thermal water systems. Geochem J 7:123–151

Roedder E (1984) Fluid inclusions, Review in mineralogy, vol 12. Mineralogical Soc. America, Washington D. C, USA

Sato K, Ito N, Tsukuda K (2019) Geothermal geological survey and production well target (Original title in Japanese). In: E-pix, Ofunato Printing Inc., Iwate, 155, ISBN978-4-901602-67-9 (in Japanese)

Sato K, Ito N, Tsukuda K (2022) Geothermal geological survey and production well target: Supplementary edition focusing on reinjection wells. In: Iwate: E-pix, Ofunato Printing Inc., Iwate, 125. ISBN978-4-901602-74-7 (in Japanese)

Shoji T (2004) Statistical estimation of deep level temperatures based on shallow level temperature profiles. J Geotherm Res Soc Japan 26(1):11–24 (in Japanese with English abstract)

Tamanyu S, Masao Y, Nomura K (1995) Deep subsurface temperature distribution pattern estimated from many temperature logging data: example of Hohi geothermal area, Kyushu Japan. Bull Geol Surv Japan 46(7):313–331 (in Japanese with English abstract)

Yoshida Y (1982) Study on the hydrothermal system in the Nigorikawa basin, Hokkaido. Onsenkagaku 33:24–36 (in Japanese with English abstract)

Yoshida Y (1991) Geochemistry of the Nigorikawa geothermal system, southwest Hokkaido Japan. Geochem J 25:203–222

Yoshida Y (2000) Sampling and chemical analysis in geothermal Development and utilization. J Jpn Geotherm Energy Assoc 37:28–42 (in Japanese with English abstract)

Chapter 11
Understanding of High Permeable Zones in Geothermal Reservoirs

Ko Sato and Narimitsu Ito

Abstract Fracture permeability, governed by the Cubic Law, plays a key role in geothermal energy production by enabling high steam output. The permeability-thickness product (*kh*), defined as the permeability multiplied by the thickness of a permeable zone, is widely used to estimate well productivity through borehole testing. Field observations suggest that a *kh* value of at least 10^{-12} m^3 is required to sustain stable steam production as illustrated in Figs. 11.5 and 11.7 generating approximately 1 MWe of electricity. Such permeability is primarily associated with open fractures, typically observed in: (1) normal faults, (2) strike-slip and pull-apart fault systems, and (3) fractures caused by volcanic processes.

Keywords Permeability-thickness · Cubic law · Permeable fracture · Steam flow rate · Magma forceful intrusion

11.1 Permeability and Permeability-Thickness (*Kh*)

11.1.1 Permeability Related to Porous Media

Permeability (k) quantifies the ease with which fluids flow through porous media and is classically defined by Darcy's law:

$$k = \frac{\mu QL}{\Delta PA}$$

K. Sato (✉)
Socio Engineering Co., Ltd., Tokyo, Japan
e-mail: satoukou6450@gmail.com

N. Ito
NEWJEC Inc., Osaka, Japan
e-mail: itonr@newjec.co.jp

J. Kasahara et al. (eds.), *Challenges in Geothermal Development*, Advances in Geological Science, https://doi.org/10.1007/978-981-95-0553-1_11

where μ is viscosity of fluid flow [Pa · s], Q is total flow rate [m^3/s], L is the flow path length [m], ΔP is the pressure drop [Pa], A is the cross-sectional area [m^2].

In the CGS system, this yields:

$$k = \frac{\mu QL}{\Delta PA} = \frac{1[\mathrm{cp}] \times 1[\mathrm{cm}^3/\mathrm{s}] \times 1[\mathrm{cm}]}{1[\mathrm{atm}] \times 1[\mathrm{cm}^2]} = 1[\mathrm{darcy}]$$

where [cp] is centipoise in viscosity unit, 1cp $=10^{-3}$Pa · s $=10^{-3}$kg /m · s.

Conversion into SI unit is as follows:

$$k = \frac{10^{-3}[\mathrm{Pa}\cdot \mathrm{S}] \times 10^{-6}[\mathrm{m}^3/\mathrm{s}] \times 10^{-2}[\mathrm{m}]}{10^{-5}[\mathrm{Pa}] \times 10^{-4}[\mathrm{m}^2]} = 10^{-12}\left[\mathrm{m}^2\right]$$

Thus, in practice, 10^{-12} m^2 is commonly referred to as 1 darcy.

Permeability is often expressed differently across engineering disciplines. For instance, civil engineering commonly uses the coefficient of permeability in cm/s or m/s, which relates to permeability through the kinematic viscosity of water. Table 11.1 summarizes typical conversion values between these units.

Laboratory measurements reveal a general trend of decreasing permeability in drill cores with increasing depth, as illustrated in Fig. 11.1. Although the original data were classified into 27 lithological units, they have been grouped into four representative rock facies for clarity. Near-surface core samples typically exhibit permeabilities on the order of 1 darcy or less, indicating that matrix permeability contributes little

Table 11.1 Approximate conversion between permeability units used in geothermal and civil engineering applications

Permeability			Permeability coefficient	
cm^2	m^2	darcy	cm/sec	m/sec
10^{-3}	10^{-7}	10^{5}	10^{2}	1
10^{-4}	10^{-8}	10^{4}	10	10^{-1}
10^{-5}	10^{-9}	10^{3}	1	10^{-2}
10^{-6}	10^{-10}	10^{2}	10^{-1}	10^{-3}
10^{-7}	10^{-11}	10	10^{-2}	10^{-4}
10^{-8}	10^{-12}	1	10^{-3}	10^{-5}
10^{-9}	10^{-13}	10^{-1}	10^{-4}	10^{-6}
10^{-10}	10^{-14}	10^{-2}	10^{-5}	10^{-7}
10^{-11}	10^{-15}	10^{-3}	10^{-6}	10^{-8}
10^{-12}	10^{-16}	10^{-4}	10^{-7}	10^{-9}
10^{-13}	10^{-17}	10^{-5}	10^{-8}	10^{-10}
10^{-14}	10^{-18}	10^{-6}	10^{-9}	10^{-11}
10^{-15}	10^{-19}	10^{-7}	10^{-10}	10^{-12}

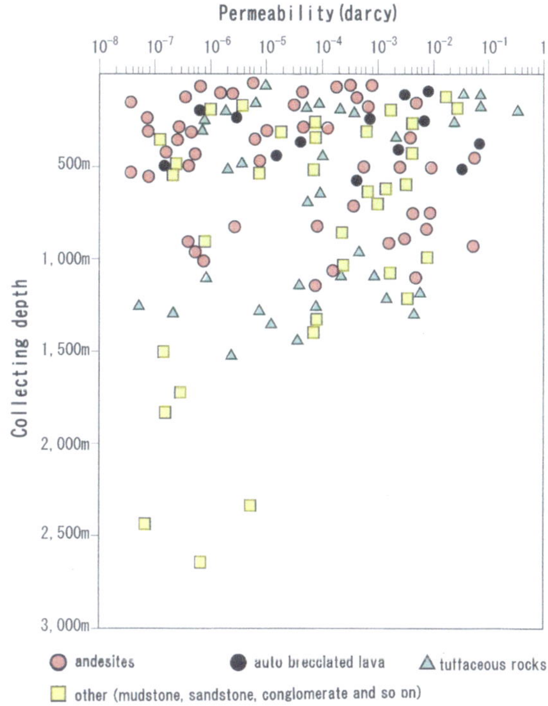

Fig. 11.1 Permeability variation of drill cores with depth (Sato et al. 2019)

to steam productivity. Instead, steam production is primarily governed by fracture permeability, which is discussed in detail in Sect. 11.1.2.

Drill core measurements from the geothermal reservoir reveal the relationship between permeability and porosity in the pre-Kusu altered volcanic rocks of late Pliocene age in the Hohi VTD region, Kyushu, Japan (Fig. 11.2). In this figure, Group A demonstrates a positive correlation between permeability and porosity, indicating that larger pore spaces facilitate fluid infiltration. In contrast, Group B shows minimal variation in permeability despite changes in porosity, likely due to differences in grain sorting within the rock. In both groups, measured permeability values do not exceed 1 darcy (10^{-12} m^2).

11.1.2 Cubic Law Related to Fracture Permeability

The Cubic Law describes fluid flow through narrow planar fractures, stating that the volumetric flow rate is proportional to the cube of the fracture aperture. This relationship is particularly important in geothermal systems, where fracture permeability dominates fluid movement. The key parameters used in the Cubic Law and as illustrated in Fig. 11.3 are defined below:

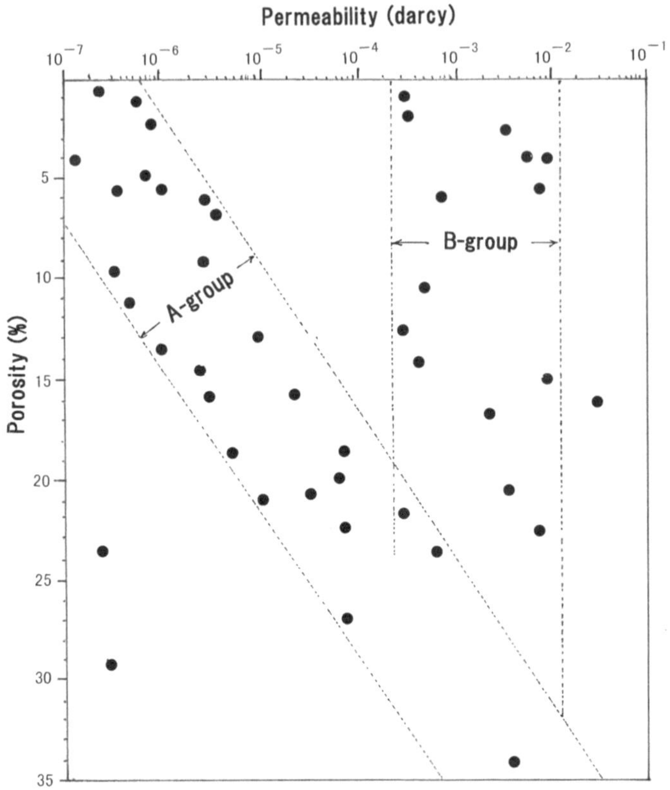

Fig. 11.2 Relationship between permeability and porosity in drill cores of the pre-Kusu altered volcanic rocks (Sato et al. 2022)

Fig. 11.3 Conceptual illustration of fluid flow through a narrow fracture plane (fissure or crack). Axes X, Y, and Z represent the spatial coordinates. Parameters C (length in Y-direction), d (fracture aperture), and L (length in X-direction) are explained in the main text

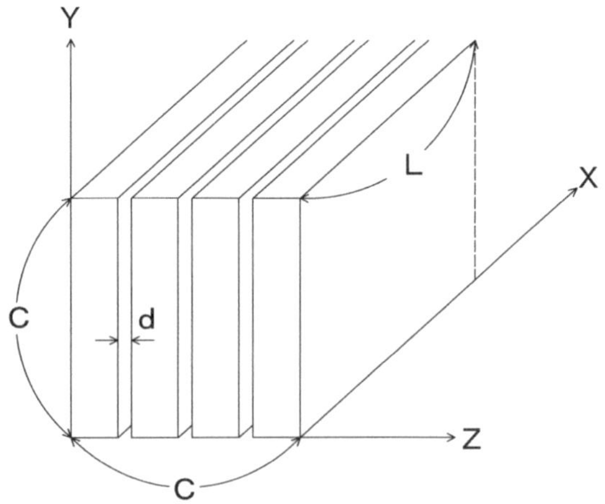

C	length along the Y-direction
d	fracture aperture width
k	permeability
L	length in the X-direction
n	number of fractures
P	pressure
Q	total flow rate
q	flow rate per fracture
V	velocity of fluid flow
LC^2	total volume of the layer
$ndLC$	total volume of porosity
φ	porosity
μ	viscosity of the fluid flow

The following equation describes the system:

$$\frac{\partial P}{\partial x} = \mu \frac{\partial^2 V}{\partial x} \tag{11.1}$$

Twice integrating the Eq. (11.1) with respect to z, we get:

$$V = \frac{d}{2\mu} z^2 \frac{\partial P}{\partial x} + AZ + B \tag{11.2}$$

Under the boundary conditions $z = 0$, $z = d$ and $V = 0$, we find that $B = 0$ and $\frac{1}{2\mu} d^2 \frac{\partial P}{\partial x} + Ad = 0$,

$$A = -\frac{d}{2\mu} \cdot \frac{\partial P}{\partial x} \tag{11.3}$$

The velocity equation becomes:

$$V = \frac{1}{2\mu} Z^2 \frac{\partial P}{\partial x} - \frac{d}{2\mu} \frac{\partial P}{\partial x} Z = -\frac{1}{2\mu} (d - Z) Z \frac{\partial P}{\partial x} \tag{11.4}$$

To find the flow rate per fracture (q), we integrate V from $z = 0$ to $z = d$:

$$q = \int_0^d V dz = -\frac{1}{2\mu} \frac{\partial P}{\partial x} \int_0^d (d - z) Z dz = -\frac{1}{2\mu} \frac{\partial P}{\partial x} \cdot \frac{d^3}{6} = -\frac{d^3}{12\mu} \cdot \frac{\partial P}{\partial x} \tag{11.5}$$

This demonstrates that the flow rate per fracture is proportional to the cube of the fracture aperture width.

The total flow rate (Q) is defined as:

$$Q = nqC = -\frac{\varphi d^2}{12\mu} C^2 \frac{\partial P}{\partial x} \tag{11.6}$$

This can be compared with Darcy's law, rewritten as:

$$Q = -\frac{kC^2}{\mu}\frac{\partial P}{\partial x} \tag{11.7}$$

Compared with Eqs. (11.6) and (11.7), the bulk permeability according to Darcy's law can be expressed as:

$$k = \frac{d^2}{12} \tag{11.8}$$

where d is the fracture aperture. Accordingly, the fracture permeability validated by the actual fluid velocity, defined as $V = U/\varphi$ (where U is the Darcy velocity), becomes consistent with the Cubic Law formulation.

For example, using Eq. (11.7), a fracture aperture of $d = 1$ mm (10^{-3} m) yields a permeability of $k = \frac{(10^{-3})^2}{12} = 8 \times 10^{-8}$ (m^2), which is sufficiently high for sustainable steam production in geothermal reservoirs. However, this estimate assumes an idealized fracture with a smooth, continuous surface. In practice, fracture surfaces are often rough and uneven, with flow controlled by the narrowest constrictions within the aperture (see Fig. 11.4). Moreover, geothermal fluids commonly precipitate or crystallize vein minerals within these fractures. While such mineralization can reduce permeability, it may also produce drusy cavities that preserve pore space. Therefore, maintaining effective steam production typically requires a wider and more open fracture aperture.

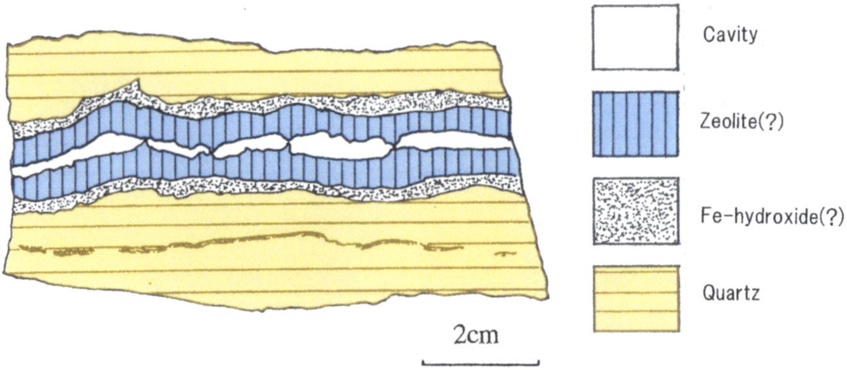

Fig. 11.4 Sketch of a hydrothermal mineral vein observed in an outcrop from the Kakkonda geothermal area, illustrating irregular fracture surfaces and vein mineral precipitation. (adapted from Sawaki et al. 1999)

11.1.3 Threshold Values of Permeability-Thickness (kh)

The permeability k of geothermal reservoir rock can be determined through laboratory analysis on wellbore cores. However, in fracture-dominated geothermal reservoirs, direct evaluation of k is difficult, as the permeability cannot be easily separated from the cumulative thickness of permeable zones (h). Therefore, permeability-thickness (kh) or transmissivity (kh/μ, where μ is fluid viscosity) is more commonly used and is typically estimated through well tests—such as build-up, fall-off, production, and injection tests. The value of kh provides a practical and integrated measure of reservoir productivity, representing the combined effect of permeability and the thickness of contributing strata.

Fractures play two essential roles in geothermal development (Hanano 2000). First, they initiate and sustain natural convection within geothermal systems. Second, they act as conduits that connect wells to the reservoir, enabling fluid extraction. In a typical convective system at reservoir temperatures of 200–250 °C, the upward velocity of geothermal fluid driven by natural convection is approximately 10^{-9} m/s, corresponding to a bulk permeability on the order of 10^{-15} to 10^{-16} m^2. The first role is particularly important for maintaining the convective flow, which is generally supported by a network of fine fractures.

An example calculation using Eq. (11.7) shows that a permeability of $k = 10^{-15}$ m^2 corresponds to a fracture aperture of $d = 1.1 \times 10^{-7}$ m (0.1 μm), a typical value observed in geothermal systems. Conversely, build-up and fall-off tests conducted in single-phase, liquid-dominated geothermal reservoirs typically yield kh values in the range of 10^{-12} to 10^{-10} m^3. If the fracture thickness is assumed to be $h = 0.011$ m, the estimated permeability falls within $k = 10^{-12}$ to 10^{-8} m^2. These results suggest that only fractures with relatively high permeability make a substantial contribution to geothermal fluid production—corresponding to the second functional role of fractures.

To clarify the minimum kh value for economically viable steam production, Fig. 11.5 presents the relationship between kh (darcy · m) and steam flow rate, based on data from the Kirishima geothermal area in southern Kyushu. If the target steam flow rate is approximately 10 t/h ($\approx$1 MWe), the data suggest that kh must be at least 1 darcy·m ($\approx 10^{-12}$ m^3). This threshold is further corroborated by the results shown in Fig. 11.7. For a higher target of 35 t/h ($\approx$3.5 MWe), a kh value of at least 20 darcy·m is required. However, this criterion can vary depending on several factors, including formation temperature, fluid pressure, depth, and aperture of feed zones, and flushing conditions during drilling.

In the Kirishima geothermal field, formation temperatures range from 214 °C to 263 °C, with an average of 231 °C. The data cluster located in the lower left portion of Fig. 11.5, characterized by both low kh and low steam flow rates, is partly attributed to small borehole diameters and fluid flushing within formations of low permeability. In many cases, increasing the borehole diameter and enhancing permeability via hydraulic fracturing has successfully shifted these low-performing wells into the higher kh and higher steam flow rate region of the graph.

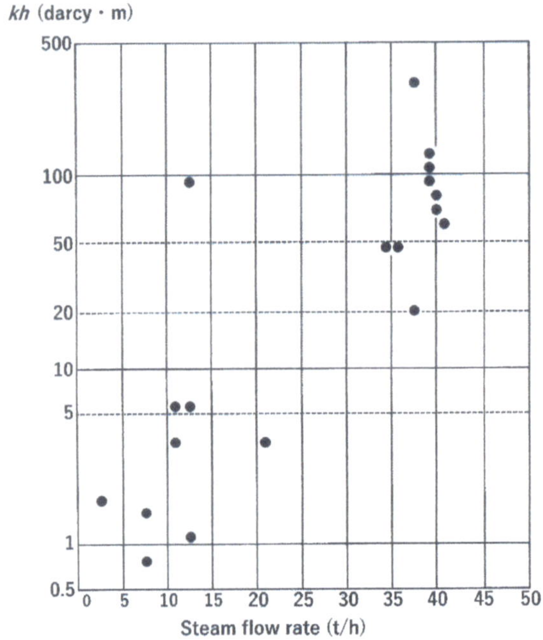

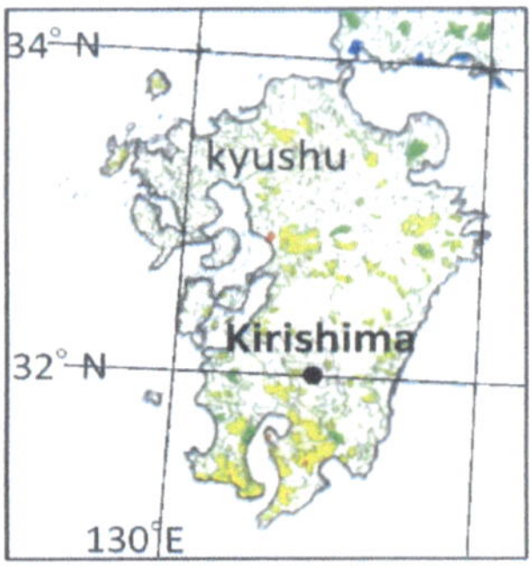

Fig. 11.5 Relationship between steam flow rate and permeability-thickness (*kh*) in the Kirishima geothermal area, southern Kyushu (modified from Kodama 1999)

11.1.4 kh Values from Borehole Data in Japan

From the 444 boreholes drilled during NEDO's nationwide Geothermal Development Promotion Survey (NEDO, 1983–2010), this study independently selected 57 wells across 23 geothermal areas based on the following three criteria:

(1) Feed points located at depths deeper than 500 m.
(2) Formation temperatures at the feed points exceeding 150 °C.
(3) Permeability-thickness (*kh*) values greater than 10^{-13} m^3.

Figure 11.6 shows the locations of 23 geothermal areas across Japan. Table 11.2 summarizes data from 57 wells independently selected in this study, clarifying the relationships among permeability-thickness (*kh*), geological conditions, and production test results. The column descriptions "a" through "g" are as follows:

a. Geothermal area and well name
b. Feed point (m): Depth ranges are shown when multiple or unclear feed points are present, or when feed zones are identified from production tests.
c. L/C (L/min): Abbreviation of Lost Circulation. Fluid loss rate during drilling, measured in liters per minute.
d. Temp. (°C): Formation Temperature at feed point

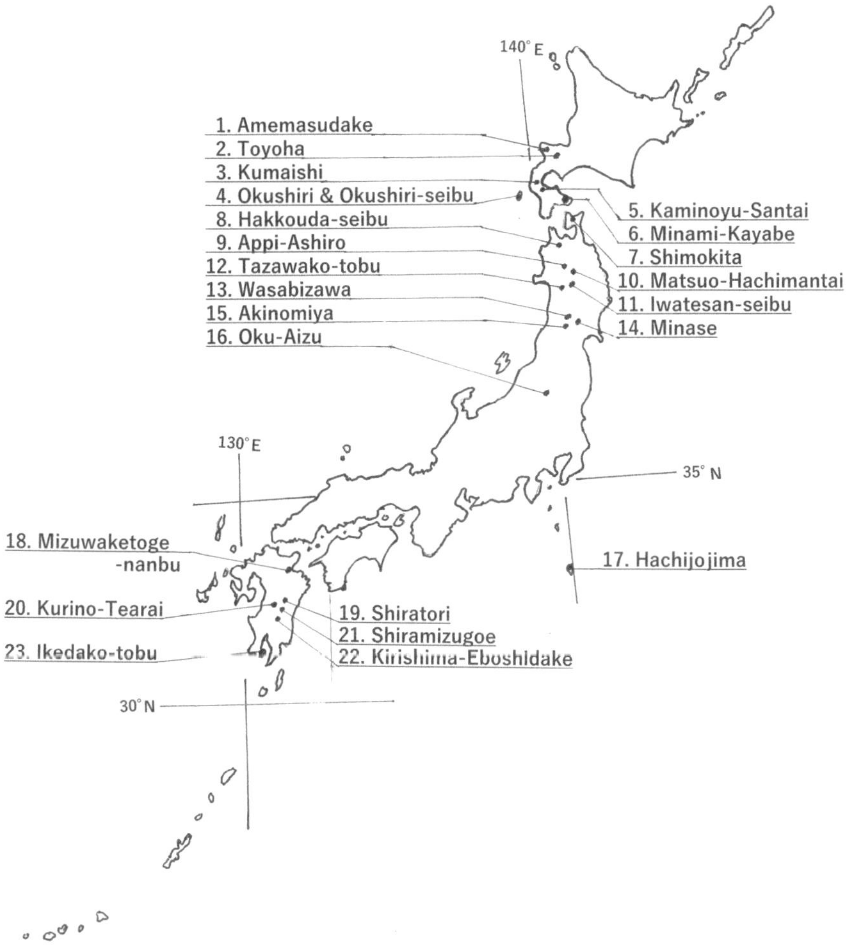

Fig. 11.6 Locations of 23 geothermal areas in Japan analyzed for permeability-thickness (*kh*), geology features, production performance, and drilling records

e. *kh* (m^3): Permeability-thickness (*kh*); Permeability-thickness estimated from well test data. Where no thickness is defined, 1 m is assumed.
f. Feed point geology: Geological features of of feed points. A = Distinctive altered minerals, B = Lithological boundary, D = Druse, F = Fault and fracture zone, V = vein.
g. Hole dia.: the final outer diameter of the borehole (e.g., NQ ≈ 75 mm, HQ ≈ 98 mm; tricone bits shown in inches) P0: Wellhead pressure from production test (MPaG). Gs(t/h): Steam flow rate during production tests. Gw (t/h): Hot water flow rate during production test.

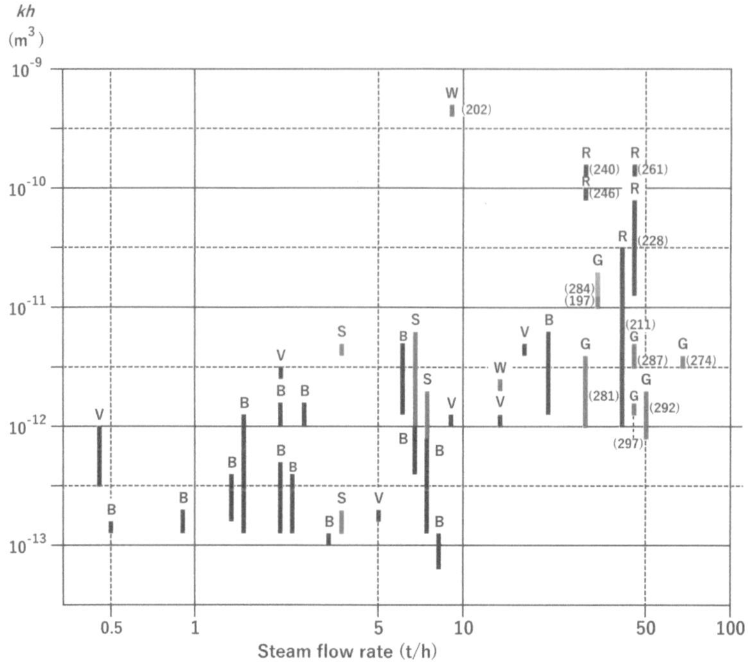

Fig. 11.7 Relationship between permeability-thickness (*kh*) and steam flow rate (Gst). Bars are color-coded by region: green and others for northern and northeastern Japan, red for southwestern Japan. Borehole diameter categories are denoted as follows: G, R: 8–1/2 inch, S, W: 5–7/8 and 6–1/4 inch, B, V: NQ and HQ core diameters (approx. 75 and 98 mm). Parentheses indicate average formation temperature (°C) for each well

We interpret the results presented in Table 11.2 as follows. At first glance, the permeability-thickness (*kh*) values tend to be higher in southwestern Japan compared to northern and northeastern Japan. The reasons for this contrast are discussed in detail in Sect. 12.2, but a brief summary is provided below:

(1) The main geothermal reservoirs in southwestern Japan are composed of relatively young formations, ranging from the Pliocene to the Middle Pleistocene. In contrast, those in northern and northeastern Japan are formed within older rocks, such as pre-Tertiary basement, Miocene, and early Pleistocene strata. These lithological differences influence rock mechanical properties, such as strength, hardness, and toughness.
(2) Since approximately 3.0 Ma, northeastern Japan has been affected by east–west compressional stress caused by the subduction of the Pacific Plate. Conversely, southwestern Japan has experienced transtensional tectonics (pull-apart structures) due to the oblique subduction of the Philippine Sea Plate.
(3) Volcanic activity has been more intense and continuous in southwestern Japan since about 6.0 Ma, enhancing geothermal potential through sustained magmatic heat input and fracturing.

Table 11.2 Relation between *kh* and geology, production data, and drilling records (continued on the next page). See text for full explanation of items a–g

a		b	c	d	e	f	g
Area, Well name		Feed point	L/C	Temp	*kh*	Feed point geology	Hole dia. P0(MPaG)
		m	L/min	°C	m^3		Gs(t/h), Gw(t/h)
Northern Japan, Hokkaido							
1 Amemasudake area							
(1)	N4-AM-4	1,040 ~ 1,050	60	172 ~ 198	$1.2 \sim 2.4 \times 10^{-13}$	(B) granodiorite/hornfels	NQ, P0 = 0.1
		1,130 ~ 1,149				(F) inside granodiorite(V) (D) quartz	Gs = 0.91
2 Toyoha area							
(2)	N61-TH-6	561, 585 ~ 632	40 → 120	193 ~ 206	$1.33 \sim 7.1 \times 10^{-12}$	(F) fracture in tuff breccia, brecciated	HQ, P0 = 0.12
		551 ~ 714, 830				andesite, (V)(D) quartz, calcite, anhydrite	Gs = 6.5
3 Kumaishi area							
(3)	N7-KI-1	1,405 ~ 1,539	5 → 94	157	$1,1 \sim 3.5 \times 10^{-13}$	(B) granite/basalt	NQ
						(V) calcite, dolomite, reticular clay	

(continued)

Table 11.2 (continued)

a		b	c	d	e	f	g
Area, Well name		Feed point	L/C	Temp	*kh*	Feed point geology	Hole dia. P0(MPaG)
		m	L/min	°C	m^3		Gs(t/h), Gw(t/h)
(4)	N8-KI-3	1,388 ~ 1,614	5 → 60	167	1.5 ~ 5.1 × 10^{-13}	(B) sandstone/ mudstone, andesite dyke	HQ
						(V) calcite	
4 Okushiri and Okushiri-seibu areas							
(5)	N4-OS-5	1,013 ~ 1,048	26 → 80	183	0.55 ~ 1.40 × 10^{-12}	(B) quartz porphyry/ granodiorite (F)	NQ, P0 = 0.062
		1,470				fracture (V) calcite, dolomite, anhydrite	Gs = 2.6
(6)	N19-OW-3	1,403 ~ 1479	690 → 800	175	1.05 ~ 1.90 × 10^{-12}	(F) fracture inside garnodiorite	6–1/4”, P0 = 0.54
							Gs = 7.02
5 Kaminoyu-Santai area							
(7)	N3-KS-4	1,280 ~ 1,292	16 → 40	165 ~ 180	2.2 ~ 7.0 × 10^{-13}	(B) tuff/quartz porphyry (A) laumontite	NQ

(continued)

Table 11.2 (continued)

a		b	c	d	e	f	g
Area, Well name		Feed point	L/C	Temp	*kh*	Feed point geology	Hole dia. P0(MPaG)
		m	L/min	°C	m^3		Gs(t/h), Gw(t/h)
						(F) vertical fracture(V) calcite	
(8)	N3-KS-6	1,509 ~ 1,800	94 → 168	185 ~ 195	$1.36 \sim 6.34 \times 10^{-13}$	(F) fracture inside limestone	HQ, P0 = 0.4 ~ 0.5
						(V) calcite, anhydrite	Gs = 1.3 ~ 2.7
6 Minami-Kayabe area							
(9)	N59-MK-2	768 ~ 1,084	10 → 85	177	1.48×10^{-13}	(F) fracture inside basalt	NQ, P0 = 0.03
						(V) quartz, calcite	Gs = 0.5
(10)	N61-MK-6	960 ~ 1,028, 1.332	35 → 110	163 ~ 180	$1.74 \sim 5.13 \times 10^{-13}$	(F) fracture inside quartz porphyry	NQ, P0 = 0.065
						(V)(D) quartz, quartzite	Gs = 1.48
Northeastern Japan, Honshu							
7 Shimokita area							

(continued)

Table 11.2 (continued)

a		b	c	d	e	f	g
Area, Well name		Feed point	L/C	Temp	*kh*	Feed point geology	Hole dia. P0(MPaG)
		m	L/min	°C	m^3		Gs(t/h), Gw(t/h)
(11)	N59-SK-6	1,686	40	232	10^{-13}	(D) druse inside quartz porphyry	NQ, P0 = 0.04
						(V) calcite, anhydrite	Gs = 3.4
8 Hakkoda-seibu area							
(12)	N3-HD-7	1,247 ~ 1,288	6 → 70	185 ~ 192	1.21 ~ 10.4 × 10^{-13}	(B) tuff breccia/dolerite (F) brecciated	NQ, P0 = 0.04
		1,457 ~ 1,480				andesite (V) quartz, calcite	Gs = 1.5
9 Appi-Ashiro area							
(13)	N7-AR-1	1,743	221	289	1.2 ~ 3.0 × 10^{-13}	(B) tuff/mudstone/andesite	6–1/4”, P0 = 0.11
						(A) actinolite	Gs = 4.3
(14)	N12-AP-1	1,025, 1,130	49 → 1,700	180 ~ 260	1.6 ~ 1.8 × 10^{-13}	(F) farcture inside dacite, fault	8–1/2”

(continued)

Table 11.2 (continued)

a		b	c	d	e	f	g
Area, Well name		Feed point	L/C	Temp	*kh*	Feed point geology	Hole dia. P0(MPaG)
		m	L/min	°C	m^3		Gs(t/h), Gw(t/h)
		1,471, 1,950				(B) mudstone/dacite	
(15)	N13-AP-2	1,701 ~ 1,887	617 → 1,533	287 ~ 294	$4.3 \sim 7.2 \times 10^{-12}$	(B) mudstone/andesite	8–1/2”, P0 = 2.2
						(A) actinolite	Gs = 44
(16)	N14-AP-4	1,618 ~ 1,660	250 → 1,231	297	$1 \sim 2.16 \times 10^{-12}$	(B) mudstone/dacite (F) fault	8–1/2”, P0 = 1.1
						(A) actinolite	Gs = 44
(17)	N15-AP-6	1,431	897 → 1,025	249 ~ 280	$0.73 \sim 1.87 \times 10^{-13}$	(B) mudstone/dacite	6–1/4”
		1,501 ~ 1,606				(V) quartz, calcite	
10 Matsuo-Hachimantai area							
(18)	N19-HA-1	1,377	130 → 820	232	6.05×10^{-12}	(B) tuff/andesite	6–1/4”, P0 = 0.04
		1,610				(D) tuff (V) calcite, crystal, anhydrite	Gs = 3.5

(continued)

Table 11.2 (continued)

a		b	c	d	e	f	g
Area, Well name		Feed point	L/C	Temp	*kh*	Feed point geology	Hole dia. P0(MPaG)
		m	L/min	°C	m^3		Gs(t/h), Gw(t/h)
(19)	N19-HA-2	1,150, 1,370	20 → 24	242	1.08×10^{-13}	(B) tuff/hornfels	6–1/4”
		1,455				(D) crystal (V) crystal, anhydrite	
11 Iwatesan-seibu area							
(20)	N1-IT-2	884	10 → 70	200	1.07×10^{-13}	(B) upper part and inside andesite dyke	NQ
		912				(D) (V) quartz, calcite	
12 Tazawako-tobu area							
(21)	N1-TZ-4	924 ~ 978	608	281	$1.56 \sim 3.61 \times 10^{-12}$	(B) tuff/inside dacite dyke	5–5/8”
						(V) calcite (A) wairakite	
13 Wasabizawa area							

(continued)

Table 11.2 (continued)

a		b	c	d	e	f	g
Area, Well name		Feed point	L/C	Temp	*kh*	Feed point geology	Hole dia. P0(MPaG)
		m	L/min	°C	m^3		Gs(t/h), Gw(t/h)
(22)	N7-WZ-5	1,010, 1,110	100	200	$1.03 \sim 1.63 \times 10^{-12}$	(B) dacite/ mudstone/ sandstone (F) fault	HQ, P0 = 0.1
		1,220, 1,250				(V) anhydrite (A) wairakite	Gs = 2.2
(23)	N7-WZ-6	1,203 ~ 1,261	27 → 130	263	$0.64 \sim 1.08 \times 10^{-12}$	(F) fault inside granite	HQ, P0 = 0.21
						(V) quartz, calcite (A) wairakite	Gs = 6.9
(24)	N7-WZ-7	1,165 ~ 1,210	500 → 1,538	281	$1.04 \sim 5.33 \times 10^{-12}$	(F) fault zone, inside granite, porphyrite	8–1/2”, P0 = 0.33
		1,361				(A) Laumontite	Gs = 21, Gw = 70
(25)	N8-WZ-9	1,450, 1,494	33 → 1,538	274	$5.02 \sim 6.0 \times 10^{-12}$	(F) fault zone, inside granite	8–1/2”, P0 = 0.84
		1,508 ~ 1,520				(A) Laumontite	Gs = 70, Gw = 115

(continued)

Table 11.2 (continued)

a		b	c	d	e	f	g
Area, Well name		Feed point	L/C	Temp	*kh*	Feed point geology	Hole dia. P0(MPaG)
		m	L/min	°C	m^3		Gs(t/h), Gw(t/h)
(26)	N57-YO-3	1,098 ~ 1,203	60 → 90	282	$0.13 \sim 1.36 \times 10^{-12}$	(B) granite/ metamorphiic rock	NQ, P0 = 0.275
						(F) fault zone (D) (V) opal	Gs = 9.3
14 Minase area							
(27)	N16-MS-1	917	34 → 950	179 ~ 197	1.16×10^{-11}	(B) andesite/ dacite, tuff/ dacite	5–7/8", P0 = 0.513
		1,410 ~ 1,487				(D) quartz, calcite (A) wairakite	Gs = 31, Gw = 90
(28)	N17-MS-2	939 ~ 1,487	950	160	$2.38 \sim 7.75 \times 10^{-12}$	(B) lapilli tuff/ mudstone/ tuff	5–7/8", P0 = 0.104
						(D) (V) calcite, laumontite	Gs = 6.5
15 Akinomiya area							
(29)	N8-AY-1	1,308	100	288	$1.3 \sim 7.4 \times 10^{-12}$	(B) mylonite/ andesite dyke	HQ, P0 = 1.1

(continued)

Table 11.2 (continued)

a		b	c	d	e	f	g
Area, Well name		Feed point	L/C	Temp	*kh*	Feed point geology	Hole dia. P0(MPaG)
		m	L/min	°C	m^3		Gs(t/h), Gw(t/h)
							Gs = 13.8
(30)	N8-AY-2	1,041 ~ 1,044	15 → 18	239	$1.3 \sim 5.4 \times 10^{-13}$	(F) fault inside granite (B) andesite	HQ, P0 = 0.32
		1,566				dykes (V) quartz, calcite	Gs = 2.3
(31)	N9-AY-3	1,530 ~ 1,680	32 → 1,533	292	$0.87 \sim 3.8 \times 10^{-12}$	(F) fault zone inside granite	8–1/2", P0 = 1.50
							Gs = 65, Gw = 90
(32)	N9-AY-4	1195 ~ 1.505	1,200	284	$1.2 \sim 3.0 \times 10^{-11}$	(F) fracture zone (B) granite/tuff	5–7/8", P0 = 3.14
						(A) wairakite	Gs = 32.3, Gw = 91
(33)	N10-AY-6	1,065	58 → 72	212	$4.8 \sim 7.2 \times 10^{-13}$	(B) granite/tuff, andesite dykes	8–1/2"
		1,477 ~ 1,697				(A) laumontite	

(continued)

Table 11.2 (continued)

a		b	c	d	e	f	g
Area, Well name		Feed point	L/C	Temp	*kh*	Feed point geology	Hole dia. P0(MPaG)
		m	L/min	°C	m^3		Gs(t/h), Gw(t/h)
(34)	N10-AY-8	1,036, 1,143	47 → 433	256	$0.09 \sim 1.6 \times 10^{-12}$	(B) inside granite, andesite dykes	6–1/4", P0 = 0.2 ~ 0.3
		1,244				(A) wairakite	Gs = 7.3, Gw = 8.8
16 Oku-Aizu area							
(35)	N57-OA-4	1,037 ~ 1,280	5 → 580	234 ~ 251	$0.72 \sim 1.33 \times 10^{-13}$	(F) fracture zone (B) dacite/ lapilli tuff	NQ, P0 = 0.3
						(D) quartz, calcite (V) pyrite, sphalerite	Gs = 6.5, Gw = 5.2
(36)	N58-OA-7	706 ~ 795	15 → 60	194 ~ 200	$1.28 \sim 3.58 \times 10^{-12}$	(B) tuff/mudstone (F) fracture zone	NQ
		878 ~ 938				(D) calcite, anhydrite (V) siderite	
Central Japan, Izu-Bonin volcanic zone							

(continued)

Table 11.2 (continued)

a		b	c	d	e	f	g
Area, Well name		Feed point	L/C	Temp	*kh*	Feed point geology	Hole dia. P0(MPaG)
		m	L/min	°C	m^3		Gs(t/h), Gw(t/h)
17 Hachijojima							
(37)	N2-HJ-4	1,050 ~ 1,092	40 → 160	154 ~ 160	$0.5 \sim 2.6 \times 10^{-13}$	(F) fracture zone of andesite, lapilli tuff	NQ
		1,108 ~ 1,500				(V) (D) quartz, zeolite	
(38)	N3-HJ-8	584, 890	40 → 120	241 ~ 288	$0.7 \sim 4{,}2 \times 10^{-13}$	(F) open fracture in andesitic tuff breccia	HQ
		1,176 ~ 1,200				(D) (V) quartz, anhydrite, zeolite	
Southwestern Japan, Kyushu							
18 Mizuwake-toge-nanbu area							
(39)	N2-MW-2	624, 635	60 → 70	198	$4.14 \sim 4.49 \times 10^{-12}$	(B) andesite/tuff breccia	NQ, P0 = 0.095
						(D) (V) calcite	Gs = 2.4
(40)	N3-MW-6	1,102, 1,120	30 → 70	199	$4.71 \sim 9.70 \times 10^{-13}$	(B) tuff breccia/andesite	NQ, P0 = 0.02

(continued)

Table 11.2 (continued)

a		b	c	d	e	f	g
Area, Well name		Feed point	L/C	Temp	*kh*	Feed point geology	Hole dia. P0(MPaG)
		m	L/min	°C	m^3		Gs(t/h), Gw(t/h)
		1,145				(D) (V) calcite (A) wairakite	Gs = 0.4
19 Shiratori area							
(41)	N4-ST-1	1,740	180	195 ~ 203	$1.66 \sim 2.40 \times 10^{-10}$	(F) fracture in lapilli tuff	7–5/8"
						(D) (A) wairakite	
(42)	N7-ST-6	1,041 ~ 1,300	370 ~ 900	196 ~ 207	6.45×10^{-10}	(F) fault zone, fracture in andesite	5–5/8", P0 = 0.025
						(V) anhydrite (A) wairakite	Gs = 8.6, Gw = 40
20 Kurino-Tearai area							
(43)	N55-KT-5	728, 840	50 → 120	280	2.66×10^{-13}	(F) fracture in autobrecciated andesite	HQ, P0 = 0.03 ~ 0.6
		920, 1,026				(D) (A) quartz, wairakite, prehnite	Gs = 3.92 ~ 6.19

(continued)

Table 11.2 (continued)

a		b	c	d	e	f	g
Area, Well name		Feed point	L/C	Temp	*kh*	Feed point geology	Hole dia. P0(MPaG)
		m	L/min	°C	m^3		Gs(t/h), Gw(t/h)
(44)	N56-KT-7	835, 1,006	50 → 200	200	2.0×10^{-12}	(F) fault breccia, autobrecciated andesite	HQ
		1,281 ~ 1,427				(D) quartz, calcite (A) mordenite	
(45)	N56-KT-8	379 ~ 934	185 → 600	280	6.3×10^{-12}	(B) andesite/tuff breccia (F) brecciated	HQ, P0 = 0.6
		1,045, 1,311				andesite (D) quartz, anhydrite	Gs = 18, Gw = 30
21 Shiramizugoe area							
(46)	N12-SZ-1	1,040 ~ 1,060	150 → 220	232 ~ 287	1.2×10^{-12}	(B) andesites (F) fault	HQ, P0 = 0.05 ~ 0.64
		1,085 ~ 1,500				(A) wairakite	Gs = 11.9 ~ 12.9
(47)	N12-SZ-2	1,325 ~ 1,487	1,000 1,800	261	1.67×10^{-10}	(B) andesites (F) fault	8–1/2", P0 = 1.5 ~ 2.4
		1,755 ~ 1,867					Gs = 39.1, Gw = 181

(continued)

Table 11.2 (continued)

a		b	c	d	e	f	g
Area, Well name		Feed point	L/C	Temp	*kh*	Feed point geology	Hole dia. P0(MPaG)
		m	L/min	°C	m^3		Gs(t/h), Gw(t/h)
(48)	N12-SZ-3	1,430, 1,490	1,000 2,000	242	3.87×10^{-11}	(B) andesites (F) fault	8–1/2”
		1,529					
(49)	N13-SZ-4	1,485, 1,635	1,250 1,500	242 ~ 256	1.682×10^{-10}	(B) andesites (F) fault	8–1/2”, P0 = 0.90 ~ 1.36
		1,751 ~ 1,776					Gs = 47, Gw = 130
(50)	N13-SZ-5	1,040, 1,470	2,300	235 ~ 258	9.7×10^{-11}	(B) andesites (F) fault	8–1/2”, P0 = 0.80 ~ 1.42
						(A) wairakite	Gs = 30, Gw = 165
22 Kirishima-Eboshidake area							
(51)	N15-EB-3	930 ~ 940	250 → 1,500	193 ~ 229	$1.17 \sim 9.78 \times 10^{-12}$	(B) andesites (V) calcite (A)(D) siderite	8–1/2”, P0 = 0.63
		1,453 ~ 1,478				corrensite	Gs = 38.8, Gw = 115

(continued)

Table 11.2 (continued)

a		b	c	d	e	f	g
Area, Well name		Feed point	L/C	Temp	*kh*	Feed point geology	Hole dia. P0(MPaG)
		m	L/min	°C	m^3		Gs(t/h), Gw(t/h)
(52)	N16-EB-4	774, 959. 973	300 → 2,500	216 ~ 221	1.15 ~ 7.89 × 10^{-12}	(B) andesites (V) calcite (A)quartzyte	8–1/2"
		1,430					
(53)	N16-EB-5	926, 941 ~ 950	400 → 2,000	228	1.88 ~ 8.68 × 10^{-11}	(B) andesites (F) fault	8–1/2", P0 = 0.79
		1,272 ~ 1,289					Gs = 43.4, Gw = 135
23 Ikedako-tobu area							
(54)	N19-IK-1	1,223	1,000	230	3.9 × 10^{-12}	(B) tuff breccia (D) anhydrite (A) wairakite	6–1/4", P0 = 0.2
							Gs = 12.7, Gw = 83
(55)	N20-IK-2	1,243	75 → 700	180 ~ 195	1.3 ~ 1.6 × 10^{-12}	(B) tuff breccia/ lapilli tuff (F) fracture zone	6–1/4"
						(D) anhydrite (A) alunite	

(continued)

Table 11.2 (continued)

a		b	c	d	e	f	g
Area, Well name		Feed point	L/C	Temp	*kh*	Feed point geology	Hole dia. P0(MPaG)
		m	L/min	°C	m^3		Gs(t/h), Gw(t/h)
(56)	N20-IK-3	1,233	870 → 1,110	180 ~ 205	1.9×10^{-12}	(F) fracture zone, inside rhyolite	6–1/4"
						(D) anhydrite (A) alunite	
(57)	N21-IK-4	1,114	150 → 1,000	203	$2.8 \sim 3.4 \times 10^{-12}$	(B) tuff breccia/ lapilli tuff (F) fracture zone	6–1/4"
		1,199 ~ 1,590				(D) anhydrite (A) wairakite	

Subsequently, Fig. 11.7 presents the relationship between permeability-thickness (*kh*) and steam flow rate (Gst) on a semi-logarithmic scale, showing an overall trend of proportionality. However, some wells in northern and northeastern Japan, such as N7-ST-6 with a large borehole diameter (8–1/2″), achieve high Gst values despite having relatively low *kh*. Conversely, wells in southwestern Japan typically exhibit high Gst only when *kh* is also high.

To facilitate regional comparison, the average formation temperatures (FT) are noted in parentheses within Fig. 11.7. In northern and northeastern Japan, FT ranges from 274 °C to 297 °C, except for one lower value (197 °C). In southwestern Japan, FT ranges from 202 to 261 °C. These observations suggest that in southwestern Japan, targeting high-temperature zones is critical to achieving high steam output, while in northern and northeastern Japan, targeting highly permeable zones is more effective.

Geological analysis of feed points indicates that most productive intervals are associated with fault and fracture zones (F) or stratigraphic boundaries (B), often accompanied by features, such as druse (D), mineral veins (V), and distinctive hydrothermal alteration (A).

Moreover, Fig. 11.7 supports the previously identified criterion: a *kh* value of at least 10^{-12} m^3 ($\approx$1 darcy · m) is generally required to achieve approximately 10 t/h of steam flow ($\approx$1 MWe), consistent with the result shown in Fig. 11.5.

11.1.5 Actual Open Fractures (Fissures and Cracks)

Actual open fractures—such as fissures and cracks—play a decisive role in steam production within geothermal systems. To understand what constitutes an "actual open fracture," we must integrate insights from multiple approaches: (1) field observations of geothermal surface manifestation, (2) analysis of drill cores, (3) thermodynamic calculations of mineral solubility, and (4) laboratory experiments on rock and mineral dissolution.

The most direct example is provided by the Oyasu Gorge (locally known as the "Oyasu Great Fountain") in Akita Prefecture, northeastern Japan, where abundant thermal fluid discharge is observed at the surface (see Fig. 11.8, photo and sketch). The outcrop location is shown in Fig. 14.39 (labeled "Of"). Steam and hot water issue from natural fissures within the thick-bedded mudstone of the Sanzugawa Formation (Pliocene age), where the aperture width reaches approximately 10 cm, and the spacing between fissures ranges from 20 to 50 cm.

This formation, consisting of mudstone, tuff, and tuff breccia, has a total thickness of about 120 cm, strikes N25°W, and dips 12°W in the area. The observed fluid flushing is attributed to vapor pressure within the reservoir exceeding groundwater pressure, which forces thermal fluids to discharge through fractures—mimicking the mechanism of fluid expulsion from a production well.

Fig. 11.8 Photograph and interpretive sketch of thermal fluid emission from open fractures at Oyasu Big Fountain, Akita Prefecture, northeastern Japan. Thick-bedded mudstone of the Sanzugawa Formation (Pliocene) hosts the fissures. (photo and sketch by the author, October 18, 2017)

The second approach involves detailed observation of drill cores. Figure 11.9 presents an example of a core retrieved from beneath a zone of total lost circulation. The core, composed of Miocene-age tuff breccia, exhibits several druses and mineral veins. Prominent pore spaces, approximately 4 cm in length and 1 cm in width, are supported by quartz and idiomorphic crystals. These druses likely formed as geothermal fluids infiltrated along pre-existing open fractures, where minerals precipitated or crystallized from the fluid and were later partially dissolved.

The third approach involves calculating quartz solubility based on deep drilling well data (Saishu et al. 2016). The key findings are as follows:

(1) Quartz mineralization is driven by pressure changes in ascending deep-crustal fluids, transitioning from lithostatic to hydrostatic conditions.
(2) Descending fluids undergo phase transitions—from liquid to vapor or supercritical states—causing a sharp reduction in quartz solubility, which promotes quartz precipitation.
(3) The mixing of CO_2 or NaCl-rich deep fluids with meteoric water locally alters quartz solubility, further influencing mineral precipitation patterns.

Fig. 11.9 Druse (black) and mineral veins (white) in a drill core composed of Miocene tuff breccia. The druse cavities are supported by quartz and idiomorphic crystals, while the veins consist of quartz and anhydrite. (photo by the author, September 3, 2019)

The final approach involves laboratory experiments on the dissolution behavior of rocks and minerals. While mineral precipitation from geothermal fluids can reduce fracture permeability by clogging pore spaces, it may also play a role in sustaining voids, such as druses. This duality—between pore occlusion and structural support—is addressed through dissolution experiments on granitic rocks (Suto et al. 2002; Wang et al. 2003), which yielded the following findings:

(1) Dissolution rates of granitic rocks increase with both rising temperature and fluid velocity.
(2) At lower temperatures (~20 °C), feldspar governs the dissolution process, whereas quartz dissolution dominates under high-temperature subcritical conditions (≥300 °C).
(3) Precipitation and dissolution reactions are strongly influenced by salinity, aluminum concentration, and CO_2 content in the fluid.
(4) Quartz dissolution increases with temperature and pressure up to the critical point (~330 °C along the saturated vapor pressure curve). Beyond this point, dissolution rates decrease with temperature but remain pressure-dependent.

In conclusion, the development and maintenance of high-permeability fractures require the following conditions. As geothermal fluids continuously flow through fractures, newly formed minerals may gradually dissolve, leaving behind undissolved components that structurally support the pore spaces. Ohtani et al. (1997) proposed that geothermal reservoirs are composed of fracture networks interconnected with miarolitic cavities—initially formed voids—based on three-dimensional X-ray CT imaging of granitic cores from the Kakkonda geothermal area. Persistent regional tectonic stress is believed to preserve these cavity structures.

11.2 Permeable Fractures in Geological Structures

The development and persistence of actual open fractures are controlled by sustained regional tectonic stress fields and the continuous inflow of geothermal fluids. Such permeable fractures are typically associated with specific structural features, including normal faults, pull-apart structures linked to strike-slip fault systems, reverse faults, fractures induced by volcanic activity, and permeable zones formed along folds and joint sets.

11.2.1 Examples of Normal Faults Decided by Geological Structure

Normal faults generated under extensional stress fields promote the development of open fractures. The associated permeability is supported by features, such as fault breccia, druse, dissolved veins, and slicken-step displacements. Cross-sectional views from the Otake-Hatchobaru geothermal areas illustrate conjugate normal faults dipping at approximately 70°, becoming gentler at greater depths (Tsutsumi et al. 2021; Fig. 6.6).

A comprehensive review of 250 geothermal fields in the Great Basin, western USA (Faulds and Hinz 2015), identified favorable structural configurations within normal fault systems, including breached step-overs or relay ramps, horsetail splays at fault terminations, and intersections of normal faults or normal faults with oblique-slip faults. These findings underscore the importance of fault-fracture assemblages as promising targets for production well placement.

11.2.2 Pull-Apart Structure Related to Strike-Slip Fault

Strike-slip faults (also known as wrench faults) generally have a near-vertical dip, as the intermediate principal stress axis (σ_2) is oriented vertically. These faults often exhibit dilational jogs or spreading centers known as pull-apart structures. Such features enhance fracture apertures and serve as key permeable pathways extending downward into the deeper subsurface due to their vertical geometry. These pathways frequently act as upflow zones for geothermal fluid circulation.

Three schematic models of strike-slip fault systems are illustrated in Fig. 11.10, with brief descriptions below:

a. A single strike-slip fault develops a pull-apart structure at bends within mechanically heterogeneous rocks due to variations in internal friction angles.
b. A right-lateral (dextral) fault system forms a pull-apart basin between *en échelon* strike-slip segments.
c. A negative flower structure, depicted in three dimensions, is created by a strike-slip duplex where steeply dipping faults extend downward into a graben-like zone under extensional stress conditions.

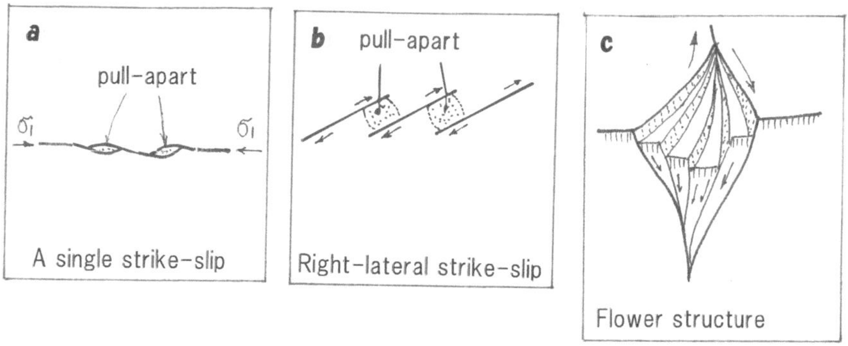

Fig. 11.10 Representative schematic illustrations of strike-slip fault systems with associated pull-apart structures. Panels a through c correspond to different geological configurations, as described in the main text

Pull-apart structures associated with strike-slip faults have been documented in major geothermal regions worldwide. For example, Lomnitz et al. (1970), Elders et al. (1972), and Sibson (1987) describe a spreading center extending from the Salton Sea in the USA through the Cerro Prieto geothermal field to the Wagner Basin in Mexico, composed of the dextral San Andreas, Imperial, and San Jacinto faults. A well-developed pull-apart basin is also present in the Hohi volcano-tectonic depression (VTD) in central Kyushu, Japan. Further details of the VTD are discussed in Sect. 12.2.2 (Fig. 12.7).

11.2.3 Reverse Fault

Reverse faults are generally formed under compressional stress fields, which tend to close fracture apertures and inhibit permeability. However, several exceptional geological conditions allow high permeability to persist even along reverse fault planes:

(1) Inversion tectonics: In northern and northeastern Japan, intense compressional stresses from the Pacific Plate can cause normal faults to be reactivated as reverse faults. These reactivated faults may retain druse, mineral veins, and other open spaces inherited from their original listric normal fault structures.
(2) Erosional unloading and uplift: The reduction of overburden pressure due to erosion and uplift in shallow subsurface layers can lead to the formation of low-angle reverse faults. In such settings, vertical stress (σ_3) becomes the least principal stress, enabling localized extensional conditions and potential fracture opening.
(3) Igneous intrusion: As described in Sect. 11.2.4, magma intrusion can mechanically fracture surrounding rocks. Reverse faults may develop just outside these fractured zones due to lateral compressive stresses caused by the forceful emplacement of magma (see Fig. 11.11b, d).

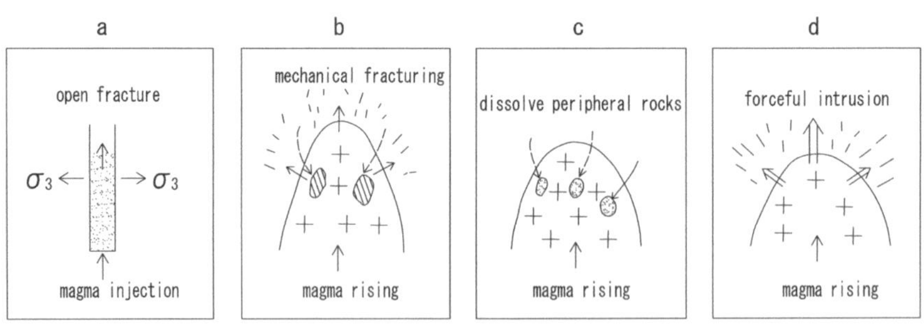

Fig. 11.11 Conceptual diagrams illustrating four mechanisms of igneous intrusive complexes associated with volcanic activity. Panels “a” through “d” are described in the main text. (Sato et al. 2019)

11.2.4 Fractures Associated with Volcanic Activity

The development of fractures in association with igneous intrusions can be categorized into four principal mechanisms based on the style of volcanic activity (Fig. 11.11):

a. Dike and sill intrusion: Magma intrudes pre-existing rock fissures under extensional stress regimes, forming dikes and sills. These intrusive bodies are commonly observed in geothermal fields throughout Japan.
b. Stoping: Magma mechanically fractures surrounding host rocks and ascends by engulfing xenoliths. This process resembles hydraulic fracturing and is facilitated by the fluid nature of the magma.
c. Assimilation: During upward migration, magma may dissolve fragments of the surrounding country rock, contributing to the evolution of its composition and permeability characteristics.
d. Forceful intrusion: The buoyant force of rising magma forcibly uplifts and displaces overlying strata, deforming the surrounding rock mass.

Forceful intrusion leads to contrasting effects on the permeability of surrounding rocks. Permeability tends to increase under the following conditions:

(1) Pore spaces develop between the surrounding rocks and the solidified magma.
(2) Cooling joints are generated as magma solidifies.
(3) A large temperature gradient between magma and surrounding rocks induces fissures due to thermal stress.
(4) Mineral dissolution around the intrusion promotes the development of drusy cavities and mineral veins.

In contrast, permeability is likely to decrease under the following circumstances:

(1) Pore spaces are compacted or sealed by overburden pressure or roof collapse.
(2) Mineral precipitation or crystallization occludes cooling joints.
(3) High temperatures result in fewer or poorly developed fractures due to thermal metamorphism or ductile deformation.
(4) Drusy cavities and veins become filled by mineral precipitates.

11.2.5 Permeable Portion with Folding and Joint

Structural complexity introduced by folding enhances permeability through the generation of joints, pore spaces, and fissures, as illustrated in Fig. 11.12. In flexural slip folding (buckling; Fig. 11.12a), open joints and conjugate minor shear faults typically develop near fold hinges, such as crests and troughs. Similarly, in flexural flow folding (Fig. 11.12b), pore spaces also tend to accumulate in these zones. The interplay between folding and permeable zones was a critical factor in the early stages of geothermal development at the Kakkonda field (Sato 1982). A notable example is the A-2 well, which yielded a substantial steam discharge of 125 t/h at 0.97 MPa G, drilled into the trough of a fold.

In addition, the uplifted margins of rock blocks caused by folding (as illustrated in Fig. 11.12c) can also form permeable zones. Beyond folding-related permeability, boundaries between lithologically distinct rock units can exhibit high permeability due to differences in rock strength—more precisely, fracture toughness—which results in mechanical weaknesses favorable for tectonic deformation and fluid migration. These boundaries are susceptible to subcritical crack propagation mechanisms, such as stress corrosion and fatigue cracking. Other joint systems, including columnar joints, cooling joints, and weathering joints, also contribute significantly to permeability in geological settings.

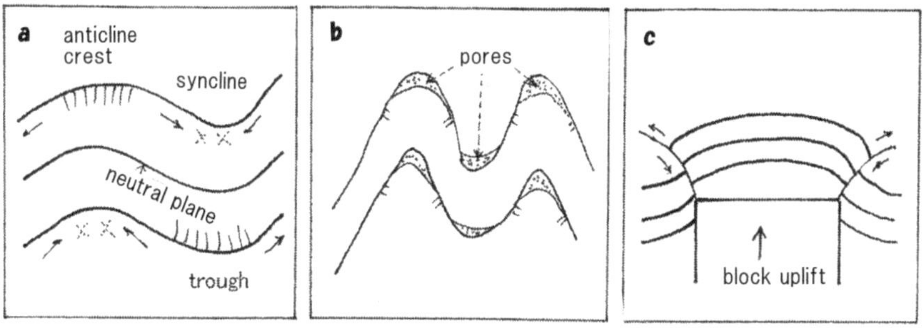

Fig. 11.12 Representative examples of permeable zones associated with folding. Panels "a" through "c" are explained in the main text (Sato et.al. 2019)

11.2.6 Genesis of Shallow Fractures

Large-scale lost circulation and blowout-type fluid discharges frequently occur in the shallow zones of geothermal reservoirs in Japan, often resulting in significant drilling and borehole damage. The following six geological scenarios are considered likely causes for the development of shallow fractures:

(1) Interfaces between shallow, low-angle, or horizontally bedded strata.
(2) Boundaries between multiple Quaternary lava flow units.
(3) Fracture zones overlain by alluvial deposits or impermeable debris layers.
(4) Horizontally developed joints in near-surface formations (e.g., joints in welded tuff or lava).
(5) Fractures, including low-angle reverse faults, are present in shallow subsurface zones.
(6) Exposure of originally deep-seated fractures near the surface due to erosion or uplift, reducing overburden pressure.

References

Elders WA, Rex RW, Meidav T, Robinson PT, Biehler S (1972) Crustal spreading in Southern California. Science 178(4056):15–24

Faulds JE, Hinz NH (2015) Favorable tectonic and structural settings of geothermal systems in the great basin region, western USA: proxies for discovering blind geothermal systems. In: Proceedings of the world geothermal congress 2015, Melbourne, Australia, 19–25 April 2015, 1–6

Hanano M (2000) Two roles of fractures in geothermal development: contribution to occurrence of geothermal resources and well production. Jour Geotherm Res Soc Japan 27(2):107–116 (in Japanese with English abstract)

Kamata H (1997) Geology of the Miyanoharu district. With Geological Sheet Map at 1:50,000. Geol Surv Japan 127 (in Japanese with English abstract 5p)

Kodama M (1999) Geothermal development in the Kirishima geothermal field, Japan -Development, present state and future prospect. Jour Geotherm Res Soc Japan 21(3):277–289 (in Japanese with English abstract)

Lomnitz C, Mooser F, Allen C, Brune JN, Thatcher W (1970) Sismicidad y Tectonica de La Region Norte del Golfo de California, Mexico Resultados Preliminaries. Geofisica Internacional, Revista Union Geofisica Mexicana, UNAM 10(2):37–48 (in Spanish with English)

NEDO (New Energy Industrial Technology Development Organization) (1983–2010) Geothermal Development Promotion Survey. (in Japanese) Kurino-Tearai area, 1983, Report No. 3, 621p. Yuzawa-Ogachi area, 1985a, Report No. 7, 814p. OkuAizu area, 1985b, Report No. 8, 811p. Shimokita area, 1986, Report No. 9, 688p. Toyoha area, 1988a, Report No. 12, 1156p. Minami kayabe area, 1988b, Report No. 13, 1170p. Minase area, 2006, Report No. 57 C-2-3, 616p. Tazawako-tobu area, 1992, Report No. 27, 1021p. Hakkoda-seibu area, 1993a, Report No. 30, 934p. Iwatesan-seibu area, 1993b, Report No. 31, 1289p. Hachijo-jima area, 1993c, Report No. 31, 1289p. Okushiri area, 1994a, Report No. 33, 1108p. Kaminoyu-Santai area, 1994b, Report No. 34, 1337p. Mizuwake- toge-nanbu area, 1994c, Report No.35, 1281p. Amemasu-dake area, 1995, Report No.36, 1139p. Shiratori area, 1997, Report No. 39 C-1, 379p. Wasabizawa area, 1998, Report No. 41 C-2, 712p. Ashiro area, 1999a, Report No. 45 B-2, 752p. Kumaishi area, 1999b, Report No. 46 B-3, 962p. Akinomiya area, 2001, Report No. 48 C-3, 710p. Shiramizugoe area, 2003, Report No. 53 C-4, 923p. Appi area, 2004, Report No. 54 C-5, 307p. Kirishima-Eboshidake area, 2005, Report No. 55 C-6, 780p. Minase area, 1990, Report No. 20, 1281p. Okushiri-seibu area, 2007, Report No. 62 C-2-9, 560p. Hachimantai area, 2009, Report No. 63 C-2-8, 562p. Ikedako-tobu area, 2010, Report No. 64 C-2-10, 310p.

Ohtani T, Nakashima Y, Muraoka H (1997) Three-dimensional imaging of miarolitic cavities in the Kakkonda Granite using X-ray CT. Jour Geotherm Res Soc Japan 19(4):209–216 (in Japanese with English abstract)

Saishu H, Okamoto A, Tsuchiya N (2016) The relationship between the permeable-impermeable boundary of the Earth's crust and the precipitation of mineral due to the change in a state of water at the geothermal fields. Jour Geotherm Res Soc Japan 38(1):17–25 (in Japanese with English abstract)

Sato K (1982) Analysis of geological structure in the Takinoue geothermal area. Jour Geotherm Res Soc Japan 3(3):135–148

Sato K, Ito N, Tsukuda K (2019) Geothermal geological survey and production well target (Original title in Japanese). E-pix, Ofunato Printing Inc., Iwate, 155p, ISBN978-4-901602-67-9 (in Japanese)

Sawaki T, Sasaki M, Komatsu R, Muramatsu Y, Sasada M (1999) Gas compositions of fluid inclusions from the shallow geothermal reservoir in the Kakkonda geothermal system, northeast Japan. Jour Geotherm Res Soc Japan 21(2):127–141 (in Japanese with English abstract)

Sibson RH (1987) Earthquake rupturing as a mineralizing agent in hydrothermal systems. Geology 15:701–705

Suto Y, Tanifuji K, Watanabe K, Hashida T, Shoji T, Takahashi H (2002) Dependence of the rate of Granite dissolution on temperature and fluid velocity under simulated geothermal reservoir environments. Jour Geotherm Res Soc Japan 24(1):47–56 (in Japanese with English abstract)

Tsutsumi S, Wulaningsih T, Soeda Y, Kiyota Y, Matsuda K, Oshima S, Taguchi S, Tsukada Y (2021) Geology, mineralogy and fluid geochemistry of wells discharging acid fluids in Hatchobaru geothermal field, Kyushu Island, southwestern Japan. In: Proceeding of the World Geothermal Congress 2020, Reykjavik, Iceland, April–October 2021,8

Wang Y, Hirano N, Tsuchiya N, Nakatsuka K, Yamasaki N, Ishida T (2003) Initial dissolution behavior of Granite by hydrothermal solution in the temperature range between 200 °C and 350 °C. Jour Geotherm Res Soc Japan 25(2):129–137. (in Japanese with English abstract)

Chapter 12
Production Well Targeting Based on Geological Analyses

Ko Sato

Abstract Targeting production wells requires geological investigations that integrate current thermal manifestations, recent hydrothermal rock alteration, and geomorphological features. The orientation of the regional stress field and the degree of rock consolidation through geological time significantly influence the development of permeable fractures. The geothermal reservoir in the Hohi region (Kyushu) is primarily composed of volcanic rocks deposited from the Pliocene to the Late Pleistocene. In contrast, the Sengan region (Tohoku) comprises both volcanic and sedimentary rocks ranging in age from the pre-Tertiary to the Middle Pleistocene.

Keywords Hydrothermal rock alteration · Thermoluminescence age determination · Chishima (Kuril) forearc sliver · Volcano-tectonic depression · Superimposed fracture system · Structural movement

12.1 Recent Surface Manifestations in Geothermal Reservoir

Various phenomena occurring in subsurface geothermal reservoirs are not directly observable but are expressed at the surface as manifestations such as hot springs, fumaroles, hydrothermal alteration, and morphological anomalies. To effectively target production wells, it is essential to focus on these surface manifestations as indicators of subsurface activity.

K. Sato (✉)
Socio Engineering Co., Ltd., Tokyo, Japan
e-mail: satoukou6450@gmail.com

J. Kasahara et al. (eds.), *Challenges in Geothermal Development*, Advances in Geological Science, https://doi.org/10.1007/978-981-95-0553-1_12

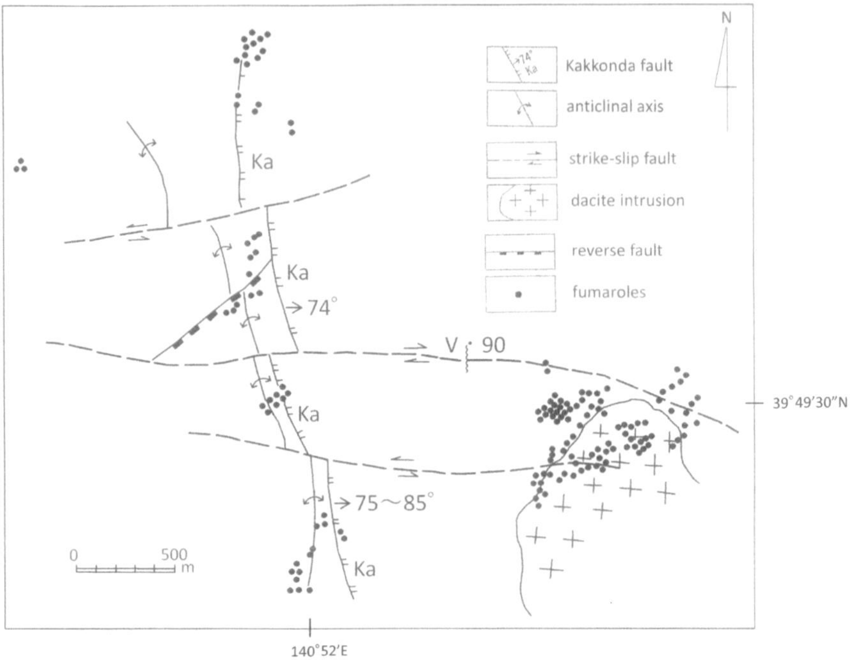

Fig. 12.1 Distribution of fumaroles, anticlinal axis, faults, and dacite intrusion in the Kakkonda geothermal area (Sato et al. 2019)

12.1.1 Hot Springs and Fumaroles

In the Kakkonda geothermal area, tracking alignments of fumaroles has proven useful in identifying fault traces, anticlinal axes, and the margins of dacitic intrusive bodies (Fig. 12.1). Based on field experience, however, we have encountered cases where even a first-class active fault lacked any associated thermal manifestations—such as hot springs or fumaroles—indicating no geothermal fluid flow along the fault. In such cases, drilling targeted at the fault proved unsuccessful.

12.1.2 Hydrothermal Rock Alteration

Impermeable, clay-rich acidic alteration zones play a vital role in geothermal systems by forming sealing layers that suppress upward heat dispersion and block the inflow of cold meteoric water. Once such alteration is identified, dating the altered rocks becomes essential. Thermoluminescence (TL) age determination is a commonly applied method, and its basic principle is illustrated in Fig. 12.2.

When crystal minerals are exposed to natural radiation over time, electrons in the mineral structure are excited to higher energy states. These electrons may become

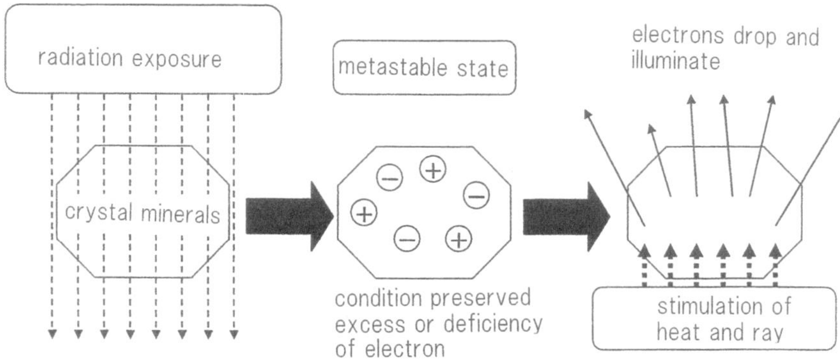

Fig. 12.2 Schematic figure of the principle of thermoluminescence imaging

trapped in metastable states, characterized by an excess or deficiency of electrons. Upon stimulation by heat or ionizing radiation, the trapped electrons return to their original state, releasing photons at specific wavelengths. The number of these trapped charges increases with the mineral's age, leading to greater luminescence intensity during measurement.

Thermoluminescence dating involves four essential steps:

1. Paleodose (PD) measurement—Luminescence intensity is recorded following stimulation to determine the total absorbed radiation dose over geologic time.
2. Sensitivity calibration—Artificial irradiation using γ-rays or X-rays is performed to evaluate the sample's response.
3. Annual dose (AD) calculation—The dose rate is estimated by measuring concentrations of uranium (U), thorium (Th), and potassium-40 (K-40).
4. TL age computation—The age is calculated using the equation proposed by Shimao et al. (1999).

$$TLage(ka) = \frac{PD}{AD}\left\{1 \pm \sqrt{\sigma^2 + \delta^2}\right\}$$

where *PD*: Paleo Dose, *AD*: Annual Dose, σ: Standard deviation of the dose–response growth curve, δ: Uncertainty in AD (uniformly set to 0.1, same as the analytical error for radioactive element measurements).

The annual dose (AD) of a rock sample represents the total radiation it receives per year. It is calculated using the following equation:

$$AD = kD_\alpha + D_\beta + D_\gamma + D_c$$

where $D\alpha$, $D\beta$, $D\gamma$, and Dc represent the annual doses from alpha, beta, gamma, and cosmic rays, respectively. The coefficient *k* accounts for the lower thermoluminescence (TL) efficiency of alpha radiation relative to beta and gamma radiation, typically ranging from 0.08 to 0.25.

For example, if a soil sample contains 3 ppm uranium, 12 ppm thorium, and 1% potassium, the corresponding values are:

$$D_\alpha = 17150\ \mu\text{Gy/a}$$

$$D_\beta = 1510\ \mu\text{Gy/a}$$

$$D_\gamma = 1250\ \mu\text{Gy/a}$$

Additionally, the cosmic ray dose D_c is generally about 150 μGy/a (Ichikawa and Hiraga 1988).

The relationship between alteration halos and TL dating of altered rocks has been demonstrated in the Hachimantai–Appi geothermal field (Takashima et al. 2001). The spatial distribution of alteration zones and sampling points (A1–A12) is shown in Fig. 12.3. Dotted lines trending WNW–ESE connect alteration zones of the same type and correspond to identified lineaments and fracture trends. Large solid circles indicate the presence of pyrophyllite, and colored symbols represent different alteration zones (see Table 12.1).

Table 12.2 shows the TL ages and related data of altered minerals. More than two hydrothermal events are recognized. Two TL ages of 499 ± 52 ka (locality A6) and 510 ± 121 ka (locality A10) indicate an ancient hydrothermal event corresponding to the formation of pyrophyllite. According to the formation conditions of hydrothermally altered minerals (Yoshimura 2001, and Izawa 2010), pyrophyllite forms at temperatures ranging from approximately 250 °C to over 300 °C. However, Table 12.1 indicates that high-temperature pyrophyllite coexists with the low-temperature minerals such as smectite and α-cristobalite. The high-temperature minerals could not have formed under current surface conditions (1 ata, 100 °C), suggesting that they originated in deep, high-temperature zones. Their present exposure is due to erosion and uplift caused by magmatic activity. Subsequently, they appear as relict minerals due to landslide modification and overprinting by later hydrothermal events. Therefore, exploration should prioritize younger hydrothermally altered zones that better reflect current formation temperatures.

Furuya et al. (2000) presented schematic cross-sectional profiles of hydrothermally altered zones of the Takigami geothermal system in Kyushu after completing several drillings to understand the sealing effect and upflow zones. A thick alteration zone is observed from sea level down to elevations of −200 m to −1,500 m. The alteration minerals—montmorillonite and mixed-layer clay minerals—act as a cap rock, preventing shallow groundwater from percolating into the geothermal reservoir. A sharp increase in formation temperature occurs between these alteration zones, corresponding to impermeable layers. A similar alteration zone was identified in the Matsukawa geothermal area (see Sect. 14.3). Both areas are considered hidden geothermal systems with few surface manifestations. These thick alteration zones pose challenges for geological and geophysical surveys because they produce a masking effect over underlying structures.

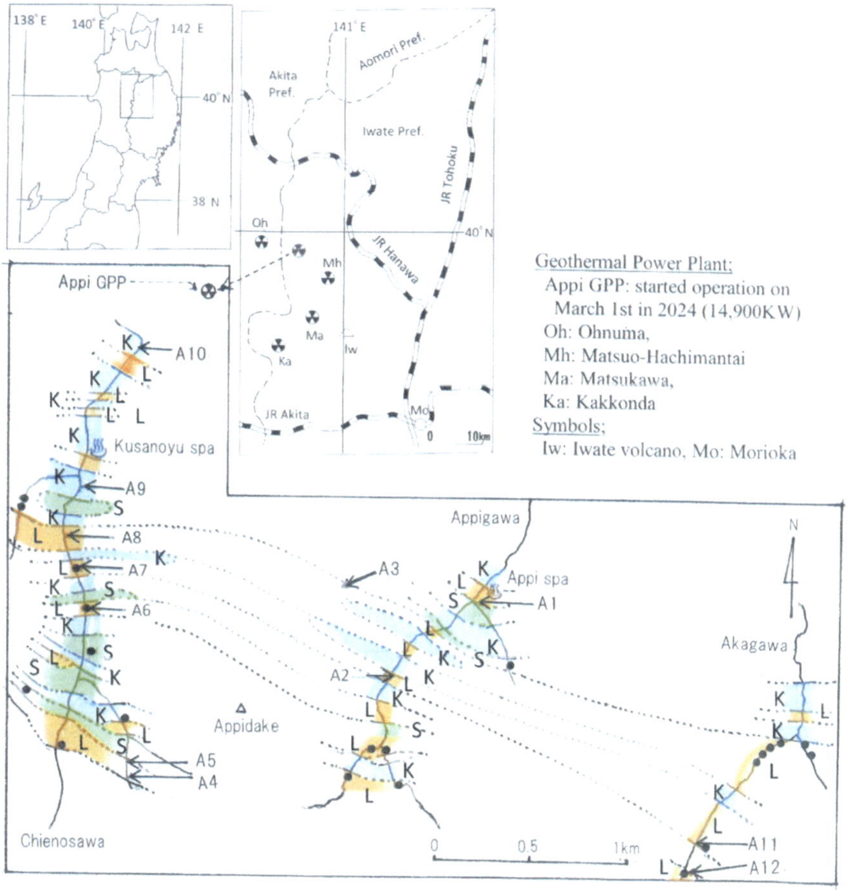

Fig. 12.3 Distribution of alteration zones and thermoluminescence (TL) dating sample locations (A1–A12) in the Hachimantai–Appi geothermal field. Alteration zones are classified as S: Smectite, K: Kaolinite, L: Alunite, and large solid circles indicate the presence of Pyrophyllite. Mineral zoning is summarized in Table 12.1 (modified from Takashima et al. 2001)

12.1.3 Fault and Fracture Decision by Geomorphological Analysis

Geomorphological analyses involve 3D terrain interpretation using stereoscopic topographic maps synthesized from DEM (Digital Elevation Model) data, as well as satellite imagery and LiDAR (Laser Imaging Detection and Ranging) analysis. In addition to these methods, the following three traditional analyses were also conducted to identify faults and fractures in geothermal systems.

Table 12.1 Zoning of hydrothermally altered minerals in the Hachimantai–Appi geothermal field (after Takashima et al. 2001)

Altered Minerals	Smectite zone	Kaolinite zone	Alunite zone
Quartz	———	———	———
α-Cristobalite	- - -	- - -	———
Tridymite	- - -	- - -	- -
Pyrophyllite	- - -	- - -	- - -
Alunite			———
Kaolinite		———	- - -
Smectite	———	- - -	- -
M/S Mixed layer	- - -	- - -	———
Sericite		- -	- -
Pyrite	- - -	———	- - -
Zunyite			- -
Dickite		- -	
Minamite			- - -

Table 12.2 Thermoluminescence (TL) ages and related data for hydrothermally altered rocks (after Takashima et al. 2001)

No.	Main altered minerals	U (ppm)	Th (ppm)	K_2O (%)	Annual Dose (mGy/a)	Paleodose (Gy)	TL age (ka)
A1	Alunite zone	1.21	3.03	0.62	1.54	315	205 ± 23
A2	Alunite zone	1.00	1.87	0.20	0.963	95	99 ± 13
A3	Alunite zone	0.20	0.10	0.01	0.223	50	224 ± 25
A4	Alunite zone	1.29	3.32	2.33	3.11	540	174 ± 37
A5	Alunite zone	1.86	1.99	0.91	2.06	390	189 ± 24
A6	Pyrophyllite	0.78	3.07	0.01	0.781	390	499 ± 52
A7	Alunite zone	1.14	3.03	0.78	1.65	215	130 ± 14
A8	Alunite zone	1.50	1.68	1.21	2.10	205	98 ± 10
A9	Kaolinite zone	1.44	4.55	0.01	1.26	140	111 ± 14
A10	Pyrophyllite	0.48	1.16	0.01	0.461	235	510 ± 121
A11	Alunite zone	1.75	2.85	0.20	1.44	390	271 ± 90
A12	Alunite zone	1.05	2.79	0.44	1.28	175	137 ± 14

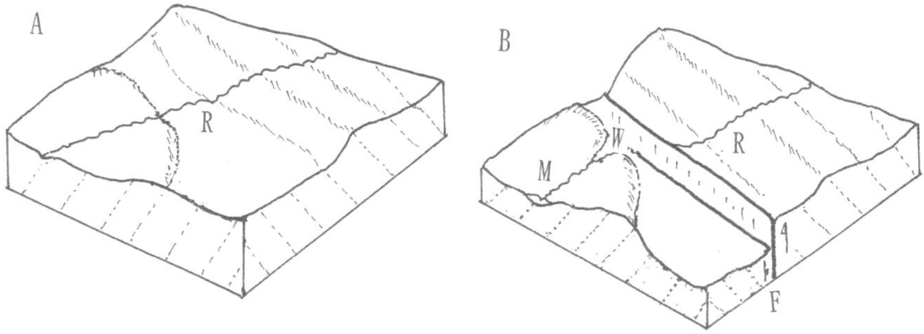

Fig. 12.4 Schematic diagram illustrating the formation of a wind gap by faulting. A: Topography before faulting. B: Topography after faulting and wind gap formation. W: Wind gap. R: River or tributary. M: Maladapted (inactive) river or tributary. F: Fault

(1) Subsurface geomorphological fractures reflect fault systems. Masuda and Okada (1968) presented a representative example of identifying active faults using topographic analysis in Japan.
(2) Fig. 12.4 shows the fault identification based on wind gap features. A wind gap formed by faulting is defined as a pass in a mountain ridge with no active stream. Ikeda (1979) identified several active faults using wind gaps in the eastern area of Beppu Bay, within the Hohi VTD (volcano-tectonic depression) region, Kyushu.
(3) A characteristic topographic feature in geothermal areas is collapsed terrain associated with the distribution of hydrothermally altered rock along faults, particularly in valleys or riverbeds (Fig. 12.5). Hydrothermal alteration erodes fractures, creating weak strata in these areas. This type of topography is discussed in Sect. 14.4.5 as a specific example of well targeting in geothermal field (Fig. 14.38).

12.2 Geothermal Systems Related to Structural Movements

12.2.1 Structural Movement of Geology in Japan

The formation of the geothermal reservoir is strongly influenced by the structural evolution of the geological framework. Permeable fractures within the reservoirs have developed in response to changes in the regional stress field, which were caused by variations in the movement direction and velocity of the Pacific and Philippine Sea plates over geological time. In addition, volcanic activity during each geological period controlled both the thickness of volcanic rock deposits and localized stress conditions. As a result, the orientation of fractures and the mechanical behavior of sediments—such as strength, hardness, and toughness—have played a role in determining reservoir permeability.

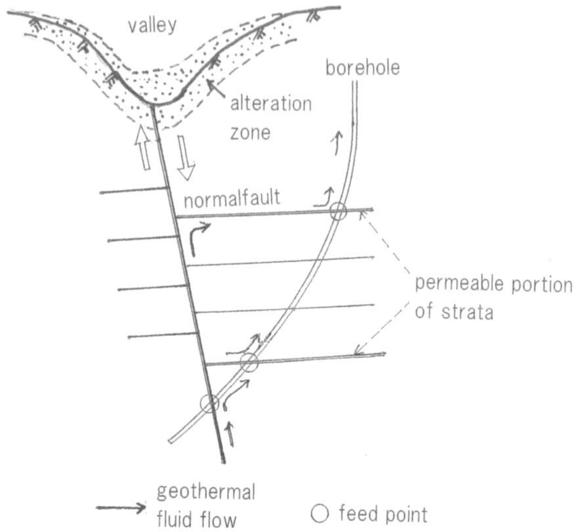

Fig. 12.5 Cross-sectional profile of a directly exposed normal fault along a valley (modified from NEDO 2004)

Furthermore, regional variations in structural geology across Japan can be traced back to the opening of the Japan Sea, which occurred between approximately 21 and 15–13.5 Ma (Fig. 12.6). In Fig. 12.6, the broken lines represent tectonic rotations—clockwise in the Tohoku region and counterclockwise in the southwestern region of Japan. The solid black line is MTL (Median Tectonic Line). Symbol A indicates the location before the Japan Sea opening, while Symbol B is their present positions. Symbol C is the Chishima (Kuril) forearc sliver.

Arrow "**a**" represents the westward motion of the Pacific Plate. Arrow "**b**" shows the northward motion of the Philippine Sea Plate prior to approximately 3.0 Ma. Arrow "**c**" indicates the southwestward motion of the Chishima (Kuril) forearc sliver. The arrow "**d**" denotes the northwestward motion of the Philippine Sea Plate after approximately 3.0 Ma.

This section examines the structural movements in two geothermal regions of Japan. The first is the Hohi Volcano-Tectonic Depression (VTD) region in Kyushu, which includes the Otake, Hatchobaru, Takigami, Waita, and Suginoi geothermal areas, as well as over thirty small-scale binary and hot spring power stations (Fig. 12.7). The second is the Sengan region in Tohoku, which comprises the Sumikawa, Ohnuma, Matsuo-Hachimantai, Matsukawa, Kakkonda, Tazawako-Tobu, and Appi geothermal areas (Fig. 12.8).

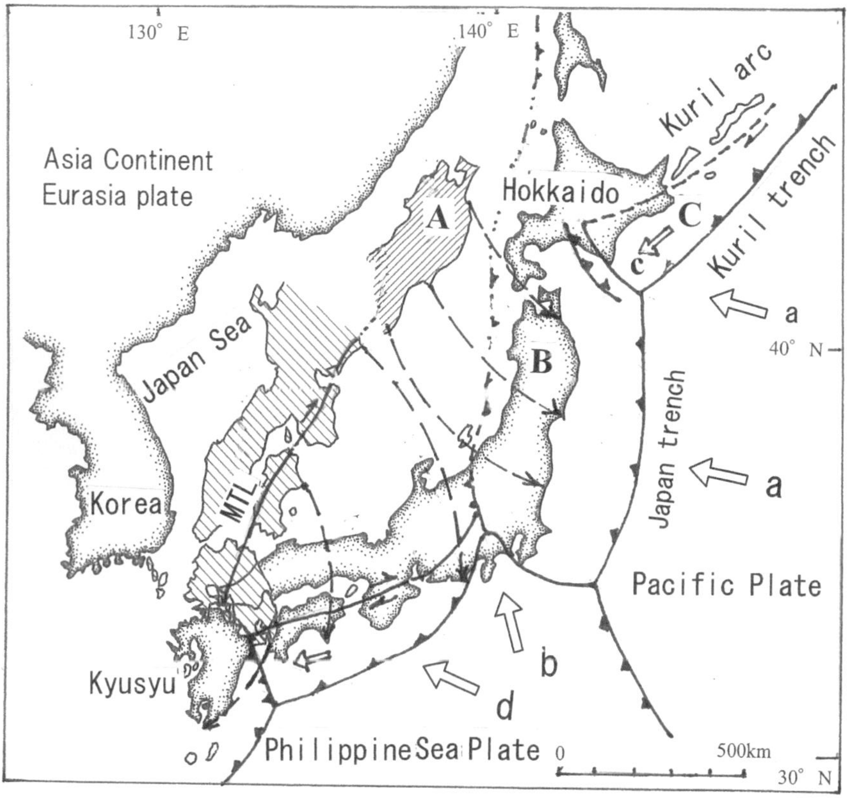

Fig. 12.6 Conceptual representation of tectonic rotary motion associated with the opening of the Japan Sea (approximately 21 to 15–13.5 Ma), and modern plate movements after approximately 3 Ma. Symbols MTL, A, B, C, a, b, c, and d are explained in the main text

12.2.2 The Comparison of Structural Movements of Geology Between the Sengan Region and the Hohi VTD (Volcano-Tectonic Depression) Region

Differences in geological structure and geothermal reservoir characteristics between the Hohi VTD and Sengan regions are illustrated in Fig. 12.9. One of the key distinctions lies in the geological age of the predominant geothermal reservoirs. In the Hohi VTD region, the reservoir mainly comprises relatively young sequences from the Pliocene to the Middle Pleistocene. In contrast, reservoirs in the Sengan region are typically formed within Miocene and Pleistocene strata, and in some cases, within pre-Tertiary basement rocks. These stratigraphic differences affect the mechanical properties of the reservoir rocks—such as strength, hardness, and toughness—due to variations in diagenetic processes like consolidation and cementation. As a result,

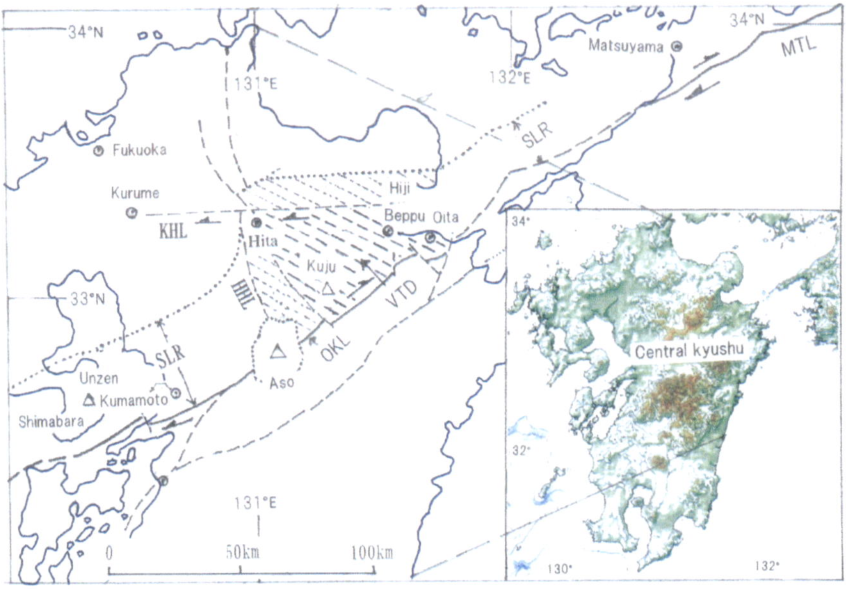

Fig. 12.7 Schematic illustration of the genesis of the Hohi Volcano-Tectonic Depression (VTD) in Central Kyushu, associated with a pull-apart structure. Symbols are described in Sect. 12.2. (Sato et al. 2022)

the Hohi VTD region contains more porous strata (e.g., A-group in Fig. 11.9), which are associated with higher permeability, compared to the Sengan region.

Volcanic activity also differs between the two regions. In the Hohi VTD, volcanism has been consistently vigorous from the Pliocene to the present, whereas in the Sengan region, volcanic intensity increased more recently, starting in the late Pleistocene. These differences have influenced both tectonic movements—such as uplift and subsidence—and the thickness and distribution of volcanic deposits.

Crustal dynamics further distinguish the two regions. The Hohi VTD developed through oblique subduction of the Philippine Sea Plate toward the northwest during the Pliocene, in conjunction with right-lateral movement along the Median Tectonic Line (MTL). In the Pleistocene, the subduction angle steepened, resulting in intensified volcanic activity, increased vertical principal stress (σ_1), and a more pronounced east–west-trending regional stress field. By contrast, the Sengan region is influenced by more complex tectonic interactions among the Philippine Sea Plate, the Pacific Plate, and the Chishima (Kuril) forearc sliver. These interactions have caused the orientations of permeable faults and fractures in geothermal reservoirs to vary in accordance with plate motions, as discussed further in the following section.

As described in Fig. 12.6, fault and fracture development in both regions reflects the broader tectonic evolution associated with the opening of the Japan Sea, including the counterclockwise rotation of the Tohoku district and the clockwise rotation of Kyushu. The MTL continues to influence crustal structure in both areas. Its

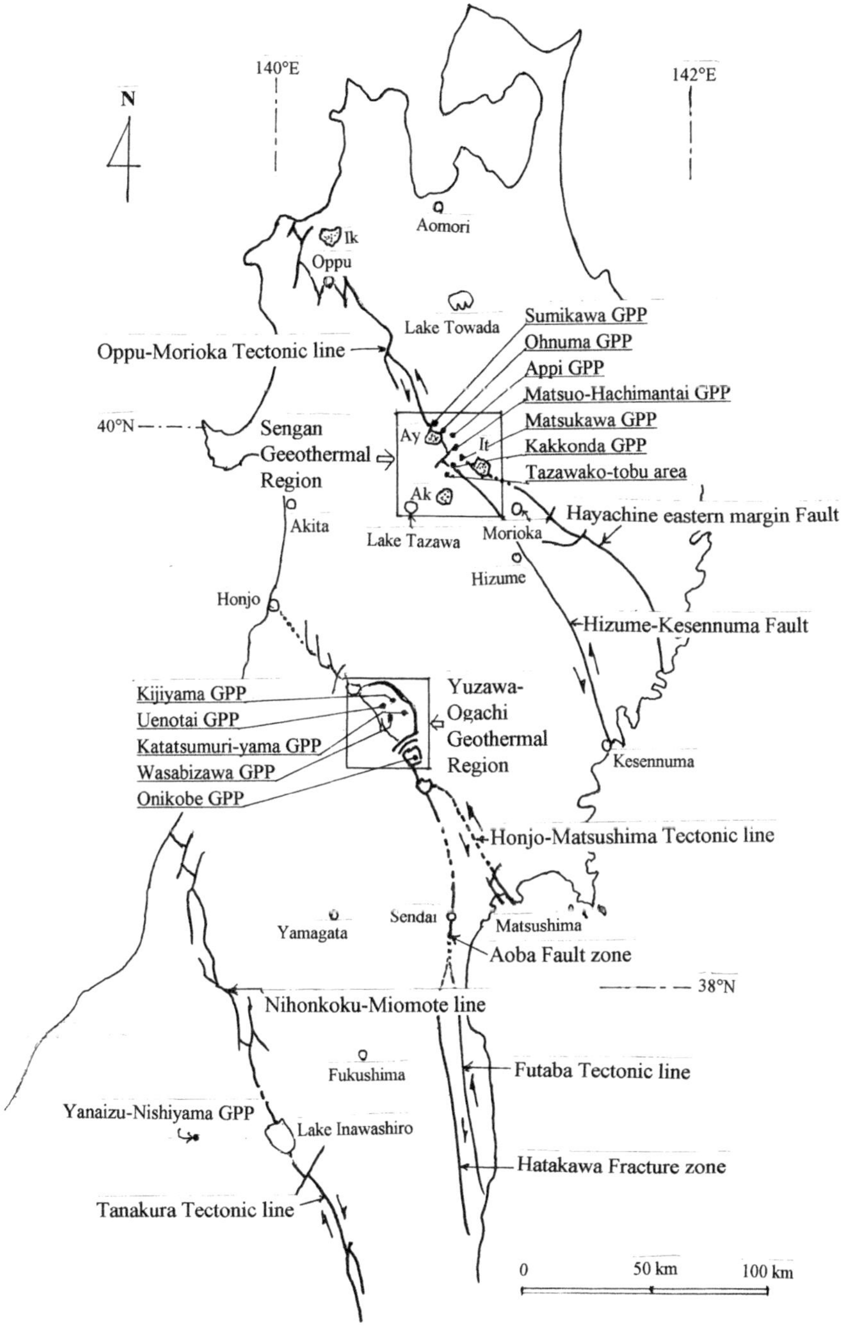

Fig. 12.8 Map showing the geological structure, geothermal power plants, and geothermal areas in the Tohoku district of northeastern Japan. Volcano symbols: Ik = Iwaki, It = Iwate, Ay = Akita-Yakeyama, and Ak = Akita-Komagatake

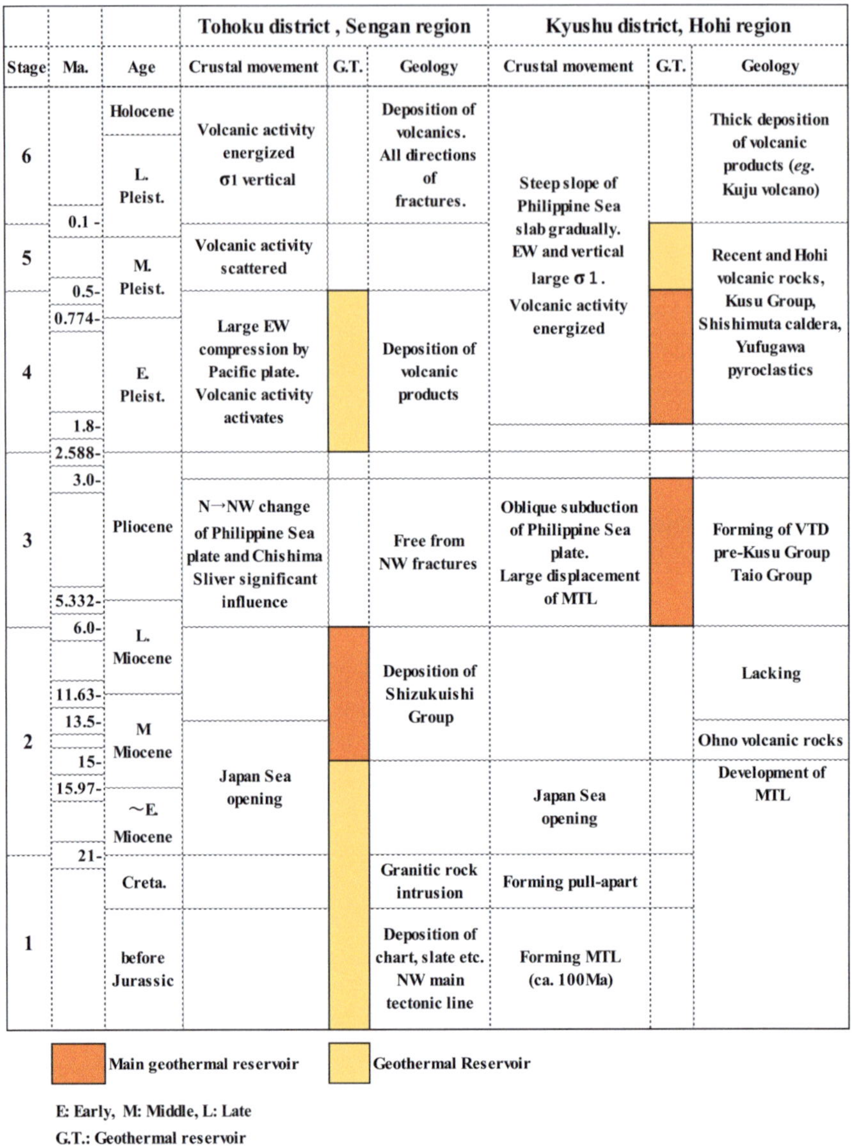

Fig. 12.9 Comparison of geological structural movements and geothermal reservoir characteristics between the Sengan region (Tohoku district) and the Hohi region (Kyushu district)

recent motion is characterized by right-lateral strike-slip movement, as detailed in Sect. 3.2.2, whereas from the Early Cretaceous (~100 Ma) to the Late Paleogene (~25–30 Ma), the MTL exhibited left-lateral strike-slip behavior (Ichikawa 1980). This earlier movement aligns with the Futaba Tectonic Line and the Hatagawa Fracture Zone in Tohoku, and extends northwest–southeast along the Honjo–Matsushima Tectonic Line (Fig. 12.7), passing through the Aoba Fault Zone in Sendai, where mylonitic and metamorphic zones have also been identified (Osozawa 2002; Sasada 1985; Sasaki et al. 2018).

(1) Tectonic Evolution and Structural Framework in the Sengan Region (Tohoku District)

According to Fig. 12.9, the differences in structural movements between the Sengan region and the Hohi VTD region are described in detail below, organized by stages from Stage 1 to Stage 6.

Stage 1 (pre-Tertiary period): The geothermal system is characterized by fractures extending into deep-seated pre-Tertiary basements as illustrated in Fig. 9. 1. These fractures suggest that the structure of the pre-Tertiary basement—overlain by Miocene and Quaternary formations—plays a significant role in geothermal system development, particularly in relation to large-scale tectonic lines in the upper crust. A prominent example is the NW–SE-trending Oppu–Morioka Tectonic Line, which bevels the Sengan region and is intersected by NE–SW-trending transverse faults (Fig. 12.8). A similar structural feature, the Honjo–Matsushima (Sendai) Tectonic Line, also bevels the Yuzawa–Ogachi geothermal region. The pre-Tertiary basements consist of a range from Paleozoic to Mesozoic formations intruded by Cretaceous granitic rocks, and they constitute geothermal reservoirs predominantly and subordinately despite their hard and compact rocks in the Tohoku district.

Stage 2 (ca.21 Ma to ca.6.0 Ma): The opening of the Japan Sea occurred between approximately 21 Ma and 13.5 Ma. Following this event, from around 13.5 Ma to 6.0 Ma, a thick sequence of marine clastic sediments was deposited, interbedded with frequent effusive volcanic rocks. In the Sengan region, geothermal reservoirs are primarily composed of formations from this period.

Stage 3 (ca.6.0 Ma to ca.3.0 Ma): During this period, tectonic disturbance occurred as a result of the southwestward movement of the Chishima (Kuril) forearc sliver. This movement originated from the oblique subduction of the Pacific Plate along the Kuril Trench (Fig. 12.6; Kimura 2002). In response, the maximum principal stress (σ_1) in the Tohoku district was oriented NE–SW, leading to the development of NE–SW-trending extensional fractures and strike-slip fractures trending N–S and E–W. In contrast, NW–SE-trending fractures did not develop under this stress regime. These tectonic changes laid the foundation for the basic geological structures observed in the present-day Sengan region.

Stage 4 (ca.3.0 Ma to ca.0.5 Ma): During this stage, the subduction direction of the Philippine Sea Plate shifted from northward to northwestward (Takahashi 2006). This change caused an increase in the westward movement rate of the

Pacific Plate, while the tectonic influence of the Chishima (Kuril) forearc sliver diminished. As a result, the east–west-trending principal stress (σ_1) intensified in the Tohoku district. Under this stress regime, dominant E–W-trending extensional fractures developed, with subordinate N-S-trending fractures occurring in association with magma intrusions (Fig. 9.3, CASE B and CASE C). Around the same time, volcanic activity associated with the Tamagawa Welded Tuffs commenced and continued through the Quaternary period. The volcanic rocks that accumulated in basin structures contributed to the subordinate development of geothermal reservoirs in the Sengan region.

Stage 5 (ca.0.5 Ma to ca 0.1 Ma): During this stage, igneous activity became more sporadic and gradually transitioned to less intense volcanic eruptions.

Stage 6 (ca.0.1 Ma to the present): Quaternary volcanism intensified during this period, including activity at the Kakkonda Quaternary granite and the Akita-Yakeyama, Akita-Komagatake and Iwate volcanoes. These volcanic events were accompanied by geothermal alteration, surface manifestations, fracture development, and the establishment of convective systems within geothermal reservoirs. With regard to fracture development, multidirectional fractures emerged due to magma intrusion and the influence of both regional and local east–west-trending principal stress (σ_1), as illustrated in Fig. 9.3 (CASE A to CASE C). Among them, NW–SE-trending fractures were either newly generated or reactivated along pre-Tertiary structures first formed during Stage 1.

(2) Tectonic Evolution and Structural Framework in the Hohi VTD Region (Kyushu District)

The Hohi VTD region is defined as the area bounded by the Oita–Kumamoto Tectonic line (OKL) to the south, the Kurume-Hiji line (KHL) to the north, the Hita-Hakuso line (HHL) to the west, and Beppu Bay to the east (Fig. 12.7). In the figure, oblique dotted lines within VTD and its surrounding areas represent NW–SE-trending faults and fractures. The symbol "SLR" denotes the Setouchi subsidence zone, also known as the Beppu–Shimabara graben.

Stage 1 (pre-Tertiary period): Prior to the opening of the Japan Sea, tectonic activity along the Median Tectonic Line (MTL) was already underway. During this period, a half-graben pull-apart basin formed in the Hohi VTD region at the accretionary wedge of the Asian continent (Okada 1993).

Stage 2 (ca. 21 Ma to ca. 6.0 Ma): The opening of the Japan Sea occurred between approximately 21 Ma and 15 Ma. Following this event, from around 15 Ma to 6.0 Ma, the Hohi VTD region experienced minimal sedimentation, resulting in limited basin development—except for the sporadic deposition of volcanic strata.

Stage 3 (ca. 6.0 Ma to ca. 1.8 Ma): As discussed in Chap. 9.3, Tamanyu (1985, 1993) proposed that horst-and-graben structures were formed in the Hohi VTD region during this period. In contrast, Kamata (1985, 1989) and Kamata and Kodama (1994) presented an alternative interpretation of the genesis of the Hohi VTD, suggesting it developed as a pull-apart basin resulting from the right-lateral movement along the Oita–Kumamoto Tectonic Line, located on the western margin of the Median Tectonic Line (MTL) (Figs. 12.6 and 12.7). This tectonic movement was driven by the oblique subduction of the Philippine Sea Plate toward the northwest. The opening of the Hohi VTD was not instantaneous but rather progressed gradually from the outer rim toward the interior of the basin over time, accompanied by subsidence and fissure eruptions. Consequently, the basin became filled primarily with lava and pyroclastic deposits, with only sparse intercalations of sedimentary rocks. The geothermal reservoir in the region is predominantly hosted within these volcanic strata. During this time, NE–SW-trending right-lateral movement along the MTL was associated with a NE–SW-trending minimum principal stress (σ_3). The combined effect of vertical maximum principal stress (σ_1), induced by magma intrusion, and the σ_3 stress field produced NW–SE-trending conjugate normal faults, which are widely distributed throughout the Hohi VTD region.

Stage 4 and Stage 5 (ca. 1.8 Ma to ca. 0.1 Ma): During this period, volcanic activity resumed within the pull-apart basin and persisted throughout the stage (Kamata et al. 1988). An east–west-trending extensional stress field was superimposed on the preexisting stress regime established during Stage 3, due to the intensified westward motion of the Philippine Sea Plate. As a result, the thick volcanic sequences that now compose the geothermal reservoir in the Hohi VTD region are characterized by fractures oriented in various directions, with a predominance of E W trending extensional fissures.

Stage 6 (ca.0.1 Ma to the present): The Hohi VTD region is extensively covered by volcanic products from the Aso and Kuju volcanic complexes (Kamata 1997). According to the depth contour map of the upper boundary of the Philippine Sea slab, a steep gradient is observed beneath the eastern half of the Kyushu district, which is associated with intensified volcanic activity in the region (Hirose et al. 2008). This combination of vigorous volcanism and the strong westward movement of the Philippine Sea Plate has activated fault systems and structural anomalies that serve as favorable pathways for geothermal fluid flow. Based on the current tectonic setting and volcanic activity, fractures in the Hohi VTD region are predominantly characterized by east–west- and north–south-trending normal and strike-slip faults (Fig. 9.4, CASE B and CASE C).

12.3 A Brief on Targeting for Production Wells

Building upon the structural and geological considerations discussed in the preceding section, this part outlines the principal factors relevant to the siting and targeting of production wells. These factors are synthesized from the insights presented in Chap. 9 through Sect. 12.2 and represent the foundational criteria for identifying productive geothermal zones.

12.3.1 Key Targeting Considerations from Prior Chapters

(1) Geothermal systems in Japan typically develop on the gentle slopes at the base of andesitic volcanoes, where groundwater tends to accumulate. These geological settings are conducive to long-term hydrothermal alteration and mineral dissolution, which contribute to the formation of cap rocks (sealing zones). As discussed in Chap. 9.1, these cap rocks are generally not associated with the latest volcanic activity. Therefore, effective well targeting must focus on areas that include upflow zones related to volcanic systems and horst–graben structures resulting from tectonic deformation.

(2) Once upflow zones are delineated, predicting subsurface temperature becomes a critical step in geothermal reservoir assessment. Regions with higher temperature distributions are more favorable for geothermal power generation. The accuracy of deep temperature estimation can be improved by distinguishing between convective and conductive heat transfer regimes, using geophysical methods such as magnetotellurics (MT), seismic reflection, and gravity surveys (see Chap. 10).

(3) Geothermal fluid production is governed by both high temperature and high permeability within the reservoir. The latter is primarily attributed to fractures, rather than pore spaces in porous media, and is commonly described by the Cubic Law. Productive open fractures—often identifiable by the presence of drusy cavities or vein minerals—should exceed a permeability-thickness (*kh*) threshold of approximately 10^{-12} m^3 to enable economically viable steam production of around 1 MWe (Figs. 11.5 and 11.7).

(4) In targeting production wells, both the permeability-thickness (*kh*) and reservoir temperature must be considered. In southwestern Japan, zones of high temperature are essential for obtaining large steam flow rates (Gs), whereas in northern and northeastern Japan, zones of high permeability play a more significant role. These regional differences are attributed to variations in reservoir formation age and rock mechanical properties—including strength, hardness, toughness, consolidation, and cementation—which collectively influence rock density (Sect. 11.1.4).

(5) High-permeability geological structures are typically associated with extensional stress regimes. These structures include normal faults, pull-apart basins formed along strike-slip fault systems, fractures induced by volcanic activity, and permeable zones generated by folding and jointing (see Sect. 11.2).

(6) The identification of high-permeability geological structures relies on comprehensive structural analysis of the targeted geothermal area. Understanding the geological history provides critical insight into fracture orientations and favorable structural configurations for geothermal fluid flow (see Sect. 12.2).

12.3.2 Advanced Approaches to Targeting Precision

(1) Recent thermal manifestations—such as hot springs, fumaroles, and zones of hydrothermal alteration—serve as surface indicators of geothermal upflow zones. Drilling conducted on faults without corresponding thermal evidence has repeatedly proven unsuccessful (Sect. 12.1.1). Productive open fractures are typically associated with drusy cavities, mineral veins, and cracks in volcanic or intrusive rocks. Therefore, thermal manifestation surveys are essential components of production well targeting (Sect. 11.1.5).

(2) Superimposed fracture systems are a key determinant of high steam productivity in geothermal fields. Productive zones often occur where fault networks and other fractures intersect and exceed a permeability-thickness (*kh*) threshold of 10^{-12} m^3. In the Hohi VTD region (central Kyushu), favorable production zones are associated with the superposition of fault systems and structural basins, such as pull-apart or caldera structures (see Chap. 12.2). In the Appi geothermal area, productive thermal fluid pathways align with both fault planes and lithological boundaries. This interpretation is supported by hydrothermal alteration mapping, geomorphological analysis, DTF (drilling-induced tensile fracture) analysis, and FMI (formation micro-imager) logs (NEDO 2004). These studies suggest that fluid flow channels are formed by the overlap of tectonic normal faults and low-angle weak planes within stratified sequences.

(3) Coherence among geological, geochemical, and geophysical datasets is essential for effective well siting. Ideally, fault and fracture interpretations align with detected geophysical anomalies and structural discontinuities. When such correlations are lacking, supplementary data—such as temperature profiles or geomorphological surveys—can provide critical support for well targeting decisions.

(4) Prior to drilling, production well targeting plans should include predicted geological cross-sections and a decision tree for avoiding operational issues. Vertical wells drilled parallel to steeply dipping fault planes—such as strike-slip or high-angle normal faults—are likely to fail. In such cases, directional drilling is preferable, with wellbores oriented as close as possible to perpendicular with respect to the dominant fault or fracture planes. Oblique or parallel drilling trajectories should be avoided, even if constrained by the drilling pad layout.

12.3.3 Drilling-Phase Considerations and Interventions

During drilling, complete or partial lost circulation frequently occurs before reaching the target depth. Standard countermeasures include treating the lost zone as a productive interval, sealing the fracture zone through cementing or casing adjustments, or continuing drilling with aerated or reduced-density mud—though the latter poses risks such as borehole instability and sloughing of weakly bonded reservoirs.

In the event of lost circulation or low steam productivity, the following four response strategies may be applied:

CASE 1: Accept the lost zone as a feed point if the estimated steam flow rate—based on calculations such as those illustrated in Fig. 12.10—is sufficient.
CASE 2: Conduct additional drilling via a side-track approach to access a secondary well target if initial steam output is inadequate.
CASE 3: Extend the borehole toward a deeper target if the original depth does not yield sufficient productivity.
CASE 4: Improve reservoir permeability through well-stimulation techniques, including hydraulic fracturing, acid stimulation, forced discharge, or cold-water injection.

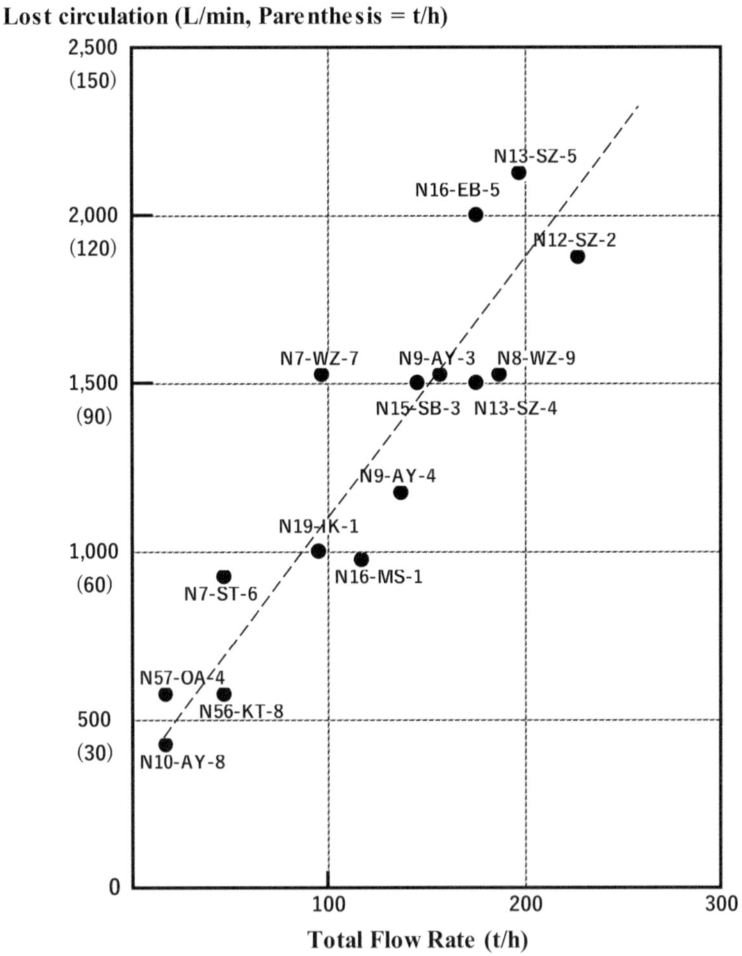

Fig. 12.10 The relationship between lost circulation and total flow rate

A notable case of successful hydraulic stimulation is reported by Kizaki and Sato (1994) at the Kakkonda geothermal field, where 9 out of 12 wells showed enhanced steam productivity following treatment. Key success factors included high-temperature conditions, low-permeability zones surrounding more permeable channels, the use of proppant, and extended stimulation duration.

12.3.4 Real-Time and Post-Drilling Flow Rate Estimation

During drilling, the total geothermal fluid flow rate can be approximately estimated by comparing lost circulation volumes with empirical data. According to NEDO's dataset from 15 geothermal fields (Table 11.2), lost circulation rates (in L/min or t/h) show a proportional relationship with total fluid discharge. This correlation is illustrated in Fig. 12.10.

Steam flow rate can be further approximated using constant enthalpy equations. The fluid temperature is typically estimated from the minimum homogenization temperature of fluid inclusions (Sect. 10.5), or, when unavailable, from neighboring wells with comparable geological conditions.

Example Calculation:

- Fluid temperature: 250 °C → specific enthalpy $h' = 1{,}089.69$ kJ/kg
- Wellhead pressure: 0.41 MPa G (≈0.5 MPa)
- Specific enthalpy: h' (hot water) = 670.50 kJ/kg, h'' (steam) = 2,748.11 kJ/kg.

Let dryness fraction be X, then: $1{,}089.69 = 2{,}748.11X + 670.50(1–X)$.

Solving yields: X ≈ 0.20 → Steam-to-water mass ratio ≈ 1:4

If lost circulation is 1,500 L/min (90 t/h), Fig. 12.10 suggests a total flow rate of 150 t/h. Based on the dryness fraction, the estimated steam flow rate is 30 t/h, equivalent to approximately 3 MWe.

Note: This is a provisional estimation used for on-site decision-making. Accurate determination of steam flow requires post-drilling two-phase flow measurements using a steam-water separator or James's lip pressure method.

References

Furuya S, Aoki M, Gotoh H, Takenaka T (2000) Takigami geothermal system, northeastern Kyushu, Japan. Geothermics 29:91–211

Geological Soc. Japan, ed. (2017) The geology of Japan—2, Tohoku, *Asakurashoten,* 693p. Geological structure referenced from pp. 428 and 462

Hayashi M (1973) Hydrothermal alteration in the Otake geothermal area. Jour Japan Geotherm Ener Assoc (Chinetsu) 10:9–46

Hirose F, Nakajima J, Hasegawa A (2008) Three-dimensional seismic velocity structure and configuration of the Philippine Sea slab in the southwestern Japan estimated by double-difference

tomography. Jour Geophys Res 113:B09315. https://doi.org/10.1029/2007JB005274. 2008, 1–26
Ichikawa K (1980) Geohistory of the median tectonic line of Southwest Japan. Memoirs Geol Soc Japan 18:187–212 (in Japanese with English abstract)
Ichikawa Y, Hiraga S (1988) Thermoluminescence dating. Mem Geol Soc Japan 29:73–82
Ikeda Y (1979) Active fault systems of the Quaternary volcanic region in the central part of Oita Prefecture, Kyushu district, southwest Japan. Geograph Review Japan 52(1):10–29 (in Japanese with English abstract)
Izawa E (2010) Chapter 8: Exploration of resources. Iwanami Earth Planet Sci 4 Observation of the Earth, 330p. Iwanami-Shoten, 291–323 (in Japanese)
Kamata H (1985) Volcanic activity in relation to the geologic structure in the central-north Kyushu Japan. Rept Geol Surv Japan 264:33–64 (in Japanese with English abstract)
Kamata H (1989) Volcanic and structural history of the Hohi volcanic zone, central Kyushu Japan. Bull Volcanol 51:315–332
Kamata H, Kodama K (1994) Tectonics of an arc-arc junction: an example from Kyushu Island at the junction of the Southwest Japan Arc and the Ryukyu Arc. Tectonophysics 233:69–81
Kamata H, Uto K, Uchiumi S (1988) Geochronology and evolution of the post-Shishimuta caldera activity around the Waitasan area in the Hohi volcanic Zone, Kyushu, Japan. Bull Volcanol Soc Japan, Ser 2 33(4):305–320
Kamata H (1997) Geology of the Miyanoharu district. In: With geological sheet map at 1:50,000, Geol Surv Jpn, 127. (in Japanese with English abstract 5p)
Kizaki Y, Sato K (1994) Increase in steam productivity by hydraulic fracturing stimulation in the Kakkonda geothermal field Jour. Jpn Geotherm Ener Assoc (Chinetsu) 31(1):1–17 (in Japanese with English abstract)
Matsuda T, Okada A (1968) Studies of active faults in Japan. Quaternary Res 7(4):188–199 (in Japanese with English abstract)
NEDO (New Energy and Industrial Technology Development Organization (2004) *Geothermal development promotion survey, survey report No.C-5, Appi area,* Geothermal resources research, supplementary sourcebook, 845p (in Japanese)
Okada H (1993) Relationship between sedimentary basin on the shelf area in the East China Sea and the central Kyushu Rift Valley. Mem Geol Soc Japan 41:163–173 (in Japanese with English abstract)
Osozawa S (2002) Formative process of the Kagitori-Ayashi Line (Monocline) and the Aoba Fault Zone, at the Moniwa and Aobayama hills, Sendai, northeast Japan. Jour Geol Soc Japan 108(12):781–793 (in Japanese with English abstract)
Sasada M (1985) The pre-Neogene basement rocks of the Kamuro Yama-Kurikoma Yama area, Northeastern Honshu, Japan,-part 2, Boundary between the Abukuma and Kitakami Belts. Jour Geol Soc Japan 91(1):1–17 (in Japanese with English abstract)
Sasaki K, Tezuka S, Takizawa K, Kaneko T, Asano K, Asai H, Nakanishi S, Akasaka C (2018) Mineralogical and hydraulic characteristics of the Wasabizawa geothermal field, Akita prefecture, Japan. In: Proceedings of 43rd Workshop Geothermal Reservoir Engineering, Stanford University, Stanford, California, February 12–14, SGP-TR-213, 1–12
Sato K, Ito N, Tsukuda K (2019) Geothermal geological survey and production well target (Original title in Japanese). E-pix, Ofunato Printing Inc., Iwate, 155. ISBN978-4-901602-67-9 (in Japanese)
Sato K, Ito N, Tsukuda K (2022) Geothermal geological survey and production well target: Supplementary edition focusing on reinjection wells (Original title in Japanese). Iwate: E-pix, Ofunato Printing Inc., Iwate, 125 pp, ISBN978-4-901602-74-7 (in Japanese)
Shimao T, Takashima I, Watanabe K, Izawa E (1999) Verification of errors and reliability of thermoluminescence age for volcanic rocks—precise beta-ray correction age of pyroclastic flow deposits of Unzen volcano, SW Japan. Jour Mineral Petrol Sci 94(4):109–119 (in Japanese)
Takahashi M (2006) Tectonic development of the Japanese Islands controlled by Philippine Sea plate motion. Jour Geography 115(1):116–123 (in Japanese with English abstract)

Takashima I, Ohkawa K, Miura K, Saiki K, Murakami H (2001) Alteration and thermoluminescence ages of altered and volcanic rocks in the Hachimantai Appi geothermal field, NE Japan -Proposal for selection of promising geothermal areas by alteration survey, dating and sensing. Jour Geotherm Res Soc Japan 23(4):251–261 (in Japanese with English abstract)
Tamanyu S (1985) Stratigraphy and geologic structures of the Hohi geothermal area, based mainly on the borehole data. Rept Geol Surv Japan 264:115–142 (in Japanese with English abstract)
Tamanyu S (1993) Geotectonic developing history in the Beppu-Kuju graben, central Kyushu Japan. Mem Geol Soc Japan 41:93–106 (in Japanese with English abstract)
Yoshimura T (2001) Sample preparation and identification of clay minerals using polarization microscope. In: Clay mineral and alteration process (Original title in Japanese). Chigaku Sousho (Series of geological division), 32, Associate Geological Collaborator Japan. 202–250 (in Japanese)

Chapter 13
Reinjection Wells

Ko Sato

Abstract In Japan, geothermal hot water containing arsenic concentrations of 0.5 ppm or higher must be reinjected into the subsurface, as mandated by the Prime Minister's Office Ordinance enacted on June 21, 1971. This chapter examines the mechanisms and behaviors of reinjected fluids within geothermal reservoirs. Topics include the cold sweep process, effects of fracture spacing, breakthrough phenomena, spatial distribution of reinjected fluids, and temperature-related influences. Furthermore, six typical reinjection well placement strategies are presented with schematic illustrations.

Keywords Hot water reinjection · Cold sweep process · Breakthrough · Spread out · Temperature effects

13.1 Cold Sweep Process and Fracture Spacing

13.1.1 Cold Sweep Process

In a liquid-dominated geothermal reservoir, pressure decreases as steam and hot water are produced. This pressure decline is offset by the recharge of colder water, which flows into the reservoir to compensate for the extracted volume. As the recharging fluid travels through the reservoir, it absorbs heat from the surrounding hot rocks, allowing the reservoir to maintain elevated temperatures during production.

This mechanism is known as the cold sweep process, illustrated in Fig. 13.1. The same concept applies to the reinjection of geothermal hot water, which is relatively cooler than the surrounding formation temperature.

A key challenge in modeling reinjected fluid behavior is quantifying the heat exchange between the flowing fluid and adjacent rock through fractures. One effective modeling approach is the Multiple Interacting Continua (MINC) method (Pruess

K. Sato (✉)
Socio Engineering Co., Ltd., Tokyo, Japan
e-mail: satoukou6450@gmail.com

J. Kasahara et al. (eds.), *Challenges in Geothermal Development*, Advances in Geological Science, https://doi.org/10.1007/978-981-95-0553-1_13

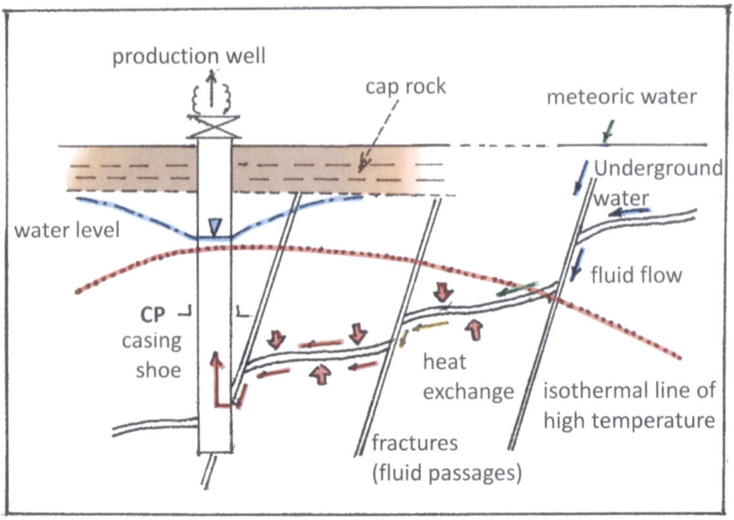

Fig. 13.1 Conceptual model of the cold sweep process (Sato et al. 2022)

and Narasimhan 1985), which simulates heat and multiphase fluid flow in complex, fractured porous media.

This method adopts a double-porosity framework, assuming three orthogonal sets of infinite, planar, and parallel fractures with uniform aperture (δ) and spacing (λ).

13.1.2 Fracture Spacing

Fracture spacing, governed by cubic law and porosity, is a critical factor in understanding the cold sweep process. Panel "A" in Fig. 13.2 presents a conceptual model of long-term formation temperature changes in production zones, depending on fracture spacing. Panel "B" illustrates hypothetical fracture arrangements in plan view.

a. Wide spacing between permeable fractures results in high flow velocity with minimal heat exchange from the rock matrix. Early-stage reinjection may cause rapid cooling of the production zone, but heat degradation stabilizes over time as the matrix gradually transfers stored heat to the fluid.
b. Represents intermediate spacing, with behavior between that of "a" and "c."
c. Narrow spacing between fractures in low-permeability rocks leads to slower flow and more effective heat exchange. Initially, reinjected fluid retains a high temperature due to prolonged rock contact. However, once the matrix heat is depleted, fluid temperature drops rapidly.

Condition "c" includes the following subsurface characteristics:

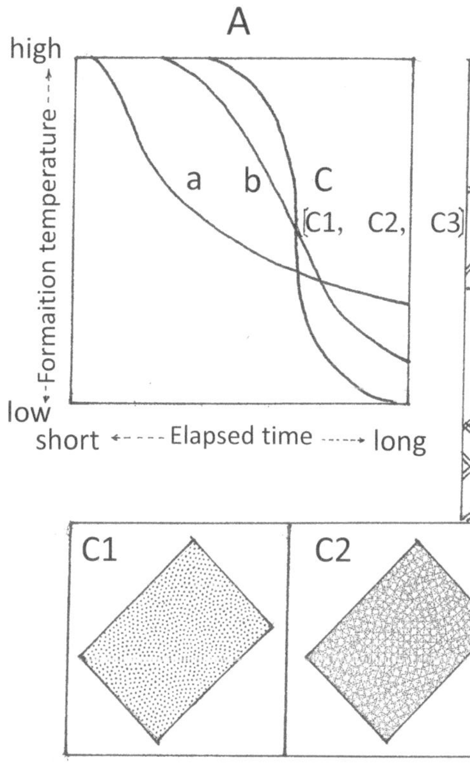

Fig. 13.2 Conceptual diagrams of fracture spacing. Symbols A, B, a, b, c, C1, C2, C3, and λ are described in the main text (Sato et al. 2022)

C1: Low-permeability rocks with <5% porosity and $kh \approx 10^{-17}$ m^3
C2: Porous units such as breccia or conglomerate with >5% porosity and $kh \approx 10^{-15}$–10^{-13} m^3
C3: Densely fractured rocks with minimal permeability, such as cleavage planes, schistosity, jointed veins, and small-scale faults or folds

To reduce the degradation of fluid flow temperature, "c" is better than "a" in the short term. On the contrary, "a" is better than "c" in the long term. The problems related to fracture spacing arise as follows:

(1) Estimating fracture spacing in the early stage of geothermal development is impossible.
(2) "c" is advantageous for the cold sweep process in the short term. However, production wells are not targeted at such zones because they have little to no permeability.

(3) Production wells select "a" for the long-term thermal recovery to avoid heat cooling by reinjection water. However, if condition "a" yields excellent production, operators tend toconcentrate well targets around such zones as a hedge for unsuccessful wells.

13.2 Breakthrough and Spread Out

13.2.1 Breakthrough

Breakthrough refers to the interaction between reinjected water and production fluids, affecting reservoir pressure and temperature. Four types of breakthroughs are commonly recognized:

B1. Pressure breakthrough: A rapid pressure response during reservoir exploitation. Pressure interference tests are used to evaluate this phenomenon and to estimate transmissibility and storativity. However, pressure changes alone do not confirm direct mixing between reinjected and produced fluids.

B2. Fluid flow breakthrough: Tracer tests are widely used to assess fluid communication between reinjection and production wells. The movement of tracers is governed by the cubic law and Darcy's laws, indicating fluid breakthrough.

B3. Thermal breakthrough (fluid temperature): A drop in production fluid temperature due to mixing with cooler reinjected water. This is identified through changes in SiO_2 concentrations and direct temperature measurements in wells.

B4. Thermal breakthrough (formation temperature): A delayed response involving a decrease in reservoir rock temperature. This follows the reduction in fluid temperature and is governed by thermal diffusivity. Formation temperature is typically measured using the Horner method after well shut-in.

The time lag between B2 (fluid flow breakthrough) and B4 (formation temperature breakthrough) can be roughly estimated using the following expressions:

Real flow velocity (representing the tracer arrival time in B2):

$$U_{B2} = \frac{U}{\phi} \tag{13.1}$$

Apparent heat transfer velocity (representing the delayed thermal response in B4):

$$U_{B4} = \frac{\rho_w - C_w}{\rho_r - C_r} \times U \tag{13.2}$$

where U: Darcy velocity, ϕ: Porosity, ρw: Fluid density, ρr: Rock density, Cw: Specific heat capacity of fluid, Cr: Specific heat capacity of rock.

This shows that thermal response lags behind tracer response due to the heat capacity and density differences between rock and fluid. Example calculation using Eqs. (13.1) and (13.2) as follows:

Given parameters:

Porosity $\varphi = 10\%$
Fluid density $\rho_w = 864.58$ kg/m^3 (at 200 °C)
Rock density $\rho_r = 2{,}500$ kg/m^3
Fluid specific heat capacity $C_w = 4{,}497$ J/kg °C (at 200 °C)
Rock specific heat capacity $C_r = 810$ J/kg °C

Using Eqs. (13.1) and (13.2):

$U_{B2} = 10U$
$U_{B4} \approx 1.92U$

Ratio of the breakthrough velocities:

$U_{B2}/U_{B4} \approx 5.21$

This result indicates that the thermal breakthrough (B4) occurs approximately five times later than the fluid flow breakthrough (B2). In other words, there is a significant time lag in detecting the formation temperature change compared to the fluid arrival, due to the thermal diffusivity effect in the rock matrix.

Subsequently, the time lag between B2 and B3 was reported based on measurements in the Palinpinon geothermal area, Philippines (Malate and O'Sullivan 1991). The breakthrough time for B2 (fluid flow) was approximately 40 h, whereas B3 (thermal fluid temperature change) occurred after 18 days (432 h).

In any case, if no countermeasures are taken and operators wait until the confirmation of B2 before responding, it becomes difficult to recover both fluid and formation temperatures in the reservoir. Figure 13.3 presents a conceptual diagram showing the difference between fluid flow breakthrough and the advancing thermal cooling front. While the cooling front lags behind fluid flow at the beginning of reinjection, it eventually catches up. Once it does, production temperature drops significantly and, in the worst-case scenario, fluid production may cease entirely.

13.2.2 Spread Out

Spread out refers to a sufficient separation between production and reinjection zones to prevent thermal interference from cooler reinjected fluids. It is a critical strategy to minimize temperature decline in the production reservoir. At the Hatchobaru geothermal power station, Tokita and Haruguchi (1998) reported a significant recovery case: after the commissioning of Unit No.2 (55 MWe) in 1990, power output dropped from 110 to 70 MWe within four years due to cooling from reinjected water. Through reservoir management—including gravity monitoring, geochemical analysis (tracers, Cl, SO_2), temperature profiling, and pressure interference tests—power

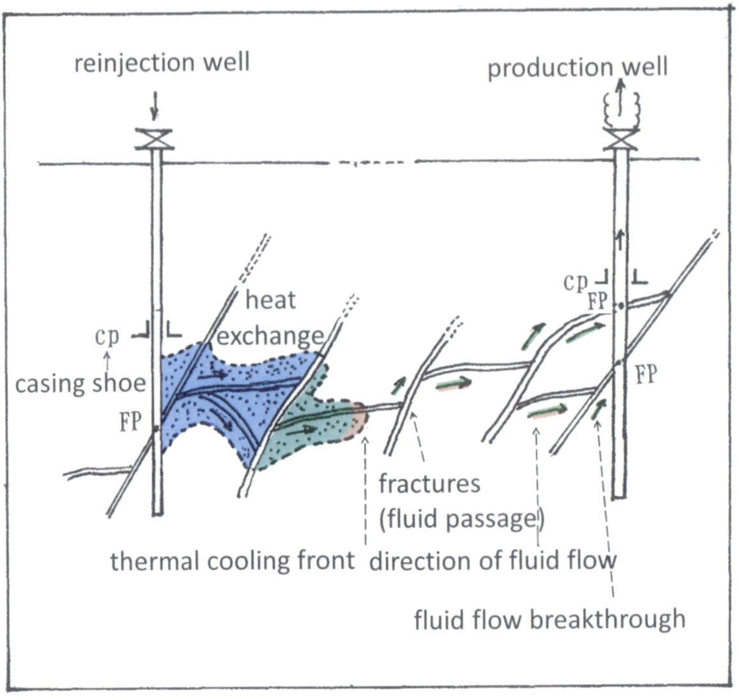

Fig. 13.3 Conceptual figure illustrating fluid flow breakthrough and the thermal cooling front. The thermal front lags behind initially but eventually catches up, leading to production temperature decline (Sato et al. 2022)

output was restored to 110 MWe by 1999. Furthermore, seven replacement wells were drilled at least 500 m north of the production zone, based on numerical simulation.

Vertical spread out is another effective reinjection strategy. When reinjection wells are completed at shallower depths than production wells, the cooler, denser reinjected water tends to sink and reach the production zones more easily, accelerating thermal interference.

For example, pure water at 250 °C has a density of approximately 800 kg/m^3, while at 150 °C it increases to about 915 kg/m^3. This density difference can induce downward density-driven flow, allowing colder water to intrude into hotter reservoir zones (Bodvarsson 1972).

Therefore, although setting reinjection targets at shallower depths is logistically simple, it may compromise reservoir temperature and long-term productivity.

13.3 Influence of Temperature in Reinjection

The effectiveness of geothermal reinjection is significantly influenced by the temperature of the reinjected water, the reservoir temperature, the reinjection volume, and the system's capacity to extract heat efficiently.

13.3.1 Temperature of Reinjected Water

High-temperature reinjection is generally preferred, as a smaller temperature difference between the reinjected water and the reservoir fluid promotes smoother heat extraction; however, some studies have presented opposing views regarding the optimal reinjection temperature.

(1) Lower-Temperature Reinjection Case:

Lower reinjection temperature increases reinjection capacity by inducing thermal contraction of surrounding rocks, enhancing fracture aperture, promoting gravity-driven flow, and increasing storativity. This has been confirmed through laboratory experiments and MINC-model simulations in the Sumikawa geothermal field (Kitao et al. 1990; Ariki and Akibayashi 2001, 2002). Additionally, low reinjection temperature supports a low wellhead pressure, which is advantageous for double-flash systems requiring higher steam output. However, cold reinjection promotes reservoir cooling through thermal breakthrough and may lead to silica scaling inside wells and equipment when wellhead pressure drops below 0.12 MPa (approx. 123 °C).

(2) Higher-Temperature Reinjection Case:

To mitigate silica precipitation and reservoir cooling, a closed reinjection system that avoids depressurization after separation is favorable. While this approach minimizes thermal breakthrough, it also has drawbacks:

- It reduces the steam yield due to higher reinjection temperatures.
- It decreases reinjection capacity, necessitating either additional reinjection wells or mechanical pumping support.

13.3.2 Formation Temperature

Geothermal areas characterized by high formation temperatures offer a distinct advantage for development. In such areas, the proportion and volume of reinjected water are reduced due to the high reservoir temperature. Additionally, a small temperature difference between reinjected water and formation fluid facilitates efficient heat exchange. Whether a high-temperature field can be effectively developed depends on the presence of permeable fractures, which must be confirmed at an early exploration stage.

13.4 *In-Situ* Boiling Caused by the Replacement of Reinjection Area

Withdrawal of fluid through production and the relocation of reinjection wells can stimulate in-situ boiling within the reservoir. This process creates a vapor-dominated zone above a liquid-dominated zone, forming what is known as a steam cap. Typically, fluid extraction leads to a decline in both pressure and temperature. However, in reservoirs enclosed by impermeable boundaries (i.e., shielding structures), high temperatures can be maintained. In such settings, the remaining fluid draws heat from the surrounding high-temperature rocks, and boiling continues via latent heat evaporation (Fig. 13.4). When the reservoir pressure decreases due to production and reinjection area shift—from "b" to "a" in Fig. 13.4—boiling persists because of the thermally insulated, closed reservoir environment.

The relocation of the reinjection area effectively "kills two birds with one stone"—enhancing power generation while preventing reservoir cooling. A representative example is the Palinpinon geothermal area in the Philippines, where the Palinpinon I power plant has been operating at 112.5 MWe since 1983 (Malate and Aqui 2010).

Ten years later, the 80 MWe Palinpinon II plant was commissioned in 1993, bringing the total installed capacity to 192.5 MWe. However, reinjection breakthroughs led to cooling of the production zones, causing mass flow reduction and

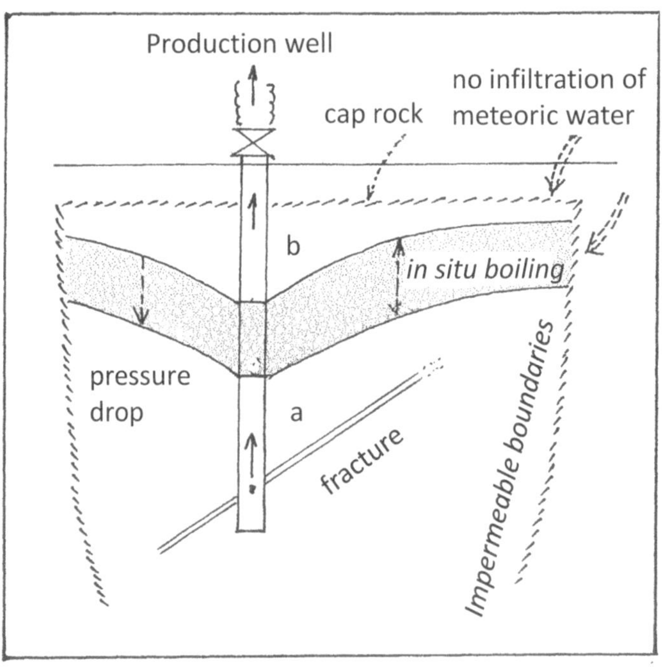

Fig. 13.4 Conceptual model of *in-situ* boiling

enthalpy decline, which resulted in power output losses from 1983 to 1989. In response, the bulk of the reinjection was shifted 2 to 4 km northward beginning in 1989. This relocation allowed the two-phase zone to expand both horizontally and vertically. As a result, an additional 35.5 MWe power plant was constructed, utilizing steam from the expanded two-phase zone. This zone remains sustainable due to a thick cap rock near the surface and high temperatures exceeding 250 °C at shallow depths.

13.5 General Plan of Reinjection Well-Setting

Figure 13.5 presents six general strategies for reinjection well placement, focusing on spatial separation ("spread out") from production zones. Since fracture spacing and breakthrough behavior are difficult to predict in the early stages of geothermal development, these factors must be considered in later planning.

Case A: Separate field with production potential. The reinjection well is located in a different field separated by an impermeable layer but with steam production potential. Although direct breakthroughs may not occur, this setting could still reduce steam output due to interference with the reservoir's pressure or heat.
Case B: Out-of-field, no steam productivity. The reinjection site lies outside the productive zone and has lower formation temperatures. The vertical and horizontal distances between reinjection and production wells are large, reducing the risk of breakthrough.
Case C: Edge of reservoir (infield). The reinjection well is set within the reservoir boundary, but far from the main upflow zone. To avoid thermal breakthrough, the feed point of the reinjection well is placed deeper than that of the production wells. However, the fracture density and temperature may be lower.
Case D: Within reservoir (infield, closer proximity). Similar to Case C but with shorter spacing. The reinjection and production zones are closer, so thermal management becomes more critical.
Case E: Shallow infield. Reinjection occurs at a shallow depth, separated laterally from production zones. Cold reinjected water can sink into the hotter zone due to density differences, increasing breakthrough risk. Monitoring with tracers and temperature logging is essential.
Case F: Shallow and near production zone. This is the least favorable scenario. The shallow reinjection zone is close to the production zone, making thermal breakthrough almost immediate and unavoidable.

The upflow zone often does not exist in reinjection areas, which generally contain fewer fractures and exhibit lower formation temperatures—excluding Case A. As a result, it is difficult to clearly identify discontinuities using geophysical methods such as electrical resistivity, magnetotelluric (MT), and gravity surveys. These limitations highlight the dual challenge of ensuring reinjection capacity while avoiding

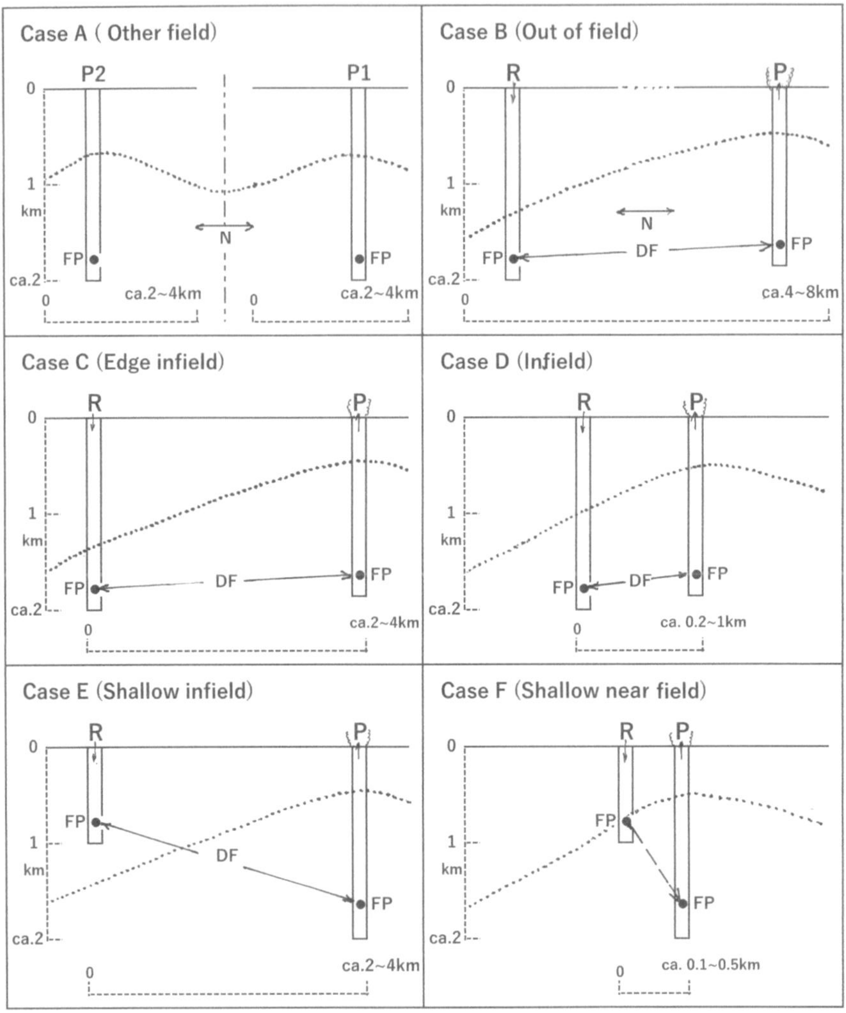

Fig. 13.5 Six general reinjection well-setting plans. Symbols: P = Production well, P1 = Production field No.1, P2 = Production field No.2, R = Reinjection well, FP = Feed point, DF = Spread out (distance between production and reinjection), N = No or low-permeability strata. Dotted lines represent conceptual temperature profiles—elevated segments indicate relatively high formation temperatures (Sato et al. 2022)

adverse thermal effects on the production zone, and they underscore the importance of planning reinjection zones with an ideal configuration from the early stages of development.

To achieve effective reinjection, countermeasures such as drilling multiple wells, using injection pumps, or enlarging borehole diameters may be required. Key characteristics of a desirable reinjection system include:

(1) Feed points of reinjection wells are deeper than those of production wells.
(2) Fracture spacing near the reinjection zone is wide, and permeability is high.
(3) Layers with low-permeability and narrow fracture spacing are distributed between production and reinjection zones, delaying thermal breakthrough while enhancing heat extraction.
(4) Vertical and horizontal spread out between production and reinjection zones is sufficient.
(5) Reservoir temperature increases gradually toward the production zone.
(6) The aquifer water level is initially low during the early phase of geothermal development.

13.6 Altered Plan of Reinjection After the Start of Operation

Breakthrough from reinjected water often occurs after power plant operations begin, leading to a decline in production fluid temperature and a gradual drop in power output. To address this, enhanced monitoring is essential in the following areas:

(1) Temperature monitoring: Downhole temperature measurements during plant maintenance and continuous monitoring from observation wells track changes in both fluid and formation temperatures.
(2) Chemical monitoring: Variations in SiO_2 concentration reflect changes in reservoir fluid temperature. Cl concentration, an inert component, helps determine the mixing ratio of reinjected water. Simultaneous reductions in both SiO_2 and Cl may indicate meteoric water inflow.
(3) Tracer testing: Fluid flow breakthrough can be tracked by injecting KI or KBr into reinjection wells. The return ratio of tracers is calculated by sampling production fluids.
(4) Pressure monitoring: Pressure variations within and around the reservoir are assessed via pressure interference tests and downhole measurements. High-sensitivity observation enables early detection of pressure-related breakthroughs.
(5) Gravity monitoring: Repeated high-resolution gravity surveys (μgal level) can detect mass changes in the reservoir due to reinjection and production. These changes also indicate reservoir recharge. This method has been applied in Hatchobaru and Takigami (Ehara and Nishijima 2004; Saito et al. 2006; Oka et al. 2014).
(6) Resistivity monitoring: Using the Audio-frequency Magnetotelluric (AMT) method (1–10,000 Hz), stable resistivity and phase changes can be monitored, offering insight into subsurface fluid behavior (Yamaya et al. 2018).
(7) Micro-earthquake monitoring: Microearthquakes, triggered by fluid movement, help locate active fractures and flow pathways. Hypocenter mapping reveals fracture networks within the reservoir.

If monitoring clearly indicates degradation of reservoir temperature, the following countermeasures may be considered:

- Reinjection wells showing significant breakthrough are decommissioned.
- New reinjection wells are drilled at greater distances from the production zones.
- Feed points that negatively affect production are sealed within the borehole.
- The reinjection strategy is revised with respect to the cold sweep process, fracture spacing, breakthrough risk, spatial separation (spread out), and temperature influence.

In all cases, these measures and associated drilling and monitoring efforts can involve substantial costs.

References

Ariki K, Akibayashi S (2001) Effects of the injection temperature on injection-capacity of geothermal wells—numerical study. J Geotherm Res Soc Japan 23(2):141–156 (in Japanese with English abstract)

Ariki K, Akibayashi S (2002) Effects of the injection temperature on injection- capacity of geothermal wells—numerical simulation on the measured downhole pressure. J Geotherm Res Soc Japan 24(2):161–175 (in Japanese with English abstract)

Bodvarsson G (1972) Thermal problems in the siting of reinjection wells. Geothermics, L. 2:63–66

Ehara S, Nishijima J (2004) Study on sustainable development of geothermal energy from geophysical monitoring of geothermal reservoir—a case study of repeat Gravity measurement. Jour Geotherm Res Soc Japan 26(2):181–193 (in Japanese with English abstract)

Kitao K, Ariki K, Watanabe H, Wakita K (1990) Well stimulation using cold-water injection experiments in the Sumikawa geothermal field Japan. J Japan Geotherm Ener Assoc (Chinetsu) 27(4):32–48 (in Japanese with English abstract)

Malate RCM, Aqui AA (2010) Steam production from the expanded two-phase region in the Southern Negros geothermal production field, Philippines. In: Proceedings of the world geothermal congress 2010, Bali Indonesia, 25–29 April 2010, 1–8

Malate RCM, O'Sullivan MJ (1991) Modelling of chemical and thermal changes in well PN-26 Palinpinon geothermal field Philippines. Geothermics 20(5/6):291–318

Oka D, Fujimitsu Y, Nishijima J (2014) Interpretation of the gravity changes pattern observed at the Takigami geothermal field, Oita Prefecture Japan. J Geotherm Res Soc Japan 36(2):75–84 (in Japanese with English abstract)

Pruess K, Narasimhan TN (1985) A practical method for modeling fluid and heat flow in fractured porous media. Soc Petrol Engi Jour February 14–26

Saito H, Tagomori K, Honda M, Kasagi J, Fujise Y, Ehara S (2006) Reservoir behavior inferred from Quasi-3D analysis of Gravity changes in the Hatchobaru geothermal field. J Geotherm Res Soc Japan 28(2):181–197 (in Japanese with English abstract)

Sato K, Ito N, Tsukuda K (2022) Geothermal geological survey and production well target: supplementary editions focusing on reinjection wells (Original title in Japanese). Iwate: E-pix, Ofunato Printing Inc., 125 pp, ISBN978-4-901602-74-7 (in Japanese)

Tokita H, Haruguchi K (1998) Strategy of reservoir management based on numerical simulation in the Hatchobaru geothermal field. In: Proceedings of an Asian geothermal energy 98, Subic Bay Metropolitan Auth. Subic Philippines, October 22, 1988, 89–96

Yamaya Y, Takakura S, Asanuma H (2018) Resistivity monitoring of a geothermal reservoir at the Okuaizu geothermal field, Japan using the AMT method. J Geotherm Res Soc Japan 40(3):159–173 (in Japanese with English abstract)

Chapter 14
Case Studies II: Well Targeting Field Examples in Japan

Ko Sato, Junzo Kasahara, and Narimitsu Ito

Abstract Five geothermal areas in northern and northeastern Japan were selected for well-targeting case studies. In the Noboribetsu-Carls area, drilling targets were identified through comprehensive analyses of geological correlations, MT, and electrical surveys. In the Nigorikawa area (Mori geothermal power plant), which lies in a funnel-shaped caldera, subsequent well targets focus on deeper and west to southwest fracture extensions based on Distributed Acoustic Sensor (DAS) data. The Matsukawa area, Japan's first commercial geothermal power site, has a natural vapor-dominated reservoir bounded by low-permeability structures, and potential targets include the boundary of intrusive rocks, the Sumikawa Fault, and deeper zones. In the Tazawako-tobu area, NW–SE to NNW–SSE trending fractures were identified as targets based on geological, geochemical, gravity, temperature, and CSAMT surveys. In the Wasabizawa area (Wasabizawa geothermal power plant), fault and fracture systems were elucidated through detailed feed point identification and Drilling-Induced Tensile Fracture (DTF) analysis.

Keywords Well targeting · Five geothermal areas · DAS analysis · Shielding structure · CSAMT survey · DTF analysis

---

K. Sato (✉)
Socio Engineering Co. Ltd, Tokyo, Japan
e-mail: satoukou6450@gmail.com

J. Kasahara
Shizuoka University, Shizuoka, Japan
e-mail: kasahara.junzo@shizuoka.ac.jp

N. Ito
NEWJEC Inc., Osaka, Japan
e-mail: itonr@newjec.co.jp

J. Kasahara et al. (eds.), *Challenges in Geothermal Development*, Advances in Geological Science, https://doi.org/10.1007/978-981-95-0553-1_14

14.1 Noboribetsu-Carls, Hokkaido, Northern Japan

14.1.1 Location Summary

Noboribetsu-Carls, the study area addressed in this section, is located in Noboribetsu City in the southwestern part of Hokkaido, northern Japan. This city is known for its geothermal activity, which has been utilized for various purposes including tourism and energy production. Although geothermal development studies have been conducted in the past, no actual development has yet been initiated (Fig. 14.1).

The northwestern part of the area features Quaternary volcanic topography with a plateau composed mainly of pyroclastic flow deposits, while the eastern part is characterized by lava flow topography and the distribution of fission craters. Figure 14.2 shows surface manifestations including surface geology, lineaments, topography, and gravity anomalies. The geological setting consists of Paleogene to Holocene volcanic rocks overlaid by pumice and alluvium, along with intrusive rocks such as granite and andesite.

The area contains two major hot springs: Noboribetsu Onsen and Carls Onsen. The chemical composition of Carls Onsen closely resembles that of a nearby geothermal survey well, suggesting that the hydrothermal water within the sedimentary rocks has been influenced by volcanic gases. Silicified zones have been identified in the deeper parts of the area, and carbon dioxide gas plumes containing hydrogen sulfide and carbon monoxide were observed in the deepest sections of former exploration wells. Quartz scale was found deposited inside the well casing.

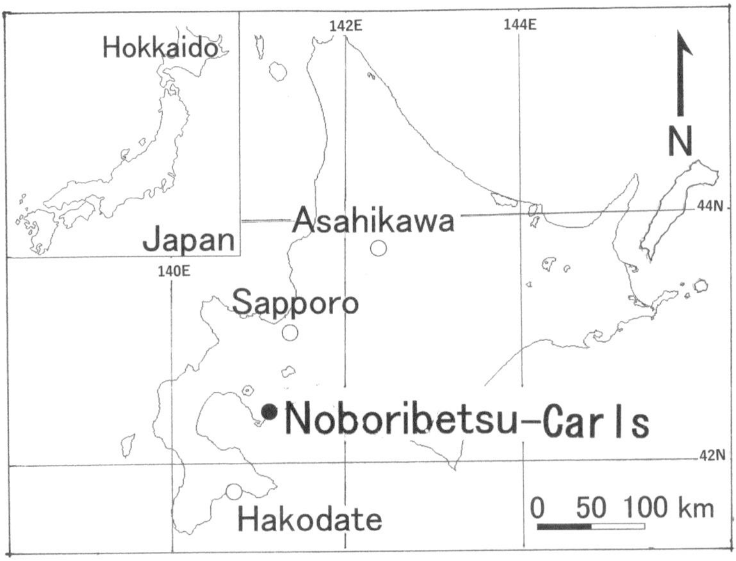

Fig. 14.1 Locality map of the Noboribetsu-Carls geothermal area

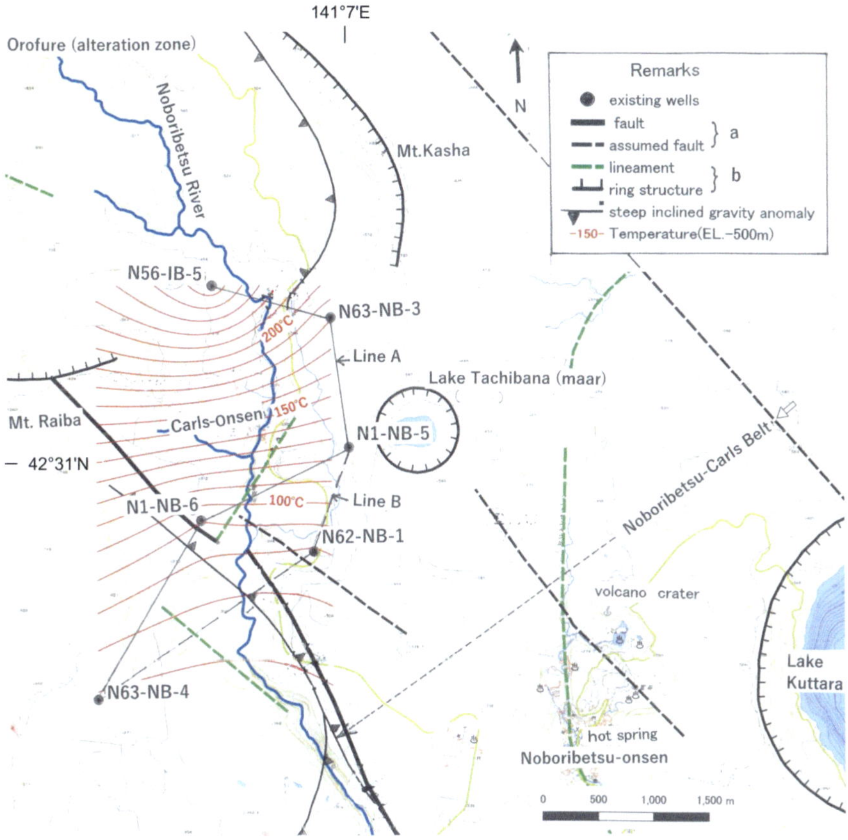

Fig. 14.2 Distribution map of locality, topography, and surface geology in the Noboribetsu–Carls geothermal area, including isothermal contours (red lines) and cross section lines (A: north–south, B: east–west; see Fig. 14.3). Remarks: "a" = surface geological survey; "b" = geomorphological analysis. The base map is from the Geospatial Information Authority of Japan

These findings indicate the presence of high-temperature thermal water saturated with silica at depth near Carls Onsen. It is inferred that hot water ascending along fractures from the deep silicified zones mixed with groundwater, resulting in Na–SO_4/HCO_3-type hydrothermal water stored within the sedimentary rock layers.

This section discusses a well-targeting study based on the results of previous geothermal surveys.

14.1.2 Previous Surveys

Surface surveys including geological, alteration zones, geochemical, magnetotelluric (MT), and electrical investigations were conducted by NEDO (1991). Electrical surveys using the Schlumberger configuration and Mise-à-la-Masse method were applied to obtain 2D cross-sectional resistivity profiles and to estimate the planar distribution of resistivity anomalies. At that time, 3D MT analysis had not yet become common practice. Gravity surveys conducted by private companies were also integrated into comprehensive analysis.

Six exploratory geothermal wells were drilled to depths ranging from 1,000 to 1,500 m. Geological logging, temperature and electrical logging, temperature recovery tests, water injection tests, and core analyses were performed. Three wells—N63-NB-3, N1-NB-5, and N1-NB-6—recorded bottom-hole temperatures of 221.2 °C, 250.7 °C, and 154.6 °C, respectively. The temperature in N1-NB-5 was measured only by a maximum registering thermometer due to deterioration of downhole conditions during drilling.

Figure 14.2 shows the well locations. From a regional perspective, a ~5 km-wide NW–SE trending Carls–Noboribetsu geothermal belt is notable, with manifestations concentrated around Lake Kuttara, Lake Tachibana, and the altered zone near Orofure Pass. The area between Mount Raiba and Kasha is interpreted as an eroded volcanic caldera (NEDO 1991). The map also shows inferred faults, lineaments, ring structures, and gravity anomalies. Cross sections A and B are presented in Fig. 14.3. Isothermal contours at −500 m elevation were created using temperature recovery test data from five wells, excluding N1-NB-5 due to its poor condition caused by high temperatures and gas blowouts.

The permeability–thickness product (*kh*) was not obtained for N1-NB-5 due to borehole collapse, but values of approximately 10^{-15} to 10^{-16} m^3 were estimated for N63-NB-3 and N1-NB-6 via pressurized water injection. Swabbing was attempted in N63-NB-3 but was unsuccessful because casing pipes covered the main feed zones.

The three wells are located near Carls Onsen and are separated by 0.5 to 1.5 km. Given the high measured temperatures, the geothermal potential in the vicinity of Carls Onsen is promising. However, since permeability is unclear, understanding the fracture system is crucial to identifying suitable targets for production and reinjection.

Two geological cross sections (Fig. 14.3) were newly created based on correlated lithologic logs. One section runs north–south through five wells (N56-IB-5 to N63-NB-4) and the other east–west through three wells (N1-NB-5 to N63-NB-4). The left side of each well log indicates drilling depth from the surface, while the right shows lost circulation rates (L/min). Major stratigraphic dip angles were calculated from tangent angles between wells. Notably, large dip angles of 21.5° and 26.5° were observed between N1-NB-5 to N62-NB-1 and N1-NB-6 to N63-NB-4, respectively, suggesting the presence of faults.

It is also worth noting that while geological cross-sectional views are commonly produced in geothermal surveys, they are often based on equally spaced wells regardless of true scale. However, for evaluating drilling targets, it is important to use a 1:1

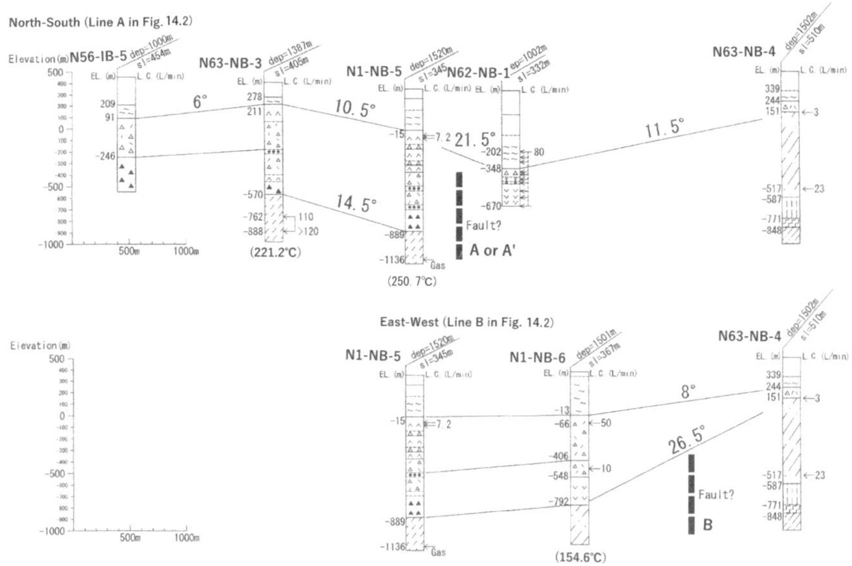

Fig. 14.3 Geological correlation and lost circulation (LC) during drilling across six boreholes in the Noboribetsu–Carls geothermal area. A, A′, and B refer to the lines shown in Fig. 14.4. Parentheses indicate bottom-hole temperatures

horizontal-to-vertical scale. This approach enables intuitive assessment of faulting by aligning formation boundaries between wells and clearly visualizing their dip angles.

14.1.3 Schematic Setting of Drilling Targets on the Plan View

Based on the above information, drilling targets for production and reinjection wells are illustrated in Fig. 14.4. The ellipses indicate the positions of resistivity structure profiles from previous two-dimensional electrical surveys; their minor axes are aligned parallel to each profile's orientation. According to the spatial distribution of these profiles, the presumed structural drop between N1-NB-5 and N62-NB-1 is likely influenced by a NW–SE trending fault (Line A in Fig. 14.4).

A NE–SW trending fracture (A′ in Fig. 14.4) appears to correspond with the rim of an eroded caldera. The fracture system between N1-NB-6 and N63-NB-4 (Line B) seems to be a result of both NW–SE and NE–SW structural trends. Another orthogonal NE–SW fracture may also be inferred between N56-IB-5 and N63-NB-3 (Line C). This fracture network is likely associated with abrupt gravity changes, ring structures, and the path of the Noboribetsu River.

The production well target is set to intersect a steep gravity gradient near N56-IB-5 (NEDO 1983), where the highest temperature at −500 m elevation was recorded.

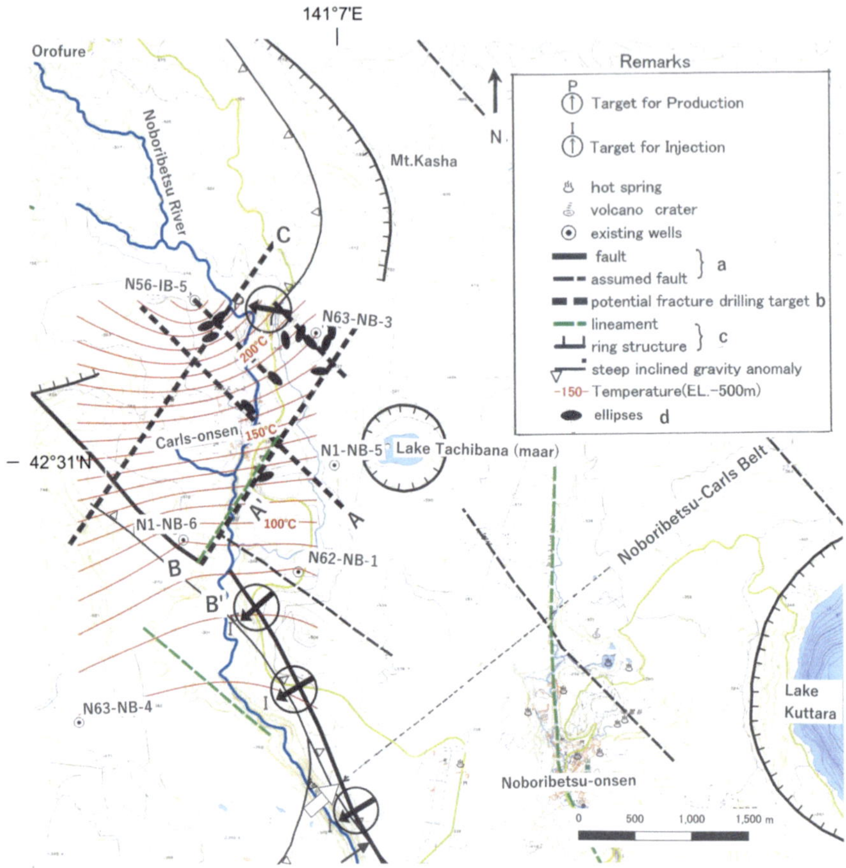

Fig. 14.4 Drilling targets for production and reinjection wells based on comprehensive analyses of geology, magnetotelluric data, and electrical surveys. Remarks: **a** surface geological survey; **b** resistivity discontinuities identified from magnetotelluric and electrical surveys; c—geomorphological survey; d—resistivity structure profiles from Mise-À-La-Masse method

The drilling trajectory is oriented to intersect both the presumed NE–SW fracture (C) and the Noboribetsu River fault line. A geological cross section comparison around N1-NB-6 shows a more than 1000 m basement drop on the river's east bank (NEDO 1991), supporting the existence of a major fault.

It should be noted that the isogravity contours are influenced not only by the belt-like structures but also by circular (ring) structures. The high temperature (250 °C) observed at 1500 m depth in N1-NB-5 is possibly due to a remnant ring structure, such as a former crater (maar) around Lake Tachibana. Although targeting this feature is feasible, it was excluded from the drilling plan due to its likely isolation, limited area for development, and poor drilling conditions (e.g., gas blowouts and borehole collapse).

Reinjection well targets were proposed in three locations south of the production zones, where the formation temperature is lower. As with production targets, these locations are selected where various structural features overlap—particularly where faults coincide with steep gravity anomalies.

14.1.4 Drilling Targets in Cross Section

Figure 14.5 shows the cross-sectional drilling targets for production and reinjection wells, assuming that the elevation differences between wells are caused by faulting. This figure was drawn with a 1:1 aspect ratio, allowing intuitive planning of inclined (directional) wells toward the target zones. The stratigraphy has been simplified into three to four major layers. Fractures between wells are assumed to be vertical due to a lack of dip data. Lost circulation depths are indicated for each well.

The production well target is set near the fracture (C in Figs. 14.4 and 14.5) between wells N56-IB-5 and N63-NB-3. Since the fracture dip is unknown, the well is designed to be sufficiently long to intersect the fracture. The bottom depth is set at EL −1000 m, with the well also intersecting the lost circulation zones between EL −762 m and −888 m in N63-NB-3. The inclined section of the well is planned to reach the target with a 20° deviation, after vertical drilling a few hundred meters from the surface. Measurement While Drilling (MWD) tools are assumed to control the trajectory from the kick-off point to the target.

Given these parameters, the wellhead of the production well is placed approximately 200 m from N63-NB-3. It is important to note that the surface conditions and site availability are not considered in this study; decisions were made solely based on subsurface geology.

The reinjection well target is designed to pass through the boundary between the second (D) and third (T) layers, and cross the nearly vertical fracture B′. The well also exceeds the lost circulation zones observed from EL −202 to −670 m in

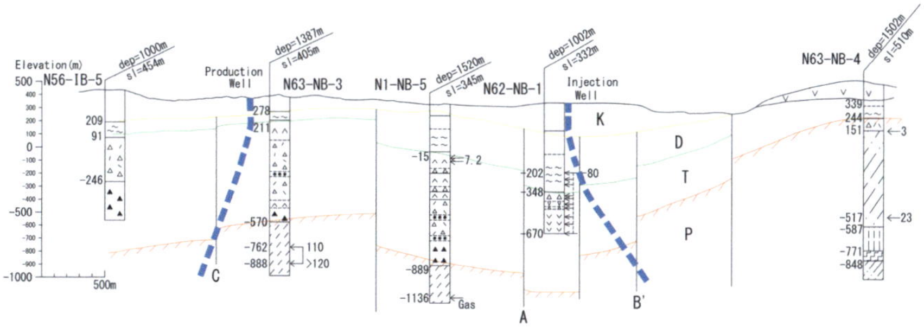

Fig. 14.5 Drilling trajectories plan by fracture analysis in cross section. K: Quaternary Kuttara volcanic rock, D: Carls débris, T: Tertiary tuff breccia and lava, P: Pre-Tertiary silicified rocks. A, B, and C correspond to fracture locations shown in Fig. 14.4

N62-NB-1. The kick-off point is below the first layer (K), and the borehole inclines toward the fracture zones.

This production-reinjection well configuration corresponds to Case B (Out of field) in Fig. 14.5. As the feed points of both wells are horizontally well-separated, the risk of thermal breakthrough is minimal.

14.2 Nigorikawa Geothermal Area, Hokkaido, Northern Japan

14.2.1 Nigorikawa Geothermal Area and Mori Geothermal Power Plant

Geological, hot spring, and oil surveys in the Nigorikawa geothermal area have been conducted since the 1910s by the Hokkaido Development Agency, the Geological Survey of Hokkaido (now the Research Institute of Energy, Environment and Geology), and the Geological Survey of Japan. Geothermal exploration began in 1972 by Japan Metals and Chemicals Co., Ltd. (JMC), which selected this area for continued development following a fundamental co-development agreement with Tohoku Electric Power Co., Inc. for the Kakkonda geothermal project.

Following the 1973 oil crisis, Hokkaido Electric Power Co., Inc. (HEP) supported geothermal development in this area to enhance energy security in southern Hokkaido and reduce dependency on oil. Development continued under Dohnan Geothermal Energy Co., Ltd. (DGE), in collaboration with JMC and HEP, leading to steam production and the construction of a 50 MWe power plant in 1977.

The Mori geothermal power plant, located within the Nigorikawa geothermal area in Mori Town, is situated in a roughly 3 km-wide circular basin (Figs. 14.6 and 14.7). The plant began operation in 1982, with DGE supplying steam and HEP responsible for power generation and plant operations.

After the start of operation, four major challenges emerged.

First, scale deposition of $CaCO_3$ and $CaSO_4$ inside production wells led to plugging. This was resolved by injecting scale inhibitors (e.g., polyacrylic acid) through coiled tubes directly into the wellbore fluid—a method that has since been adopted in other geothermal fields.

Second, silica and stevensite (a smectite-group mineral) scales accumulated in reinjection wells and pipelines. This issue was successfully addressed by pH control.

Third, the reservoir capacity proved insufficient for 50 MWe generation. Although the capacity was initially estimated using the volumetric method, significant pressure drawdown occurred shortly after operation began, leading to steam decline and inflow of low-temperature, low-chloride groundwater. A history-matching reservoir simulation conducted from 1988 to 1993 indicated a sustainable steam supply of 200–250 t/h (equivalent to 20–25 MWe). Based on this, the plant output was reduced to 25 MWe in 2010 and has since been managed solely by HEP.

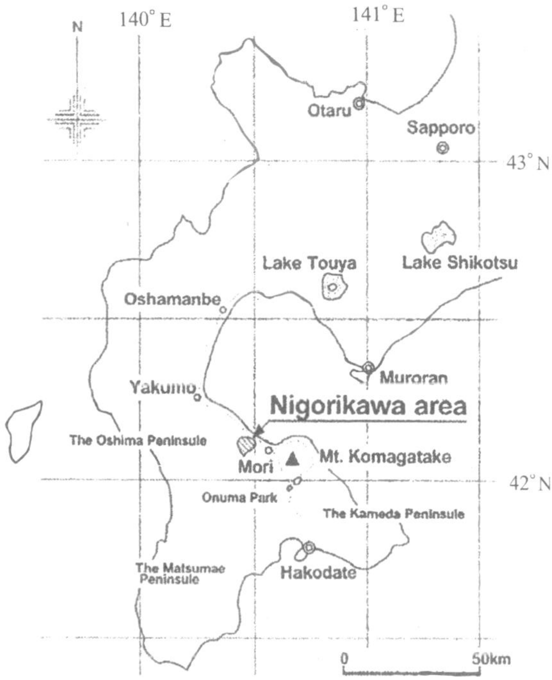

Fig. 14.6 Location map of the Nigorikawa geothermal area in southern Hokkaido, Japan

Fourth, reinjected water caused thermal breakthrough, cooling the reservoir and lowering the enthalpy of the produced fluid. This resulted in reduced steam production. To mitigate this, a spread-out strategy was implemented using tracer tests, temperature data, and chloride concentrations to increase the separation between production and reinjection zones.

The latter two issues—reservoir capacity and reinjection breakthrough—remain unresolved. Hanano et al. (2005) provided a detailed account of these problems.

14.2.2 Geological Setting in the Nigorikawa Geothermal Area

(1) Subsurface configuration of the Nigorikawa caldera

The Nigorikawa geothermal area is characterized by a basin-shaped structure interpreted as a funnel-shaped caldera. Detailed subsurface mapping reveals that the caldera has a pentagonal outline, with a broad upper section measuring approximately 3 km × 2.5 km, narrowing to a lower section of about 0.7 km × 0.5 km at depths exceeding 1,000 m below sea level. The caldera elongates in the NE–SW direction, forming a roughly rectangular shape (Fig. 14.8). The caldera wall slopes at an angle of about 60° to 70° in the upper part and becomes nearly vertical in the

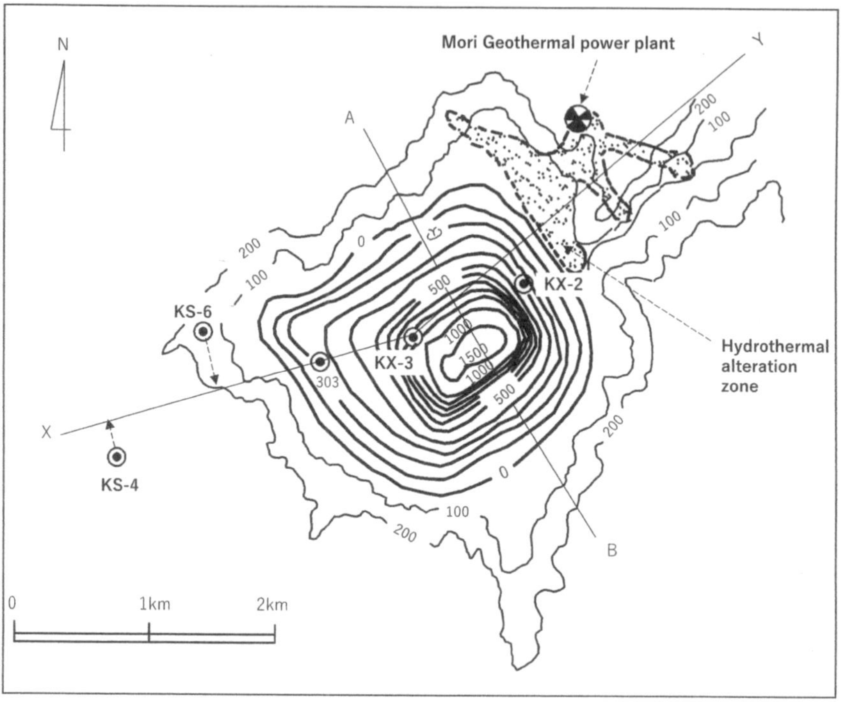

Fig. 14.7 Map of the Nigorikawa geothermal area and the Mori geothermal power plant. Solid lines represent subsurface caldera wall contours at 100 m and 500 m intervals below sea level. Wavy lines indicate surface topographic contours. Lines A–B and X–Y correspond to the cross sections shown in Figs. 14.8 and 14.21, respectively. Exploration wells KS-4 (= N3-KS-4), KS-6 (= N4-KS-6), 303 (= NT303), KX-2 (= N63-KX-2), and KX-3 (= N2-KX-3) are labeled and also appear in Figs. 14.8 and 14.21

lower part. This configuration is delineated by data from more than 42 geothermal wells, including one reaching a depth of 3,250 m.

(2) Caldera fill deposits and caldera formation process

The caldera, ranging from the surface to 824 m below sea level, is infilled with alluvial deposits, lake deposits, post-caldera andesite intrusions, and unconsolidated vent-fill materials (lapilli tuff) in descending order. The deep strata of the caldera consists of consolidated vent-fill materials and andesite to dacite intrusions which occupy the whole vent space beneath 2,300 m below sea level. The vent-fill materials are non-welded lapilli tuff and tuff breccia including accidental lithics and shattered fragments such as pumice, andesite, tuff, slate, chert, granite, and hornfels. Moreover, the vent-fill materials are characterized by stirred fragments and accretionary lapilli. The lapilli are treated as a "fallback" deposit but occupy not more than 30% of the total vent fill.

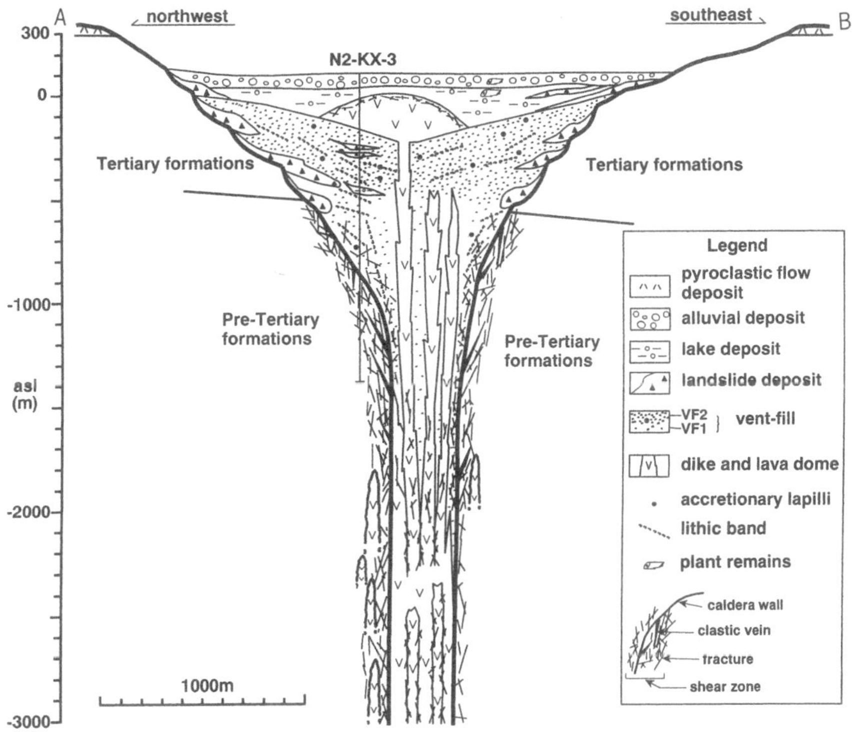

Fig. 14.8 Schematic cross section of the Nigorikawa caldera (after Kurozumi and Doi 2003). The cross section line A–B is shown in Fig. 14.7

The Nigorikawa caldera was formed about 12,000 years ago by violent volcanic activity involving the simultaneous formation of pyroclastic flow deposits and vent-fill materials. As a result, the main cause of the caldera's formation is attributed to a volcanic explosion during the pyroclastic flow eruption, accompanied by the crushing of basement rocks around the caldera wall.

The initial cause of caldera formation in this area can be inferred from the distribution of hydrothermal alteration zones—comprising opal, α-cristobalite, montmorillonite, sericite, and kaolinite—concentrated around the northeastern edge of the caldera. These alteration zones are overlain by recent pyroclastic flow deposits, indicating that a hydrothermal convection system existed prior to the caldera formation. In other words, fracturing of the basement occurred before the caldera formed, enabling the volcanic vent to open along the pre-existing fractures.

(3) Fracture appearance around the caldera wall

The geothermal reservoir in this area is entirely related to the caldera structure, especially the crushing structures of the caldera wall, the basement around the caldera wall, and caldera fill deposits. The detailed observation of the continuous core of the N2-KX-3 well makes clear the crushing structures of the caldera (Kurozumi and Doi

2003) as follows. The caldera wall is not in a straight fault line but shear zones about 30 cm wide. Abundant micro-open cracks are distributed around the shear zones in a net-like structure without displacement.

A volcanic explosion accompanied by caldera formation affects the surrounding basement strata, resulting in shear zones as illustrated in Fig. 14.8. Based on the occurrence of clastic dikes and andesite intrusions, the shear zones extends approximately 130 to 500 m from the surface rim of the caldera. Shear zones are especially prominent to the northeast and southwest of the caldera, and these areas are considered useful targets for well placement. In particular, the clastic dikes are notable for their partially open spaces and the presence of calcite veins.

During the early stage of geothermal development, a significant issue arose regarding the basement rock. It had initially been assumed that a thick Tertiary formation underlay the caldera. However, data from the first exploration well, 301, revealed that the Tertiary layer was thin and instead a thick pre-Tertiary formation was present around the caldera. This finding suggests that the fractured zones in the basement is more limited than previously expected, as the pre-Tertiary units consist of hard, compact rocks such as chert, slate, limestone, and Jurassic pyroclastic rocks collectively referred to as the Kamiiso Group (or Kamiiso Complex, as detailed later). If a thick Tertiary formation had been present, wider shear zones would likely have formed due to the lower fracture toughness of the weaker rock units during the caldera-forming explosion.

The internal fracturing of caldera fill deposits can be inferred from lost circulation observed during drilling. Approximately 90% of lost circulation occurs within the caldera fill deposits, particularly within 150 m of the caldera wall. However, the extent of the fractured zones remains limited despite the presence of open cracks. In contrast, andesite to dacite intrusions are also associated with lost circulation, as the boundaries between multiple intrusive bodies often serve as permeable zones.

14.2.3 Distributed Temperature Sensor (DTS) and Distributed Acoustic Sensor (DAS) Surveys in the Nigorikawa Caldera Basin

14.2.3.1 Introduction

Kasahara et al. (2022) carried out the geothermal exploration using the F-01 geothermal borehole of the Mori geothermal power plant of Hokkaido Electric Co., Inc. They used optical fiber DTS and DAS in the F-01 borehole. An optical fiber system that was installed down to 2,100 m depth in the F-1borehole is shown in Fig. 14.9. (Kasahara et al. 2022) The F-01 borehole is dipping from the east to the west. DTS and DAS data were obtained at 2,100 m and 2,000 m depths, respectively. The temperature data at 1 m spacing were obtained by the DTS mode using the same optical fiber in the F-01 borehole. The continuous seismic data were obtained by the optical fiber DAS and 26 three-component surface geophones (Fig. 14.8). The DAS data were obtained by the ODH-4 interrogator with a 20 kHz ping rate and a 1 kHz decimation. The gauge length for the DAS and the data acquisition were chosen as 16.4 m, and 1 m, respectively. The IVI Envirovib seismic sources at eight locations were used (Fig. 14.9). The source at SP4 is located near the F-01 wellhead. At each location, 480 times (8 h) up-sweeps from 10 to 75 Hz frequency for 30 s and 30 s of sleep were repeated.

After transient-noise reduction and cross-correlation with the source sweep signal, all 480 seismic records were stacked together. This processing enhanced the signal-to-noise ratio by approximately 20 dB.

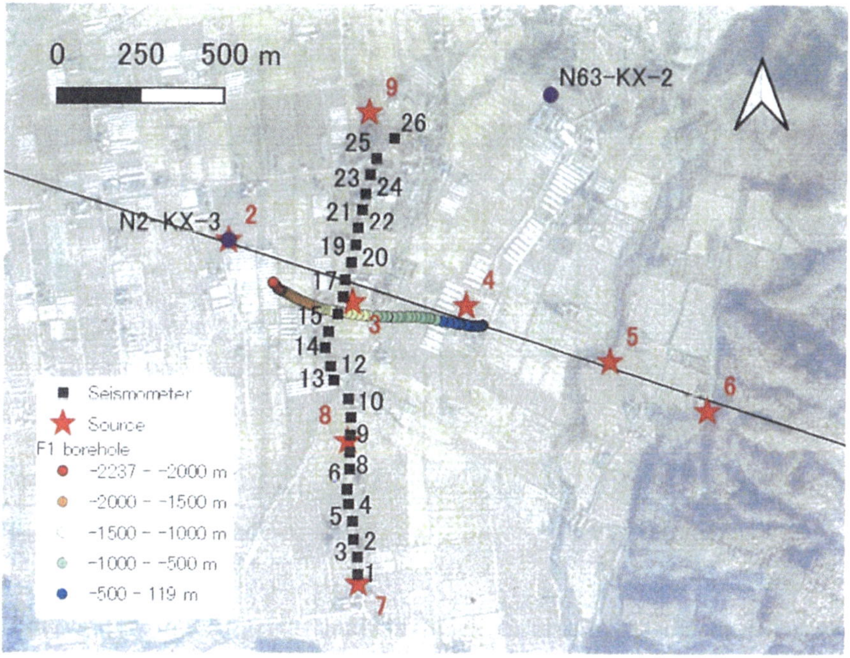

Fig. 14.9 Location map of the F-01 geothermal borehole, seismic sources (stars), and surface seismometers (square boxes) in the Nigorikawa geothermal area (Kasahara et al. 2022). The DAS along the F-01 borehole position is shown by small circles. The solid squares are surface geophones. The black line is the line for the velocity inversion. The red stars are seismic source positions. The purple circles are the locations of the wellheads of N2-KX-3, and N63-KX-2 (NEDO 1993)

14.2.3.2 Result

Using the optical fiber system in the F-01 borehole, Kasahara et al. (2022) obtained temperature and seismic data by DTS and DAS, respectively.

(1) *Temperature profile obtained by DTS measurement*

The results of the temperature measurement are shown in Fig. 14.10. The maximum temperature was approximately 204 °C at the bottom of the F-01 borehole. The boundary of air and water in the borehole is 60 m depth.

(2) *Seismic results obtained by DAS measurements*

After removing transient and spike noises and compensating source signatures, the stacked DAS seismic data were obtained from eight source locations. Figure 14.11 shows one example.

(3) *Observed DAS records*

Figure 14.12 shows the stacked DAS records for eight source locations. All DAS records show discontinuity at 1,060 m depth. The first arrivals at this depth strongly attenuate or show time delays. The trapped reflected phase within a 100 m thick zone is also seen around this depth.

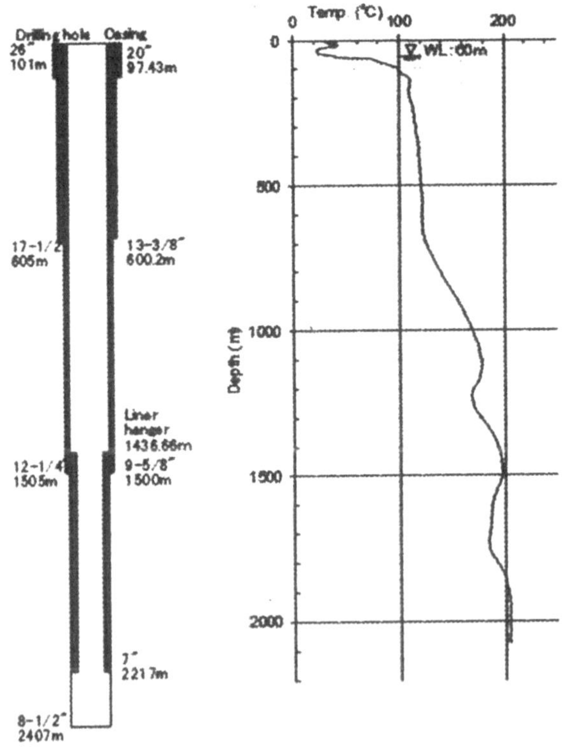

Fig. 14.10 (Left) Casing plan of F-01 borehole. (Right) Temperature profile in the F-01 geothermal borehole measured via DTS mode (Kasahara et al. 2022)

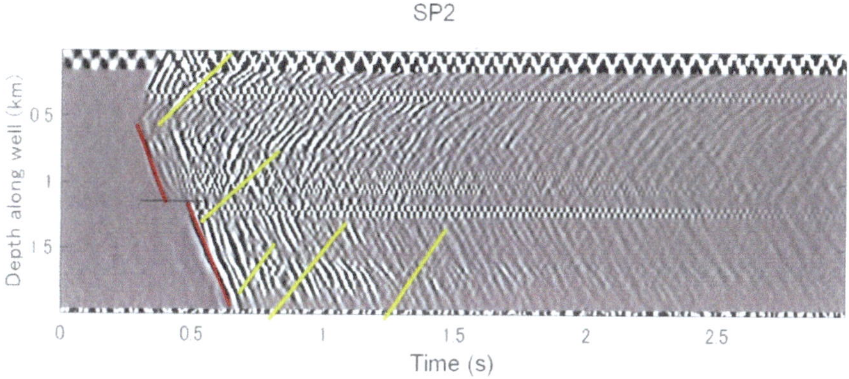

Fig. 14.11 DAS records of the source location SP2 at the western side of the F-01 wellhead (Kasahara et al. 2022). The vertical and the horizontal axes are the borehole depth in meters and the travel times in seconds, respectively. First arrivals (red lines) show travel time discontinuity at 1,060 m depth. Several reflected phases are identified (yellow lines)

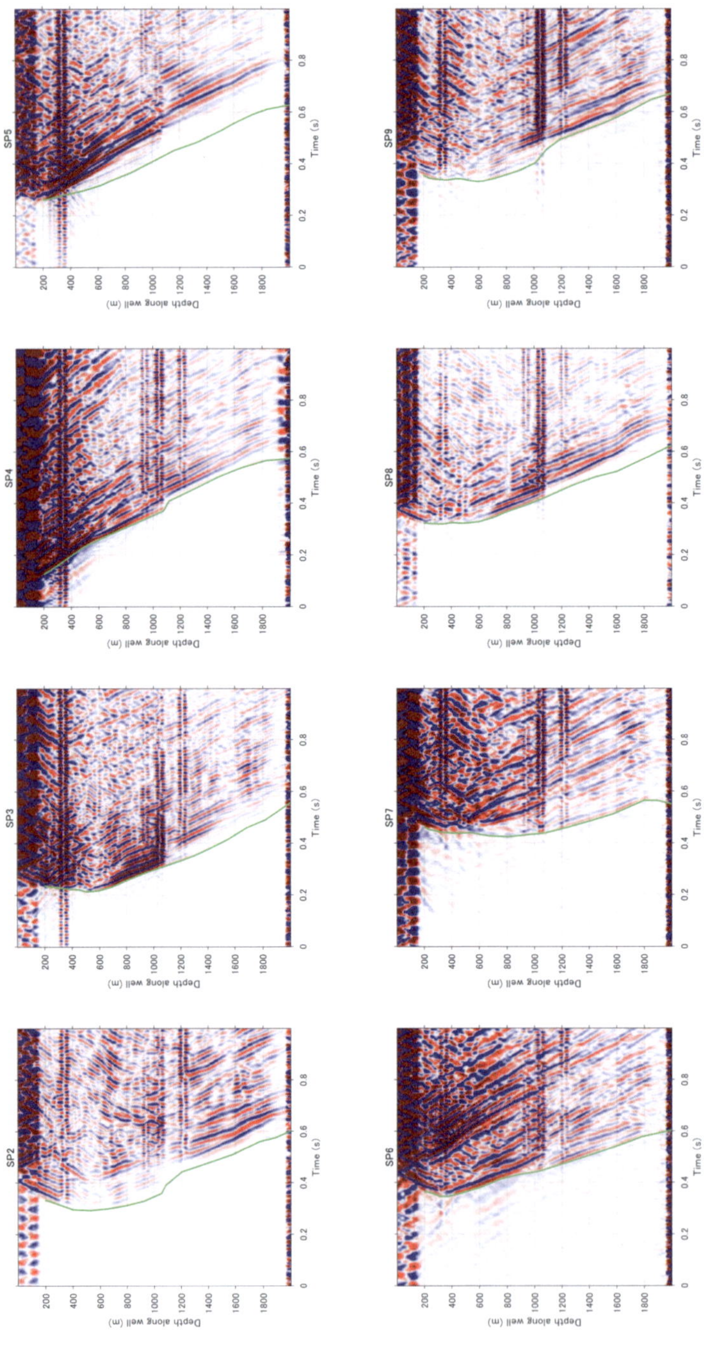

Fig. 14.12 Observed DAS waveforms for the eight-shot locations (Kasahara et al. 2022). Green lines are the first arrival readings for the tomographic travel time inversion. All DAS records show travel time discontinuity (shift) or strong attenuation of first arrivals at 1,060 m depth

14.2.3.3 Data Analysis

(1) *Separation of down-going phase and up-going phase*

The F-K velocity filter was applied to the stacked DAS records to separate the down-going and up going phases (Fig. 14.13). The up-going and down-going phases correspond to the direct arrivals and the reflected arrivals, respectively. Many reflected phases can be recognized in Fig. 14.13.

(2) *Velocity profile*

Using travel time readings of the first arrivals of DAS records, and surface seismic records, the 2D Vp profile along the east–west line was calculated by the tomographic travel time inversion method. The 2D Vp profile along the east–west line (Fig. 14.9) is shown in Fig. 14.14. Pink dotted lines show the Nigorikawa caldera wall proposed by Kurozumi and Doi (2003). The F-01 borehole shown by the black line is located in the caldera fill deposits. The Vp for the depth shallower than 200–300 m is less than 2 km/s. In the caldera below 200–300 m depth, Vp is approximately less than 4 km/s. The outer region of the caldera is comprised of the pre-tertiary layer with Vp of 4.5 km/s.

(3) *Synthetic DAS waveform calculation and comparison to observed DAS records*

Using Vp profile shown in Fig. 14.14, Vs and density profiles are assumed to calculate the synthetic DAS waveforms (Fig. 14.15). The synthetic DAS waveforms are calculated for five source locations (Fig. 14.16). Comparison of observed and synthetic DAS records, there are distinct differences for waveforms deeper than 1,060 m depth. The observed first arrivals are fitted to the ones of synthetics. However, the amplitudes of first arrivals between observed and synthetics below 1,060 m depth are quite different. Although observed ones show discontinuity of first arrivals at approximately 1,060 m depth, the synthetic ones do not show such discontinuity.

To examine the first arrival discontinuity at 1,060 m depth, alternative models are examined shown in Fig. 14.17 which places a flat layer 50 m thick, and 250 m long filled by water at 1,060 m depth. The five synthetic DAS records using this model are shown in Fig. 14.18. The results can explain travel time discontinuity at 1,060 m depth and the presence of a trapped phase within a 50 m thick layer. However, the reflected phase at 1,060 m is larger than observed DAS records.

(4) *Generation of Migration Image*

To evaluate the zones for reflections, we computed migration images along the east–west line using reflection using the F-K filtered sections shown in Fig. 14.19. The reflections are distributed across the caldera wall at this depth. The migration images are shown in Fig. 14.20. The reflectors are concentrated between 2,000 and 3,000 m depth.

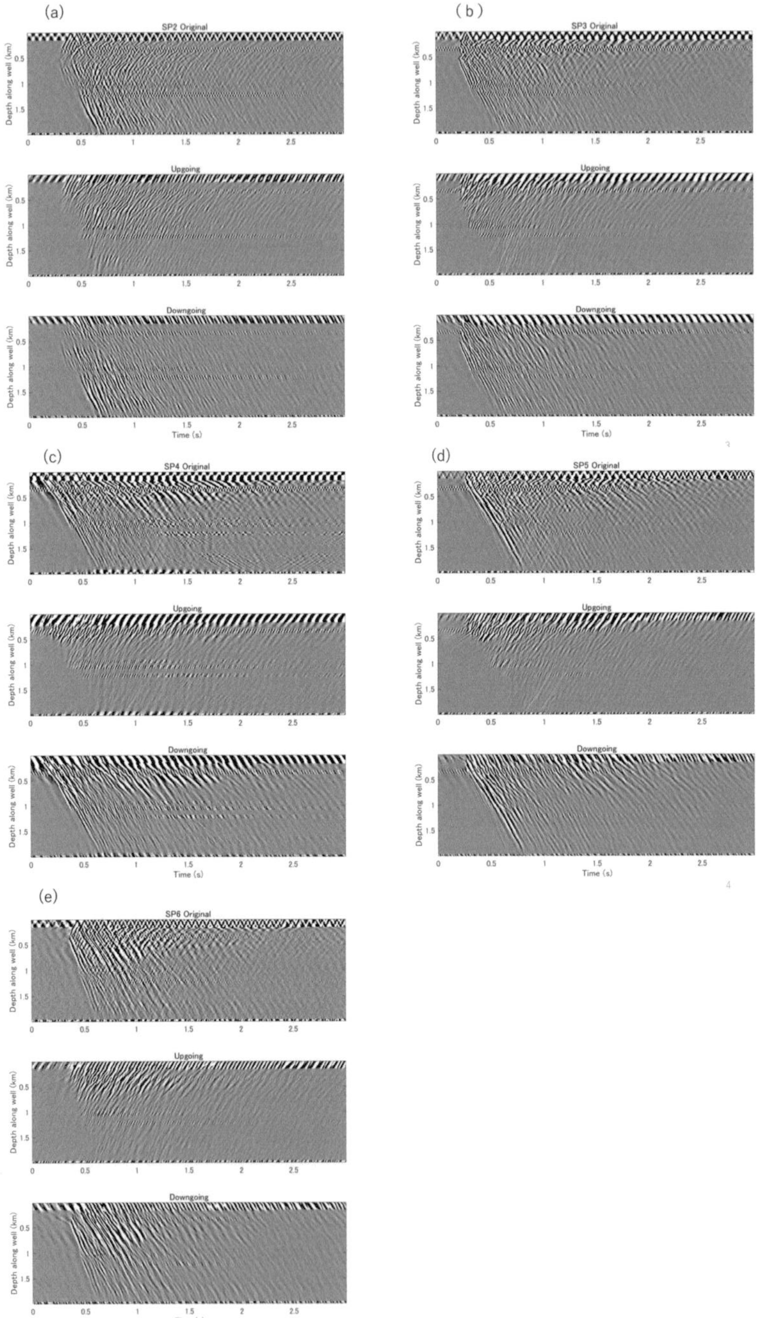

Fig. 14.13 DAS seismic sections after applying the F-K filter (Kasahara et al. 2022). The top of each (**a**)-(**e**) is the original record. The middle of each has up-going phases and the bottom has down-going phases. The reflected arrivals are shown as yellow lines

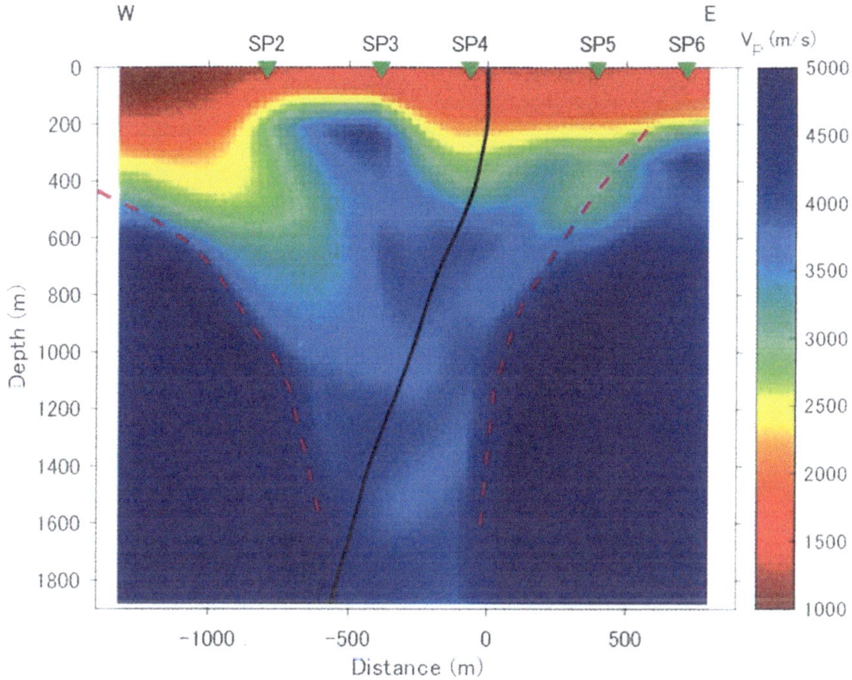

Fig. 14.14 Results of 2D Vp tomographic inversion for East–West profile (Kasahara et al. 2022). Zero in distance is the position of the wellhead. The F-01 borehole is shown as a black line. The green inverted triangles are source positions. The caldera wall model (Kurozumi and Doi 2003) is shown by pink dotted lines

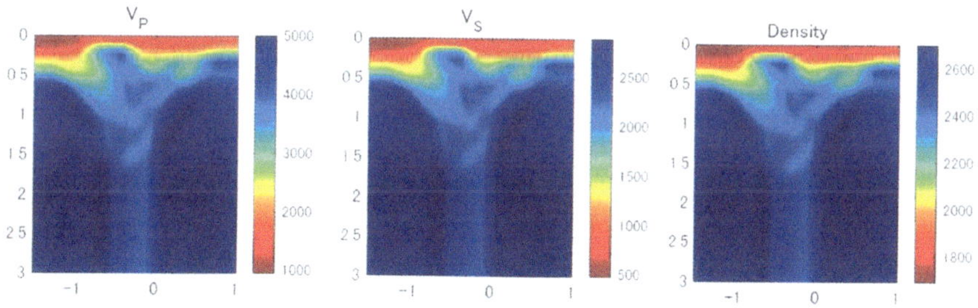

Fig. 14.15 Vp, Vs, and density model to calculate synthetic DAS waveforms. Vp model is based on the results of tomographic velocity inversion shown in Fig. 14.14. The units of Vp, Vs, and density are m/s, m/s, and kg/m^3, respectively

14.2.3.4 Discussion

The drilling operation of F-01 revealed the total circulation loss at 1,060 m depth (personal communication). The velocity profile model obtained by the tomographic

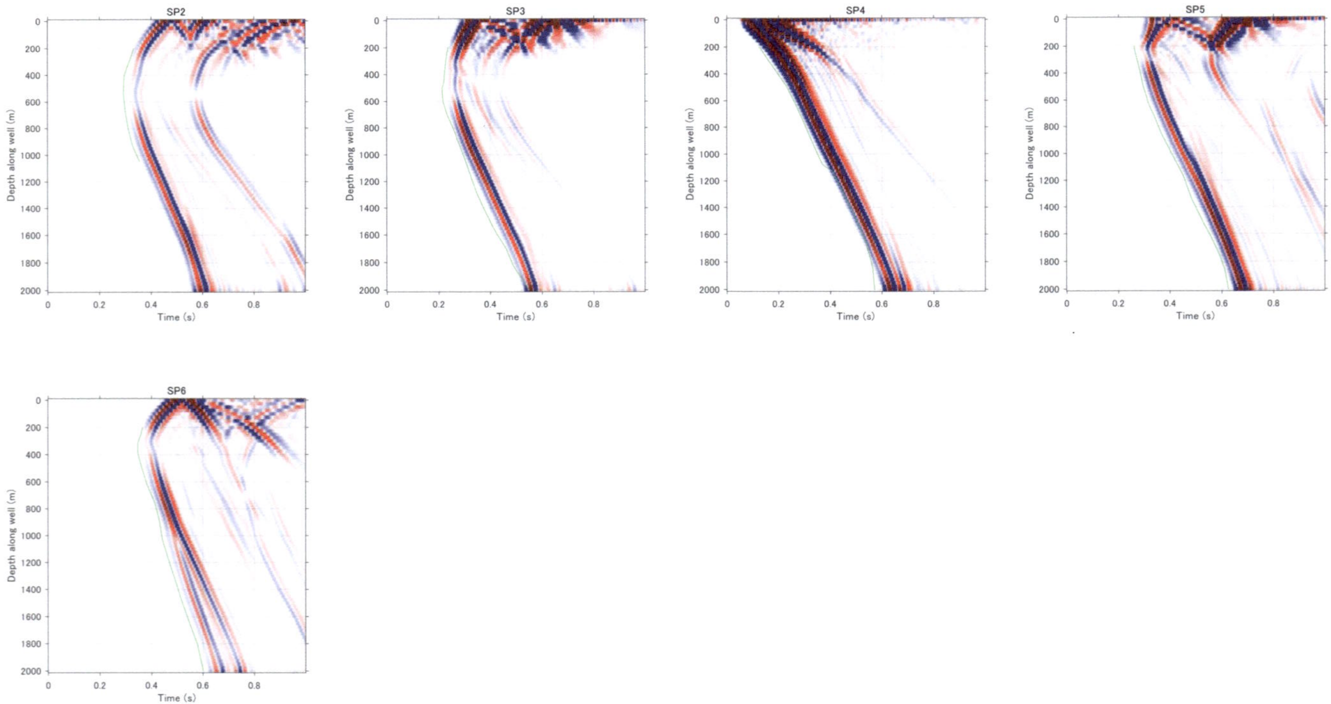

Fig. 14.16 Synthetic waveforms using the structural mode in Fig. 14.15. Green lines are calculated on first arrival and they fit to synthetic waveforms. However, the amplitude of first arrivals does not show strong attenuation at 1,060 m depth

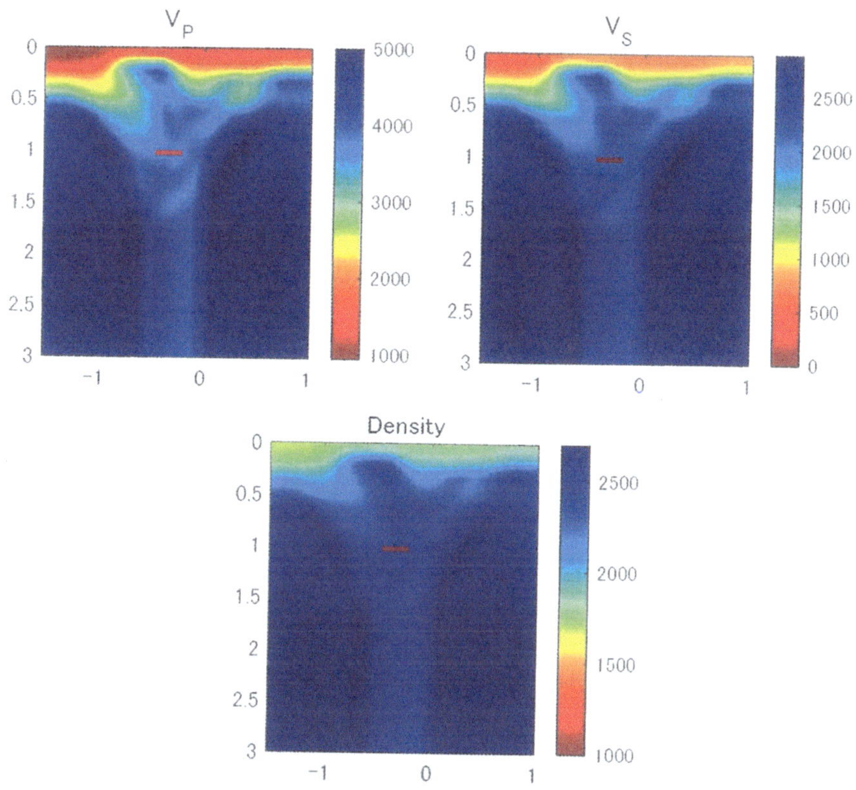

Fig. 14.17 Vp, Vs, and density models to explain the 1,060 m depth travel time discontinuity. The rectangular boxes (50 m thick and 250 m long) were filled with water place at 1,060 m depth

travel inversion does not have a discontinuous structure at 1,060 m depth although the travel time shifts at this depth are included in the travel time data. The synthetic DAS waveforms using this Vp and associated parameters do not reveal such discontinuous first arrivals at 1,060 m depth.

When we placed 50 m-thick and 250 m-long layers filled with water at 1,060 m depth, we could produce a similar waveform discontinuity at this depth. If such a layer exists at this depth, it might be caused by fractured zones filled with hot water. However, there is some discrepancy between observed and synthetics. The excess reflections at this depth were generated and there is not such strong reflection at this depth.

The Geomagnetic survey shows low resistivity for shallow deposits of the Nigorikawa Caldera, but the low penetration of electromagnetic waves did not resolve the deeper than 1,000 m (Kajiwara et al. 2000).

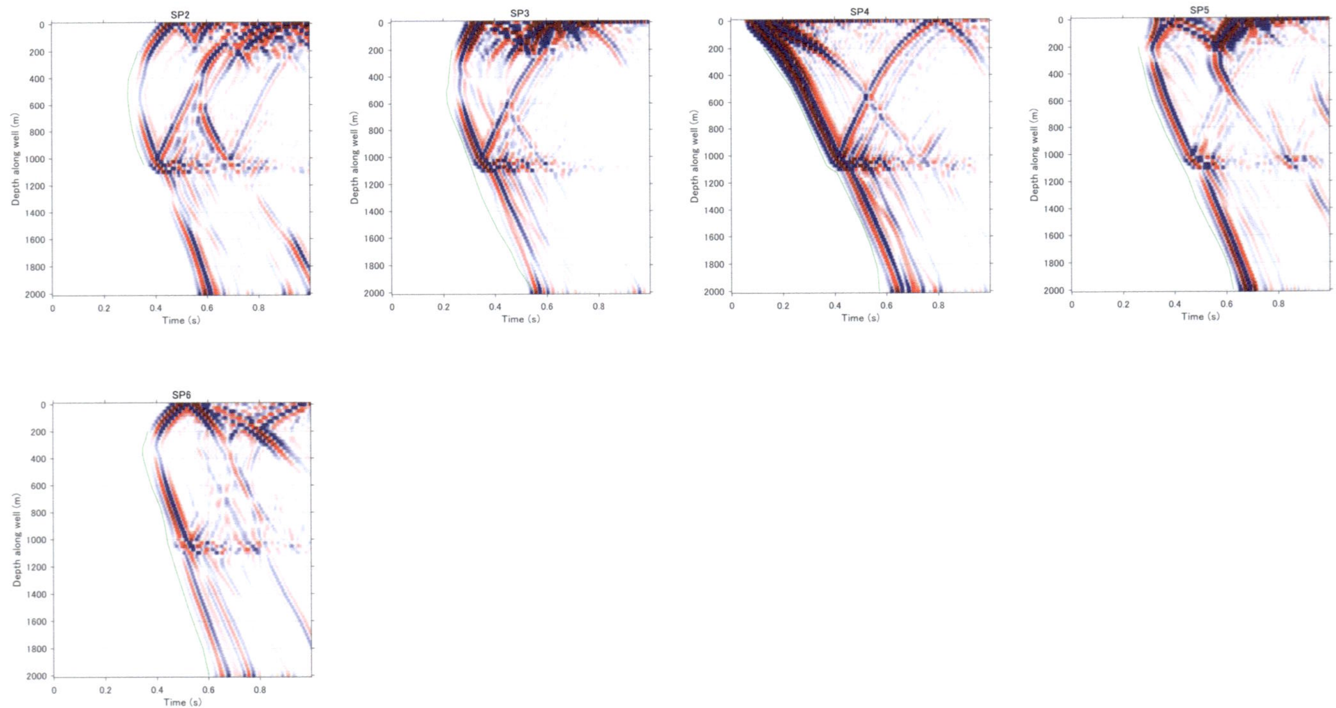

Fig. 14.18 Results of five synthetic DAS waveforms calculated by the model shown in Fig. 14.17. There is a travel time shift at 1,060 m depth and a trapped phase within a 50 m thick layer. However, the reflection at 1,060 m depth is larger than the observed ones

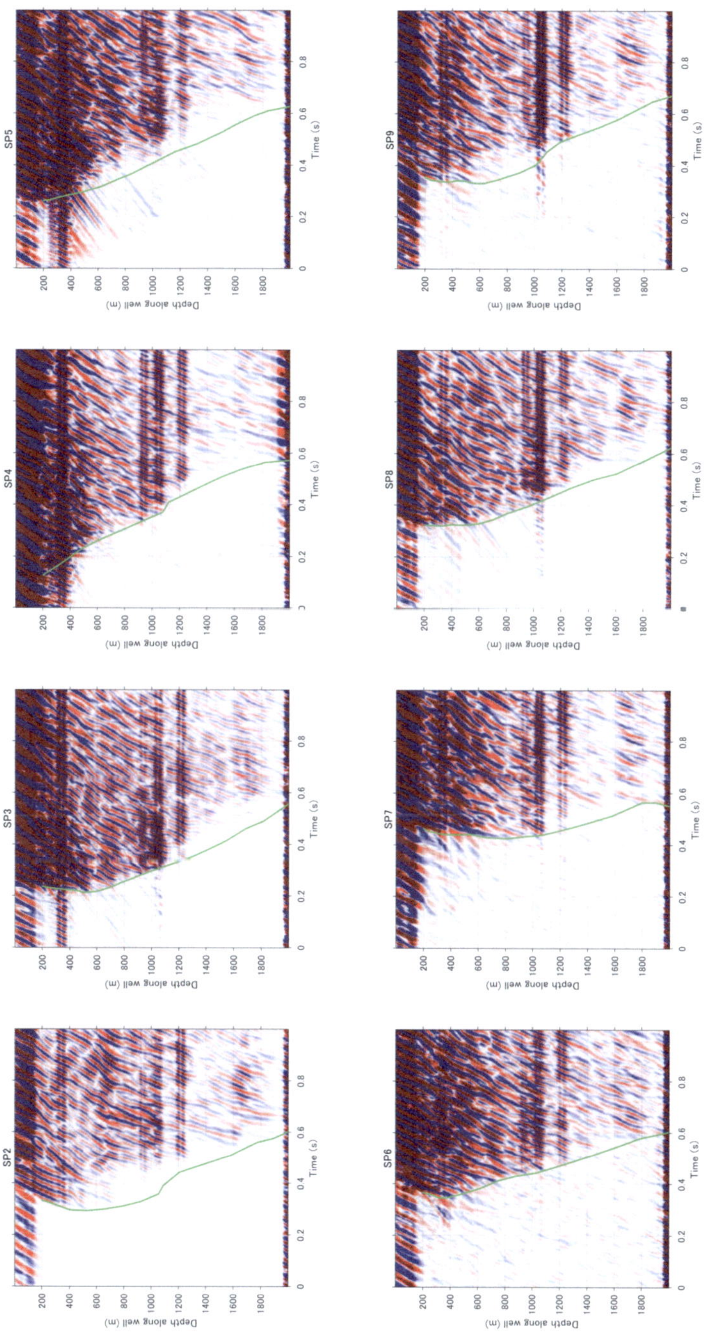

Fig. 14.19 Reflected phases obtained by the F-K filtering for eight shootings. Many reflections are identified for each section. The migration image along the east–west line shown in Fig. 14.20 is calculated using SP2–SP-7

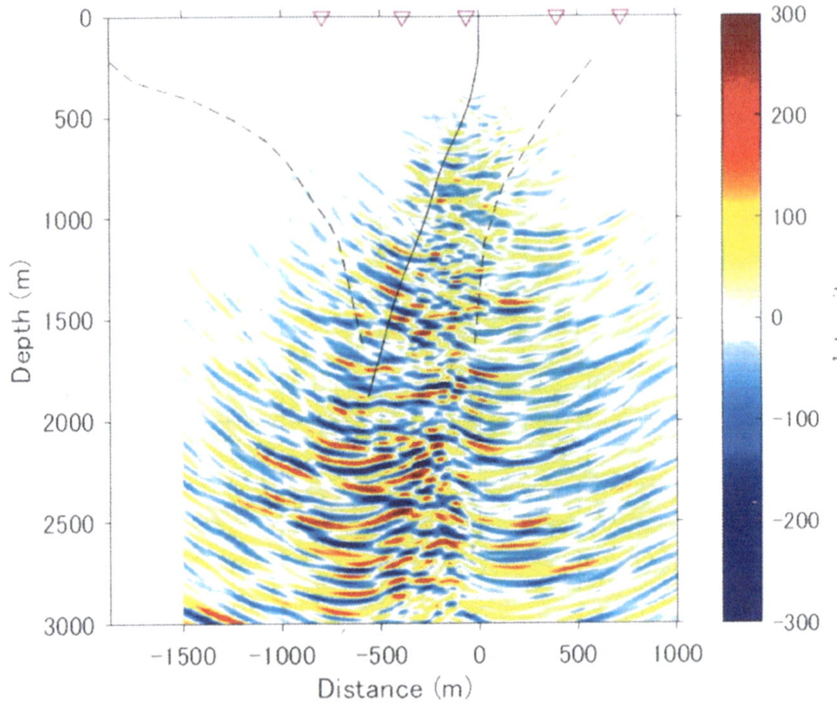

Fig. 14.20 Results of migration of reflected phases

14.2.3.5 Summary of DTS/DAS Survey

We carried out the DTS/DAS geophysical survey using the F-01 geothermal borehole of the Mori Geothermal Power Plant owned by the Hokkaido Electric Power Co. Inc., in the Nigorigawa Caldera basin. The optical fiber system was installed at a depth of 2,100 m depth. The temperature was measured by the DTS mode, and the highest temperature was obtained as 204 °C at the borehole bottom. The seismic records by eight source locations are obtained by the optical fiber DAS in the F-01 borehole and surface geophones at the ground surface. The DAS records show discontinuous (or travel time shift) first arrivals at 1,060 m depth. This might correspond to the total circulation loss at this depth encountered during the F-01 drilling.

The 2D Vp profile was obtained by the travel time tomographic inversion. The result shows the consistency of the caldera structure. We applied the F-K velocity filtering to the DAS records. By the velocity filtering, the reflected phases (up-going arrivals) are separated from the down-going arrivals. Many reflections originated at a depth deeper than 2,000 m depth.

Using the Vp profile and associated Vs and density profiles, we calculated the synthetic DAS waveforms and compared them with observed records. The synes-thetic waveforms obtained by the smooth velocity model obtained by the tomographic model do not show discontinuous first arrivals at 1,060 m depth, but the model with a layer of 50 m-thick and 250 m-long filled by water explains observed discontinuity at this depth. This might suggest the presence of vertical fracture zones filled with hot water at 1,060 m depth.

We made the migration image along the east–west profile using the five DAS sections of up-going arrivals. The result shows the gathering of reflections depth between 2,000 m and 3,000 m and this depth fits one of the production zones.

14.2.4 Summary of Permeable Fractures in the Nigorikawa Geothermal Area

(1) Three types of permeable fractures

Two major issues remain unresolved in the Nigorikawa geothermal area: limited reservoir capacity and the return of reinjected fluid. Addressing these challenges requires a comprehensive understanding of the permeable fracture systems. These fractures are classified into three main types, as shown in Figs. 14.7, 14.8, and 14.21:

Type 1: Fractures along the caldera wall, consisting of ~30 cm wide shear zones occupied by micro-open cracks.

Type 2: Fractures within the caldera fill deposits.

Type 2-1: Boundaries between multiple andesite to dacite intrusive bodies.

Type 2-2: Fractures within 150 m of the caldera wall inside the fill deposits.

Type 3: Fractures in Tertiary and pre-Tertiary formations surrounding the caldera.

Type 3-1: Shear zones located 130–500 m from the caldera rim.

Type 3-2: Hydrothermal alteration zones around the northeastern rim, where strata fractured before the caldera was formed (see Fig. 14.6).

Type 3-3: Fractures trending west to southwest from the caldera; this area has yet to be developed.

(2) Correlation between DAS analysis and geology

Numerous seismic reflectors identified through DAS analysis are concentrated outside the caldera vent at depths between 2,000 and 3,000 m (Type 3–1; see Fig. 14.20). This finding is significant for optimizing production well targets, as past developments have primarily focused on fractures along the caldera wall (Type 1) in deeper formations.

DAS waveforms show a distinct travel time discontinuity at a depth of 1,060 m near the F-01 well, which corresponds to the reinjection zones of hot water where

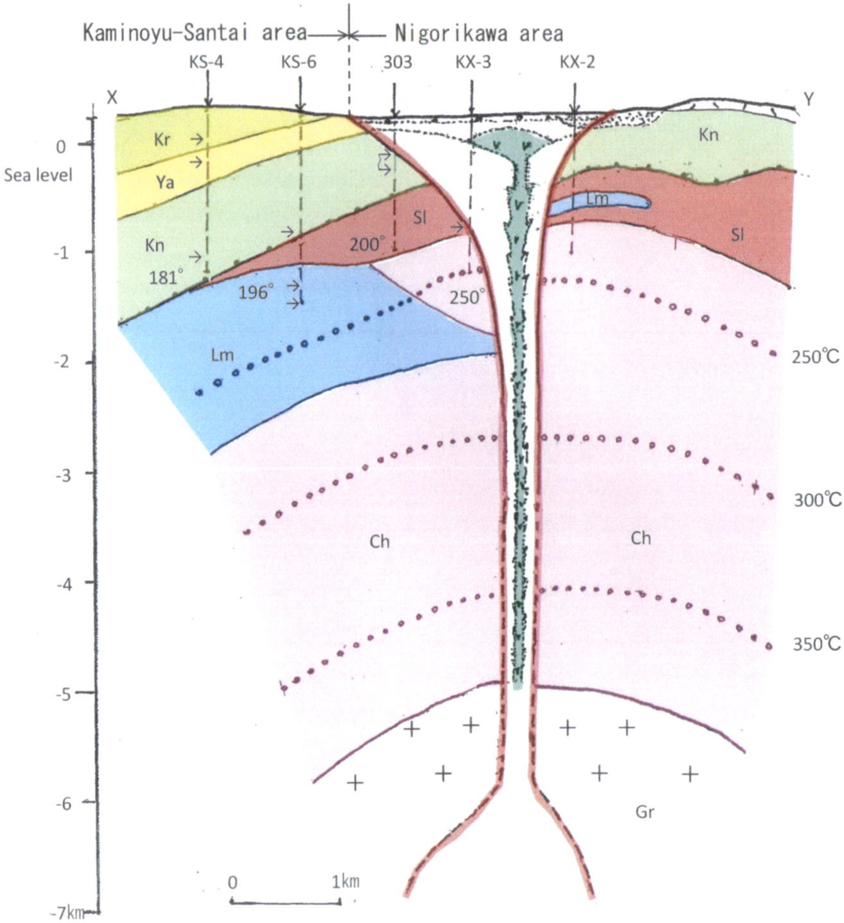

Fig. 14.21 Cross-sectional view from the Kaminoyu–Santai area to the Nigorikawa geothermal area. Line X–Y corresponds to the cross section trace shown on the plan view in Fig. 14.7. All well numbers are consistent with those in Fig. 14.7

lost circulation was observed during drilling. This discontinuity suggests the presence of vertical fracture zones. As the F-01 well penetrates the caldera fill deposits (Fig. 14.14), the lost circulation is likely associated with Type 2 fractures.

According to Fig. 14.21, multiple fracture zones of Types 1 through 3 are distributed around the 1 km depth level. However, reinjection into Type 1 and Type 2 fractures at this depth should be avoided. If the reinjection feed points are shallower than those of the production wells, the higher density of the reinjected water can accelerate cooling of the production zones. This is due to density-driven convection, where colder reinjected water sinks below the hotter reservoir fluid (see Sect. 13.2.2).

14.2.5 Well Targeting for Production and Reinjection Wells

(1) Production wells should target the following six zones:

Type 1: Caldera wall in the deeper formation

Type 2-1: Multiple andesite to dacite intrusions within caldera fill deposits at depth

Type 2-2: Fractures within caldera fill deposits in the deeper formation

Type 3-1: Fractures outside the caldera vent in the deeper formation, as revealed by DAS analysis

Type 3-2: Fractures below the hydrothermal alteration zones at the northeast edge of the caldera

Type 3-3: Fractures distributed in the west-to-southwest direction (see following section)

(2) Consideration of Type 3–3

New insights into fracture orientations in the deep-seated formation are needed. Drilling should avoid the northwest direction due to the presence of oil and gas emissions from shale, which hinder both borehole construction and power plant development. Exploration wells have also confirmed that the southeast direction exhibits low permeability. In contrast, the northeast and southwest directions are considered promising for targeting production and reinjection wells. As mentioned earlier, the northeast direction includes hydrothermal alternation zones and indicates the presence of a sheared basement extending approximately 500 m from the caldera margin (Type 3–2, Fig. 14.8).

The west-to-southwest direction for reinjection and production, where NEDO (1994) conducted surveys and drilled exploration wells, is considered a well-targeting area within the Kaminoyu-Santai geothermal zones, which partially includes the Nigorikawa area (Fig. 14.21). Miocene formations in this region include the Kr (Kuromatsunai Formation: siltstone, sandstone, and tuff), Ya (Yakumo Formation: mudstone), and Kn (Kunnui Formation: tuff and tuff breccia), as shown in Fig. 14.20. The Kamiiso Complex consists of Sℓ (slate), Lm (limestone), and Ch (mainly chert), while Gr represents a Cretaceous granodiorite (Fig. 14.21). For example, multiple instances of lost circulation occurred in the middle Miocene formations of the N3-KS-4 well, located about 1 km west-southwest of the caldera's sea-level edge. High-angle fractures in sandstone and mudstone were encountered at depths of 300 m and 505 m. A lost circulation at 571 m is attributed to auto-brecciated andesite lava. Further lost circulations at 1,280 m and 1,291 m, associated with quartz porphyry intrusions, yielded kh values ranging from 2.2 to 7.0×10^{-13} darcy·m.

The N4-KS-6 well is located approximately 300 m west of the sea-level edge of the caldera. In this well, lost circulation was observed between depths of 1,118 m and 1,180 m, attributed to alteration zones in limestone and slate containing calcite druses. Additional lost circulations occurred between 1,509–1,525 m and 1,721–1,734 m, caused by calcite and anhydrite druses in marbleized limestone. The zones yielded steam production of 1.3 to 2.7 t/h and hot water production of 7.2 to 16 t/h at a wellhead pressure of 0.4 to 0.5 MPa. These values were recorded with a borehole diameter of HQ (approximately 10 cm).

Partially lost circulation occurred between 303 and 524 m in the caldera wall and Tertiary formations in the NT-303 exploration well, as well as in the caldera wall in the N2-KX-3 well. A challenge remains in that these four wells are located far from existing infrastructure, necessitating long pipelines and extended plant systems. However, it is essential to leverage the findings from these wells to address the limited reinjection capacity and to prevent thermal breakthroughs between reinjection and production zones near the caldera wall. The current reinjection system, corresponding to Case E and Case F in Fig. 13.5, promotes reservoir cooling due to breakthrough occurrences. Therefore, the conditions represented by Case B or Case C in Fig. 13.5 are more desirable for reinjection in this area. We believe that recent technological advances—such as high-angle directional drilling, insulated polyethylene pipelines, and modern civil construction—can mitigate the risk of thermal breakthroughs and overcome reinjection limitations.

(3) Production systems to be adopted shortly

Two production systems are expected to be adopted in the near future. The first involves binary or partial single-flash power generation in the west-to-southwest direction around the caldera. Formation temperatures at this location include 180 °C at 1,500 m in the N3-KS-4 well, 196 °C at 1,800 m in the N4-KS-6, 200 °C at 1,300 m in the NT-303, and 250 °C at 1,500 m in the N2-KX-3. These temperature levels are sufficient for binary or single-flash power generation, provided the wells are not designated for reinjection. The second system targets geothermal fluid production from a deep-seated reservoir located more than 3 km below the surface, as shown in Fig. 14.21.

The geological profile from the surface down to 3 km below sea level is inferred from existing borehole data, with the deepest well reaching 3,250 m.

The section from 3 to 7 km depth is interpreted based on the following considerations.

First, the total estimated thickness of the Jurassic Kamiiso Group is approximately 5 km. This group is currently regarded as the Kamiiso Complex, an accretionary prism (i.e., chaotic deposit), making it difficult to estimate its detailed thickness. However, a supermassive limestone body measuring 5 km × 4 km has been identified within the complex, suggesting a deposit thickness of around 5 km assuming a spherical shape (Geol. Soc. Japan ed. 2010). The complex, composed of chert, slate, limestone, and pyroclastic rocks, is the most extensively exposed in southern Hokkaido, including the Nigorikawa area. The presence of chert and granodiorite is evidenced by accidental lithics and shattered fragments found in the vent-fill materials.

Second, seismic data from the Kaminoyu–Santai area show hypocenters located between 1.5 km and 7 km below sea level (NEDO 1994). Microearthquakes with S-P times of less than 2 s and magnitudes around 0.7 occurred between September 4 and October 20, 1990. These events were recorded within the chert layers of the

Kamiiso Complex and the overlying Cretaceous granodiorite above a discrete magma chamber.

Third, Mannen (2006) proposed a conceptual model for the formation of a "Nigorikawa-type" funnel-shaped caldera, estimating the depth from the surface to the magma chamber to be about 7.5 km. A similar model was presented by Kamata (1989) for the Shishimuta caldera in the Hohi VTD region of Kyushu, where the top of the residual magma chamber lies at approximately 7 km depth.

14.3 Matsukawa Geothermal Area, Iwate Prefecture, Northeastern Japan

14.3.1 A Brief History of the First Commercial Geothermal Power Plant in Japan, Matsukawa Geothermal Development

Tomioka Shigenori, the first president of JMC (Japan Metals and Chemicals Co., Ltd.; formerly Azuma-kako), established a ferroalloy manufacturing company in 1936, which required inexpensive and large quantities of electricity. One day, he happened to witness powerful steam emissions at Tateyama Hell Valley in Toyama Prefecture, central Japan. This encounter inspired him to consider geothermal power generation. He later consulted with Sato Sukekuro, president of Sato Kogyo Co., Ltd., who had visited a geothermal power plant in Italy around 1950.

On the other hand, the Matsuo local administration (now Hachimantai City) drilled seven boreholes in the Matsukawa area between 1952 and 1954 to promote tourism. However, three of them emitted only steam, not hot water, at depths ranging from 57 to 327 m. This information reached President Tomioka via the Geological Survey of Japan (GSJ). In response, JMC recognized the potential and launched a systematic exploration in 1956, including deeper drilling and borehole repair. In 1958, JMC and GSJ began a collaborative geothermal research and development project, with GSJ in charge of geological and geophysical surveys and analysis.

After that, several challenges had to be overcome before the success of the first production well, M-1, which produced a large volume of steam—60 t/h, equivalent to approximately 6 MWe per well—from a depth of 937 m in 1963. The first challenge was that two exploration wells, with depths of 350–425 m and casing shoes at 150–200 m, failed to sustain steam production due to inflow of shallow cold water. This issue was resolved by setting deeper casing and drilling down to around 1,000 m. The second challenge was JMC's financial management during the development phase. Fortunately, the Japan Science and Technology Agency provided financial support for the construction of four production wells, a power plant, and power transmission lines.

The power plant has been operated by the Tohoku Electric Power Co., Inc. group since 2003 and has continued successful power generation—currently 14,990 kWe—for more than 59 years (Figs. 14.22 and 14.23).

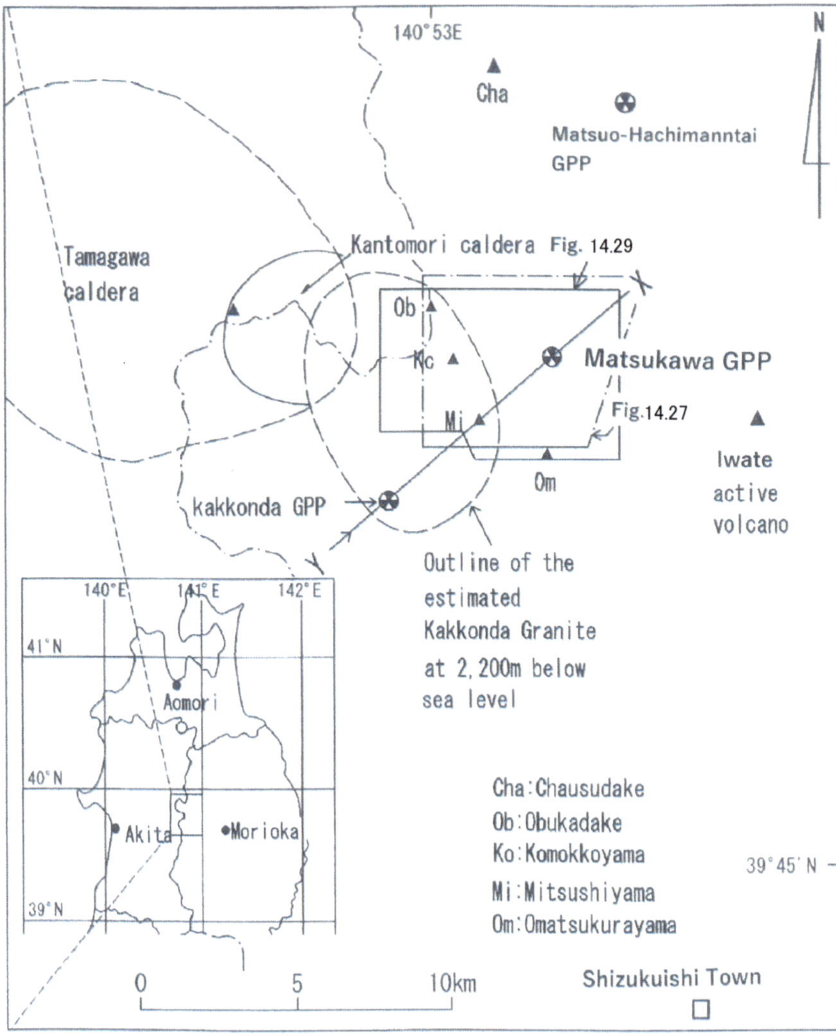

Fig. 14.22 Locality map of the Matsukawa geothermal power plant and its surrounding area. Line X–Y corresponds to the cross section shown in Fig. 14.26. "GPP" stands for geothermal power plant

Fig. 14.23 Autumn foliage scenery featuring the cooling tower of the Matsukawa geothermal station. Watercolor painting by professional artist Yoshiyuki Kudo, dated October 10, 2022

14.3.2 Characteristics of the Matsukawa Geothermal Area

The Matsukawa geothermal power plant is well known not only as Japan's first commercial geothermal power station—put into operation in 1966 with an initial output of 9.5 MWe and increased to 20–23.5 MWe in 1967—but also for its naturally occurring vapor-dominated reservoir identified during the early phase of development. Detailed studies on the vapor-dominated reservoir in the Matsukawa area have been actively conducted by Hanano et al. (1989, 1991, 1993) and Hanano (1994, 2003, 2011, 2017, 2021, 2022). Their research highlights the following three key aspects.

The first factor is the mechanism of superheated steam production from geothermal wells. Hanano (2021) proposes two mechanisms for superheated steam generation. The first involves steam produced by the boiling of immobile liquid adsorbed in microfractures, which then migrates through highly permeable fractures and enters the wellbore as superheated steam. The second mechanism occurs when two-phase fluids flash and dry out near the wellbore during discharge under exploitation, also resulting in superheated steam entering the well.

In the Matsukawa geothermal area, the superheated steam observed is attributed to the first, naturally occurring mechanism. Figure 14.24 illustrates the temperature recovery curves of the M-1 well, the first production well developed during the early exploration phase. Saturated conditions are evident from approximately 300 m to 700 m depth, overlain by an impermeable cap rock, with a pressurized liquid

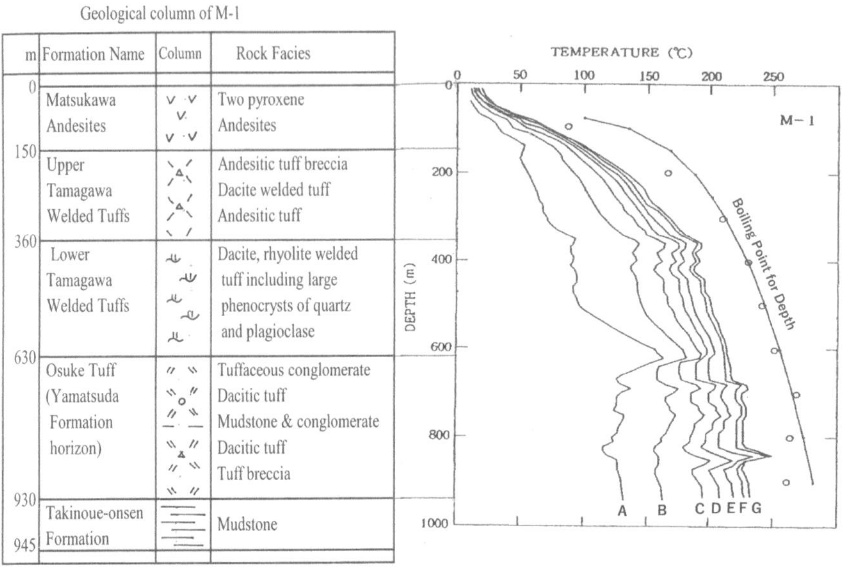

Fig. 14.24 Geological column and temperature recovery curves of the M-1 well. Temperature data are from Takagi and Tanaka (1968), and the geological column is drawn by the author. Open circles represent the inferred formation temperatures calculated using Horner's method. The standing time (ST) and water level (WL) for each temperature profile are as follows: A: ST = 8 h, WL = 147 m (Jan. 5, 1964), B: ST = 26.5 h, WL = 110 m (Jan. 6, 1964), C: ST = 39.2 h, WL = 95 m (Jan. 7, 1964), D: ST = 73.2 h, WL = 91 m (Jan. 8, 1964), E: ST = 107 h, WL = 85 m (Jan. 9, 1964), F: ST = 147 h, WL = 77 m (Jan. 11, 1964), G: ST = 167 h, WL = 75 m (Jan. 12, 1964)

condition present below 700 m (Hanano 2022). These data support the occurrence of the first mechanism.

The second factor is heat transfer by the heat pipe phenomenon occurring in the vapor-dominated geothermal reservoir. In this system, vapor rises along fractures, condenses beneath the cap rock, descends through fractures, and returns to the hydrothermal zones via latent heat transfer. The upward flow of vapor is driven by conductive heat flow (buoyancy), whereas the downward flow of condensed water is driven by gravity.

A conceptual model of this heat pipe mechanism in the Matsukawa area is illustrated in Fig. 14.25. During the vapor-dominated phase of the reservoir's initial state, vapor ascends along southwest-trending fractures, while condensed water descends near the northeastern shielding structure after 34 years of operation. Although boiling brine (or possibly no-brine zones) is speculated to exist below the vapor-dominated layer, drilling has not confirmed this.

Akatsuka et al. (2005) suggest the presence of a vapor–liquid two-phase fluid beneath the vapor-dominated zones, based on measurements of minimum homogenization temperatures in fluid inclusions (MHTF). Quartz within clastic dikes containing pyrophyllite and diaspore shows MHTF values of 261–288 °C, while quartz in dikes without these minerals shows lower values of 157–207 °C. These

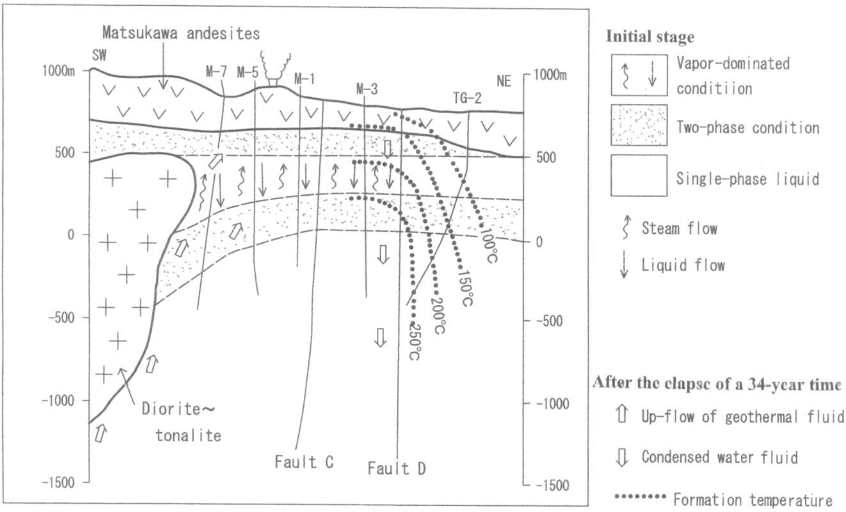

Fig. 14.25 Conceptual model combining the initial reservoir state and the condition after 34 years of operation. Note: the horizontal scale is not proportional

temperature variations indicate that both vapor and hot water are present in the subsurface of this area.

The third factor in the vapor-dominated reservoir is the presence of a low-permeability shielding structure (barrier). This structure, which encloses the reservoir at the top and lateral sides, must have permeability as low as approximately 5×10^{-17} m^2 (Hanano 2011) to prevent water intrusion, suppress vapor escape, and minimize heat loss. Such shielding confines both inflow and outflow, thereby maintaining reservoir pressure. During geothermal fluid production, pressure reduction occurs around production wells, enabling the discharge of superheated steam.

However, a fundamental question remains unanswered: Why is the Matsukawa reservoir vapor-dominated, while the nearby Kakkonda reservoir, located only about 7 km away, is liquid-dominated?

To explore this, we examine the geological conditions. A geological cross section between the Matsukawa and Kakkonda areas is shown in Fig. 14.26, and the corresponding plan view is presented in Fig. 14.27.

In Matsukawa, the reservoir is interpreted to be enclosed by a shielding structure composed of thick cap rock in the vertical direction and intrusive rocks in the horizontal extent. For instance, Ozeki et al. (2001) identified the northeastern boundary between production well M-3 and exploration well T-21 based on abrupt changes in formation temperature, pressure, and water level—despite the absence of direct geological evidence.

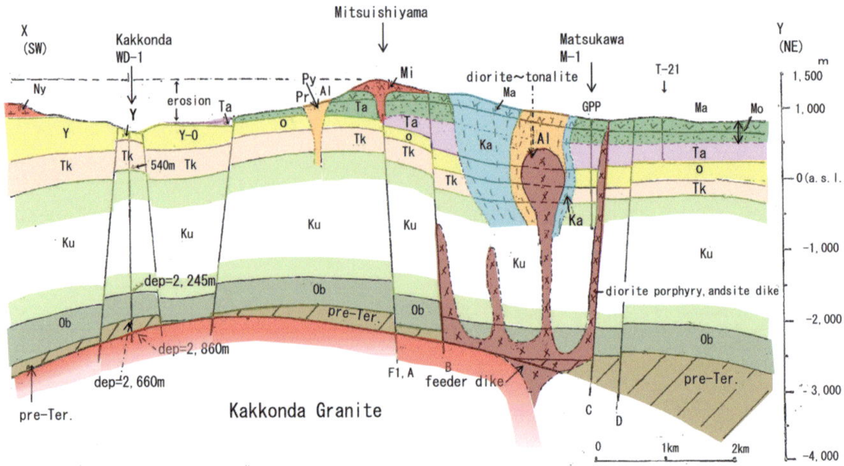

Fig. 14.26 Geological cross section from the Matsukawa area to the Kakkonda area. The location is indicated in Figs. 14.22 and 14.27. Symbols: Ny = Nyutosan Andesite; Mi = Mitsuishiyama Andesite; Ma = Matsukawa Andesite; Ta = Tamagawa Welded Tuffs; Y = Yamatsuda Formation (sandstone, mudstone, tuff, conglomerate); O = Osuke Tuff; Y–O = Yamatsuda–Osuke Tuff Formations; Tk = Takinoue-onsen Formation (mudstone, tuff); Ku = Kunimitoge Formation (green tuff, mudstone, andesite, basalt); Ob = Obonai Formation (altered volcanic rocks); Pre-Ter. = Pre-Tertiary basement. Alteration zones: Mo = Montmorillonite; Ka = Kaolinite; Aℓ = Alunite; Py = Pyrophyllite; Pr = Pyrite. Faults: A, B, C, D, F1

On the northwestern side, impermeable diorite to tonalite intrusive rocks form a natural barrier. This interpretation is supported by water level data from observation wells: well T-22, located inside the reservoir, shows water level decline corresponding to steam withdrawal, whereas well T-23, located outside the intrusive boundary to the north, has maintained stable water levels over time (Fig. 14.28; Fukuda et al. 2013).

In contrast, the southwestern and southeastern boundaries of the reservoir have yet to be clearly defined.

14.3.3 The Low-Permeability Shielding Structure in a Plane View

The geological structure of the Matsukawa reservoir in plan view is now interpreted based on several geological field observations (Figs. 14.29 and 14.30). The dominant structural trend is NE–SW, as evidenced by the alignment of hydrothermal alteration zones, the distribution of reservoir shut-in pressures from 1988 (Fig. 14.28), and gas concentration data (Yoshida and Ishizaki 1989). This NE–SW trend contrasts with the NW–SE structural trend typical of the broader Sengan region, including the Matsukawa area. The NE–SW structure is interpreted to have originated as a

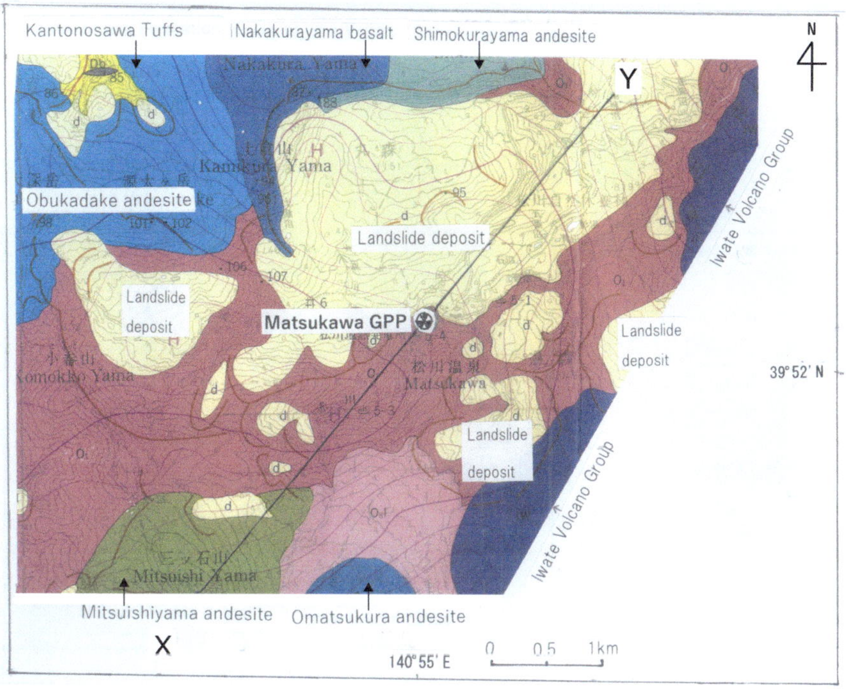

Fig. 14.27 Geological map around the Matsukawa area. (after Suto 1992)

transverse fault cutting across the Oppu–Morioka Tectonic Line (Fig. 12.8). It was later reactivated and extended during the southwestward movement of the Chishima Forearc Sliver from the late Miocene to the late Pliocene, resulting in NE–SW trending extensional fractures (Sect. 12.2.1).

(1) Southern Side

The geology of the Matsukawa area exhibits a superimposed structure resulting from overlapping NE–SW and E–W trends. The E–W trending structure is represented by Fault F1, which strikes east–west at Akidorisawa, a tributary of the Kakkonda River located approximately 4.5 km west of the Matsukawa area (Fig. 14.30). North of Fault F1, acidic alteration minerals such as kaolinite, alunite, and dickite are widely distributed. In contrast, moderately weakly alkaline alteration minerals containing chlorite and sericite are distributed to the south of Fault F1. Kimbara (1983) also confirmed this distribution in the field but described Fault F1 as trending WNW–ESE. Of particular importance is Fault A, which trends NNW–SSE and extends from Komokkoyama to Mitsuishiyama, as described by Takatsuka and Takashima (1991). Fault F1 is offset by Fault A, shifting from the Akidorisawa outcrop to the southern edge of the acidic alteration zones in the Matsukawa area.

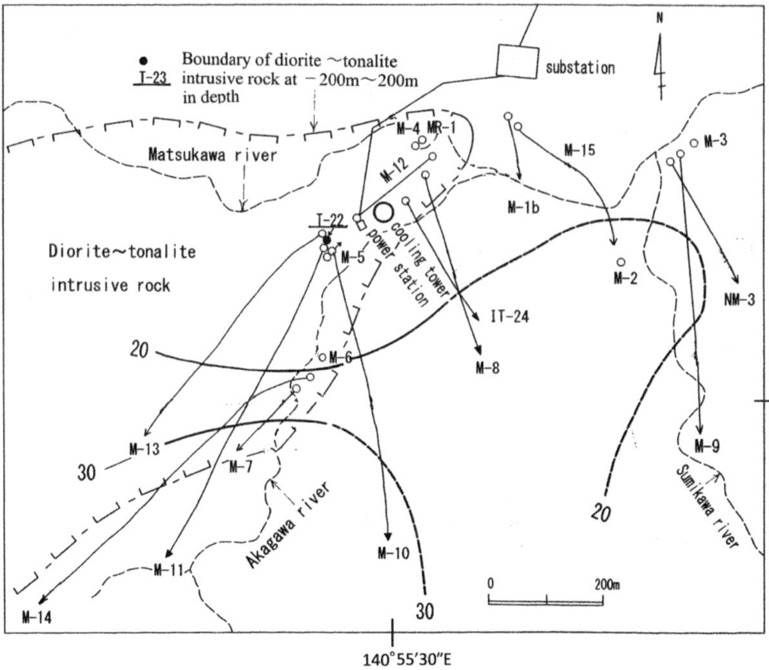

Fig. 14.28 Boundary of intrusive rock, borehole trajectories, and shut-in pressure direction. Lines and numbers indicate pressure values in bar. (Fukuda et al. 2013; modified by the authors)

(2) Northern Side

The northern boundary of the shielding structure is defined by diorite–tonalite intrusive rocks and the observed water level changes in wells T-22 and T-23, as previously discussed. Additionally, E–W trending fracture zones is exposed approximately 2 km west of the intrusive rock extension (b in Fig. 14.29). Fault F2 delineates a connecting line between the intrusive body and the exposed fracture zones.

(3) Eastern Side

The eastern boundary of the shielding structure is supported by the following three lines of evidence:

(a) The NNW–SSE trending Fault D, which crops out downstream along the Matsukawa River and forms a boundary between wells M-3 and T-21 (Fig. 14.29).
(b) An abrupt decrease in formation temperature to the east of Fault D (Fig. 14.25), suggesting that geothermal fluid does not migrate beyond the fault due to the presence of a low-permeability barrier.
(c) The TG-2 well, drilled under NEDO's 1995 project for enhanced geothermal reservoir recovery, encountered impermeable strata composed of mineral-filled fractures down to a depth of 1,300 m east of Fault D.

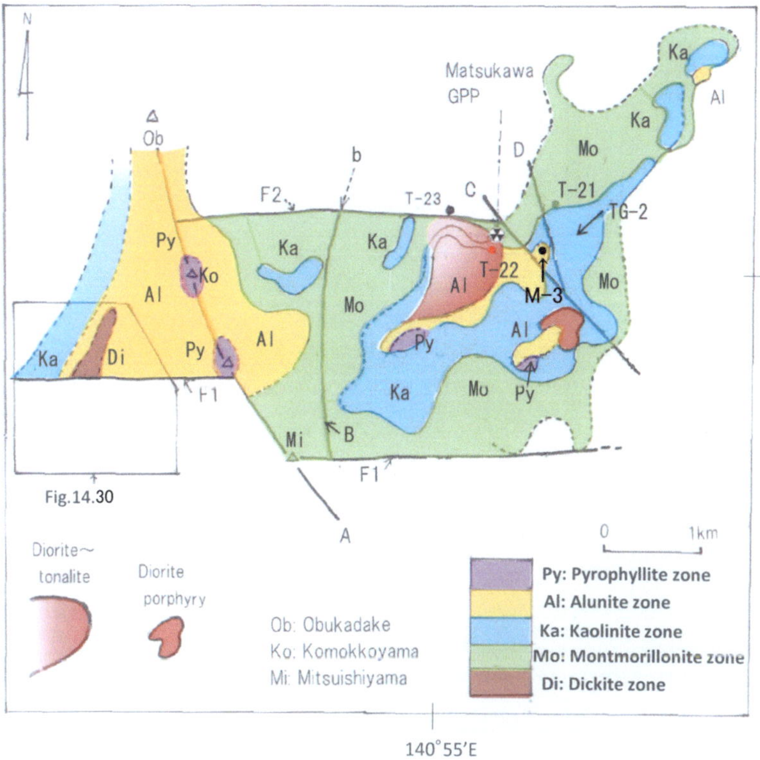

Fig. 14.29 Distribution map of acidic alteration zones, intrusive rocks, and faults (A to D, F1, and F2)

(4) Western side

The western boundary of the shielding structure is less clearly defined compared to the other sides, as it is characterized by north–south trending acidic alteration zones west of the Matsukawa area. It remains uncertain whether the alteration zones is associated with hydrothermal activity in the Matsukawa reservoir or with the Obukadake volcano, which erupted andesite dated to 0.77 ± 0.04 Ma (K–Ar age). Nevertheless, two faults identified in outcrops—Fault A (NNW–SSE trending) and Fault B (N–S trending)—are considered potential candidates for delineating the western extent of the shielding structure (Figs. 14.26 and 14.29).

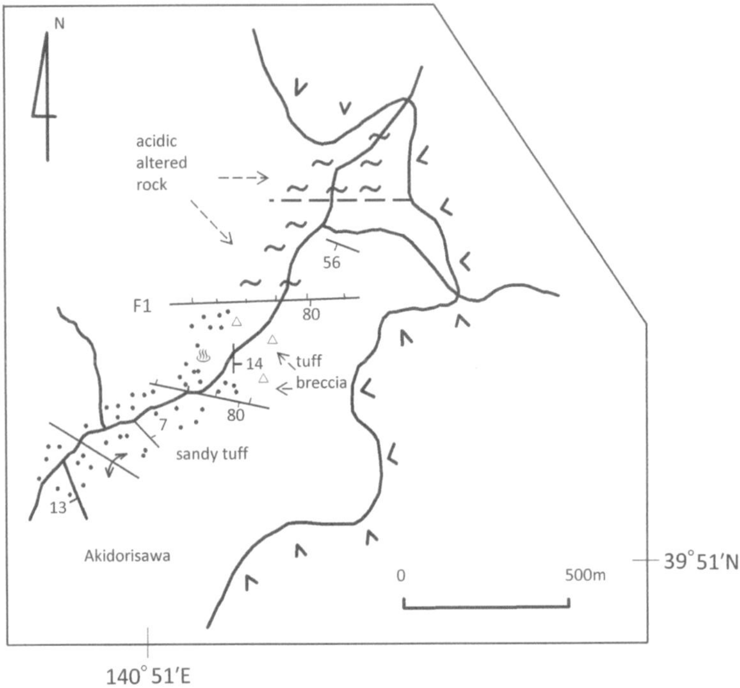

Fig. 14.30 Geological route map of the Akidorizawa area, a tributary of the Kakkonda River

14.3.4 The Low-Permeability Shielding Structure in a Vertical View

The shielding structure, acting as a cap rock layer, consists of thick clayey acidic alteration and Matsukawa andesites—hard, compact, and impermeable lavas that erupted during the early Pleistocene. The andesites are approximately 160 m thick. Clayey acidic alteration extends from the surface to depths beyond 1,200 m. This significant thickness results from high-temperature alteration of sulfur deposits and the influx of acidic gases at depth.

This alteration can be divided into two types: one is shallow and distributed above 500 m elevation (about 300 m depth), and the other has a funnel-shaped distribution (Fig. 14.26). The shallow zones contain montmorillonite and iron-rich saponite, which function as effective sealing layers. The funnel-shaped zones consists of kaolinite, alunite, and pyrophyllite, which intersect the surrounding alkaline alteration zones composed of chlorite, sericite, and mixed-layer sericite–montmorillonite clays.

The shielding structure in vertical view is further clarified by the presence of recently identified diorite porphyry and andesitic dikes intruding along the NW–SE trending Sumikawa Fault (Fault C in Figs. 14.26 and 14.29). Additionally, surveys

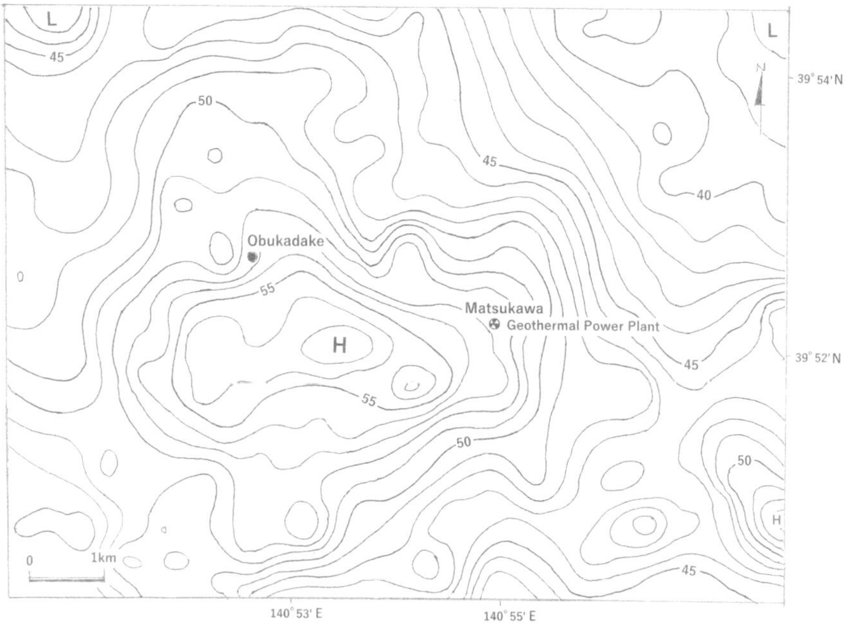

Fig. 14.31 Distribution map of Bouguer gravity anomalies around the Matsukawa geothermal area. Contour interval: 1 mGal; assumed density (ρ): 2.40 g/cm^3. (modified after Komazawa 1991)

by Akatsuka et al. (2005, 2006, 2011) suggest that additional intrusive bodies and dikes are concealed beneath the Pleistocene and Miocene formations, supplied via feeder dikes acting as conduits from the magma chamber to localized intrusions. These intrusive rocks, dikes, and feeder dikes enhance the integrity of the shielding structure. This interpretation is consistent with the observed high gravity anomaly caused by dense rock masses in and around the Matsukawa area, despite the presence of a subsiding structural setting (Fig. 14.31).

The deep-seated geological structure has been clarified by drilling data only down to approximately 700 m below sea level (corresponding to a depth of 1,500 m). The structure beneath this level is inferred from adjacent drilling data in the Kakkonda area (Fig. 14.26; locality shown in Fig. 14.22). According to WD-1 well data (Muraoka et al. 1998), the pre-Tertiary formation appears at 2,000 m below sea level (2,660 m in depth), and the Kakkonda Granite emerges at 2,200 m below sea level (2,860 m in depth). In contrast, in the Matsukawa area, the top of the pre-Tertiary is inferred to occur at 2,560 m below sea level (3,410 m in depth), due to a 560 m deeper position of the middle Miocene formation relative to the Kakkonda area. This inference assumes that the stratigraphic units—specifically the lower Tamagawa Welded Tuffs and Osuke Tuff—are comparable, as both consist of similar lithofacies (tuff breccia) and intertongue with the Osuke Tuff and the Yamatsuda Formation (composed of sandstone, mudstone, tuff, and conglomerate), which extends into the Matsukawa area (Fig. 14.26).

The Quaternary Kakkonda Granite, with an intrusion age ranging from 0.34 Ma to 0.01 Ma (Table 14.1), is recognized as the primary heat source extending from the Kakkonda area to the Matsukawa area (Figs. 14.22 and 14.26). This extensive, consolidated body of granite forms an egg-shaped spheroid approximately 8 km along its major axis and 5.5 km along its minor axis (Doi et al. 2001). Volcanic and seismic activity at Mount Iwate began in 1995, accompanied by large-scale fumarolic gas emissions. Subsequently, this activity migrated westward toward Mitsuishiyama, about 9 km from Mount Iwate (Fig. 14.22).

In 1998, ground uplift exceeding 10 cm was observed around Mitsuishiyama due to volcanic-seismic activity. Doi et al. (2001) inferred that this uplift was caused by a spherical pressure source associated with ongoing magma feeding into the Kakkonda Granite. If the spheroidal granite body extends toward the Matsukawa area, its upper boundary is estimated to lie below approximately 4,000 m, based on its depth of emergence in the Kakkonda area. These observations suggest that the vertical shielding structure is formed by multiple geological elements: the Matsukawa andesite, thick clay-rich acidic alteration zones, intrusive rocks, dikes and feeder dikes, the pre-Tertiary basement, and the Kakkonda Granite, which is also responsible for the high gravity anomaly in the region.

14.3.5 Structural Analysis in the Matsukawa Area

As described in Sect. 14.3.2, the geological structure of the Matsukawa area comprises a superimposition of NE–SW and E–W trends. The NE–SW trend initially originated as a transverse fault, which was subsequently influenced by the southwestward movement of the Chishima Forearc Sliver during the late Miocene to late Pliocene. Since the early Pleistocene, westward movement of the Pacific Plate has reactivated this fault as a strike-slip fault.

The Tamagawa caldera was formed between 2.2 and 1.2 Ma during the early Pleistocene, resulting in violent volcanic eruptions and the deposition of extensive pyroclastic materials known as the Tamagawa Welded Tuffs (Table 14.1). The eastern margin of this caldera lies approximately 6 km west of the Matsukawa power plant. This large-scale caldera experienced initial uplift due to magma intrusion, followed by subsidence in its later stages. Within this structure, the nested Kantomori caldera developed from approximately 1.6 to 1.2 Ma. These events led to a dominant vertical principal stress (σ_1) inside the caldera and the Matsukawa area due to the repeated uplift and collapse.

Meanwhile, around 3 Ma, the subduction direction of the Philippine Sea Plate shifted from northward to northwestward (Takahashi 2006), increasing the westward velocity of the Pacific Plate and reducing the tectonic influence of the Chishima Forearc Sliver. As a result, E–W trending normal faults could develop where the vertical σ_1 exceeded the horizontal E–W compressive stress (σ_2) caused by this tectonic regime.

Table 14.1 Chronological correlation of hydrothermal alteration stages, intrusive and volcanic activities in the Matsukawa geothermal area. The table integrates thermoluminescence (TL), K–Ar, and U–Pb age determinations for major geological events influencing the development of low-permeability shielding structures

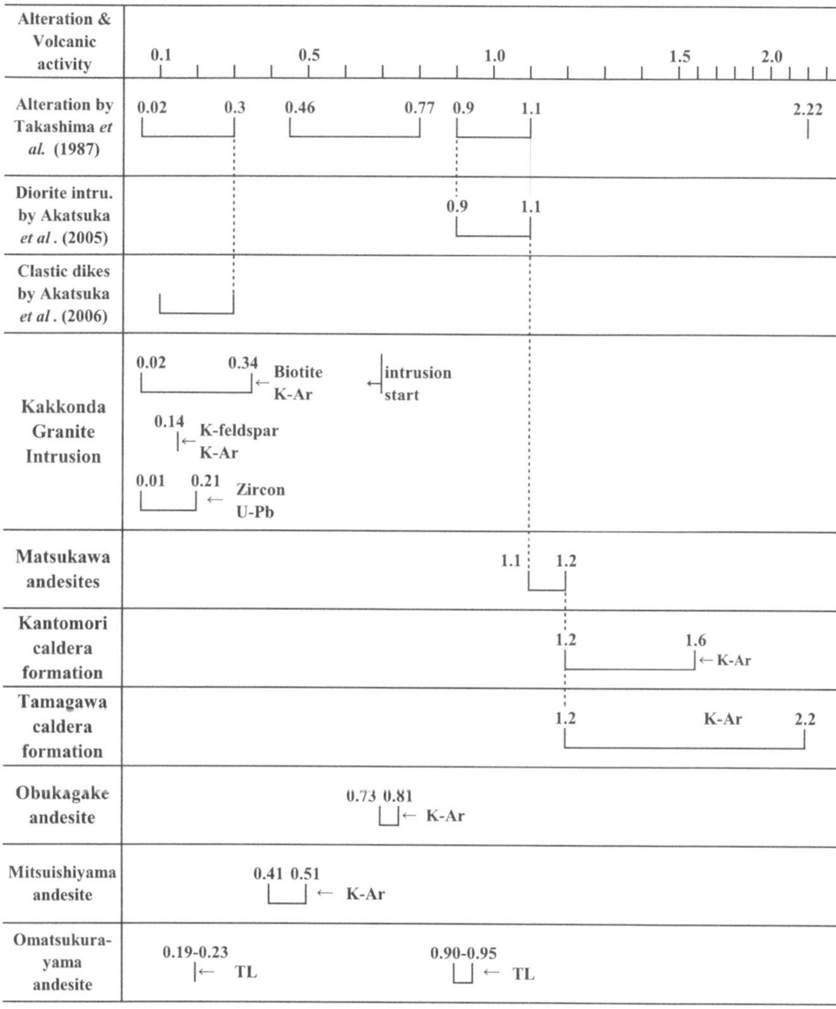

intru. : intrusion K-Ar, Zircon U-Pb and TL: Age determination

The NW–SE to N–S trending structures, including Faults A, B, C, and D as well as alteration zones west of the Matsukawa area, were formed under local stress conditions characterized by a strong vertical principal stress (σ_1) and a weaker horizontal stress (σ_3) trending NE–SW to E–W. These stress conditions are attributed to repeated uplift and subsidence associated with the formation of the Tamagawa caldera and the intrusive activity of the Kakkonda Granite from the early Pleistocene to the Holocene.

14.3.6 Thermal and Structural Transition in the Matsukawa Area

Hydrothermal rock alteration in the Matsukawa area has already been discussed in Sects. 14.3.1 and 14.3.2. Based on TL age dating, Takashima et al. (1987) and Takatsuka and Takashima (1991) divide the alteration history into four intermittent stages (Table 14.1).

The first stage dates back to approximately 2.2 Ma. The second stage, ranging from 1.1 Ma to 0.9 Ma, is supported by fission-track dating of zircon in diorite porphyry breccia found in clastic dikes, which yielded an age of 1.0 ± 0.1 Ma (Ito et al. 2005). Additionally, the intrusion of diorite to tonalite rocks in the northern part of the area occurred around 1.0 Ma, suggesting a significant contribution to thermal alteration during this period. The third stage occurred between 0.77 Ma and 0.46 Ma and corresponds to the eruption of Obukadake and Mitsuishiyama andesites. The fourth stage, from 0.3 Ma to 0.02 Ma, coincides with the activity of the Kakkonda Granite, the Omatsukurayama andesite eruption, and the formation of clastic dikes in the Matsukawa area.

The Tamagawa Welded Tuffs were formed between 2.6 Ma and 1.2 Ma. Their subsequent thermal alteration by intrusions of diorite, tonalite, and diorite porphyry—dated between 1.1 Ma and 0.9 Ma—is a key factor in the development of the low-permeability shielding structure. This is because the E–W trending structural features had already formed prior to the intrusions. In other words, the diorite, tonalite, and diorite porphyry likely intruded along pre-existing faults (fissures) associated with the Tamagawa caldera activity, thereby forming impermeable walls that now contribute to the vapor-dominated reservoir.

The age of the Matsukawa Andesites remains uncertain due to contradictions between radiometric dating and stratigraphic relationships. Although K–Ar dating results suggest ages ranging from 2.1 ± 0.6 Ma to 1.34 ± 0.05 Ma (Suto and Ishii 1987), these andesites overlie the Tamagawa Welded Tuffs, creating a geological inconsistency.

Key geological characteristics of the low-permeability shielding structure in the Matsukawa area, in comparison with the Kakkonda area, are summarized below:

(1) Acidic clayey alteration zones extend from the surface to a depth of 1,200 m in the Matsukawa area. In contrast, neutral to alkaline alteration predominates in the Kakkonda area. The boundary between these differing alteration zones is found at the Akidorisawa tributary of the Kakkonda River.
(2) In addition to the alteration zones, low-permeability shielding structures in the Matsukawa area are developed by Matsukawa andesite lavas, intrusive rocks

such as diorite and porphyry, and pre-existing fault systems. In contrast, intrusive dacite and andesite rocks are only locally distributed in the Kakkonda area.

(3) The Matsukawa area features a subsidence (graben) structure formed by overlapping NE–SW and E–W trending faults, influenced by the formation of the Tamagawa and Kantomori calderas. In contrast, the Kakkonda area is characterized by an anticlinal horst structure with NW–SE and E–W trending pull-apart features.

14.3.7 Well Targeting in the Matsukawa Area

(1) Well targeting in the shallow reservoir (to 2,000 m)

Two-phase tracer tests using alcohols have successfully predicted production interference among several wells, thereby identifying key permeable fracture zones (Fukuda et al. 2013):

- The southeastern margin and interior of diorite–tonalite intrusive rocks.
- Boundaries between intrusive rocks and surrounding wall rocks.
- Within the wall rocks themselves.
- The Sumikawa Fault (Fault C), delineated by magnetotelluric (MT) surveys (Aoyama et al. 2022), and associated andesitic intrusions in the eastern sector.

As previously stated in Sect. 14.3.2, we consider that the targetable permeable fractures are not limited to those associated with intrusive rocks but also include the NE–SW trending fractures along the Akagawa tributary. This is due to the predominant steam production observed along the Akagawa, as indicated by the reservoir pressure distribution (Fig. 14.28), hydrothermal alteration trends (Fig. 14.29), and gas concentration data.

Therefore, targeting additional wells along the NE–SW direction, farther from the M-14 well, is preferable for complementary drilling.

Nonetheless, access to potential drilling sites in the NE–SW direction is hindered by steep terrain along the Akagawa valley. Modern civil engineering techniques—such as temporary road construction, strut-supported platforms, inclined planes, and monorails—can help overcome this challenge. High-angle directional drilling is also feasible for targeting these zones.

Akatsuka et al. (2005, 2006) identified numerous clastic dikes, including diorite porphyry breccia and quartz fragments, in the southeastern upstream area of the Sumikawa Valley. The presence of diorite porphyry breccia suggests an underlying hidden porphyry body in the deeper zone. Fluid inclusions in quartz exhibit minimum homogenization temperatures ranging from 261 °C to 265 °C, indicating a high-temperature upflow that likely led to the development of permeable drusy textures. Due to the narrow valley topography in the southeastern upstream of the Sumikawa Fault (Fault C), high-angle directional drilling is the only viable method for accessing this target.

(2) Well targeting in the deep-seated reservoir (2 to 4 km)

The deep-seated reservoir remains unexplored to date, as the deepest well drilled reaches only 700 m below sea level (1,500 m depth). According to the geological profile in Fig. 14.26, the Quaternary Kakkonda Granite is expected to appear at approximately 2,800 m below sea level (3,600 m depth). Therefore, deep drilling to around 4,000 m should be planned in the next stage to penetrate the granite body. Such drilling operations must be carefully managed to avoid damaging the overlying shallow reservoir—by ensuring proper mud circulation, cementing, and control of lost circulation.

14.3.8 Sustaining Liquid Saturation in the Superheated Reservoir

Concerns have been raised about the future state of the vapor-dominated reservoir in the Matsukawa area, as continued steam production leads to rising steam temperature and increased superheat. For example, the degree of superheat at the wellhead of well M-7 increased from 10 °C to 40 °C over 19 years, suggesting a potential depletion of steam in the future.

To counter this, initial injection experiments were conducted from 1988 to 1989 to maintain liquid saturation in the superheated reservoir (Hanano et al. 1991). Building on these results, a long-term injection trial was performed intermittently over approximately 11 years from injection well MR-1 to response well M-12 (Fukuda et al. 2013). In addition, injection from well M-6 to response wells M-8 and M-13 was carried out in 2000.

However, the injection trial is not currently being conducted for the following reasons:

- Inadequate injection leads to a decrease in fluid temperature, causing the reservoir to cool. In particular, the vertical separation between injection and production zones was not considered in the Matsukawa area. The feed points of both injection and response wells are located at the same formation level (600 m to 1,120 m depth). As discussed in Sect. 13.2.2, the feed points of injection wells should be deeper than those of production wells to prevent direct thermal breakthrough.
- Injection wells should also be placed sufficiently far from production wells in the horizontal plane to avoid direct connection. However, the distance between injection well MR-1 and response well M-12 is only 210 m, which is insufficient for proper thermal separation.

Therefore, to maintain liquid saturation in the superheated reservoir, deeper drilling beyond the current well levels must be planned and executed in the future, as outlined in Sect. 14.3.5.

14.4 Tazawako-Tobu Geothermal Area, Akita Prefecture, Northeastern Japan

14.4.1 Geothermal Outline of the Tazawako-Tobu Geothermal Area and the Nyuto-Onsen Geothermal Field

The Tazawako-tobu geothermal area, located on the eastern side of Lake Tazawa, includes the Akita-Komagatake active volcano and the Nyuto-onsen geothermal field. As described in Sect. 9.1, this area represents a typical geothermal system associated with an andesitic volcano exhibiting significant topographic relief (Fig. 9.1). The Nyuto-onsen geothermal field lies approximately 5–6 km north of the base of Akita-Komagatake (Fig. 14.32). The volcano has been active from around 0.1 Ma to the present, with the most recent eruption occurring between 1970 and 1971. As a result, the area is underlain by widespread pyroclastic deposits and andesitic lavas. These volcanic materials obscure the underlying geological structure, including the pre-Tertiary basement and middle Miocene formations.

Figure 14.33 shows the fumarolic and hydrothermally acid-altered area near Kuroyu Hot Spring in the Nyuto-onsen geothermal field. Figure 14.34 illustrates Lake Tazawa (Tazawako), which forms a circular, funnel-shaped caldera approximately 6 km in diameter and 423.4 m at its deepest point. Figures 9.1 and 14.32 (NEDO 1992) include gravity survey data, hydrothermal alteration zones, and borehole temperature profiles, which collectively provide a basis for interpreting the subsurface structure.

These investigations suggest that a primarily N–S trending structure extends northward from the volcano but shifts to a NW–SE to NNW–SSE orientation within the Nyuto-onsen geothermal field. The NW–SE trend corresponds with the major regional fault zones known as the Oppu–Morioka Tectonic Line (see Fig. 12.8). In the narrow valley floor of the Nyuto-onsen field, middle Miocene formations are exposed, overlain by Quaternary volcanic deposits, debris flows, landslide material, and river terrace sediments (Figs. 9.1 and 14.35).

No geothermal power plant has been developed in this area to date, the reasons for which will be discussed in Sect. 14.4.5.

14.4.2 Temperature Distribution in the Tazawako-Tobu Geothermal Area and the Nyuto-Onsen Geothermal Field

The temperature distributions presented in Figs. 9.1 and 14.32 are based on static downhole temperature measurements from five exploration wells drilled in the area (NEDO 1992; Sato et al. 2022). Figure 14.33 shows fumaroles and hydrothermal

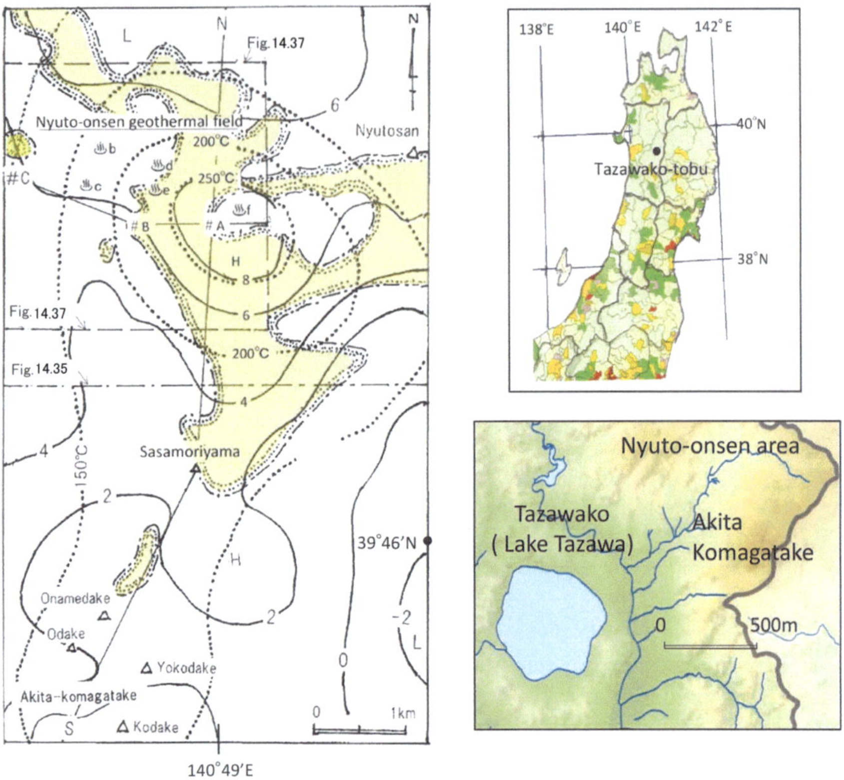

Fig. 14.32 (Left) Location and subsurface structural map of the Tazawako-tobu geothermal area. Solid lines represent residual gravity anomalies (in mgal; $\rho = 2.40$ g/cm^3), with "H" and "L" indicating high and low gravity anomalies, respectively. Dotted lines show the distribution of formation temperature at sea level, and chain double-dotted lines outline hydrothermal acidic alteration zones. Symbols b to f represent hot springs with the following geochemical temperature estimates: b (Kaniba Onsen) Na–K thermometer (Fournier) = 161 °C; c (Taenoyu Onsen) Na–K (Truesdell) = 184 °C; d (Magonoyu Onsen) Na–K (Truesdell) = 170 °C and Na–K (Fournier) = 173–186 °C; f (Ippon-Matsuzawa Onsen) Na–K (Truesdell) = 290 °C. Borehole sites are labeled as #A, #C, and #D. The N–S line corresponds to the cross section in Fig. 9.1. (Right bottom) Location of Tazawako (Lake Tazawa) and the Nyuto-onsen geothermal field

acidic alteration zones near the Kuroyu hot spring, vividly illustrating the surface expression of geothermal activity in the Nyuto-onsen field. Figure 14.34 presents the scenery of Lake Tazawa (Tazawako), a circular, funnel-shaped caldera approximately 6 km in diameter with a maximum water depth of 423.4 m.

In addition to the above well data, the Tazawako-machi Hot Spring hole No. 3 and several geochemical thermometers contribute to the thermal characterization of this area. Estimates from Na–K geothermometers suggest the presence of geothermal upflow zones near well N1-TZ-4 and the Ippon-Matsuzawa hot spring, as indicated in Fig. 14.35.

Fig. 14.33 Fumarolic activity and hydrothermal acidic alteration zones near the Kuroyu Hot Spring in the Nyuto-onsen geothermal field (photo by the author, November 16, 2023)

Fig. 14.34 Scenic view of Lake Tazawa (Tazawako), showing its caldera topography (photo by the author, November 16, 2023)

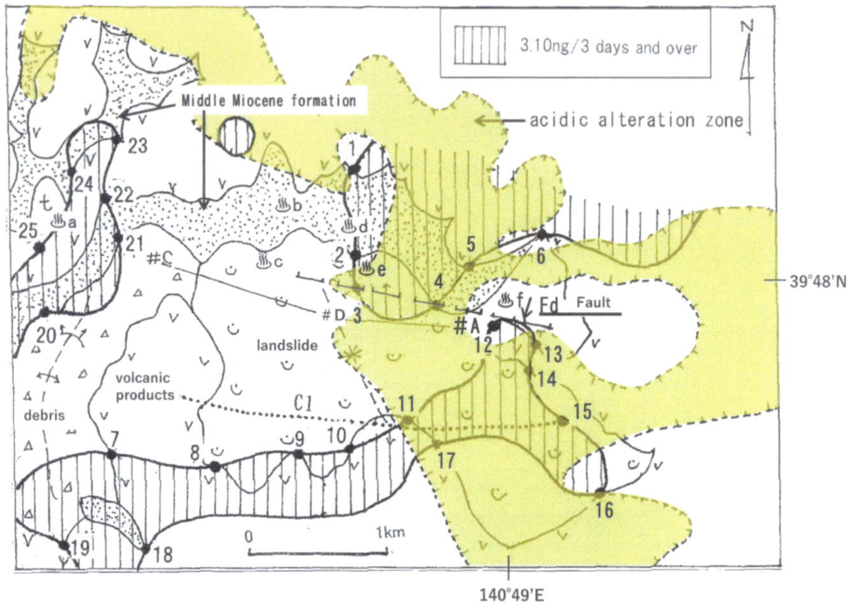

Fig. 14.35 Map showing Hg gas concentrations and corresponding geological features in the Nyuto-onsen geothermal field. Black dots numbered 1 through 25 indicate boundaries between zones of low ("light") and high ("thick") Hg gas concentrations. The locations of hot springs (b to f) and boreholes (#A, #C, and #D) are consistent with those shown in Fig. 14.32. (Sato et al. 2022)

14.4.3 Geochemical Survey in the Nyuto-Onsen Geothermal Field

Soil gas concentrations of mercury (Hg) were measured using boreholes 60 cm deep and 5 cm in diameter at approximately 650 m intervals, totaling 56 sampling points across an area of about 20 km^2. The measurement results are illustrated in Fig. 14.35, and their notable features are summarized as follows.

The boundary between low and high concentrations, with a threshold of 3.10 ng/3 days, generally intersects with the boundaries of geological formations. Black circles numbered 1, 5, 6, 7, 8, 9, 10, 11, 12, 13, 14, 17, and 19 in Fig. 14.35 mark boundaries between impermeable formations and other geological units. High Hg concentrations correspond with Fault Fd (at points 3 and 4) and the eastern segment of lineament C1 (at points 11 and 15).

14.4.4 CSAMT Method Survey in the Nyuto-Onsen Geothermal Field

The Controlled Source Audio-frequency Magnetotelluric (CSAMT) survey was conducted at 208 grid points within a 7.7 km^2 area, using 200-m spacing between stations. The current electrode separation was 2 km, and the transmitter was positioned 4 to 6 km away from the measurement area. One-dimensional inversion was performed after applying static shift corrections to extract resistivity discontinuities. These discontinuities were further clarified using additional analyses, including two-dimensional CSAMT inversion, a simplified MT method, and a regional-scale CSAMT interpretation. For stations affected by near-field and transient effects at low frequencies, apparent resistivity values derived from the simplified MT method were applied.

Figure 14.36 shows an example of a resistivity cross section used to identify discontinuity points (black circles in the figure) through one-dimensional analysis at sea level. The observed resistivity differences are attributed to variations in the degree of rock alteration, which are in turn influenced by differences in fluid flow behavior. This interpretation is supported by the fact that the stratigraphic horizon at sea level consists of the same rock facies across the section. On the western side (left side in Fig. 14.36), the resistivity structure is more complex due to the effects of electric transmission lines, fine fractures, and intrusive rocks.

The connection of discontinuity points into lines was carried out based on the following concepts. First, resistivity discontinuities with similar characteristics were linked together as indicative of fractures. Next, NW–SE to NNW-SSE trending structures were interpreted as pre-Tertiary faults that have been reactivated from the late

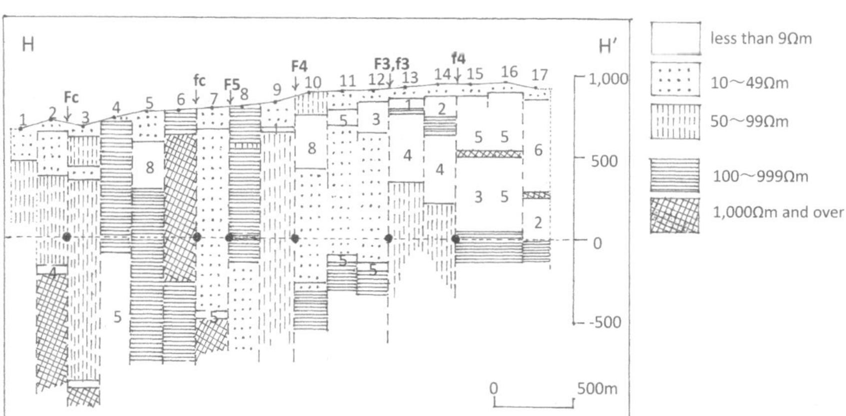

Fig. 14.36 Resistivity cross section along H–H′ from the CSAMT survey (Sato et al. 2022). The location of H–H′ is indicated in Fig. 14.37. Black dots represent resistivity discontinuity points. Numbers 1 to 17 correspond to measurement stations. Symbols (f4, f3, F3, F4, F5, fc, and Fc) indicate the identified resistivity discontinuities

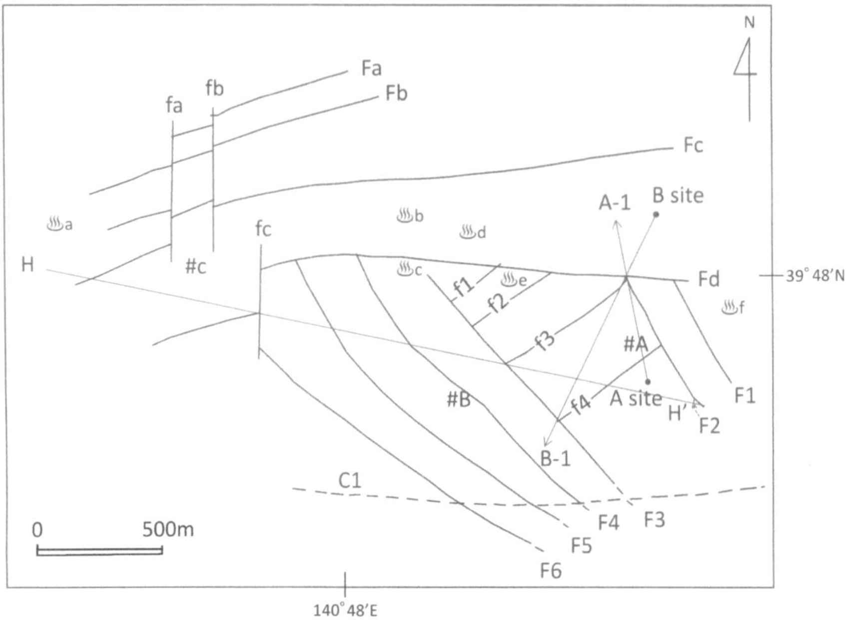

Fig. 14.37 Distribution of resistivity discontinuities identified by the CSAMT survey. Line types are defined as follows: (1) Thick solid lines indicate resistivity discontinuities; (2) The thin solid line labeled H–H′ corresponds to the cross section shown in Fig. 14.36; (3) The broken line represents the main lineament C1; (4) Arrows indicate the proposed well trajectories A-1 and B-1 (refer to Fig. 14.38). Hot spring and borehole locations correspond to those shown in Figs. 14.32 and 14.35

Pleistocene to the present. These structures are also supported by high gravity anomalies and elevated temperature zones (Fig. 14.32). On the other hand, E-W and N-S trending resistivity discontinuities appear to have formed after the late Pleistocene and intersect the earlier NW–SE to NNW-SSE structures. Figure 14.37 shows the integrated resistivity discontinuities, interpreted as a fault system in the Nyuto-onsen geothermal field.

14.4.5 Well Targeting in the Nyuto-Onsen Geothermal Field

(1) Fracture trend

It is important to consider the fracture trend in well targeting. NW–SE to NNW-SSE trending fractures—originally dominant before the late Miocene—are confirmed by gravity, temperature distribution, and CSAMT surveys. These trends were not formed during the middle Pleistocene, as the southwestward movement of the Chishima Forearc Sliver was not conducive to NW–SE to NNW-SSE trending fracture formation (Sect. 12.2.2, Stage 4). However, E-W trending regional compressive stress from the late Pleistocene to the present has reactivated pre-existing "old" fractures as NW–SE trending strike-slip faults (Sect. 12.2.2, Stages 5 and 6). As these older fracture planes are predicted to be closed by consolidation, it is advisable to design directional wells that intersect multiple resistivity discontinuities (i.e., multiple fractures) when targeting these NW–SE to NNW-SSE fractures.

The southwestward movement of the Chishima Forearc Sliver from the late Miocene to the late Pleistocene resulted in the formation of NW–SE trending extensional fractures and N–S to E–W trending strike-slip faults. These faults were subsequently influenced by magma intrusions from the late Pleistocene to the present (Fig. 9.3, Cases B and C). The E–W directed regional compressive stress associated with these strike-slip faults also contributed to the reactivation and enhancement of NE–SW trending fractures.

(2) Selection of drilling site

One of the challenges in selecting drilling sites in this area is its status as one of Japan's most famous secluded hot spring regions, necessitating that wells be located as far as possible from the hot springs. Additionally, most of the area lies within the Special Protection Zone (Class 2) of Towada-Hachimantai National Park.

In this context, Site A has been tentatively designated approximately 150 m south of well N1-TZ-4, near the inferred upflow zones (Fig. 14.37). If a well is drilled from Site A, the trajectory would pass through the footwall of normal Fault Fd. Should the fault dip be gentler than expected, the trajectory may not intersect the fault, compromising the target.

Alternatively, Site B is proposed on the northern side of Fault Fd so that the trajectory intersects the hanging wall. However, constructing Site B poses difficulties due to steep terrain.

Two issues arise in selecting a reinjection well site. First, Site C (well N2-TZ-7), corresponding to Case B or Case C in Fig. 13.5, is located 2.3 km WNW of Sites A and B, requiring a longer pipeline for reinjected water. Second, the N2-TZ-7 well exhibits low permeability, necessitating additional reinjection wells.

(3) Examples of well-targeting plans (A-1 and B-1 wells)

(a) Target of A-1

A-1 well penetrates three targets: the resistivity discontinuity f3, normal fault Fd, and a structurally weak portion of geological strata at a valley, which reflects faults and fractures (see Fig. 14.37 for the plan view and Fig. 14.38 for the cross section). In northeastern Japan, fractured Tertiary rocks are often exposed at valleys or rivers due to thin Quaternary cover and erosion caused by hydrothermal alteration (see Fig. 12.5). The total drilling depth is planned at 2,000 m (with a vertical depth of 1,650 m), and the horizontal deviation is 800 m at a depth of 2,000 m. The expected formation temperature ranges from 260 °C to 290 °C.

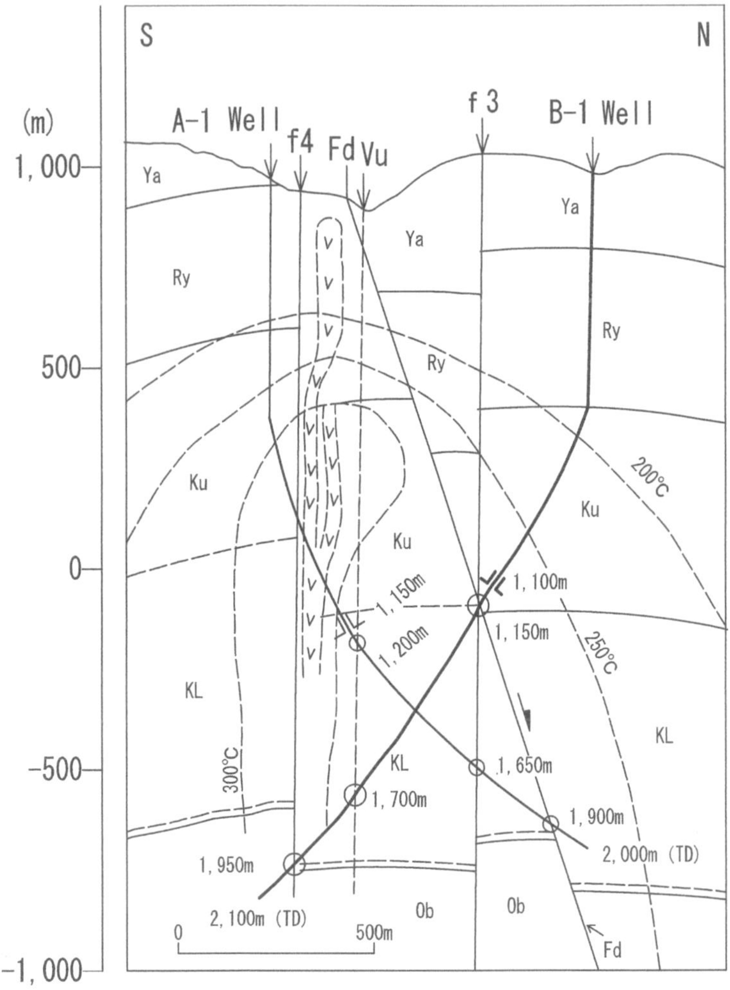

Fig. 14.38 Cross-sectional view showing the well-targeting plans for A-1 and B-1 wells. The corresponding plan view is shown in Fig. 14.37. Circles indicate well targets. Vu denotes the vulnerable portion of strata in the valley, and TD indicates the total depth of the borehole. Symbols f3, f4, and Fd represent resistivity discontinuities. Geological formation names (main rock facies in parentheses): Ya: Yamatsuda (sandstone), Ry: Koshitomaesawa and Ryukawa (mudstone), Ku: Upper Kunimitoge (tuff and mudstone), KL: Lower Kunimitoge (tuff and lava), Ob: Obonai (altered tuff). Symbol v indicates an intrusive rock

(b) Target of B-1

B-1 well penetrates three targets: (1) the intersection point of f3 and Fd, (2) the structurally weak portion of geological strata mentioned above, and (3) f4. The total drilling depth is planned at 2,100 m (with a vertical depth of 1,850 m), and the horizontal deviation is 850 m at a depth of 2,100 m. The expected formation temperature ranges from 260 °C to 300 °C.

14.5 Wasabizawa Geothermal Area, Akita Prefecture, Northeastern Japan

14.5.1 Geothermal Development in the Wasabizawa Geothermal Area

Geothermal resource development in the Wasabizawa area began in 1971, when its high potential was recognized alongside the neighboring Uenotai, Akinomiya, and Minase areas, all part of the Yuzawa-Ogachi geothermal region (Fig. 14.39).

Detailed surveys and drilling by NEDO from 1985 to 2001 revealed a close geothermal connection between the Wasabizawa and Akinomiya areas. Based on these findings, development of the Wasabizawa geothermal power plant—including the Akinomiya area—began in 2010.

The project was led by Yuzawa Geothermal Corporation, a special purpose company (SPC) capitalized by Electric Power Development Co. (J POWER), Mitsubishi Materials Corporation, and Mitsubishi Gas Chemical Company, Inc. The plant began operation in 2019 with a generation capacity of 46,199 kWe.

In the same region, the Uenotai geothermal power plant has been operating since 1994 (28,800 kWe), while the Katatsumuriyama and Kijiyama plants in the Minase area are under construction and scheduled to begin operation in 2027 and 2029, respectively (each 14,990 kWe).

14.5.2 Geological Setting in the Wasabizawa Geothermal Area

The Wasabizawa geothermal area is situated within the Sanzugawa caldera (also referred to as the Ogachi caldera by Takeno 1988, and Sasaki et al. 2018), which is an elliptical volcano-tectonic depression (VTD) covering approximately 500 km^2 and trending NW–SE along its major axis. This VTD lies along the Honjo-Matsushima Tectonic Line, which cuts across the Yuzawa–Ogachi region in a NW–SE direction (Fig. 12.8). Notably, the Onikobe caldera—home to the Onikobe geothermal power plant (14,900 kWe)—is also located along this tectonic line to the southeast.

The geology of the Wasabizawa area mainly consists of pre-Tertiary granite (Gr), mylonite (My), and schist, which are hard and compact, indicating low permeability and porosity. The area is located in a horst structure bounded by the NW–SE to WNW-ESE trending Doroyu Fault (Fig. 14.40). These pre-Tertiary rocks are unconformably overlain by Miocene, Pliocene, and Pleistocene formations. The Miocene clastic

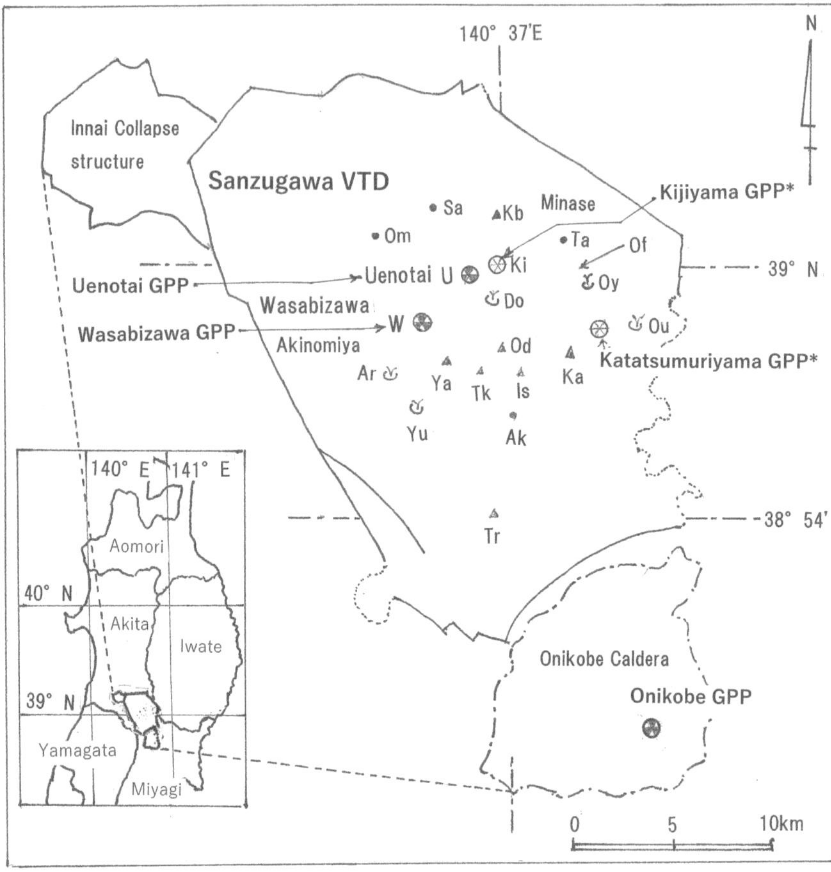

Fig. 14.39 Location map of the Yuzawa–Ogachi geothermal region, including the Wasabizawa, Akinomiya, Uenotai, and Minase areas. Symbols: GPP = Geothermal Power Plant (* = under construction); Ak = Akayunomata, Ar = Arayu, Do = Doroyu, Is = Ishigamiyama, Ka = Katatsumuriyama, Kb = Kabutoyama, Od = Oyasudake, Of = Oyasu Big Fountain (see Fig. 11.8), Om = Okumaemori, Ou = Ooyu, Oy = Oyasu-onsen, Sa = Sanzugawa, Ta = Takinohara, Tk = Takamatsudake, Tr = Torageyama, Ya = Yamabushidake, Yu = Yunomata-onsen

sediments, interbedded with volcanic rocks (excluding the latest Miocene sediments), are subdivided into the Doroyu Formation (Dr: tuff, andesite, and sandstone) and the Ootoritanisawa Formation (Ot: mudstone, tuff, and dacite) in ascending order. The latest Miocene to early Pleistocene formations consist of thick pyroclastic deposits and lake sediments derived from volcanic eruptions and depressions associated with the VTD.

The pyroclastic deposit is referred to as either the Torageyama or Minasegawa Formation (Tm), depending on the locality, due to their separated distribution and minor stratigraphic differences, despite having similar rock facies such as tuffaceous rocks and lava. The associated lake sediment is known as the Sanzugawa Formation

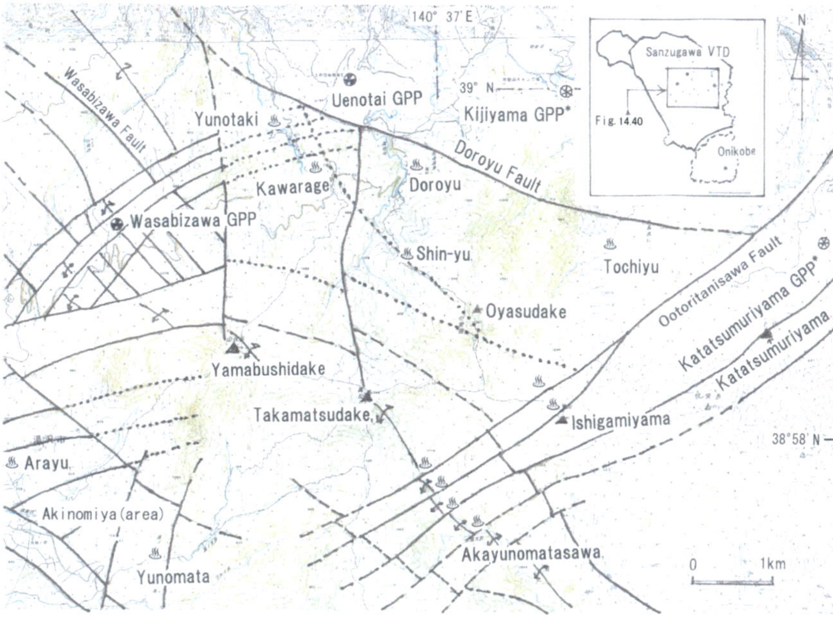

Fig. 14.40 Comprehensive diagram of fault and fracture systems surrounding the Wasabizawa area. Line types: Solid line = confirmed fault or fracture; Broken line = presumed fault or fracture; Dotted line = presumed lineament; Diverging arrows (←—|—→) = anticlinal axis. Abbreviations are explained in the main text. (sources: Kuriyama 1985; Sato et al. 2019; Nunohara et al. 2021)

(mudstone, tuff, and conglomerate), which interfingers with the upper part of the pyroclastic deposit. The surface of the Wasabizawa area is almost entirely covered by Quaternary debris and volcanic products (An), such as lava and pyroclastic flow deposits, resulting in limited exposure of pre-Tertiary to early Pleistocene formations.

14.5.3 Fracture Systems in the Wasabizawa Geothermal Area

The fracture system of the geothermal reservoir, comprising rocks of low permeability and porosity, mainly consists of large-scale fault structures, formation boundaries, intrusive rock margins, and fault intersections. These fracture systems have been identified using the methods shown in Figs. 14.40, 14.41, and Table 14.2. A geological cross section along the NE–SW direction from WZ-1 to AY-6 reveals that the Miocene and pre-Tertiary formations form horst and graben structures, showing significant elevation differences (throws), as illustrated in Fig. 14.42. Stratigraphic correlation based on borehole data indicates fault throws ranging from a few meters to approximately 500 m, as represented by faults "a" through "g" and C′ through E′

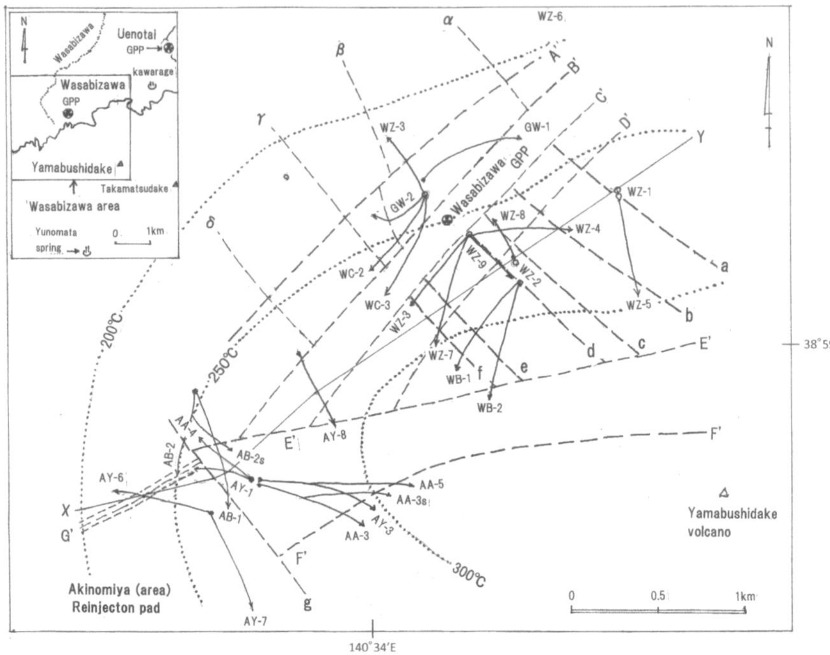

Fig. 14.41 Plan view of the fault and fracture systems in the Wasabizawa geothermal area. Solid lines represent borehole trajectories, while dotted lines indicate formation temperature contours at 500 m below sea level. WZ-1 through AY-7 are well names. Broken lines and their labels (a to g, A' to G', and α to δ) correspond to faults documented in Table 14.2

in Fig. 14.41. The complex throw of fault G′ near AY-6, located on the left side of the section, is likely due to repeated faulting and cliff formation.

The most important point when considering the fracture systems is that the Minasegawa (or Torageyama) Formation of the late Miocene to Pliocene age shows only minor elevation differences across the faults, while most of the Miocene strata and pre-Tertiary rocks exhibit significant vertical displacement.

This suggests that large-scale diastrophism occurred between the late Miocene and latest Miocene, around 6 Ma (Table 14.2). This deformation phase corresponds to Stage 4 (ca. 8 Ma to 3 Ma) described in Sect. 12.2.2, during which NW–SE trending fractures were generally unlikely to develop (Acocella et al. 2008).

The NW–SE trending faults labeled "a" through "g" in Fig. 14.41 contradict this general trend. It is inferred that the westward propagation of the Pacific Plate exerted a stronger influence than the forearc sliver effect of the Kuril Arc (Fig. 12.6), promoting the development of NW–SE trending fractures in this region, which lies along the NW–SE trending Honjo–Matsushima Tectonic Line (Fig. 12.8).

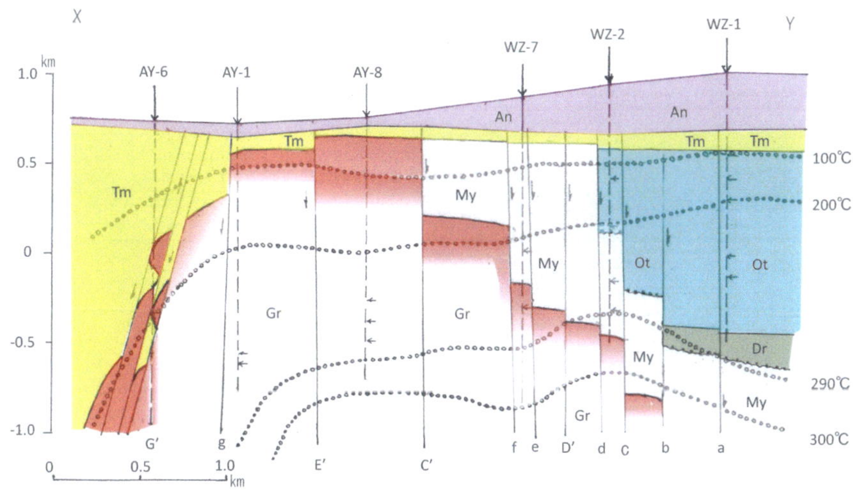

Fig. 14.42 Geological cross section of the Wasabizawa geothermal area. The X–Y section is shown in Fig. 14.41. Abbreviations for geological formation names are explained in Paragraph 14.5.2

14.5.4 Relationship Between Fracture Systems and Geothermal Activity in and Around the Wasabizawa Geothermal Area

Understanding the role of VTD formation is essential for analyzing the fracture systems and geothermal activity. Takeno (1988) identified two pyroclastic deposit units associated with the VTD formation based on K–Ar dating: one dating from approximately 5 to 6 Ma, and the other from around 3 to 4 Ma (Fig. 14.43).

Both units are associated with large-scale volcanic activity and accompanying subsidence. In contrast, Ohki and Ohba (2014) report that the Torageyama Formation comprises eight pyroclastic layers and two episodes of caldera subsidence. They suggest that the source of these pyroclastic deposits lies in the Oyasu–Takinohara zone, located about 10 km east of the Wasabizawa area (Fig. 14.39).

Sasaki et al. (2018) identify two distinct periods of VTD formation and associated geothermal activity. The first geothermal event occurred during the deposition of the Torageyama (or Minasegawa) Formation in the early to late Pliocene. At that time, geothermal fluids rose through both pre-existing and newly formed fractures, leading to the precipitation of alteration minerals such as epidote. Epidote reduces permeability because it fills the fractures.

The second geothermal event took place during the deposition of the Sanzugawa lake sediments in the late Pliocene to early Pleistocene, resulting in the formation of alteration minerals such as prehnite, which contributed to high reservoir permeability. The detailed distribution of prehnite is shown in Table 14.2.

Figure 14.40 presents a comprehensive diagram of fracture systems in the area surrounding Wasabizawa, located south of the Doroyu Fault and centered on the

Table 14.2 Identification of feed points related to faults and fractures for each well in the Wasabizawa geothermal area

Fault number	Well name	Main lost circulation (volume in l/min) and prehnite	Remarks
a	WZ-1	460 m (45), 590 m (16), 700 m (56), 719–1,000 m (100+), 1,014–1,500 m (70+)	NW-SE discontinuity in MT, A few throw of fault
b	WZ-5 WZ-4	667 m (800+), 671–763 (190–800), 1,002 m (100+) 1,198 m (60), 1,209–1,220–1,501 m (100+) 1,330–1,550 m (preh.)	ca.250 m in throw of fault
c	WZ-2	514 m (120+), 797 m (20–50), 1,126 m (66) 1,270–1,330–1,400 m (66–80)	ca.300 m in throw of fault
d	WZ-9	1,450 m (33–50), 1,511 m (1,000), 1,545 m (1,538)	NW-SE discontinuity in CSAMT, ca. 500 m in throw of fault
e	WZ-7	785 m (75)	ca.400 m in throw of fault
	WA-3	665–1,065 m (preh.)	
	WB-1	1,000–1,250 m (preh.)	
	WB-2	Unidentified in depth but indicates $kh = 2{,}3$ darcy·m	
f	WZ-7	1,164–1,345–1,701 m (38–1,538)	NW-SE discontinuity in MT and CSAMT ca.450 m in throw of fault
	WA-3	1,100–1,430 m (preh.)	
	WB-1	1,250–1,670 m (preh.)	
	WB-2	Unidentified in depth but indicates $kh = 2{,}3$ darcy·m	
g	AY-1	1,140–1,181 m (3–19), 1,254–1,308–1,480 m (33–100)	
	AB-1	1,200–1,251–1,73 m0 (preh.)	
	AB-2	1,136 m (preh.)	
A'	WZ-3	833 (50), 1,032 (30), 1,052 (40), 1,275 (20+)	ENE-WSW discontinuity in MT
B'	GW-1	1,400–2,00 m	None of lost circulation indicates NE-SW trending blank space between B' and C'
	AY-8	62 m (583-667)	
	WC-3	None of lost circulation	
C'	AY-8	1,036 m (47), 1,141–1,143 m (116–433)	
	WC-3	None of lost circulation	
D'	WA-4	700 m (preh.)	ENE-WSW discontinuity in CSAMT
	WZ-8	584 m (400–500+), 589–825 m (presumpton)	
	WZ-9	963 m (64)	

(continued)

Table 14.2 (continued)

Fault number	Well name	Main lost circulation (volume in l/min) and prehnite	Remarks
	WZ-7	1,345~(38–1,538)	
E'	WB-2	Unidentified in depth but indicates *kh* = 2,3 darcy m	
	AY-8	1,244 m (pumping pressure decreases	
	AB-2s	1,465–1,600 m (preh.)	
	AA-4	Unidentify in depth but indicates *kh* = 2,3 darcy m	
F'	AY-3	700–746 m (113–30), 846–879 m (20), 1,530–1,610–1,680 m (32–1,533+)	Presumed by surface geological survey and steep anomaly of gravity by NEDO (1998 and 2001)
	AA-3	1,165–1,565 m	
	AA-3s	1,330–1,400 m (preh.)	
	AA-5	Unidentified in depth but indicates *kh* = 16.7 darcy m	
G'	AY-6	382–412 m (33–50), 1,065 m (58), 1,477 m (101), 1,581 m (23–43), 1,643–1,697 m (12–72)	granitic rocks appear from 677 m to 1,701 m in spot (complicated profile)
α			Wasabizawa Fault confirmed by the surface surveys
β	GW-2	1,030–1,130 (preh.)	NNW extension is steep discontinuity in gravity anomaly
	WC-2	unidentified in depth but indicates *kh* = 43.3 darcy m	
γ	WC-2	unidentified in depth but indicates *kh* = 43.3 darcy m	NW-SE discontinuity in MT & CASMT
δ			NW-SE discontinuity in MT

Yamabushidake and Takamatsudake volcanoes. Solid lines represent faults and fractures confirmed by geological, geophysical, and borehole surveys, as well as by the distribution of springs and fumaroles. Broken lines indicate inferred faults and fractures. Dotted lines mark stratigraphic anomalies identified through lineament and DTF (Drilling-induced Tensile Fractures) analyses.

The lineament analyses were performed using aerial photographs, 3D topographic maps derived from DEM data, and satellite imagery. DTF analysis suggests that approximately E–W trending normal faults are commonly observed in the area around Wasabizawa.

To determine optimal production well targets, it is essential to restore the in-situ stress field in order to identify highly permeable apertures created by faults and fractures. Borehole-based measurements—such as hydraulic fracturing, core analysis using nonlinear deformation, and Drilling-induced Tensile Fracture (DTF) analysis—are employed for this purpose. Among them, DTF analysis is particularly

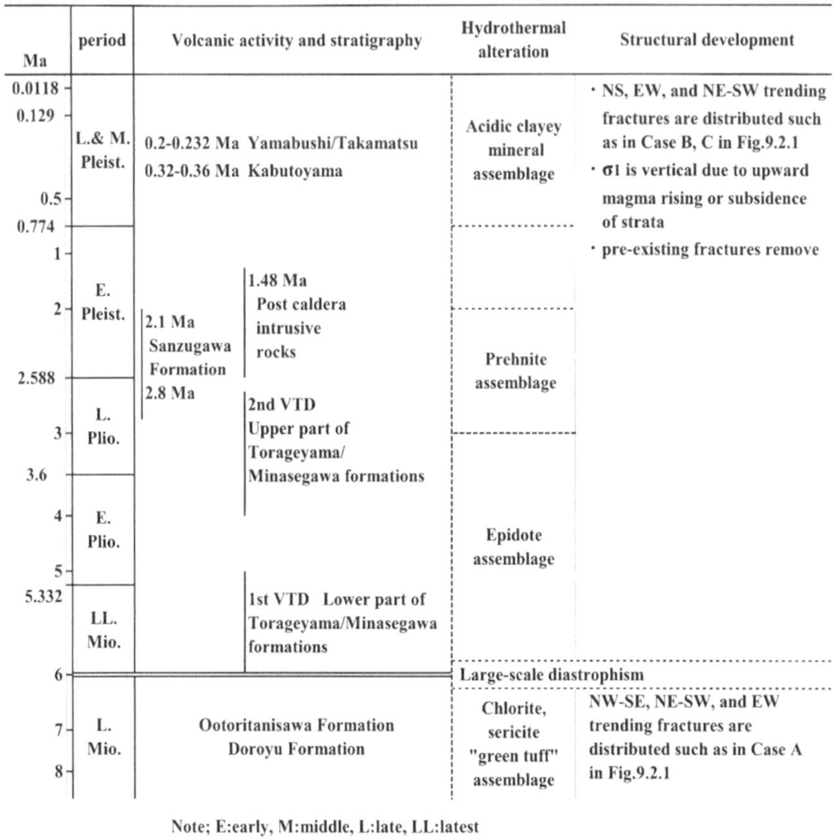

Fig. 14.43 Geological and tectonic evolution in and around the Wasabizawa geothermal area, including volcanic activity, caldera formation, and diastrophism

useful because it provides information on both the orientation and magnitude of the principal stress axes.

DTFs are longitudinal cracks composed of small, parallel echelon fractures oblique to the borehole axis. They form on borehole walls due to mud pressure and thermal stress during drilling, which concentrate in zones of stress relief. Each DTF is characterized by its circumferential position ($\theta_m d$) and inclination (γ_m) relative to the borehole axis (see Fig. 14.44).

These features are observed using Formation Micro Imager (FMI) or Borehole Televiewer (BHTV) tools at various depths along deviated wells. Three-dimensional stress fields are then reconstructed based on multiple measurements of $\theta_m d$ and γ_m taken at different orientations. The principal stress directions (σ_1, σ_2, σ_3) are estimated through statistical inversion techniques applied to a dataset from a single inclined borehole (Okabe et al. 1997, 1998).

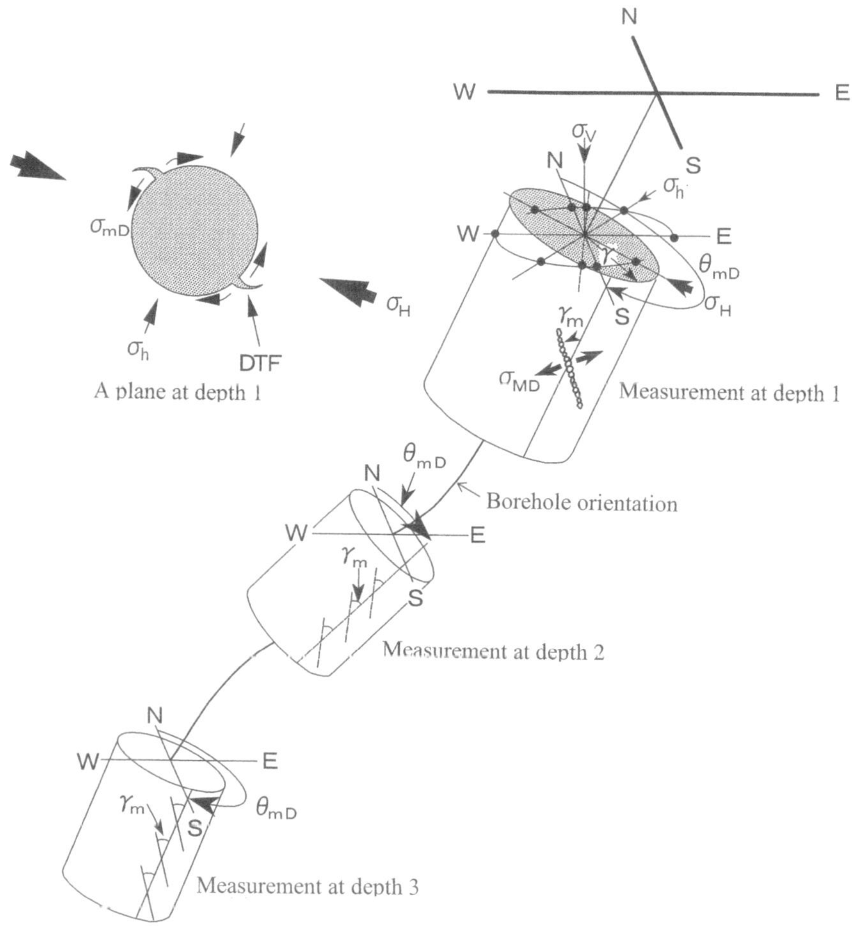

Fig. 14.44 Conceptual diagram illustrating the estimation method of a three-dimensional stress field using DTF data, under conditions of axis misalignment between the borehole orientation and the principal stress directions (modified from Okabe et al. 1997). Symbols: σ_H = maximum horizontal compressive principal stress; σ_h = minimum horizontal compressive principal stress; σ_v = vertical principal stress; σ_{MD} = maximum compressive stress in the circumferential direction; θ_{md} and γ_m = parameters defined in the main text

An application example is provided from the N9-AY-3 well at depths of 1,406 m, 1,560 m, and 1,750 m in the Akinomiya area of the Wasabizawa geothermal field, Akita Prefecture (NEDO 2001). The measured circumferential positions of DTFs (θ_md) were 106.8° (±3.9), 103.9° (±5.5), and 109.3° (±6.2), respectively, measured clockwise from true north. The inclination angles (γ_m) were 10.0° (±2.4), 10.3° (±3.0), and 11.4° (±3.3), respectively.

Mud pressure during drilling ranged from 12.5 to 13.7 MPa, and thermal stress was estimated between 12.9 and 18.5 MPa. The estimated principal stress orientations and magnitudes at 1,550 m depth were:

σ_1: Strike 285.4° (±2.04), Dip 51.1° (±6.0), Magnitude 40.1 MPa (±2.57).
σ_2: Strike 76.7° (±7.07), Dip 35.1° (±5.5), Magnitude 24.9 MPa (±1.31).
σ_3: Strike 173.7° (±35.9), Dip 14.3° (±3.5), Magnitude 21.5 MPa (±0.93).

This subsurface stress field at 1,550 m depth indicates an E–W trending normal fault with an oblique-slip component.

Two types of fracture systems associated with the Yamabushidake and Takamatsudake volcanoes are interpreted from Fig. 14.40. One is an elliptical system encircling the volcanoes, comprising NE–SW, NW–SE, and E–W trending fractures likely formed during large-scale tectonic activity around 6 Ma (see Sect. 14.5.3). Their formation mechanism is illustrated in Fig. 9.3 (Case A). The other system consists of radial fractures (N–S, E–W, and NW–SE) generated by local stress fields with vertical σ_1 due to magma upwelling, as shown in Fig. 9.3 (Cases B and C). The E–W and NW–SE trending fractures may result from both mechanisms.

Dacite from the Yamabushidake volcano is dated at 0.2 Ma (fission-track method), and andesite from Takamatsudake at 0.232 Ma (TL method); both volcanoes remain active. These tectono-volcanic settings foster high-permeability fracture systems, although some pre-existing fractures may have reduced permeability due to clay infill and consolidation.

14.5.5 Current Temperature Distribution in the Wasabizawa Geothermal Area

The temperature distribution at 500 m below sea level and the temperature profile along the X–Y cross section are illustrated in Fig. 14.41. The high-temperature zones are centered around the Yamabushidake and Takamatsudake volcanoes. High-temperature geothermal fluid flows laterally and extensively toward the Uenotai, Akinomiya, and Minase areas. The formation temperature (Tf) measured in wells drilled toward the volcanoes is higher than the minimum homogenization temperature (Tr) measured from fluid inclusions.

This observation implies that current volcanic and geothermal activities have intensified. For example, in the AY-3 well of the Akinomiya area, Tf is 286 °C and Tr is 251 °C at 500 m below sea level (NEDO 2001).

14.5.6 Thermal Transition Based on Hydrothermal Rock Alteration in the Wasabizawa Geothermal Area

Hydrothermal rock alteration has occurred periodically over five distinct periods in the Wasabizawa area (Fig. 14.43). The first period, during the Miocene, was characterized by neutral alteration minerals such as chlorite and sericite, generally referred to as "green tuff alteration." The second period took place from the latest Miocene to the late Pliocene, during which epidote assemblages were predominant. The third period occurred from the late Pliocene to early Pleistocene, when hydrothermal fluids ranging from 200 °C to 300 °C locally intruded into the area, resulting in the precipitation of vein minerals such as prehnite and zeolite. As previously noted, prehnite is associated with highly permeable zones (Sasaki et al. 2018). The fourth period, dated from approximately 0.36 Ma to 0.13 Ma (based on TL age dating), corresponds to the middle Pleistocene and is characterized by acidic clay minerals including montmorillonite, kaolinite, and alunite. The fifth period extends from the late Pleistocene to the present, during which acidic clay minerals such as kaolinite, alunite, cristobalite, and tridymite are widely distributed across the Yuzawa-Ogachi geothermal region.

Currently, hydrothermal alteration is actively occurring, accompanied by acidic hot springs and fumaroles containing H_2S. Notably, TL age dating of kaolinite collected from dacite tuff in the Arayu section of the Akinomiya area indicates an age of 0.0505 ± 0.0172 Ma.

14.5.7 Well Targeting in the Wasabizawa Geothermal Area

Geothermal power generation is already in operation in this area, and many fracture systems and formation temperature profiles have been identified. In the future, if power output declines due to insufficient steam production or reinjection capacity, additional wells will need to be drilled. Furthermore, in order to sustain long-term geothermal development, consideration of deeper reservoirs is essential. Accordingly, well-targeting strategies for production and reinjection wells are proposed for the following three geothermal reservoir zones, which are categorized by depth.

(1) Shallow reservoir (up to ~2 km)

The deepest well drilled so far reaches approximately 2 km. If drilling is confined to the shallow reservoir, it is necessary to expand the drilling area laterally. Potential production well targets include the following:

(a) Target known fracture systems using high-angle directional drilling from existing sites. For example, the intersections of NW–SE trending faults (c to g) with the EW trending fault E′ in Fig. 14.41 represent promising targets.
(b) Exploratory wells drilled by NEDO are concentrated in the Wasabizawa and Akinomiya areas (NEDO 1998; 2001). The N7-WZ-6 well, located north of Wasabizawa, recorded a temperature of 233.6 °C at a depth of 1,345 m and a

permeability-thickness product (*kh*) of 1.03 darcy-m. Furthermore, a significant steam flow of 46.6 t/h under atmospheric pressure was observed in N9-AY-4, approximately 1.3 km southeast of the bottom of N10-AY-7 in the Akinomiya area (Fig. 14.41), which recorded a *kh* of 1.2–3.0 × 10^{-11} m^3 and a fluid temperature of 286 °C.

(c) The alteration mineral prehnite, as noted in Sect. 14.5.4 (Sasaki et al. 2018), indicates permeable zones. However, because prehnite forms under subsurface high-temperature conditions (approximately 200–300 °C), it is unlikely to be detected at the surface unless significant erosion and uplift have occurred. In this area, prehnite has been identified at a depth of 650 m (approximately 250 m above sea level).

(2) Deep-seated reservoir (~2 to 3 km)

Geothermal fluid from the deep-seated reservoir is expected to be obtained from deeper portions of granitic rocks, which have yielded geothermal fluid in previous drilling efforts. In particular, in the northeastern part of the area (right side in Fig. 14.42), these granitic rocks lie beneath schist and Miocene formations at depths greater than 2 km. In any case, geophysical prospecting will be essential to guide successful deep drilling.

(3) Great deep-seated reservoir (below 3 km in depth)

Surveys and research on a deep-seated reservoir at depths greater than 3 km were conducted by NEDO (2021) as part of the Supercritical Geothermal Resource Leading Survey—Yuzawa-nanbu Region, which includes the Wasabizawa area. The results of the magnetotelluric (MT) survey indicated deep low-resistivity zones (<10 Ω·m) around Ishigamiyama, with an approximate diameter of 5 km at a depth of 4.5 km below sea level.

Additionally, broader low-resistivity zones (<40 Ω·m) with a diameter of about 8.75 km was identified beneath Takamatsudake and Katatsumuriyama at the same depth. Furthermore, the distribution of seismic epicenters supports the interpretation of a heat source in this area (Nunohara et al. 2021).

(4) Reinjection well

The reinjection wells (AA-4, AA-5, AY-1, AY-3, and AB-1) are located in the Akinomiya area, southwest of the Wasabizawa geothermal field (Fig. 14.41). This reinjection pad corresponds to Case B in Fig. 13.5. However, because AY-1 and AY-3 also show production potential, they may fall under Case A in Fig. 13.5. For future development, it would be advisable to drill additional reinjection wells in the peripheral areas of Akinomiya to supplement reinjection capacity.

References

Acocella V, Yoshida T, Yamada R, Funiciello F (2008) Structural control on late Miocene to Quaternary volcanism in the NE Honshu arc, Japan. Tectonics 27:1–24. TC5008. https://doi.org/10.1029/2008TC002296

Akatsuka T, Arai F, Fukuda D, Ito H, Sarudate M (2005) Geological model based on the relationship between clastic dikes and the advanced argillic alteration in the Matsukawa geothermal field, Iwate Prefecture. Geotherm. Res. Soc. Japan, 2005 Annual Meeting Abstracts, Wel Heartpia Obama, 12. (in Japanese)

Akatsuka T, Fukuda D, Arai F, Sarudate M (2006) Hydrothermal alteration process from advanced argillic to neutral-alkaline alteration in the Matsukawa geothermal field, northeast Japan. Geotherm Res Soc Japan, 2006 Annual Meeting Abstract. B02. (in Japanese)

Akatsuka T, Sugimoto T, Obara K (2011) Geological model of the Matsukawa geothermal field: magmatic contributions to the formation of the geothermal system. Geotherm Res Soc Japan, 2011 Annual Meeting Abstract. B1. (in Japanese)

Aoyama K, Mogi T, Suzuki K, Sasaki J, Yamaya Y (2022) Magnetotelluric study on a vapor-dominated geothermal reservoir in the Matsukawa area, Japan, Geothermics, 102362, 1–14

Doi N, Kasai K, Sakagawa Y, Shigehara S, Kato O, Takahashi M, Koide K, Tosha T (2001) Geothermal reservoir of the Kakkonda, Iwate Prefecture, Japan. In: a volcanic structure in the shallow part and volcanic fluid, 12K-3. Confer. Report of Disaster Prev Res Inst Kyoto Univ. 89–100 (in Japanese)

Fukuda D, Kuwano T, Itoi R (2013) Two-phase tracer tests using alcohols at the Matsukawa vapor-dominated geothermal reservoir: Evaluation of responses to water injection and prediction of production interference. J Geotherm Res Soc Japan 35(2):69–81 (in Japanese with an English abstract)

Geological Soc. Japan, ed (2010) The geology of Japan -1, Hokkaido, Asakura-shoten, 631p. Kamiiso Complex referenced from 18–25 (in Japanese)

Hanano M (1994) Reservoir engineering studies at the Matsukawa geothermal field. J Geotherm Res Soc Japan 16(3):255–284 (in Japanese with an English abstract)

Hanano M (2003) Sustainable steam production in the Matsukawa geothermal field, Japan. Geothermics 32:311–324

Hanano M (2011) Improvement of understanding of vapor-dominated geothermal resources. J Geotherm Res Soc Japan 33(2):79–85

Hanano M (2017) Success of the first commercial-scale geothermal power development in Japan: Matsukawa geothermal power station. Mining Mater Process Inst Japan 133(9):214–222 (in Japanese with an English abstract)

Hanano M (2021) Two mechanisms of production of superheated steam from geothermal wells. J Geotherm Res Soc Japan 43(4):133–144

Hanano M (2022) Vapor-dominated geothermal resources: (1) Resource characteristics. J Geotherm Res Soc Japan 44(1):37–53 (in Japanese with an English abstract)

Hanano M, Kotanaka K, Ohyama T (1989) Operation and reservoir management of the Matsukawa geothermal power station. J Japan Geotherm Ener Assoc (Chinetsu) 26:67–91 (in Japanese with an English abstract)

Hanano M, Ohmiya T, Sato K (1991) Reinjection experiment at the Matsukawa vapor-dominated geothermal field: Increase in steam production and secondary heat recovery from the reservoir. Geothermics 20:279–289

Hanano M, Kotanaka K, Ohyama T (1993) A quarter century of geothermal power production at Matsukawa, Japan. Geotherm Res Coun Bull, 32–47

Hanano M, Kajiwara T, Hishi Y, Arai F, Asanuma M, Kei S, Takanohashi M (2005) Overview of Production at the Mori Geothermal Field, Japan. Proc World Geotherm Congr, Antalya, Turkey, 24–29 April 2005, 1–19

Ito H, Arai F, Sarudate M (2005) Zircon fission-track dating on clastic dykes in the Matsukawa geothermal area. Geotherm Res Soc Japan, 2005, Annual Meeting Abstract. P11 (in Japanese)

Kajiwara T, Mogi T, Fomenko E, Ehara S (2000) Three dimentional modeling of geoelectrical structure based on MT and TDEM data in Mori geothermal field, Hokkaido, Japan. Proc World Geotherm Congr 2000, Kyushu-Tohoku, Japan, May 2000, 1313–1318

Kamata H (1989) Shishimuta caldera, the buried source of the Yabakei pyroclastic flow in Hohi volcanic zone. Japan. Bull Volcanol Soc Japan 51:41–50

Kasahara J, Hasada Y, Mikada H, Ohnuma H, Fujise Y (2022) Geophysical exploration of subsurface structure in six geothermal fields using distributed acoustic and temperature sensing methods. Grand renewable energy 2022 international conference (2):47

Kimbara K (1983) Hydrothermal rock alteration and geothermal systems in the eastern Hachimantai geothermal area, Iwate Prefecture, northern Japan. J Mineral Petrol Econ Geol 78:479–490 (in Japanese with an English abstract)

Komazawa M (1991) Gravity anomaly map of the center in the Sengan geothermal region. Resear Refer Materials of Geol Surv Japan (168)

Kuriyama T (1985) Geothermal system in the Yuzawa-Ogachi area, Northern Honshu. J Geotherm Res Soc Japan 7(3):311–328 (in Japanese with an English abstract)

Kurozumi H, Doi N (2003) Inner structure of the Nigorikawa caldera, Hokkaido. Japan. Bull Volcanol Soc Japan 48(3):259–274

Mannen K (2006) A review of the geology of "Nigorikawa-type" (funnel-shaped) caldera. Chikyu Monthly 28(4):257–264 (in Japanese with an English figures)

Muraoka H, Uchida T, Sasada M, Yagi M, Akaku K, Sasaki M, Yasukawa K, Miyazaki S, Doi N, Saito S, Sato K, Tanaka S (1998) Deep geothermal resources survey program: Igneous, metamorphic and hydrothermal processes in a well encountering 500 °C at 3,729 m depth, Kakkonda. Japan. Geothermics 27(5/6):507–534

NEDO (New Energy Industrial Technology Development Organization) Geothermal Development Promotion Survey. (in Japanese) Iburi area, 1983, Report No.4, 506p. Noboribetsu area, 1991, Report No.22, 845p. Tazawako-tobu area, 1992, Report No.27, 1021p. Mori area, 1993, Summary, 444p. Wasabizawa area, 1998, Report No. C-2, 712p Akinomiya area, 2001, Report No.C-3, 710 p

NEDO (New Energy Industrial Technology Development Organization) (1994) The advanced survey for geothermal development, the Kaminoyu-Santai geothermal area. (in Japanese)

NEDO (New Energy and Industrial Technology Development Organization) (1995) Project of enhancement technology exploitation of geothermal reservoir recovery. 479p (in Japanese)

NEDO (New Energy and Industry Technology Development Organization) (2021) The supercritical geothermal resource leading survey: Yuzawa-nanbu region, 396p. (in Japanese)

Nunohara K, Okano H, Yamada R, Hirano N, Tsuchiya N (2021) Geothermal potential in southern part of the Sanzugawa Caldera, Akita Prefecture, Northeast Japan. J Geotherm Res Soc Japan 43(2):65–78 (in Japanese with an English abstract)

Ohki I, Ohba T (2014) Volcanic activity and geological structure of the Sanzugawa Caldera, Yuzawa City, Akita Prefecture. Japan Geosci. Uni. Meeting 2014, in Pacifico Yokohama, 29 April, [S-VC54–1PO1]. 1–2. (in Japanese)

Okabe T, Shinohara N, Takasugi S, Hayashi K (1997) Estimation of in-situ three-dimensional stress field using drilling-induced tensile fracture. J Geotherm Res Soc Japan 19(1):1–15 (in Japanese with an English abstract)

Okabe T, Hayashi K, Shinohara N, Takasugi S (1998) Inversion of drilling-induced tensile fracture data obtained from a single inclined borehole. Int J Rock Mech Min Sci 35(6):747–758

Ozeki H, Fukuda D, Okumura T (2001) Historical changes of geothermal reservoir models at Matsukawa. Japan. J Japan Geotherm Ener Assoc (Chinetsu) 38:27–56 (in Japanese with an English abstract)

Sasaki K, Tezuka S, Takizawa K, Kaneko T, Asano K, Asai H, Nakanishi S, Akasaka C (2018) Mineralogical and hydraulic characteristics of the Wasabizawa geothermal field, Akita Prefecture, Japan. Proc. 43rd Workshop Geotherm. Reserv. Engi. Stanford Univ. Stanford Calif. Feb. 12–14, SGP-TR-213, 1–12

Sato K, Ito N, Tsukuda K (2019) Geothermal geological survey and production well target (Original title in Japanese). E-pix, Ofunato Printing Inc., 155 p, ISBN: 978-4-901602-67-9 (in Japanese)
Sato K, Ito N, Tsukuda K (2022) Geothermal geological survey and production well target: focusing on Reinjection Wells (Supplemented ed.) (Original title in Japanese). E-pix, Ofunato Printing Inc., 125 p, ISBN: 978-4-901602-74-7 (in Japanese)
Suto S (1992) Explanatory text of the geological map of the central part of the Sengan Geothermal Area, scale 1:50,000. Miscellaneous Map Series (21–5), Geol Surv Japan. 73p. (in Japanese with an English abstract 6 p)
Suto S, Ishii T (1987) Geology of the Shizukuishi district. With geological sheet map at 1:50,000, Geol. Surv. Japan, 142p. (in Japanese with an English abstract 5p)
Takagi S, Tanaka S (1968) Electrical logging and temperature surveys at Matsukawa geothermal wells. Monthly Report Geol Surv Japan 19(8):13–24
Takahashi M (2006) Tectonic development of the Japanese Islands controlled by Philippine Sea plate motion. J Geogr 115(1):116–123 (in Japanese with an English abstract)
Takashima I, Honda S, Wada T (1987) Geothermal history and hydrothermal system of the Kakkonda-Omatsukurazawa and adjacent areas, Hachimantai geothermal field, Northeast Japan—Application of thermoluminescence dating of altered rocks to geothermal exploration. J Geotherm Res Soc Japan 9(1):43–55 (in Japanese with an English abstract)
Takatsuka H, Takashima I (1991) Alteration zone and thermoluminescence (TL) dating at the Matsukawa area, Iwate Prefecture. Geotherm. Res. Soc. Japan, 1991 Annual Meeting Abstract, A8 (in Japanese)
Takeno N (1988) Geology of the North Kurikoma geothermal area, Akita Prefecture. Northeast Japan. Rept Geol Surv Japan 268:191–210 (in Japanese with an English abstract)
Yoshida Y, Ishizaki H (1989) Geochemical model of the Matsukawa geothermal field. Proc Internat Symp Geotherm Ener Kumamoto and Beppu, Japan, 128–131

Correction to: Challenges in Geothermal Development

Junzo Kasahara, Ko Sato, Hirofumi Muraoka, Koichi Tagomori, and Narimitsu Ito

Correction to:
J. Kasahara et al. (eds.), Challenges in Geothermal Development, Advances in Geological Science, https://doi.org/10.1007/978-981-95-0553-1

The original version of the book was updated with the belated corrections in the chapters. The book and its chapters have been updated with the changes.

FM Correction:

- Page xviii, HEP mentioned incorrectly twice, hence the first HEP has been corrected to $\hbar$ Reduced Planck constant which is the Planck constant divided by 2π, read as "h-bar".
- Page xx, μ Velocity of fluid flow corrected to μ Viscosity of fluid flow.
- Page xxi, σ1 (1 is lowercase) Maximum compressive principal stress in a regional stress field.

Chapter 1:

- Page 14-21. Table 1.3 alignment has been updated.
- Page 44, Line 18, 24 and 26 the citation of Fig. 1.31a has been updated to Fig. 1.31b.

Chapter 2:

- Page 84, Figure caption has been updated from Fig 2. to Fig. 2.8.

The updated version of this book can be found at https://doi.org/10.1007/978-981-95-0553-1

J. Kasahara et al. (eds.), *Challenges in Geothermal Development*, Advances in Geological Science, https://doi.org/10.1007/978-981-95-0553-1_15

- Page 99, Line 36 and 37 the citation of Fig. 24c, Fig. 24d, Fig. 24e has been updated to Fig. 2.24c, Fig. 2.24d, Fig. 2.24e.
- Page 100, Line 4, 5, 6, 18, 20 and 21 the citation Fig. 24a, Fig. 24b, Fig. 24c, Fig. 24e, Fig. 24f, Fig. 25a-f, Fig. 25a-f, Fig. 24a-c, e has been updated to Fig. 2.24a, Fig. 2.24b, Fig. 2.24c, Fig. 2.24e, Fig. 2.24f, Fig. 2.25a-f, Fig. 2.25a-f, Fig. 2.24a-c, e.
- Page 102, Line 22, 24 and 32 the citation Fig. 26a, Fig. 26b, has been updated to Fig. 2.26a, Fig. 2.26b.
- Page 103, Line 8 and 10 the citation Fig. 26c, Fig. 26b has been updated to Fig. 2.26c, Fig. 2.26b.
- Page 105, Line 23, 25 and 26 the citation Fig. 25c, Fig. 25a,e, Fig. 25b,d,f has been updated to Fig. 2.25c, Fig. 2.25a, e, Fig. 2.25b, d, f.

Chapter 5:

- Page 191, 192, 193. Fig. 5.6, Fig. 5.7, and Fig. 5.8 has been moved in the section "5.2.2".
- Page 193, One line space has been updated between 'f: frequency(Hz)'and the next paragraph.
- Page 194, "*H*: orthogonal magnetic field *(nT)(Ex/Hy, Ey/Hx)*" below Fig. 5.9 in page 194 should be moved to the next line of "*E*: electric field *(mV/km)* (electric field in 50–100 meter distance)" in page 193.

Chapter 6:

- Page 245, the word rehydration has been updated to dehydration.

Chapter 9:

- Page 337, Athree-dimensional has been updated to A three-dimensional.
- Page 340, see Chapter 11.1.5 has been updated to see Section 11.1.5.

Chapter 10:

- Page 353, Moved the figure to the end of Section 10.3, immediately before the start of Section 10.4.
- Page 354, Moved the figure to the end of Section 10.3, immediately before the start of Section 10.4, and placed it directly after Fig. 10.5.

Chapter 11:

- Page 365, kh/μ, where μ is (italicized)
- Page 367, This line has been deleted "Color coding: Sky blue = 10–13m^3 … Yellow = 10–11 to 10–10m^3".
- Page 368, Bars are color-coded by region: green and others for northern and northeastern Japan.

- Page 371, In Table 11.2 (item 10) the value has been updated from 960~1,0281.332 to 960~1,028, 1,332.
- Page 387, The figure citation Fig. 6.40 has been updated to Fig. 6.6.
- Page 388, The text citation in Chap. 12.2.2. has been updated to in Sect. 12.2.2.
- Page 389, Placed the fig 11.12 immediately before section 11.2.6.

Chapter 12:

- Page 396, The text citation see Chap. 14.3 has been updated to see Sect. 14.3.
- Page 399, The text citation in Chap. 14.4.5 has been updated to Sect. 14.4.5.
- Page 402, In the caption of Fig 12.7, the citation Chap. 12.2. has been updated to Sect. 12.2.
- Page 405, The text citation Chap. 3.2.2. has been updated to Sect. 3.3.2.
- Page 406, Revised the text formatting to match item (1) for Section Title "(2) Tectonic Evolution and Structural Framework in the Hohi VTD Region (Kyushu District)".
- Page 408:

  The text citation Chap. 11.1.4 has been updated to Sect. 11.1.4

  The text citation Chap. 11.2 has been updated to Sect. 11.2.

Chapter 13:

- Page 418:

  Equation (13.1) has been updated from UB2=U/Q to UB2=U/φ

  In Section 13.2.1, U, φ, ρw, ρr, Cw, Cr (not italicized) has been updated to U, φ, ρw, ρr, Cw, Cr (italicized).

Chapter 14:

- Page 429, Footer text- SOCIO ENGINEERING has been updated to Socio Engineering.
- Page 454, The Sentence: "Type 3-3: Fractures trending west to southwest from the caldera; this area has yet to be developed." has been moved at the end of Section (1), immediately before the start of Section (2).
- Page 473, As discussed in Chap.13 (13.2.2) has been updated to As discussed in Sect. 13.2.2.
- Page 474, Fig. 14.32 (Right bottom) 500m (on scale bar) changed to 5000m.
- Page 478, The text citation has been updated from (Chap. 12.2.2, Stage 4). (Chap. 12.2.2, Stage 5 and 6) to (Sect. 12.2.2, Stage 4). (Sect. 12.2.2, Stage 5 and 6).
- Page 481, Fig. 14.38 has been moved to the end of Section 14.4.5, that is, immediately before the start of Section 14.5.
- Page 487, In Table 14.2, Γ(uppercase, used as a fault number) has been updated to γ (lowercase).

Epilogue

Dr. Junzo Kasahara, Professor Emeritus of The Univ. of Tokyo, geophysicist (seismologist), visited Morioka City on November 15, 2022, to explain to me an invitation for this book publication. Dr. Kasahara emphasizes the following three purposes for publication.

First, few worldwide books on Japanese geothermal and development have been published although only oral and poster presentations have been performed in workshops and symposiums. So, the current investigation and development in Japan introduce case studies. Second, the research on the brittle-ductile transition in the crust must contribute to fracture development influenced by geothermal fluid flow. Third, the distributed acoustic sensor (DAS) and distributed temperature sensor (DTS) surveys are valid for greater precision of fracture decisions of geothermal reservoirs.

Dr. Narimitsu Ito, NEWJEC Inc. liquid resource engineer, and I published books in Japanese titled "Geothermal Geological Survey and Production Well Target, Japan" in 2019, and its supplement in 2022, based on our experiences of the geothermal survey, drilling, and development from 1979 to the present in Matsukawa, Kakkonda, Mori, Oguni, and many other areas. we therefore accepted Dr. Kasahara's proposal to pulish a book ongeothermal well targeting for an international audience.

After that Dr. Hirofumi Muraoka, a former Professor of Hirosaki Univ., geothermal geologist, Dr. Koichi Tagomori, and Dr. Naotsugu Ikeda, both geophysicists of WEST JEC Inc. approved this book publication. A kick-off meeting was held on March 22, 2023. Ten regular meetings were held to improve and examine the writing details at the Tokyo Office of WEST JEC. Inc. Through the writing and discussion of the meetings, the following items impress me.

The permeable fractures in the geothermal reservoir were formed by subjecting to the directions of regional and local stress field changes in Japan. The major regional changes are given below:

(1) The opening of the Japan Sea from ca. 21 Ma to ca. 15 or 13.5 Ma affects different rotary motions which are clockwise rotations in the Tohoku district and counterclockwise rotations in the southwest district in Japan.

J. Kasahara et al. (eds.), *Challenges in Geothermal Development*, Advances in Geological Science, https://doi.org/10.1007/978-981-95-0553-1

(2) The Pacific, the Philippine Sea, the Eurasia plates, and the Chishima (Kuril) forearc sliver were besieged and interacted with the Japanese island arc. Especially, a change of direction in the Philippine Sea plate occurred ranging in age from ca. 6 Ma to ca. 3 Ma affecting the formation of the geothermal reservoir.

According to the above tectonic deformation (diastrophism), the main reservoir in the Hohi VTD region in the southwest district consists of relatively younger sequences ranging from the Pliocene to the middle Pleistocene. In contrast, the geothermal reservoir in the Sengan region of the Tohoku district is mainly composed of the Pleistocene strata, the Miocene strata, and the pre-Tertiary basement depending on the place.

Ko Sato
Geothermal geologist
Morioka, Iwate Pref.
December 2024

Index

A

Accretionary prism, 456
Acidic geothermal fluid, 339
Acidic SO_4 type hot water, 338
Acidic water, 131
Acid stimulation, 410
Active fault, 394, 399
Active volcano, 337, 338, 340, 473
Activity index, 351
After-flow effect, 279
Age dating, 184
Alkaline geothermometer, 347
Alteration halos, 396
Alteration zoning, 183
Ambient hot water, 290
Anion index, 338, 339
Anti-Stokes scattering, 296
Apparent resistivity, 193, 194
Applied geophysical exploration, 225
Ar-Ar age dating, 184
Arrhenius equation, 114
Artificial fracture systems, 118
Aso Caldera, 204
Assemblage setting, 387
Asymptotic trend, 268, 269
Ata Caldera, 226
Attenuation of light, 287

B

Back-arc spreading, 76, 78, 80, 83
Back propagation, 297
Batholith, 127, 128
Bayesian criterion ABIC, 225
Beppu geothermal well, 202
Beppu-Shimabara Graben, 141, 143, 406
Binary power generation, 456
Binary power plant, 211, 226
Boiling point curve, 351, 352
Borehole survey, 488
Borehole Televiewer (BHTV), 489
Borehole tragectory, 296
Breached step-over, 387
Breakthrough, 415, 418, 419, 422, 423, 425, 426, 436, 437, 456, 473
Brillouin, 295
Brine evolution, 63
Brittle-ductile transition, 168, 169, 171, 172
Brittle fracturing, 161
Buckling, 390
Build-up, 365
Bulk-moduls, 304
Buoyancy, 139, 142, 145, 149–153, 337, 459
Byerlees's law, 161, 162

C

Cable material, 290
$CaCO_3$ and $CaSO_4$ scale, 436
Caldera, 182, 185
Caldera fill deposit, 438–440, 445, 453, 455
Caldera formation, 439, 440, 488
Caldera structure, 409, 439, 452
Cap rock, 339, 340, 396, 408, 423, 458, 462, 467
Carbon neutrality target (2020), 110
Cation exchange equilibria, 185, 346
Cation geothermometer, 345, 346
Cerro Prieto geothermal field, 37
Chaotic deposit, 457
Chemical bonds, 288
Chemical equilibrium, 346
Chemical reaction characteristics, 114

J. Kasahara et al. (eds.), *Challenges in Geothermal Development*, Advances in Geological Science, https://doi.org/10.1007/978-981-95-0553-1

Chemical tracer tests, 277
Chishima (Kuril) forearc sliver, 400, 402, 405, 406
Chloride concentration, 347–350, 437
Chronological depth variation of the brittle-plastic transition, 100, 105
Circumferential position, 489
Clastic dike, 440, 460, 470, 472
Clay minerals, 191, 194–196
Clay minerals in cap rock, 208
Clockwise movement, 402
Cold sweep process, 415–417, 426
Collision tectonics of the Izu-Bonin arc, 76
Columnar joints, 390
Commercial power generation feasibility, 120
Composite volcano, 340
Conceptual model, 116, 119, 120
Condensed water, 459
Conduction and convection, 351
Conductive heating, 338
Conductive heat transfer, 267
Confining pressure, 161, 162
Conjugate normal fault, 387
Consolidation, 393, 401, 408, 478, 490
Continental crust, 4, 7
Continuous temperature profiles, 267
Controlled Source Audio-frequency Magnetotelluric (CSAMT), 429, 477–479
Convection system, 337, 365, 406, 439
Convective circulation cycle, 119
Convective flow, 269, 270
Cooling from water injection, 286
Cooling joint, 389
Core, 4, 5, 7, 9
Cost-effectiveness and efficiency, 112
Counterclockwise movement, 402
Crack, 362, 383, 390, 409, 440, 453, 489
Critical point, 56, 63, 65, 66, 68
Crust, 4, 6, 8, 9
Cubic law, 359, 361, 364, 408, 416
Curve correlation techniques, 270
Curve-matching method, 280
Cyclic temperature pattern, 285

D

Damping factor, 326
Darcy's law, 121, 359, 364, 418
DAS seismic migration, 218, 220
Data sampling, 296
Deep drilling trends, 234, 235, 236, 246, 254, 255, 385, 472, 492
Deep low-frequency earthquakes, 97, 98, 100, 101, 105
Deep subsurface temperature, 351, 352
Density current, 193
Depletion of geothermal reservoirs, 108
Descramble project in Italy, 234, 252
Detailed gravity survey, 189
Diastrophism, 485, 488
Differential diastrophism, 485
Diffusivity equation, 121
Digital Elevation Model (DEM), 397, 488
Dike, 127, 128, 131
Directional drilling, 409, 456, 472, 492
Discrete magma chamber, 337, 457
Dissolution hydrogen, 289
Dissolution of minerals, 383, 386, 389
Distributed Acoustic Sensor (DAS), 208, 429, 440–447, 449, 450, 452–455
Distributed Fiber Optic Sensing (DFOS), 295
Distributed Temperature Sensor (DTS), 264, 283–286, 296, 440, 442, 452
Double flash system, 204, 205
Double-porosity concept, 416
Downflow, 352, 420
Downgoing phase, 445
Downgoing waves, 300
Downhole Geyser-like phenomena, 283
Downhole measurements, 425
Downstream stage of geothermal development, 114
Drawdown, 436
Drill cores, 360–362, 383, 385
Drilling-induced Tensile Fracture (DTF), 409, 488–490
Drilling into the Kilauea lava lake in the USA, 249
Drilling into the solidified crust of the lava lake, 250
Drilling trajectory, 409, 434
Driving force of magma ascent, 139, 140
Druse, 340, 367, 383, 385, 386, 388, 408, 409, 456
Dry hole issue and fluid balance, 117
Dryness, 411
Dry-up (depletion) of steam, 472
DTS flow profiles, 280
DTS logs, 264, 270, 275, 283
DTS measurement principle, 264, 287
DTS monitoring of aging geothermal production well, 283
DTS survey QC process, 292
DTS transient data, 279

Ductile deformation, 162, 163
Dynamic range, 297

E
Early history of Kakkonda, 234
Earth's interior, 3–7, 9, 10
Eased regulations for national parks, 110
Economic assessment, 114
Eigen function, 324, 326, 328, 329
Eigen value, 326
Electrical logging, 432
Electrical survey, 429, 432–434
Electromagnetic fields, 192–194
Elliptical fracture system, 489
Elliptical VTD, 482
Enthalpy, 346, 349
Environmental impact assessment, 114
Eötvös, 325
Equilibrium temperature, 238, 268
Estimating source materials of geothermal fluids, 185
Exceeding 500 °C at the well bottom, 238, 240, 247, 251
Exceeding 1000 °C at the well bottom
Exploration well WD-1a, 236
Extensional fractures, 340
Extensional stress, 386, 387, 389, 407, 408
Extensional tectonic field, 202
Extension yielded by the lateral tight bending fold, 144

F
Fall-off test, 365
Fast reading interval, 283
Fault, 128–130
Fault and fracture zone, 367, 383
Fault evaluation, 182
Faults and tectonic lines, 230
Fault systems, 189, 195, 196, 399, 407–409
Fault systems in Hatchobaru, 206
Feeder dike, 467, 468
Feed-in tariff enactment (2012), 107
Feed-In Tariff (FIT), 106–108, 204
Feed point, 352, 354, 366, 367, 383, 423–426, 429, 436, 455, 473, 486
Financial viability in geothermal projects, 114
Finite difference method, 120, 121
Fission track age determination, 470
Fissure, 362, 383, 384, 389, 390, 407, 471
FIT scheme effectiveness, 108, 109
F-K filtering, 300, 301
F-K velocity filter, 445, 452
Flashing and condensing phenomena, 286
Flexural slip folding, 390
Flower structure, 387
Flow law of rocks, 164, 165
Fluid characterization, 185
Fluid flow breakthrough, 418–420, 425
Fluid inclusion, 355, 356, 460, 472, 491
Fluid level oscillation, 283
Fluid loss, 270
Fluid movement, 270
Fluid speed profile, 270
Fluid temperature, 348, 349, 352, 365, 386, 411, 416, 418, 419, 425, 473, 492
Fluid temperature estimation, 275
Fluid zone, 283
Fluorine doping, 292
Flushing depth, 365
FMG, 323
Folding and joint, 390, 408
Formation Micro Imager (FMI), 409, 489
Formation temperature, 353–356, 365, 366, 368, 383, 396, 415, 416, 418, 419, 421, 423–425, 435, 456, 461, 462, 466, 474, 480, 491
Formation temperature pattern, 341
Fossa Magna, 76, 78
Fossa Magna Syntaxis, 76, 78
Fountain, 383, 384
Fracture characterization, 273, 276
Fracture correlation, 275
Fracture pair correlation, 276
Fracture permeability, 275, 359, 361, 364, 386
Fracture planes, 275
Fracture property analysis, 280
Fractures associated with dacite, 230
Fracture spacing, 415–417, 423, 425, 426
Fracture system, 409, 429, 432, 433, 453, 483–485, 488, 489, 491, 492
Fracture toughness, 390, 440
Frequency and depth relation, 192
Frictional sliding, 161, 162
Fuji volcano (Fuji San, Mt.Fuji), 77, 79
Fumaroles, 393, 394, 409, 476, 488, 491
Funnel-shaped caldera, 429, 437, 457, 473, 476
Fushime area, 226

G
Gauge length, 296, 297
Geochemical exploration, 182, 185, 186
Geochemical survey, 345

Geological age of geothermal resernoirs, 128, 132
Geological basement geometry, 189
Geological column mapping, 461
Geological correlation, 429
Geological exploration, 182
Geological strata, 390, 480
Geologic remote sensing, 182
Geomorphological analysis, 397, 409, 431
Geomorphological evaluation of volcanic edifices, 182
Geomorphological survey, 434
Geophysical anomalies, 345, 351
Geophysical exploration, 186
Geoscientific surveys and conceptual model, 116
Geostatic temperature computation, 267
Geostatic temperature profile, 269
Geostatistical approach, 351
Geothermal area, 339, 342, 356, 364–367, 386, 387, 394, 396, 399, 400, 403, 408, 409, 419, 421, 422, 429–431, 436–438, 441, 453, 454, 458, 468, 469, 473, 474, 482, 485, 486, 488
Geothermal conceptual model, 209
Geothermal energy potential, 106, 111
Geothermal exploration, 345, 346, 348, 436, 440
Geothermal fluid, 337, 339, 340, 342, 345, 346, 348, 349, 364, 365, 385–387, 394, 407, 408, 411, 456, 460, 466, 487, 491, 492
Geothermal fluid origin, 211
Geothermal fluids, 127–129
Geothermal gradient, 7
Geothermal grid capacity in Kyushu
Geothermal heat sources, 127, 128
Geothermal power development stages, 114, 115
Geothermal power generation, 10
Geothermal power plant, 10, 12, 14, 349, 403, 429, 436, 438, 440, 452, 457–459, 475, 480, 481, 483
Geothermal power plant distribution, 106–108
Geothermal power plants in Kyushu, 204–206
Geothermal reservoir, 190, 191, 194–197, 337, 339, 340, 349, 351–353, 355, 361, 364, 365, 368, 386, 390, 393, 396, 399, 401, 402, 404–408, 415, 439, 458, 466, 484, 492
Geothermal reservoir conceptual model, 206
Geothermal reservoir model, 225
Geothermal reservoir properties, 120
Geothermal reservoirs, 127, 128, 132
Geothermal reservoir structure, 230
Geothermal resource depths and parks, 108, 110
Geothermal resource development barriers, 106
Geothermal resources, 3
Geothermal station, 460
Geothermal structure, 338
Geothermal survey, 366, 430–432
Geothermal system, 337–339, 347, 349, 350, 352, 361, 365, 383, 394, 396, 397, 405, 408, 473
Geothermometer, 345–348
Geysers geothermal field, The, 34–36
Ginyu and Shiramizugoe faults, 222, 225
Government's carbon-neutral policy, 106
Graben formation, 202
Graded-index fiber, 292
Gravimeter, 188
Gravitational tensor, 324–326, 328, 329, 331
Gravity and convection, 119
Gravity anomaly, 467, 468
Gravity filtering analysis, 189
Gravity lineaments, 189
Gravity monitoring, 419, 425
Gravity monitoring stations, 189, 193
Gravity surveys, 188, 189
Great deep-seated reservoir, 492
Great East Japan Earthquake, 101, 246
Green tuff alteration, 491
Grid arrangement and cell parameters, 121
Grid search method, 319
Groundwater recharge, 339, 340

H

Hakone caldera, 77, 79
Hatchobaru, 204–210, 299
Hatchobaru geothermal field conceptual model, 116
Hatchobaru geothermal wells, 205, 206
HCO_3-type water, 338
Healing of micro-crack, 355
Heat exchange, 415, 416, 421
Heat input from reservoir, 286
Heat pipe phenomenon, 458
Heat source, 337, 467, 493

Heat source effect by the depth, 154
Heat source (Magma Chamber), 116, 117, 119
Hellisheidi geothermal field, 28
Hg gas concentration, 476
Hida Marginal Belt, 86, 89
High-angle directional drilling, 456, 472, 492
High-chloride contents in silicic crust
High host rock density in contraction tectonic fields, 139
High-relief topography, 339
High-temperature and high-pressure deformation experiments, 161
High-temperature hydrothermal systems, 127, 136, 137, 154
History matching simulation, 121, 122
Hohi Volcano-Tectonic Depression (VTD) region, 340, 342
Holocene volcano, 10
Homogenization temperature, 355, 411, 460, 472, 491
Honjo-Matsushima (Sendai) Tectonic line, 405, 482, 485
Horner analysis, 268
Horner's method, 353, 354, 356, 461
Horse-tailing, 387
Horst-graven structure, 340–342
Hot fingers hypothesis, 85, 88, 89, 91, 93
Hot spring, 3, 9, 338, 339, 345, 348, 349, 393, 394, 400, 409, 430, 436, 473–476, 479, 491
Hot springs and community cooperation
Hot water reinjection, 415
Hydraulic fracturing, 365, 389, 410, 489
Hydrogen diffusion, 289
Hydrogen issues, 288
Hydrogen sources, 289
Hydrological convection, 337
Hydrostatic pressure computation, 271
Hydrothermal alteration, 393, 399, 408, 409, 439, 453, 455, 463, 469, 472, 474, 480, 491
Hydrothermal alteration zone, 119, 439, 453, 455, 463, 474
Hydrothermal cold spot in central Japan, 93, 96
Hydrothermal convection, 346, 439
Hydrothermally altered mineral, 396, 398
Hydrothermal mineralogy, 209
Hydrothermal mineral vein, 364
Hydrothermal rock alteration, 339, 340, 393, 394, 470, 491
Hyperparameter, 324, 326–329
Hypersaline water in the plastic zone, 63, 68
Hypocenter determination, 319

I

Iceland Deep Drilling Project, 234, 254, 255
IDDP reached 4659 m
Idemitsu Kosan, 210, 212
Idiomorphic crystal, 385
Impermeable cap rock, 119
Impermeable layer (aquiclude), 352
Impurity absorption, 288
Inflow of shallow cold water, 458
Influence of water content on the flow law, 174
Infrared absorption, 287
Injection-induced temperature changes, 273
Injection test, 355, 365, 432
Injectivity index, 275
In-situ boiling, 422
In situ stress measurement, 177, 178
Installed capacity and power generation, 109
Insulating by the Philippine sea slab in central Japan, 83, 86–93
Insulating polyethylene pipeline, 456
Interfinger, 483
Interrogator, 296, 297
Inversion tectonics, 388
Investment plans and resource risk, 114
Isostasy, 4
Isothermal contours, 431, 432
Isotopic analysis, 119
Isotopic compositions, 348
Itoigawa-Shizuoka Tectonic Line, 76, 86, 92

J

James's lip pressure method, 411
Japan's Geothermal Power Plants (2024), 109
Japan's Largest Geothermal Power Plant, 208
Japanese Islands, 76, 77, 81, 91, 97, 99
Japanese public consensus on energy, 106
Japan Sea opening, 400
JAPEX exploration, 228
JOGMEC, 328, 330–332
JOGMEC's role in geothermal development, 108, 109

Jyoban-Choshi Forearc Extension Zone, 141

K
Kagoshima Graben, 135, 141, 143, 226
Kakkonda deep geothermal resources Survey Project, 236
Kakkonda geothermal area, 234
Kakkonda Granite, 235, 238–244, 467, 468, 470, 472
Kakkonda Unit 1, 235
Kakkonda Unit 2, 235, 236
K-Ar age dating, 184
K-Ar dating, 471, 485
K-horizon, 253, 254
Kijiyama, 299
Kuju, 328, 330
Kyushu Electric Power Company (Kyushu EPC), 106, 204, 210, 211

L
Landslide, 396, 475
Laplacian operator, 326
Larderello geothermal field, 23, 29, 252–254
Large elevation difference, 435
Laser Imaging Detection and Ranging (LiDAR), 397
Lateral flow, 354
Lava dome, 182
Least squares method, 277
Light attenuation formula, 290
Lineament, 430, 432, 477, 479, 483, 488
Linear regression of Horner plot, 268
Listric normal fault, 388
Lithostatic condition, 69, 385
Local stress field, 340, 489
Local temperature features, 270
Logarithmic function, 351
Logging temperature in Hohi
Logging temperature in Northern Honshu
Long-term trend of geothermal power, 108
Looped fiber-optic cable, 290
Lost circulation, 341, 353, 366, 385, 390, 409–411, 432, 435, 440, 455, 456, 472
Low-chloride contents in mafic crust, 56
Low host rock density in extension tectonic fields
Low-resistivity anomalies, 225
Lubricant, 289

M
Maar, 434
Magma buoyancy controlled by the host rock density, 139
Magma density estimated from the major compositions, 146
Magma-feeding, 468
Magma forceful intrusion, 389
Magma intrusion, 340, 388, 406, 470, 478
Magma originated water
Magma rising, 389
Magmatic deep low-frequency earthquakes, 97, 105
Magnetotelluric (MT), 423, 432, 471, 477, 492
Magneto-Telluric (MT) method, 191, 194, 195
Magnetotelluric survey, 471, 477, 492
Mantle, 4–9
Mariana-type subduction, 76, 78, 80
Mataloko project, 48, 49
Matrix permeability, 360
Matsukawa geothermal area, 457, 458, 468, 469
Matsukawa geothermal field, 106
Maximum compressive principal, 490
Maximum enthalpy point, 52, 54–56
Median Tectonic Line (MTL), 400
Medipolis, 297, 299, 300, 305, 314–316, 319
Metamorphic reaction under stress, 177
Meteoric-origin geothermal fluids, 119
Meteoric water, 337–340, 385, 394, 425
Miarolitic cavity, 386
Microearthquake hypocenters, 209
Micro-earthquake monitoring, 425
Microgravity changes, 189
Microgravity correlation, 189
Micro thermometry, 356
Mid-ocean ridge, 6, 8, 9
Mid-ocean ridge basalt, 8
Migration image, 445, 451, 453
MINC, 415, 421
Mineralization in geothermal reservoir, 230
Minimum compressive principal, 195
Minimum homogenization temperature, 355, 411, 460, 472, 491
Mise-À-La-Masse, 432, 434
Misfit kernels, 304
Mitigation methods, 290
Mixing model, 186
Mixing processes, 348, 355
Model fitting, 267

Moho discontinuity, 4
Mohorovičič discontinuity, 4
Mori, 299, 302
Mori geothermal power plant, 429, 436, 438, 440, 452
MT, 408
MT inversion in Takigami, 216
MT (Magnetotelluric) data, analysis, 208, 225–227
Mud circulation, 354, 472
Multiple feed point, 366

N
Natural convection, 365
Natural crossflow, 273, 280
Natural Geyser mechanism, 286
Natural state simulation, 121, 122
Natural vapor-dominated reservoir, 429
Near-neutral pH alkali chloride water, 338
Nesjavellir geothermal field, 28
Nested caldera, 470
Net-like structure, 440
Neutral buoyancy depth of magma estimated in Japan, 150
Neutral chloride water, 129, 131
Neutral pH alkali chloride water, 338
1959 summit eruption in the Kilauea volcano, The, 249
Nigorikawa geothermal area, 436–438, 441, 453, 454
Nitrogen flushing, circulation, 265, 290
Noboribetsu-Carls geothermal area, 430
Noise floor, 296, 297
Non-square grids, 120
Normal fault, 340, 342, 359, 386–388, 400, 408, 409, 470, 479, 480, 488, 489
Normal fault field, 140, 141, 143, 150
Northeastern fluid, 213
Number of wells available for data, 120
Numerical methods and reservoir behavior, 121
Numerical simulation, 419
Numerous fractures by dehydration reactions, 239–241
Numerous fractures in the contact metamorphic aureole, 239–242

O
Oblique subduction, 368, 402, 405
O-C, 319, 322, 323
Oceanic crust, 4, 8
Ogiri, 206, 328, 331, 332
OH group formation, 289
Ohnuma, 299–303
Oil shocks and Japan's energy policy, 106
Ophiolites, 4
Oppu-Morioka Tectonic line, 405, 464
Optical fiber, 295–299, 304, 440, 442, 452
Oscillating fluid level, 286
Oscillating temperature fluctuations, 286
Otake geothermal field, power plant, 204, 205
Overburden pressure, 388–390

P
Pacific Plate, 76–78, 84, 88, 89, 91, 94, 95, 103, 104, 368, 388, 400, 402, 405, 406, 469, 470, 485
Palau-Kyushu Ridge, 202
Parallel fracture planes, 275
PD, 395
Penetrating the plastic zone, 242
Penetrating the subsolidus magma chamber, 235
Peridotite, 4, 6, 8
Permeability, 359–365, 386, 388–390, 399, 402, 408–411, 417, 424, 425, 429, 432, 455, 460, 466, 469, 471, 480, 483, 484, 487–489
Permeability calculation, 279
Permeability-thickness (kh), 359, 365–368, 383
Permeable formation and impermeable cap rock, 117
Permeable fracture, 337, 352, 386, 393, 399, 416, 421, 453, 458, 471, 472
Permeable fracture planes structure, 277
Permeable fractures, 273, 276
Permeable layer, 337, 352, 425, 453
Phase transition, 286
PH control, 436
Philippine Sea plate, 76, 80, 84, 89, 91, 97, 103, 368, 399, 400, 402, 405, 407, 470
Philippine Sea slab-induced magmatism in cenral Japan, 96
Ping rate, 296, 297
Plastic flow in rocks, 163
Plastic zone EGS power generation, 246, 247
Poisson's ratio, 319
Polyacrylic acid, 436
Pore fluid temperature difference, 275
Pore pressure, 275

Pore space, 361, 364, 385, 386, 389, 390, 408
Porosity, 361–363, 416–418, 483, 484
Porous matrix, 417
Power plant construction and operation, 120
PP, 306, 310
Prehnite, 488, 491, 492
Pressure breakthrough, 418
Pressure deviations, 271
Pressure gradient, 285
Pressure interference test, 418, 419, 425
Pressure profiles, 285
Pressure, Temperature, Spinner logging (PTS logging), 355, 462
Pressure transient phenomena, 280
Pretreatment, 292
Production well, 340, 383, 387, 393, 407–409, 417–419, 423–425, 433, 435, 436, 454, 455, 458, 460, 462, 473, 489, 492
PS, 306, 310
P-to-S conversion, 307, 310, 314–318
Pull-apart basin, 39–48
Pull-apart structure, 368, 386–388
Pumping mechanism, 139, 153
Pyroclastic flow deposits, 430, 439, 484
Pyroclastic flow eruption, 439

Q
Quartz gauge data, 272
Quartz precipitation, 385
Quartz solubility, 385

R
Radar imagery, 182
Radiation exposure, 394
Radioactive tracer logs, 270
Raman, 295, 296
Raman backscattering, scattering, 264, 291
Raman scattering, 295, 296
R&D focus areas for geothermal development, 118
R&D for geothermal resource exploitation, 111
Rapid temperature fluctuation, 285, 286
Rate-variable convolution technique, 279
Rayleigh, 295, 296
Rayleigh scattering, 265, 266, 287, 295, 296
Recharge area, 337, 339
Reciprocal principle of Green's function, 303
Reconnaissance survey, 114, 122
Reconnaissance survey and desktop activities, 115
Reflected waves, 297, 300
Regional stress field, 340, 393, 399, 402
Regional surveys, 189
Reinjection strategy, 419, 426
Reinjection well, 415, 419, 421–426, 433–436, 455, 479, 492, 493
Reinjection well-setting, 423, 424
Relaxation method, 351
Relay ramp, 387
Remote reference analysis, 225
Renewable energy push, 106
Repeatability check, 293
Reservoir characteristics, 119
Reservoir cooling, 421, 422, 456
Reservoir location and geometry, 113, 119
Reservoir management, 419
Reservoir mass changes, 189
Reservoir simulation programs, simulator, 119
Reservoir temperature (240-280°C), 119
Reservoir temperatures, 230
Reservoir type, 230
Residual gravities, 189
Residual magma chamber, 457
Resistivity anomalies, 194, 195
Resistivity discontinuity, 478, 480
Resistivity distribution, 225
Resistivity imaging, 208
Resistivity method, 190, 191, 195
Resistivity structure, 433, 434, 477
Resource risk in geothermal development, 114
Reverse fault, 386, 388, 390
Reverse fault field, 141, 143, 150
Reverse time migration, 297, 298, 303
Reykjanes geothermal field, 256
Right lateral (dextral) strike-slip fault, 387, 388
Rigidity, 304
Ring structure, 432, 433
Rising basaltic magma conduits, 103, 104

S
Salinity measurements, 355
Sanzugawa (Ogachi) caldera, 482
Saturated water pressure, 285, 286
Scale abatement system, 213
Scale-inhibiting chemicals, 436
Scale precipitation issues, 229
Scattering mechanism

Schlumberger configuration, 432
Sealing, 339, 394, 396, 408, 409
Secondary inclusion, 355
Seismicity, 172
Seismic reflection, 209
Seismic survey, 230
Seismic tomography, 5, 6, 81–85, 87, 91, 99, 104
Sengan region (in Tohoku district), 339, 393, 400, 404, 405, 463
Serpentinite mélange, 86
Setouchi subsidence, 406
Shallow high-resistivity zone, 225
Shear zone, 440, 453
Shielding structure, 422, 459, 460, 462, 465–469, 471
Shishimuta caldera, 213
Shut-in pressure, 463, 464
Side-track method, 410
Si-F bonds, 292
Silica geothermometer, 346, 347, 349
Silica scale precipitation, 421
Silicified zones, 430
Simulation calibration with wellbore data, 120
Simulation error levels, 120
Simulation quality and engineer's role, 121
Single-flash power generation, 456
Single-mode fiber, 292
Skin depth, 193
Smaller, sustainable geothermal units, 108
Smectite cap rock, 195
Smith and Shaw Diagram, 184
S/N, 297, 311
Sociological impact assessment, 114
Solubility equilibria, 185, 346
Solubility of quartz, 66–68
Solute geothermometer, 185
Southwestern reservoir, 213
Spatial smoothness factor, 326
Specific enthalpy, 348, 350, 411
Spike noise, 297
Spinner measurements, 275
Spread out, 418, 419, 423, 424, 426, 437
SS, 306, 310
Stable isotope analysis, 185
Standing Time (ST), 353, 354, 356, 461
Static correction offset, 477
Static shift effect, 194
Steam cap, 422
Steam flow rate, 365–368, 383, 408, 410, 411
Steam-liquid separation, 348, 349
Steam loss, 63, 66
Steam production, 361, 364, 365, 383, 408, 423, 436, 437, 456, 458, 472, 492
Stevensite (smectite group) scale, 436
Stock, 127
Stokes, 295, 296
Stoping, 389
Strain rate, 163–167
Strategy of ultra-high temperature drilling, 247
Stress difference, 161, 162
Stress field, 340, 341, 386, 388, 393, 399, 402, 489, 490
Stress profiles of crust, 170, 171
Strike and dip angle, 278
Strike-slip fault, 340, 341, 386–388, 407, 408, 469, 478, 479
Strike-slip fault field, 141
Structural anomaly, 407
Structural geology, 182, 183
Structural movement, 400, 404, 405
Structural relaxation, 292
Subcooled liquid water, 285
Subduction angle, 78, 80
Subduction regime, 76–78, 80
Subduction zone, 4, 6, 9
Subsurface heat exchange network, 113
Subsurface resistivity structure, 208
Subsurface survey, 341
Subsurface temperature, 345, 349, 352, 353
Successive transitional temperature profiles, 275
Sulfur deposit, 467
Sumatran fault, 39–41, 43–48
Sumikawa, 299
Sunshine Project, 106
Supercritical Geothermal Project, 246
Supercritical phase, 385
Superheated steam production, 458
Superimpose fracture system, 409
Surface manifestation, 351, 393, 396, 406, 430
Sustaining liquid saturation, 472
Swabbing, 432
Synthetic DAS waveform calculation, 445
Systematic approach to geothermal projects, 115

T

Takigami, 299, 302
Takigami geothermal field, reservoir, geothermal power plant, 202, 210, 211, 214

Tamagawa caldera, 470, 471
Tazawako-tobu geothermal area, 473, 474
Tectonic, 340–342, 368, 390, 399–402, 405–409, 464, 470, 474, 483, 488, 489
Tectonic deep low-frequency earthquakes, 97, 98, 100, 101, 105
Tectonic line, 400, 464, 474, 482, 483, 485
Tectonic stress, 386
Temperature effects on geoscientific complexity, 114
Temperature gradient, 351, 352
Temperature gradient extrapolation, 351
Temperature influence, 426
Temperature inverted zone, 230
Temperature monitoring, 425
Temperature profile, 345, 351–353, 356, 424, 442, 461, 474, 491
Temperature recovery test, 263, 267
Terrestrial heat flow, 7–10
Thermal breakthrough, 418, 419, 421, 423, 425, 436, 437, 456, 473
Thermal conduction, 7, 9
Thermal conductivity, 7
Thermal convection, 4, 7, 9
Thermal cooling front, 419, 420
Thermal expansion on plastic deformation, 176
Thermal fluid, 349, 383, 384, 409, 456, 460, 466, 487, 491
Thermal infrared imagery, 182
Thermal manifestation, 393, 394, 409
Thermoluminescence age dating, 184
Thermoluminescence age determination, 394, 398
Thin layers of fluid flow, 275
Three-component surface geophone, 441
Three-dimensional flow law, 175
3-D inversion analysis, 196
3-D inversion code, 225
3D MT inversion analysis, 209
3D numerical model (3D-model), 119
3D numerical reservoir simulation
3 Key Elements for Geothermal Systems (3KEYGS), 115
Three types fault fields, 140, 143
Throw, 484
Time Domain Electromagnetic (TEM), 194
Tomographic travel time inversion, 444, 445
Total flow rate, 360, 363, 410, 411
Tracer test, 418, 425, 437, 471
Tracer test results, 277
Transformation superplasticity, 176
Transient phenomena, 275
Transient temperature phenomena, 264, 270
Transitional temperature profiles, 270, 280
Transmissibility, 418
Transmission line, 458, 477
Transverse oblique-slip fault, 387
Travel time discontinuity, 443–445, 449
Trench-volcanic front gap, 80
Trinity of geothermal systems, 127, 128
2D and 3D model comparison, 225
2D forward modeling, 208
Two-phase condition, 423
Two-phase fluid, 284
Two-phase fluid flash, 458, 460
Two-phase fluid simulation, 121
Two-phase tracer test, 471

U

Ultramafic rocks, 4
Unaltered lava layer, 225
Upflow zone, 337, 341, 342, 387, 396, 408, 409, 423, 476, 479
Upgoing phase, 445
Upgoing waves, 300
Upstream stage of geothermal development, 114, 116
Upward flow between fractures, 283
Upward-flow region, 283, 458

V

Vapor-dominated geothermal system, 52, 54–56
Vapor-dominated reservoir, 352, 429, 458, 471, 472
Vapor-dominated zone, 422, 460
Vapor zone, 283
Vein mineral, 340, 364, 408, 491
Vent-fill material, 438, 439, 457
Vertical faults and fractures, 119
Vertical fracture, 435, 453, 455
Vertical resistivity sections, 225, 227
Vertical spread out, 419, 437
Vertical stress, 388, 470, 489, 490
Volcanic activity, 339, 368, 386, 388, 389, 399, 402, 406–408, 439, 487, 488
Volcanic activity and geothermal activity, 113, 117
Volcanic front, 78, 81–84, 94, 96, 99, 103, 108
Volcanic geology, 182
Volcano-seismic activity, 468

Volcano-Tectonic Depression (VTD), 340, 342, 361, 388, 399–402, 405–407, 409
Volumetric method, 436
Volumetric water-vapor ratio, 285
Vp, 300, 305, 306, 309, 314, 318, 319
Vp/Vs, 307, 310, 314–320
Vs, 305, 306, 309, 311, 314, 317–319
VSP, 316
Vulnerable portion, 481

W

Wadachi diagram, 319
Wadati-Benioff zone, 77, 78, 81, 82, 88, 89, 92
Wairakei geothermal power plant, 30
Wasabizawa geothermal area, 480, 482, 484, 485, 488, 491
Water-dominated geothermal system, 52, 55
Water injection, rate, 263, 283
Water-rock interactions, 338
Water saturation pressure, 284
Water-vapor saturation pressure, 285
Waveform adjoin, 304
Waveguide imperfections, 288
Wellbore traces, 277
Well collapse, 432
Wellhead pressure, 367, 411, 421, 456
Well targeting, 399, 408, 409, 429, 471, 472, 478, 491
Widespread consolidated heat source, 473
Wind gap, 399
Wrench fault, 387

Y

Yamagawa geothermal field, 206, 226
Young intrusions penetrated by geothermal wells, 142

MIX
Papier aus verantwortungsvollen Quellen
Paper from responsible sources
FSC® C105338

If you have any concerns about our products,
you can contact us on
ProductSafety@springernature.com

In case Publisher is established outside the EU,
the EU authorized representative is:
Springer Nature Customer Service Center GmbH
Europaplatz 3, 69115 Heidelberg, Germany

Printed by Libri Plureos GmbH
in Hamburg, Germany